AA007066

International Seminar on Chemical Engineering Soehadi Reksowardojo (STKSR 2019)

Bioenergy and Biobased Chemical Product for National Sovereignty and Self-Reliance

IOP Conference Series: Materials Science and Engineering
Volume 823

Kupang, Indonesia
7 - 9 October 2019

Editors:

Antonius Indarto
Graecia Lugito
Khoirudin

ISBN: 978-1-7138-3608-7
ISSN: 1757-8981

Printed from e-media with permission by:

Curran Associates, Inc.
57 Morehouse Lane
Red Hook, NY 12571

Some format issues inherent in the e-media version may also appear in this print version.

Institute of Physics
Dirac House, Temple Back
Bristol BS1 6BE UK

Phone: 44 1 17 929 7481
Fax: 44 1 17 920 0979

techtracking@iop.org

Additional copies of this publication are available from:

Curran Associates, Inc.
57 Morehouse Lane
Red Hook, NY 12571 USA
Phone: 845-758-0400
Fax: 845-758-2633
Email: curran@proceedings.com
Web: www.proceedings.com

TABLE OF CONTENTS

BIO-BASED CHEMICALS & PHARMACEUTICALS

BIOENERGY

BIO-BASED MATERIALS

BIOMASS UPGRADING & UTILIZATION

SUSTAINABLE PROCESS & TECHNOLOGY DEVELOPMENT

IOP Conf. Series: Materials Science and Engineering **823** (2020) 011001 doi:10.1088/1757-899X/823/1/011001

International Seminar on Chemical Engineering Soehadi Reksowardojo (STKSR) 2019

"Bioenergy and Bio-based Chemical Product for National Sovereignty and Self-Reliance"

7th-9th October 2019
Swiss-Belinn Kristal Kupang Hotel
Kupang, Nusa Tenggara Timur
Indonesia

Editor:

Dr. Antonius Indarto
Dr. Graecia Lugito
Dr. Khoirudin

IOP Conf. Series: Materials Science and Engineering **823** (2020) 011001 doi:10.1088/1757-899X/823/1/011001

Preface

The first higher education in Indonesian Chemical Engineering was established at ITB in 1941. International Seminar on Chemical Engineering Soehadi Reksowardojo (STKSR) is an annual seminar held by the Chemical Engineering Study Program of ITB, in collaboration with other institutions, and usually takes place in Bandung. The International Seminar on Chemical Engineering Soehadi Reksowardojo (STKSR) 2019, however, was held in Kupang, the Province of East Nusa Tenggara (NTT) for three days, starting from October 7th to October 9th. Brought forward the theme of "Bioenergy and Bio-based Chemical Products for National Sovereignty and Self-Reliance", the STKSR 2019 was inviting scholars, practitioners and experts to contribute to this event. The aim of STKSR 2019 is to promote the implementation of EBT (Energi Baru Terbarukan/Renewable Energy) in Timor Island, NTT supporting the government program of national sovereignty and self-reliance. Throughout the procession of this seminar, there have been collaborations between the Department of Chemical Engineering of ITB, Faculty of Agricultural Technology of Universitas Kristen Artha Wacana (UKAW) Kupang, the City Government of NTT, and Directorate General of Renewable and Conservation Energy (EBTKE), Ministry of Energy and Mineral Resources of the Republic of Indonesia (ESDM).

IOP Conf. Series: Materials Science and Engineering **823** (2020) 011001 doi:10.1088/1757-899X/823/1/011001

Message from Chairman of STKSR 2019

On behalf of the Organizing Committee of International Seminar on Chemical Engineering Soehadi Reksowardojo, I would like to show my gratitude to all stockholders of STKSR 2019. The Seminar raises the topic of "Bioenergy and Bio-based Chemical Products for National Sovereignty and Self-Reliance", expanding the scopes into a multidisciplinary platform.

As we all aware of the global issues of climate change and energy crisis, STKSR 2019 was intended to be the first step toward the frontier of petroleum-to-bio-based movement, especially for our society in the Eastern part of Indonesia. Being blessed with great varieties of natural resources, Indonesia should be able to achieve it's national independence and self-reliance in terms of bioenergy and bio-based materials.

STKSR 2019 was enriched by plenary lectures and parallel oral presentations from scholars and industrial practitioners coming from 5 countries as well as Indonesian government agencies representatives. We have received more than a hundred abstracts, among which one fourth comes from the Eastern part of Indonesia; we are sincerely grateful for your great enthusiasm and contributions. Also, for those of you who return to STKSR again and again, thank you for your continued friendship, scientific sharing, and kind support.

Taking place in Kupang, STKSR 2019 could not be made possible without the great support from our counterparts in Universitas Kristen Artha Wacana. They have arranged several social functions (welcoming dinner, gala dinner, and excursions) so that all of you may leverage your community outreach and collaborative partnerships as well as experience the Kupang exotic beauty and warm hospitality.

Finally, the committee is most grateful to all sponsors for providing funds. And I thank all the committee members, all the plenary and invited speakers, as well as all oral presenters for their kind efforts and contributions in making this conference a grand success.

Warm regards,

Chairman of STKSR 2019,

Dr. Antonius Indarto

IOP Conf. Series: Materials Science and Engineering **823** (2020) 011001 doi:10.1088/1757-899X/823/1/011001

Sponsor

World Class University (WCU), Institut Teknologi Bandung, Indonesia

Badan Pengelola dana Perkebunan Kelapa Sawit (BPDPKS), Ministry of Finance, Indonesia

Bakti Pendidikan Djarum Foundation

Pupuk Kaltim

PT. EMI (Persero)

Universitas Katolik Widya Mandira

PP Energi

IOP Conf. Series: Materials Science and Engineering **823** (2020) 011001 doi:10.1088/1757-899X/823/1/011001

Steering Committee

IOP Conf. Series: Materials Science and Engineering **823** (2020) 011001 doi:10.1088/1757-899X/823/1/011001

Organizing Committee

Chairman: Dr. Antonius Indarto

ITB Organizing Committee

Dr. Jenny Rizkiana
Dr. Dian Shofinita
Dr. Ardiyan Harimawan
Dr. Khoiruddin
Dr. Graecia Lugito
Dr. Anggit Raksajati
Dr. Wibawa Hendra Saputra
Dr. Dianika Lestari

Harris Fikren Taufik
Zoealya Nabila Zafra
Robby Anggoro
Jihan Camilia Nabila
Ignatius David
Bernadeta Adinda I. P.
Nabila Fitriani Masrura
Lathifah Karina Putri
Fitridhia Aliyafaiqah Alis
Jorgensen
Rizky Eka Rachmatillah Ahmad
Felica Thalia
Fathur Rahman Wibisono
Clarissa Prakarsa
Owen Khosasi
Budiningtyas Puspito
Nasywa Kamilah

UKAW Organizing Committee

Ir. Melkianus Nuhamara, MP
Drs. Silverius Yohanis, M.Si
Marthen Makaborang, S.TP., M.Sc
Mery Rambu Djoru, S.P., M.Si
Arista M. Tamonob, S.Si., M.Si
Dr. Elvis A. Bintoni, S.Pd., M.A
Ir. I Dewa Agung. A.R.R. Adi, MP
Herry Aprilia Manubulu, S.Sos., M.Si
David Bobo
Br. Angelinus Nadut, S.Pd
Wilson Lodewik Tisera,SPi,Msi. P.hD
Regina. I. M. BanoEt, S.P., MP
Dr. Anggraeni D.N. Rupidara, S.Si., M.Si
Dr. Andriani Rafael, S.Si., M.Si
Drs. Marianus Kleden, M.Si
Ir. Jemmy. J.S. Dethan, MP
Ena Rede Bire, S.Sos
Lai Valentin Riwu Kaho
Marten Luther Lano, S.TP., MP
Ir. F. J. Haba Bunga, MP
Arlindo Umbu S. Kette, S.P., M.Si
Lodowik Landi Pote, S.Si., M.Sc
Jonny Dalla
Gabriela E. Hetharia, S.P., M.Sc
Wenny Kana, S.H
Gradict Semiun, S.Si., M.Si
Ir. Zet Malelak, M.Si
Ir. Jery. H. Mahubesy, M.T
Dr. Nikodemus P.P.E. Nainiti, S.TP., MP

IOP Conf. Series: Materials Science and Engineering **823** (2020) 011002 doi:10.1088/1757-899X/823/1/011002

Peer review statement

All papers published in this volume of *IOP Conference Series: Materials Science and Engineering* have been peer reviewed through processes administered by the proceedings Editors. Reviews were conducted by expert referees to the professional and scientific standards expected of a proceedings journal published by IOP Publishing.

IOP Conf. Series: Materials Science and Engineering **823** (2020) 012001 doi:10.1088/1757-899X/823/1/012001

Molecular modelling of antioxidant agent by QSAR study of caffeic acid derivatives

Istiqomah Rahmawati[1], Helda Wika Amini[1], and Rizki Fitria Darmayanti[1]

[1]Department of Chemical Engineering, Faculty of Engineering, Universitas Jember, Jember, 68261, Indonesia

E-mail: istiqomah.rahmawati@unej.ac.id

Abstract. Molecular modeling using Quantitative Structure and Activity Relationship (QSAR) has been performed on caffeic acid derivatives which is previously studied as an effective antioxidant agent. This research focuses on a set of experimentally IC_{50} value data of 4 caffeic acid derivatives. The mathematical method (i.e., multilinear regression calculation) was used to build the QSAR model. QSAR analysis was employed on fitting subset using log ($1/IC^{50}$) as a dependent variable and atomic net charges aromatic carbons, dipole moment and partition coefficient in n-octanol/water as independent variables. The PM3 method was used to calculate the quantum chemical descriptors, chosen to represent the electronic descriptors of molecular structures. The relationship between log ($1/IC_{50}$) and the descriptors was described by resulted in the QSAR model. The resulted QSAR model for caffeic acid derivatives as an antioxidant is presented below:

$$-7.858+1.149\text{dipol}+0.485\log P-61.68C5$$

$$R=1;\ R^2=0.999;\ SE=0.008;\ F=342$$

QSAR model for caffeic acid derivatives showed the enhancement of antioxidant activity due to the decrease of electronic properties (derived from the dipole moment value and C5 atomic charge), Log P representing hydrophobicity did not show a significant effect on antioxidant activity while increasing the chain length of antioxidant molecules indicate an increase in steric hindrance causes a decrease in antioxidant activity. The calculated PRESS (Predicted Residual Error Sum of Square) value was 6.69E-05, which indicates the calculated log ($1/IC_{50}$) using QSAR Hansch Model of caffeic acid derivatives is similar with experimental data.

1. Introduction

Environmental pollution, unhealthy lifestyles, and current community habits can spur the growth of free radicals that can damage the body. Bad habits such as consuming fried foods, fatty foods, cholesterol foods, and low-fiber foods can cause deadly diseases such as cancer, heart disease, hypertension, and stroke. Various diseases are caused by free radicals [1].

The efforts to prevent or reduce the degenerative diseases caused by free radical activity are by consuming foods that contain antioxidants. Antioxidants are substances that can delay and prevent the occurrence of free radical reactions. Based on the source, antioxidants are divided into two groups, namely synthetic antioxidants and natural antioxidants. Synthetic antioxidants are such as butyl hydroxyanisole (BHA), butylhydroxytoluene (BHT) and propyl gallate (PG). Many foods can be a source of natural antioxidants, such as spices, tea, chocolate, and vegetables. Most sources of natural antioxidants are plants and generally are phenolic compounds that are spread throughout all parts of plants both in wood, seeds, leaves, fruit, roots, flowers and pollen [2-6].

Phenolic compounds have been widely reported of their activity as natural antioxidants. Caffeic acid is one example of phenolic compounds that are mostly contained in coffee, apples, artichokes, and pears. Caffeic acid is working to press cancer and pro-inflammatory gene expression, as well as inhibiting enzymes inducing free radical and oxygen peroxidation. Caffeic acids and their derivatives have various mechanism which determines their bioactivity [7-11].

The new antiradical compounds can be found by molecular designs, by direct synthesis, or by a modeling approach using computational chemistry concepts. One application of computational chemistry that can be applied is the study of Quantitative Structure-Activity Relationship (QSAR) or quantitative relations of activity structure. QSAR studies quantitative correlations between molecular structure and the value of experimentally measured biological activities [12-13].

The research will analyze the relationship between the chemical structure and the bioactivity of 5 caffeic acid derivatives. In this study, QSAR analysis of caffeic acid derivative was carried out by using antiradical activity studies as a function of steric, hydrophobic, and electronic variables.

2. Methods

The material used in this study is antioxidant activity (expressed in IC_{50}) caffeic acid derivative compounds obtained from the literature. The compound antioxidant activity data are presented in Table 1. The structure of the caffeic acid derivative compound is shown in the following Figure 1.

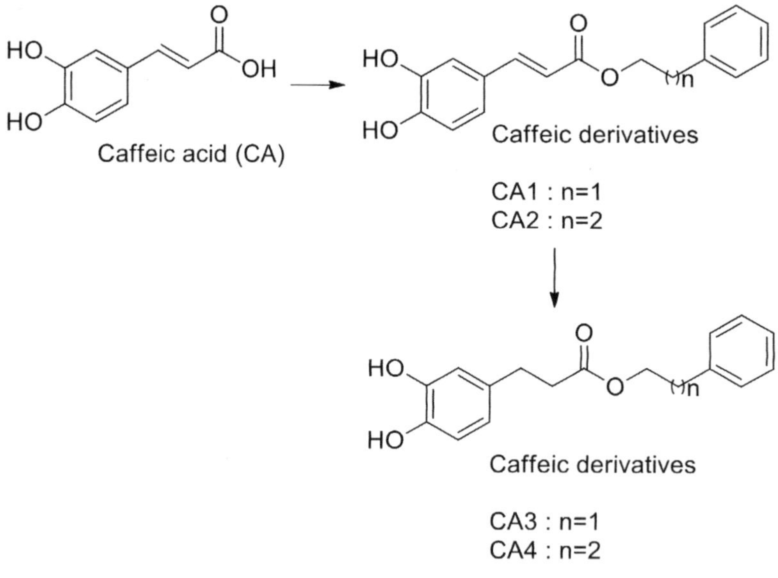

Figure 1. Caffeic acid derivatives [14].

Table 1. The experimental antioxidant activity of caffeic acid derivatives [14]

Compound	Antioxidant Activity	
	IC_{50}ˣ	Log ($1/IC_{50}$)ˣ
CA	15.3	-1.18
CA1	16.5	-1.22
CA2	11.9	-1.08
CA3	24.6	-1.39
CA4	23,0	-1.36

ˣDPPH assay was used to assess the free radical scavenging activity

IOP Conf. Series: Materials Science and Engineering **823** (2020) 012001 doi:10.1088/1757-899X/823/1/012001

2.1. Equipment
This research used Intel (R) Celeron (C) processor with 204 MB RAM as computer hardware. Hyperchem 7.02 computational chemistry software for computational chemical calculations and SPSS 16.020 for statistical analysis, and Chemdraw Ultra 8.0 to create 2D molecular structures.

2.2. Research procedure

2.2.1. Collection of descriptors. Descriptor determination is important for determining the best QSAR equation. Descriptors used in this study are atomic charge, dipole moment (μ) and logarithm of partition coefficient (log P). The data can be seen in Table 2. For electronic descriptors, calculations are carried out covering the stages of computational chemistry modeling with the geometry optimization procedure of each compound structure. Each compound used in this study was made into a two-dimensional structure model using the Chemdraw Ultra 8.0 application. After that, the model is equipped with hydrogen atoms on each atom and is formed into a three-dimensional (3D) structure with the Build (Add H and Model Build) menu. The descriptors are atomic net charge, dipole, and log P (coefficient partition n-octanol/water). We analyze the atomic net charge of carbon numbers 11, 10, 9, and 5 for caffeic acid derivatives. The position of the carbon is shown in Figure 2. Atomic net charge is determined by comparing the number of proton (positive charge) and electron (negative charge) in an atom. If an atom gains electron then it has more electrons, which give it a negative charge, and vice versa. The atomic charge indicates the density of electron in atom. Dipole moment describes the polarity of molecules. Besides, log P represents hydrophobicity properties of a molecule.

Table 2. List of descriptors and how to optimize them

No	Symbol	Descriptor	Unit	Calculation Method
1	qC_5, qC_9, qC_{10}, qC_{11}	The atomic charge of C_5, C_9, C_{10}, and C_{11}	Coulomb	Semiempirical method of PM3, Hyperchem, compound optimization
2	μ	dipole moment	Debye	Semiempirical method of PM3, Hyperchem, compound optimization
3	Log P	Partition coefficient of n-octanol/water	-	QSAR Properties, Semiempirical method of PM3, Hyperchem, compound optimization

The next procedure is to perform structural geometry optimization by minimizing molecular energy in order to obtain the most stable molecular conformation. The calculation used the semiempirical method of PM3 with the convergence limit was 0.01 kcal / Å.mol for caffeic acid derivatives. The optimization method was based on the Polak-Ribiere algorithm.

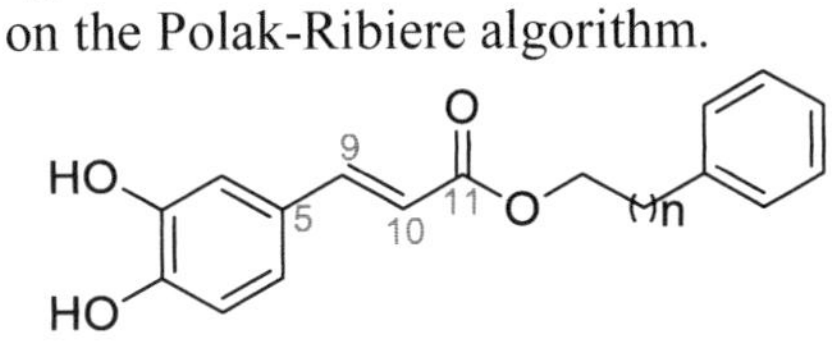

Figure 2. The label of selected carbon net charge

2.2.2. QSAR study. The best QSAR equation used to predict IC_{50} was determined by multilinear regression statistical analysis. The analysis was performed using SPSS application with the backward method. In this study, the original data using four compounds for caffeic acid derivatives were optimized by select the independent and dependent variables that influence the QSAR equation. Log

IOP Conf. Series: Materials Science and Engineering **823** (2020) 012001 doi:10.1088/1757-899X/823/1/012001

($1/IC_{50}$) data was set as the independent variable and combination value between dipole moment, atomic charge, and partition coefficient as independent variables. In general, the final relationship of the QSAR approach is expressed by the regression equation as follows

$$\text{Log}\ (1/IC^{50}) = k1\log P + k2qC1 + k3qC2 + k4qC3 + k5qC4 + k6qC5 + k7qC12 + k8\mu + k9$$

Multilinear regression analysis is performed on the output data by statistical parameters as correlation coefficient R, R^2, standard deviation (SE) and significance value. The model is chosen based on the consideration of the statistical results obtained.

3. Results and Discussion

3.1. The result of descriptors calculation on caffeic acid derivatives

In this study, we investigate caffeic acid (CA) and four caffeic acid derivatives (CA_1, CA_2, CA_3, and CA_4) using QSAR method between dependent variable antioxidant activity (Table 2) and eight independent descriptors. In addition, four caffeic acid derivatives, synthesized by LeBlanc and co-workers (2012), are shown in Figure 2. CA_1 is synthesized by the incorporation of a phenyl ethyl ester to the carboxylic moiety on caffeic acid structure. Besides, incorporation phenyl propyl ester to the carboxylic moiety on caffeic acid structure generated CA_2. CA_2, a well-known compound, named CAPE (caffeic acid phenyl ethyl ester), is the most active compound in propolis [15]. Furthermore, CA_1 and CA_2 were converted to CA_3 and CA_4, respectively by hydrogenation reaction of caffeic acid. Table 2 listed the experimental result of antioxidant activity using DPPH assay. It is seen that CA, CA_1, and CA_2 have ethylene moiety which affords higher antioxidant activity (IC_{50}=12-15 μM) than their counterpart CA_3 and CA_4 (IC_{50}=23-25 μM).

Table 3 showed that modification of carboxylic moiety in caffeic acid structure generated higher log P in relative to those of CA_1-CA_4. Also, it is observed that the dipole of caffeic acid decreased, the CA_4 has the lowest dipole value. From CA to CA_4, the net charges of C_{11}, C_{10}, C_9 are decreased. Besides, the net charge in C_5 is only slightly decreased from CA-CA_4.

Table 3. Descriptor data as independent variables for caffeic acid derivatives

No	Compound	Atomic Net Charge (Coulomb)				Dipole (deybe)	Log P
		C_{11}	C_{10}	C_9	C_5		
1	CA	0.420043	-0.200263	0.012367	-0.083385	1.611	-0.66
2	CA_1	0.414012	-0.203953	0.006267	-0.079643	1.236	0.64
3	CA_2	0.412319	-0.201729	0.001791	-0.0786	1.246	1.03
4	CA_3	0.369905	-0.104317	-0.041991	-0.08621	0.759	0.56
5	CA_4	0.368428	-0.103922	-0.042344	-0.086329	0.622	0.95

The results of antioxidant activity are shown as IC_{50} (μmolar) of caffeic acid, and the derivatives are listed in Table 2. Generally, all compounds have good inhibitor activity (IC_{50} = 10-200 μmolar). CA_2 shows the highest activity compared to other CA derivatives. CA and CA_1 produce low cytotoxic activity. It could occur because the structures of CA and CA_1 produce electron deficiency conditions that interfere oxidation/reduction reactions of the catechol ring. The electron deficiency condition was caused by the high polarity of CA and CA_1 where it induced the electron density headed to the ester group. Meanwhile CA_3 and CA_4 produce low antioxidant activity. It could occur because steric hindrance conditions of the structure of CA_3 and CA_4 disrupt oxidation/reduction reactions of the catechol ring [16]. This condition caused low dipole value and high log P value on CA_3 and CA_4 does not affect antioxidant activity on CA_3 and CA_4.

3.2. Analysis of QSAR on caffeic acid derivatives

Table 4 denotes the best QSAR equation for caffeic acid derivatives. The descriptors such as dipole, logP, and atomic charge of C_5 have a strong correlation to the antioxidant activity. The results exhibit antioxidant activity influenced by a dipole, log P and atomic charge of C_5. Dipole values and the atomic charge of C_5 represent electronic variables, log P represents hydrophobicity, and chain length represents steric variables. The results showed the highest antioxidant activity value was obtained from the CA3 structure which showed the effect of low dipole values and atomic charge of C_5 as shown in table 3 and table 4. This shows an increase in antioxidant activity due to decreased electronic properties of antioxidants. The P log representing hydrophobicity did not show a significant effect on antioxidant activity. The steric effect shows a significant effect on antioxidant activity which can be seen from the decreased value of antioxidant activity with increasing chain length which indicates that steric barriers are getting higher.

Table 4. The result of the best correlation between descriptors and antioxidant activity for caffeic acid derivatives

Equation	$-7.858+1.149\text{dipol}+0.485\log P-61.68C_5$
R	1
R^2	0.999
SE	0.008144
Sig	0.04
F	342

Table 5 showed the residual errors between the experimental log ($1/IC_{50}$) and the calculated Log ($1/IC_{50}$) is ignorable with the value of 6.69E-05. It indicates the calculation of the calculated log ($1/IC_{50}$) using QSAR Hansch Model for caffeic acid derivatives has excellent agreement with experimental data of Log ($1/IC_{50}$).

Table 5. Experimental log ($1/IC_{50}$), calculated log ($1/IC_{50}$), and PRESS value for caffeic acid derivatives

Compound	Experimental Log ($1/IC_{50}$)	Calculated Log ($1/IC_{50}$)	Residual error	[Residual error]^2
CA	-1.185	-1.183	-0.001	1.96E-06
CA1	-1.217	-1.214	-0.003	8.91E-06
CA2	-1.076	-1.078	0.003	7.03E-06
CA3	-1.391	-1.396	0.005	2.85E-05
CA4	-1.362	-1.357	-0.005	2.05E-05
		PRESS		6.69E-05

4. Conclusion

This research analyzed the relationship between the chemical structure and the bioactivity of 5 caffeic acid derivatives. In this study, the enhancement of antioxidant activity occurred due to decreased electronic properties (derived from the dipole moment value and C5 atomic charge), Log P representing hydrophobicity did not show a significant effect on antioxidant activity, while increasing the chain length of antioxidant molecules indicate an increase in Steric hindrance causes a decrease in antioxidant activity. The resulted QSAR model for caffeic acid derivatives as an antioxidant is presented below:

$$-7.858 + 1.149 \text{ dipole} + 0.485 \log P - 61.68 \text{ C5}$$
$$R=1;\ R^2=0.999;\ SE=0.008;\ F=342$$

In the future study, this research suggests designing good partition coefficient of n-octanol/water and dipole moment caffeic acid derivatives giving excellent bio-activity.

References

[1] Bogucka-Kocka A, Zidorn C, Kasprzycka M, Szymczak G and Szewczyk K 2018 Phenolic acid content, antioxidant and cytotoxic activities of four kalanchoë species *Saudi Journal of Biological Sciences* **25** 622-30
[2] Đurović S, Nikolić B, Luković N, Jovanović J, Stefanović A, Šekuljica N, Mijin D and Knežević-Jugović Z 2018 The impact of high-power ultrasound and microwave on the phenolic acid profile and antioxidant activity of the extract from yellow soybean seeds *Industrial Crops and Products* **122** 223-31
[3] Irakli M, Chatzopoulou P and Ekateriniadou L 2018 Optimization of ultrasound-assisted extraction of phenolic compounds: Oleuropein, phenolic acids, phenolic alcohols and flavonoids from olive leaves and evaluation of its antioxidant activities *Industrial Crops and Products* **124** 382-88
[4] Shen Y, Sun H, Zeng H, Prinyawiwatukul W, Xu W and Xu Z 2018 Increases in phenolic, fatty acid, and phytosterol contents and anticancer activities of sweet potato after fermentation by Lactobacillus acidophilus *J. Agric. Food Chem.* **66** 2735–41
[5] Li R, Narita R, Nishimura H, Marumoto S, Yamamoto S P, Ouda R, Yatagai M, Fujita T and Watanabe T 2018 Antiviral activity of phenolic derivatives in pyroligneous acid from hardwood, softwood, and bamboo *ACS Sustainable Chem. Eng.* **6** 119–26
[6] Sun Y, Hui Q, Chen R, Li H, Peng H, Chen F and Deng Z 2018 Apoptosis in human hepatoma HepG2 cells induced by the phenolics of Tetrastigma hemsleyanum leaves and their antitumor effects in H22 tumor-bearing mice *Journal of Functional Food* **40** 349-64
[7] Sricharoen P, Lamaiphan N, Patthawaro P, Limchoowong N, Techawongstien S and Chanthai S 2017 Phytochemicals in Capsicum oleoresin from different varieties of hot chilli peppers with their antidiabetic and antioxidant activities due to some phenolic compounds *Ultrasonics Sonochemistry* **38** 629-39
[8] Khan S, Khan T and Shah A J 2018 Total phenolic and flavonoid contents and antihypertensive effect of the crude extract and fractions of Calamintha vulgaris *Phytomedicine* **47** 174-83
[9] Martinez J R M 1988 Phenolic compounds in coffee pulp: Quantitative determination by HPLC *J. Sci. Food Agric.* **43** 135–44
[10] Coronel M, Marnet N, Kolli V, Roussos S, Guyot S and Augur C 2004 Characterization and estimation of proanthocyanidins and other phenolics in coffee pulp (Coffea arabica) by thiolysis-high-performance liquid chromatography *J. Agric. Food Chem* **52** 1344–49
[11] Muranaka K 2001 Anticancer activity of estradiol derivates:a quantitative structure-activity relationship approach *J. Cem. Educ* **78** 1390-93
[12] Chen J C, Li Q, Yong S, Mei L C and Zheng K C 2008 A QSAR study and molecular design of benzothiazole derivatives as potent anticancer agents *Science in China Series B: Chemistry* **51** 111-19
[13] Datar A P 2014 2D-QSAR Study of indolylpyrimidines derivative as antibacterial against Pseudomonas aeruginosa and Staphylococcus aureus: A comparative approach *Journal of Computational Medicine* 1-9
[14] LeBlanc L M, Pare A F, Jean-Francois J, Hebert M J G, Surette M E and Touaibia M 2012 Synthesis and antiradical/antioxidant activities of caffeic acid phenethyl ester and its related propionic acetic, and benzoic acid analogues *Molecules* **17** 14367-650
[15] Nie J, Chang Y, Li Y, Zhou Y, Qin J, Sun Z and Li H 2017 Caffeic Acid Phenethyl Ester

(Propolis Extract) Ameliorates Insulin Resistance by Inhibiting JNK and NF-κB Inflammatory Pathways in Diabetic Mice and HepG2 Cell Models *Journal Agricultural Food and Chemistry* **65** 9041–53

[16] Ketabforoosh S, Amini M, Shafiee A, Azizi E and Kobarfard F 2013 Synthesis, evaluation of anticancer activity and QSAR study of heterocyclic esters of caffeic acid *Iran J Pharm Res* **12** 705–19

IOP Conf. Series: Materials Science and Engineering **823** (2020) 012002 doi:10.1088/1757-899X/823/1/012002

Antioxidant extraction based on black rice (*Oryza Sativa L. Indica*) to prevent free radical

Gabriela E Hetharia[1], Anjar Briliannita[2], Mary Astuti[3], Y Marsono[3]

[1] Department of Agriculture Product's Technology, Universitas Kristen Artha Wacana, Kupang
[2] Department of Nutrition, Politeknik Kesehatan Kemenkes Sorong, Papua
[3] Department of Science and Food Technology, Universitas Gadjah Mada, D.I. Yogyakarta

E-mail: gelysia1705@gmail.com

Abstract. Free radicals are molecules that in their outer orbit have one or more unpaired electrons; they are very labile and very reactive. These molecules have an important role in tissue damage and pathological processes in living organisms. Antioxidants are compounds that can inhibit oxidation reactions by binding free radicals. The antioxidants produced by the human body are very limited, so we need the intake of antioxidants from outside, especially those from food. Research on antioxidant extraction based on black rice (Oryza Sativa L.Indica) to prevent free radicals has been carried out. This research was conducted using the DPPH method to measure the ability to capture free radicals and FRAP methods to measure antioxidant capacity. This study aimed to obtain data on the level of antioxidant activity contained in black rice and its effect to counteract free radicals. The result showed that the ability to capture free radicals contained 25ppm DPPH anthocyanin extract could inhibit 55,00%. While using the FRAP method, the results of black rice anthocyanin extract have a high antioxidant capacity of 824 ± 17.24 μM. It is evident that the DPPH method used to show in anti-oxidant extracts based on black rice can capture free radicals, while the FRAP method proves that antioxidants in black rice have the capacity to prevent free radicals in the body.

1. Introduction

Free radicals are defined as atoms or molecules that have one or more unpaired electrons [1]. Molecules that lose electrons can become radical and may attract electrons from other molecules that are close to them so that a chain reaction occurs [2]. Free radicals can be formed from the body (endogenous) and outside the body (exogenous). The formation of free radicals can be inhibited by the presence of antioxidants. Antioxidants are chemical compounds that can contribute one or more electrons to free radicals so that free radicals can be suppressed [3]. The human body does not have excess antioxidant reserves, so if there is excessive radical exposure, the body needs exogenous antioxidants. Bioactive compounds in food can act in a variety of biological activities, for example, as antioxidants in the body. The role of food today is not only to meet the nutritional needs and giver of satiety but is also expected to be beneficial for health (functional food). One type of food that is very close to the people of Indonesia is rice. Rice (*Oryza sativa*) is a staple food in Indonesia and based on the color of the pericarp there are several kinds of colored rice, such as brown rice, black rice, brown rice. Rice colors such as red, black, and brown contain antioxidant anthocyanins and dyes that are found in the aleuron layer also contain vitamins, especially the group of vitamin B and minerals, especially Fe and Zn. Bioactive compounds that cause pigment in rice are anthocyanin and proanthocyanidin. Hosoda et al. [4] reported that black rice and brown rice contain anthocyanin and proanthocyanidin which are potentially used as sources of antioxidants other than as a source of starch in ruminants. Seawan et al. [5] also stated that pigmented rice has high antioxidant activity potential due to the high content of bioactive compounds. Bioactive compounds in pigmented rice can reduce oxidative stress, prevent cancer, cardiovascular, diabetes complications, and others [6]. One type of antioxidant that has revealed many physiological effects on health is anthocyanin. Anthocyanin is a pigment naturally belonging to the flavonoid group which is responsible for color red, purple and blue in food. The main anthocyanin in black rice is cyanidin-3-glucoside (C3G) which is an important source of anthocyanin in Asia. Other than that, Black rice contains active phytochemicals such as tocopherol, tocotrienol, oryzanols, B complex vitamins, and phenolic compounds [7]. Anthocyanins are a class of flavonoids that are found in many foods that are red, blue, purple and black, but also are water-soluble so they are easily absorbed into the body. The main function as an antioxidant, both inside the cell (intracellular) and outside the cell (extracellular). As an antioxidant anthocyanin can capture radical compounds by donating electrons or transferring hydrogen atoms with hydroxyl groups to free radicals. At this stage, anthocyanin acts as a catcher of ROS at the initiation stage and breaks the chain of propagation in lipid peroxidation. Therefore, it is deemed necessary to conduct research to see the ability of anthocyanin extracts from black rice to counteract free radicals. In this study, extraction was carried out for aleuron from black rice, which is the outer layer of rice containing black pigment, in contrast to previous studies that used bran or even whole grains of black rice.

2. Material and Method

The research material used was black rice from Bantul, Yogyakarta. The farmers' harvested black rice is brought to the UGM Faculty of Agricultural Technology, sorted to be separated with dirt, and dried to a moisture content <13%. Then put in a two-layer sack and stored in a dry room. Anthocyanin extraction is done by maceration using ethanol and citric acid. The selection of this method which is to combine the solvent with the addition of citric acid is expected to maximize the extraction process. Robinson (1995) in Tensiska [8] states that extraction of flavonoid compounds is recommended in an acidic environment, because acid serves to denature the growth of tana man cells, then dissolves anthocyanin pigments so that they can get out of cells, and can prevent flavonoid oxidation. While organic acids which are often used for the extraction of pigments are hydrochloric acid, citric acid, and acetic acid. The flow diagram of the black rice anthocyanin extraction process can be seen in Figure 1.

IOP Conf. Series: Materials Science and Engineering **823** (2020) 012002 doi:10.1088/1757-899X/823/1/012002

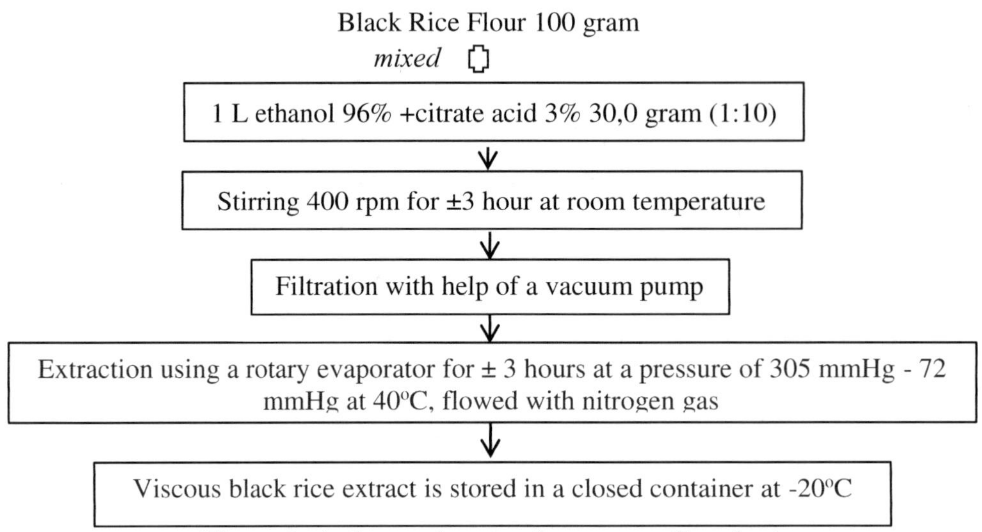

Figure 1. Anthocyanin extraction process flowchart

2.1. DPPH (1,1- diphenyl–2– picrylhydrazyl) method

The antioxidant activity test of black rice extract was carried out by the DPPH method [9]. The crude extract of black rice dissolved in methanol p.a. to obtain concentrations of 25 ppm, 50 ppm, 75 ppm and 100 ppm. Controls using vitamin C are 100 ppm. DPPH solution is prepared by dissolving DPPH crystals in a methanol solvent p.a. with a concentration of 1 mM. The process of making a 1 mM DPPH solution is carried out in low temperatures and protected from sunlight. The extract solution and the BHT antioxidant solution were each taken 4.50 ml and reacted with 500 μl DPPH 1 mM solution in different test tubes. The reaction took place at 37°C for 30 minutes then the absorbance was measured using a UV-VIS spectrophotometer at a wavelength of 517 nm. The absorbance of the blank solution was measured to calculate the percent inhibition. A blank solution was made by reacting to 4.50 ml of methanol solvent with 500 mL of 1 mM DPPH solution in a test tube. The antioxidant activity is expressed in percent inhibition, which is calculated by the formula:

$$\%\,Inhibition = \frac{absorbance\ blanks - absorbance\ of\ sample}{absorbance\ blanks} \times 100\% \qquad (1)$$

2.2. FRAP (Ferric Reducing Antioxidant Power) method

Analysis of total antioxidant activity by the FRAP method, FeSO4.7H2O as standard [10]. In the manufacture of FRAP reagents, 0.1 M acetate buffer solution (pH 3, 6), TPTZ solution (2,4,6-tripydyl-s-triazine) 10 mM in 40 mM HCl and 20 mM FeCl3.6H2O solution are prepared in advance then the solution is mixed in a ratio of 10: 1: 1. A total of 50 μL of sample solution and 150 μL of distilled water were added to the tube which contained 1.5 mL of FRAP reagent. The mixture was incubated for 8 minutes in a dark place at room temperature. The sample absorbance was measured at a wavelength of 594 nm and the results were calculated in Fe + 2 equivalent (Fe + 2mM) using a standard FeSO4.7H2O curve.

3. Result and Discussion

3.1. DPPH

In this study, antioxidant activity was tested using the DPPH method. Vitamin C and vitamin E 100 ppm were used as positive controls. Variations in the levels of extracts used were 25 ppm, 50 ppm, 75 ppm, and 100 ppm anthocyanin. Vitamins C and E are used as positive controls because this type of antioxidant vitamin is often added to the product. The results of the analysis of antioxidant activity using the DPPH method can be seen in Table 1 and Figure 2. The results showed that the decrease in absorbance occurred as a result of changes in color from purple to yellow, DPPH radical compounds trapped by anthocyanins by donating hydrogen atoms to form DPPH-H complex [11]. The higher the anthocyanin level in black rice extract, the higher the ability of DPPH radical capture will be due to the low black rice extract concentration which will reduce the number of hydrogen atoms, this condition is in accordance with the previous research [12].

The ability to capture DPPH radicals possessed by 25 ppm anthocyanin extract is higher than vitamin C 100 ppm; this condition can occur because anthocyanin extract has the ability to donate hydrogen atoms with the anthocyanin component which majority has a hydroxyl group in position 3 which is very easy to escape due to enthalpy the energy of the hydroxyl group bond on ring B is the lowest [13]. Antioxidant activity in this study showed that the inhibition ability of 25 ppm extract was 55.00% greater than the antioxidant activity using DPPH method with methanol as a solvent in a previous study by Wanti et al. [14] which was 46.20%.

Table 1. Antioxidant activity of black rice extract using DPPH method

C(ppm)	A	% Inhibition
25	0,44	55,00
50	0,27	60,86
75	0,18	73,91
100	0,15	76,81
Vitamin C (100)	0,66	4,30

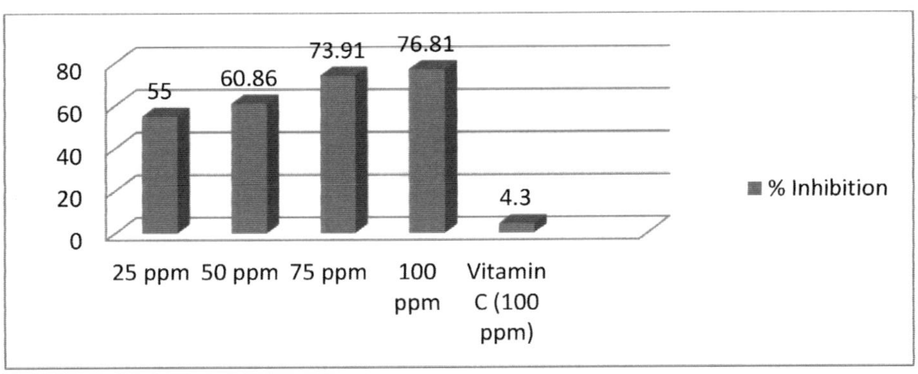

Figure 2. Ability to inhibit free radical by black rice anthocyanin extract

3.2. FRAP

In this study, testing the antioxidant capacity of black rice extract using the FRAP method. The results of black rice anthocyanin extract have a high antioxidant capacity with 3 replications of analysis of 824 μM. Determination of antioxidant activity by FRAP test measures the ability of antioxidants to reduce Fe + 3 in Fe3 + -TPTZ complexes into Fe + 2-TPTZ by donating electrons. The high antioxidant activity in the RR watering treatment is related to the high content of the phenolic compounds and the two flavonoids which are capable of reducing Fe + 3 [15].

Measurement of antioxidant activity can be done by several methods including CUPRAC, DPPH, and FRAP. The FRAP method [16] uses the Fe (TPTZ) 23+ 2,4,6-tripiridyl-triazine besilligan as a reagent. The blue Fe complex (TPTZ) 23+ will function as an oxidizing agent and will be reduced to Fe (TPTZ) 22+ which is yellow with the following reaction:

$$Fe(TPTZ)_2^{3+} + AROH \rightarrow Fe(TPTZ)_2^{2+} + H+ + AR=$$

So this shows that anthocyanin extract from black rice is a good antioxidant and can prevent free radicals in the body.

Table 2. Antioxidant activity of black rice extract using FRAP method

Sample	Concentration of Antioxidant (μM or μmol/L)	Concentration of Antioxidant (μmol/100 gram)
Extract	820,667 ± 17,243	820,667 ± 17,243

4. Conclusion

Based on research that has been done, the conclusions can be drawn that It is evident the DPPH method used to show in anti-oxidant extracts based on black rice can capture free radicals. The ability to capture DPPH radicals possessed by 25 ppm anthocyanin extract is higher than vitamin C 100 ppm, while the FRAP method proves that antioxidants in black rice have the capacity to prevent free radicals in the body.

References

[1] Florence T M 1995 The role of free radicals in disease *Australian and New Zealand Journal of Ophthalmologists* **23**(1) 3–7
[2] Phaniendra A and Jestadi D B 2015 Free radicals : Properties, sources, targets, and their implication in various diseases *Indian Journal of Clinical Biochemistry* **30**(1) 11–26
[3] Suhartono E, Fujiati and Aflanie I 2002 Oxygen toxicity by radiation and effect of glutamic pyruvat transamine (GPT) activity rat plasma after vitamine C treatment *International Environmental Chemistry and Toxicology* (Yogyakarta, Indonesia)
[4] Hosoda K, Sasahara H, Matsushita K, Tamura Y, Miyaji M and Matsuyama H 2018 Anthocyanin and proanthocyanidin contents, antioxidant activity, and *in situ* degradability of black and red rice grains *Asian-Australas Journal of Animal Sciences* **31**(8) 1213–20
[5] Seawan N, Vichit W, Thakam A, Thitipramote N, Chaiwut P, Pintathong P and Thitilertdech N 2014 Antioxidant capacities, phenolic, anthocyanin and proanthocyanidin contents of pigmented rice extracts obtained by microwave-assisted method *Suranaree Journal of Science and Technology* **21**(4) 301–6
[6] Walter M and Marchesan E 2011 Phenolic compounds and antioxidant activity of rice *Brazilian Archives of Biology and Technology* **54**(2) 371–77
[7] Jang H H, Park M Y, Kim H W, Lee Y M, Hwang K A, Park J H, Park D S and Kwon O 2012 Black rice (Oryza sativa L.) extract attenuates hepatic steatosis in C57BL/6 J mice fide a high-fat diet via fatty acid oxidation *Nutrition and Metabolism* **9**(27)
[8] Tensiska, Sukarminah E and Natalia D 2006 *Extraction of natural dyes from Arben (Rubus idaeus (Linn.)) and their applications in the food system* Available from: http://digilib.umm.ac.id
[9] Hanani E, Mun'im B and Sekarini R 2005 Identifikasi senyawa antioksidan dalam spons Callispongia Sp dari kepulauan seribu *Majalah Ilmu Kefarmasian* **2**(3) 127-33
[10] Vichitphan S, Vichitphan K and Sirikhansaeng P 2007 Flavonoid content and antioksidan activity of krachaidum (*Kaemferia parviflora*) wine *Journal of Science Technology* **7** 97-105
[11] Rufino M S M, Alves R E, Brito E S D, Perez-Jimenez J, Saura-Calixto F and ManciniFilho J 2010 Bioactive compounds and antioxidant capacities of 18 nontraditional tropical fruits

from Brazil *Food Chemistry* **121** 996-1002

[12] Hou D X 2003 Potential mechanisms of cancer chemoprevention by anthocyanins *Cur. Mole. Med.* **3** 149-59

[13] Vaya J, Mahmood S, Goldblum A, Aviram M, Volkova N, Shaalan A, Musa R and Tamir S 2003 Inhibition of LDL oxidation by flavonoids in relation to their structure and calculated enthalphy *Phtochemistry* **62** 89-99

[14] Wanti S, Andriani M A M and Parnanto N H R 2015 The effect of various types of rice on antioxidant activity in red yeast rice by Monascus purpureus *Biopharmaceutical* **13**(1) 1–5

[15] Prior R L, Wu X and Schaich K 2005 Standardized methods for the determination of antioxidant capacity and phenolics in foods and dietary supplements *Journal of Agricultural and Food Chemistry* **53** 4290-302

[16] Benzie I F F and Strain J J 1996 The ferric reducing ability of plasma (FRAP) as a measure of ''Antioxidant Power'': The FRAP assay *Analytical Biochemistry* **239** 70-7

IOP Conf. Series: Materials Science and Engineering **823** (2020) 012003 doi:10.1088/1757-899X/823/1/012003

Analysis of production kojic acid from endophytic fungi *Aspergillus flavus* isolated from *Annona squamosa* leaves using an OSMAC Approach

Antonius R B Ola[1,2], Yoseph Sugi[3], Christina A Pada Soa[1], Reiner I Lerrik[1], Bibiana D Tawa[1]

[1]Chemistry Department, Faculty of Science and Engineering, Nusa Cendana University, Kupang 85118, Indonesia
[2]Integrated Center Research (Biosains) Laboratory, Nusa Cendana University, Kupang 85118, Indonesia
[3]Faculty of Science and Engineering, Nusa Cendana University, Kupang 85118, Indonesia

E-mail: antonius.ola@staf.undana.ac.id

Abstract. An analysis production of kojic acid by endophytic fungi *Aspergillus flavus* from leaves of *Annona squamosa* has been undertaken using three different types of media including rice (Oryza sativa), sweet corn (Zea mays L.) and waxy corn (Zea mays ceritina). The fungi were extracted using ethyl acetate and quantitatively analyzed their content of kojic acid using HPLC. The intrapolation peak area of kojic acid's content in each fungal medium to the regression equation of standard kojic acid resulted in the concentration of kojic acid in each extract 3.149% for rice media, 5.998% for sweet corn media and 2.226% for waxy corn media. Pure kojic acid from an extract of fungi grown on rice media was obtained from 90% DCM fraction by VLC fractionation. The structure was determined by means of NMR and HPLC analysis data.

1. Introduction

The discovery of Taxol (paclitaxol), a high valuable anticancer drug produced by the endophytic fungi *Taxomyces andreana* [1] and also other drugs by endophytic fungi such as antifungal agents' griseofulvin from *Penicillium griseofulvum* [2] together with inhibitor of cholesterol biosynthesis lovastatin from *Aspergillus terreus* [3] have replaced the focus of the drugs discovery program from plants to endophytic microorganisms. Moreover, endophytic fungi that live in plant tissues are known to produce chemical compounds similar to those produced by their hosts. Several bioactive metabolites had been reported from endophytic fungi [4-6]. One of the bioactive fungal metabolites kojic acid was recently reported produced in large quantities by endophytic fungi *Aspergillus flavus* [6].

Kojic acid (5-hydroxy-2-hydroxymethyl-pyrone) has attracted much attention due to its wide application in cosmetic, medical, pharmaceutical, agricultural and food industries. It has been used as an anti-inflammatory drug and painkiller in the medical sector and as a skin whitening agent, antioxidant and also as protection against UV radiation [7]. Most of kojic acid is produced by fungi from the genus *Aspergillus spp.* and *Penicillium ssp.* [8] especially from the species *Aspergillus flavus* [9]. Several media have been used for production as kojic acid such as using cooked starch, yielded 0.25 g of kojic acid/g of corn starch [10].

One technique to expand and to increase the production of the secondary metabolite from endophytic fungi is by applying the concept of OSMAC (***One Strain-Many Compound)*** including using different fermentation conditions to produce a variety of bioactive compounds [11]. This simple and effective approach can be undertaken by changing cultivation parameters such as media composition, aeration, pH, temperature, time of cultivation, cultivation vessel and addition of enzyme inhibitors [12].

Several reports have revealed that changes in the media of cultivation had an effect on the metabolite production of some endophytic fungi. For example, endophytic fungi *Chaetominium Chiversii* cultivated in liquid media and solid media showed different secondary metabolites profile [13]. Moreover, endophytic fungi *Fusarium tricinctum* cultivated in solid rice media added with vegetable and fruit juices caused an increase in the number of fusarielin J by 80 times [14]. Therefore, changing cultivation media may result in the changes profile and the production of secondary metabolites from endophytic fungi.

2. Experimental Method

2.1. Fungal material

Strain endophytic fungi *Aspergillus flavus* was isolated from *Annona squamosa* leaves. The fungus was identified by observing the morphological characteristics and by molecular biological protocol based on the sequence of the ITS region as described early [4]. The fungus was transferred onto a petri dish containing potato dextrose agar. A small portion of the fungus from the petri dish was transferred to the sterile solid rice (*Oryza sativa*), waxy corn (*Zea mays ceritina*) and sweet corn (*Zea mays L.*) media in L Erlenmeyer and incubated at room temperature for 30 days. After 30 days, the ethyl acetate was added into the culture and left for 2 days. The mixture was then filtered and the solvent was removed under vacuum.

2.2. Kojic acid identification

The crude extracts and fractions were analyzed for the amount of chemical compound using HPLC with ACE 5 C18 column (150x4.6 mm id) using methanol as the mobile phase with a flow rate of 1 mL/min with detection at 235 nm and semi-preparative column Phenomenex Luna 5u C18 (250x10) with flow rate 1.5 mL/min. The EtOAc extract also subjected to vacuum liquid chromatography column eluted with the mixture of *n*-hexane : EtOAc (20, 40, 60, 80 and 100% *n*-hexane) and MeOH : DCM (20, 50, 70, 90 and 100% DCM). Each fraction was analyzed with HPLC. The structure of kojic acid was confirmed with the ^{1}H NMR spectrum performed on Agilent Varian 500 MHz with methanol (MeOD) as the solvent.

3. Results and Discussion

The media used for cultivation has been known to affect the growth of endophytic fungi and thus can influence the production and the diversity of secondary metabolites [12-14]. In this study, rice was used as a source of carbohydrates together with local yellow corn (jagung kuning) (*Zea mays L.*) and waxy (bose) corn media (*Zea mays ceritina*) found in Kupang (Timor). Rice has sufficient nutritional content, soft texture and smooth shape and thus creating more room for fungi to multiply while corn has a higher protein content (9.5%) compared to rice (7.4%). Endophytic fungi were grown in solid rice, sweet and waxy corn media and incubated for 30 days [15].

Ethyl acetate crude extracts from three different media were analyzed for the content of kojic acid using HPLC (Figure 1-4). Intrapolation of peak area for the content of kojic acid in each cultivation media to the regression equation of standard kojic acid revealed that the concentration of kojic acid in each extract were 3.149% for rice media, 5.998% for sweet corn media and 2.226% for waxy corn media. From the results of the analysis, it was observed that the highest concentration of kojic acid was from the extracts of *A. flavus* grown on sweet corn media followed by rice and then waxy corn media. However, based on the weight of the extracts, the rice media produced a higher amount of kojic acid as the main metabolite. Based on the weight of extracts from each media, the amount of kojic acid was estimated around 0.5782 grams for rice, 0.20322 grams for sweet corn and 0.1583 gram for waxy corn.

Therefore, the *Aspergillus flavus* extract on rice media was further separated based on its level of polarity with VLC (*Vacum Liquid Chromatography*). The stationary phase used in VLC was silica gel G_{60} with a size of 70-230 mesh eluted in an increased gradient from nonpolar to polar using mixture of *n*-hexane: ethyl acetate and dichloromethane (DCM): methanol (MeOH). Pure kojic acid was obtained from 90% DCM fraction as shown in Figure 5 has the same retention time as the standard (Figure 1).

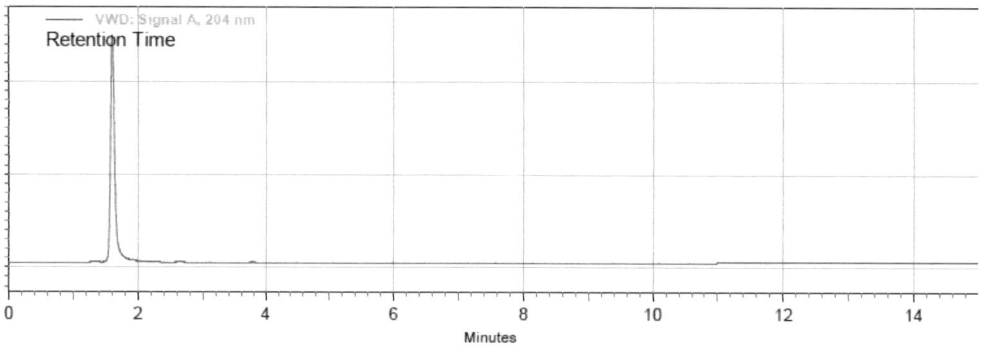

Figure 1. Standard HLPC chromatogram of kojic acid 1 mg/mL

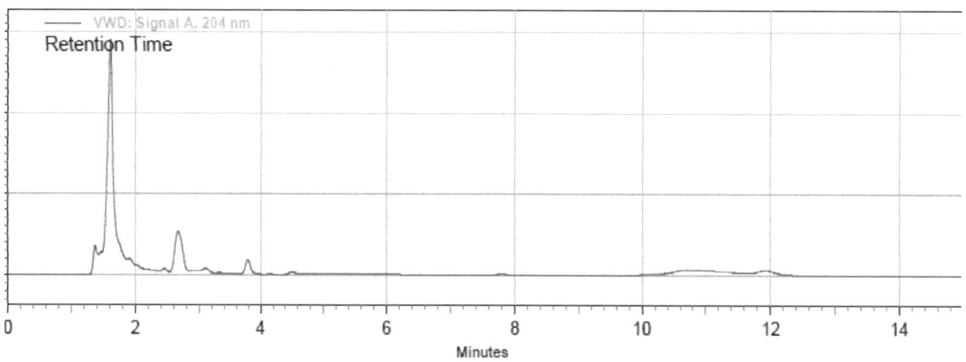

Figure 2. HLPC chromatogram of *Aspergillus flavus* extract from rice media

IOP Conf. Series: Materials Science and Engineering **823** (2020) 012003 doi:10.1088/1757-899X/823/1/012003

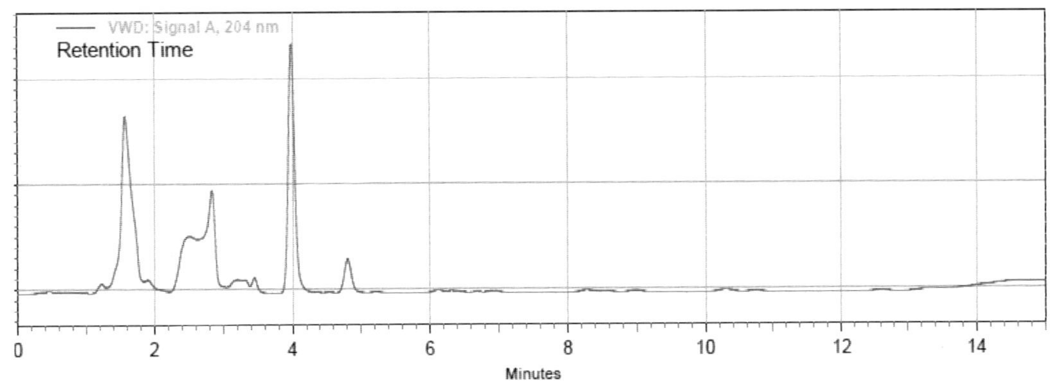

Figure 3. HLPC chromatogram of *Aspergillus flavus* extract from yellow corn media

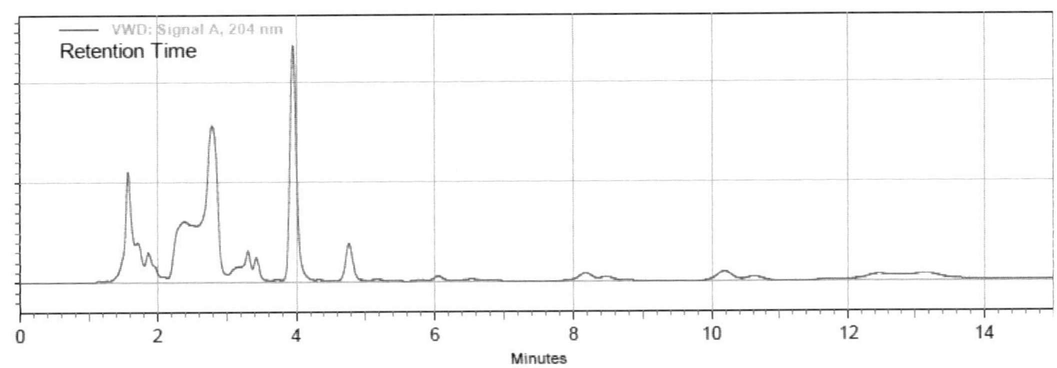

Figure 4. HLPC chromatogram of *Aspergillus flavus* extract from bose corn media.

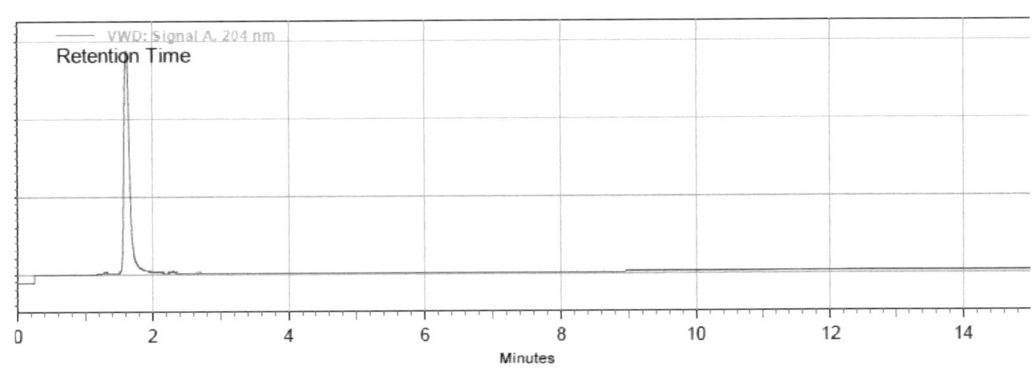

Figure 5. HLPC chromatogram of isolated kojic acid from 90% dichloromethane fraction

The structure of pure kojic acid obtained from the 90% DCM fraction was confirmed by comparison with the standard via HPLC analysis at the same condition and also supported by the ^{1}H NMR spectra, as shown in Table 1.

Table 1. ^{1}H NMR Data of kojic acid in MeOD (500 MHz)

Position	Chemical shifts, δ (ppm)	
	^{1}H-NMR	^{1}H-NMR[16]
2	-	-
3	6.45 s, 1H	6.52 s, 1H
4	-	-
5	-	-
6	7.95 s, 1H	7.97 s, 1H
7	4.41 s, 2H	4.43 s, 2H

4. Conclusion

Changes in media cultivation have led to the different chemical profiles of the endophytic fungi *Aspergillus flavus*. The use of local yellow corn and waxy corn media was able to expand the chemical profiles of *A. flavus*, but the production of kojic acid was better controlled using the rice media.

References

[1] Stierle A., Strobel G and Stierle D 1993 Taxol and taxane production by Taxomyces andreanae, an endophytic fungus of Pacific yew *Science* **206**(5105) 491-502

[2] Jimenez M, Mateo R, Querol A, Mateo J J and Hernandez E 1990 Differentiation of Penicillium griseofulvum Dierckx isolates by enzyme assays and by patulin and griseofulvin analyses *Applied and Environmental Microbiology*

[3] Hajjaj H, Niederberger P and Duboc P 2001 Lovastatin biosynthesis by Aspergillus terreus in a chemically defined medium *Applied and Environmental Microbiology*

[4] Ola A R B, Aly A H, Lin W, Wray V and Debbab A 2014 Structural revision and absolute configuration of lateritin *Tetrahedron Letters* **55**(45) 6184-7

[5] Ola A R B, Tawo B D, Belli H L L, Proksch P, Tommy D and Hakim E H 2018 A New antibacterial polyketide from the endophytic fungi Aspergillus fumigatiaffinis *Natural Product Communications* **13**(12)

[6] Ola A R B, Metboki G, Lay C S, Sugi Y, Rozari P D, Darmakusuma D and Hakim E H 2019 single production of kojic acid by *Aspergillus flavus* and the revision of flufuran *Molecules* **24**(22) 4200

[7] Chaudhary J, Pathak A N and Lakhawat S 2014 Production technology and applications of kojic acid *Review Article Annual Research & Review in Biology* **4**(21) 3165-96

[8] Saeedi M, Eslamifar M and Khezri K 2018 Kojic acid applications in cosmetic and pharmaceutical preparations *Biomedicine & Pharmacotherapy* 110 582–93

[9] El-Kady I A, Zohri A N A and Hamed S R 2014 Kojic acid production from agro-industrial by-products using fungi *Biotechnology Research International* **74**(11) 1294-300

[10] Rosfarizan M, Madihah S and Ariff A B 1998 Isolation of a kojic acid producing fungus capable of using starches as a carbon source *Lett. Appl. Microbiol.* **26** 27-30

IOP Conf. Series: Materials Science and Engineering **823** (2020) 012003 doi:10.1088/1757-899X/823/1/012003

[11] Liu M, Grkovic T, Liu X, Han J, Zhang L and Quinn R J 2017 A systems approach using OSMAC, Log P and NMR fingerprinting: An approach to novelty *Synthetic and Systems Biotechnology* **2**(4) 276–86
[12] Wei H, Lin Z, Li D, Gu Q and Zhu T 2010 OSMAC (one strain many compounds) approach in the research of microbial metabolites *Acta Microbiologica Sinica* **50**(60) 701-9
[13] Paranagama P A, Wijeratne E M K and Gunatilaka A A L 2007 Uncovering biosynthetic potential of plant-associated fungi: Effect of culture conditions on metabolite production by Paraphaeosphaeria quadriseptata and Chaetomium chiversii *Journal of Natural Products*
[14] Hemphill C F P, Sureechatchaiyan P, Kassack M U, Orfali R S, Lin W, Daletos G and Proksch P 2017 OSMAC approach leads to new fusarielin metabolites from Fusarium tricinctum *Journal of Antibiotics*
[15] Sugiyono, Soekarto S T, Hariyadi P and Supriadi A 2004 Kajian optimasi pengolahan beras jagung instan *J. Teknol. dan Industri Pangan* **15**(2) 119-28
[16] Ismail F D, Nurunnabi T, Billah M M, Sarker S, Nahar L, Rahman S M M and Sharples G 2018 Antimicrobial activity of kojic acid from endophytic fungus Colletotrichum gloeosporioides isolated from Sonneratia apetala, a mangrove plant of the Sundarbans *Asian Pacific Journal*

An improved synthesis of 1-monoolein

F O Nitbani[1*], Jumina[2], D Siswanta[2], E N Sholikhah[3], and B A Nurohmah[2]

[1]Department of Chemistry, Faculty of Science and Engineering, Universitas Nusa Cendana, Kupang, Indonesia
[2]Department of Chemistry, Faculty of Mathematics and Natural Sciences, Universitas Gadjah Mada, Yogyakarta, Indonesia
[3]Department of Pharmacology and Therapy, Faculty of Medicine, Universitas Gadjah Mada Yogyakarta, Indonesia

E-mail: febri_nitbani@yahoo.com

Abstract. The synthesis of 1-monoolein has been carried out through a two-step reaction: transesterification of ethyl oleate and 1,2-acetonide glycerol in the presence of sodium carbonate as a catalyst, and followed by deprotection using an Amberlyst-15 catalyst in ethanol. The transesterification reaction of ethyl oleate could produce 1,2-acetonide-3-oleoyl glycerol as a yellow liquid with a yield of 74%. Meanwhile, the deprotection of the intermediate compound could afford 1-monoolein as an unstable white soft solid in a yield of 59% and melting point at 35-37 °C.

1. Introduction

Monoacylglycerols or monoglycerides are a class of lipids with the hydrophilic and lipophilic functional group present in a molecule [1]. Based on their structure, monoglycerides are a non-ionic surfactant that potentially being used as an emulsifier, stabilizer, plasticizer, and conditioner. Monoglycerides could be produced conventionally from glycerol and vegetable oils or animal fats through chemical or enzymatic reactions. Industrial production of monoglyceride mainly involves a glycerolysis reaction of fats or vegetable oils that requires a high temperature reaction (220-250 °C), nitrogen atmosphere system, an inorganic alkaline catalyst, and a high vacuum distillation for purification [2]. Although the reaction requires high energy consumption, it produces monoglyceride in a lower quality, with a dark-colored and charred-smell compound, and also in a low yield.

Monoglycerides have antimicrobial activity. Monolaurin compounds, like 1-monolaurin and 2-monolaurin, have been reported to have antibacterial, anti-fungal, and antiviral activity [3-7]. Other monoglycerides such as 1-monocaprin and 1-monomyristin also have shown antibacterial and antifungal properties [8-10]. On the other hand, monoolein exhibits a high antioxidant and anti-arteriosclerosis activity [11,12]. Therefore, the synthesis of monoolein as an unsaturated monoglyceride has been of great interest to researchers.

Wang *et al.* [13] have reported the synthesis of 1-monoolein through esterification reaction of oleic acid and 1,2-acetonide glycerol, to produce an intermediate compound 1,2-acetonide-3-oleoyl glycerol, in the presence of *Novozym-435* lipase enzyme as a catalyst. In their work, the deprotection reaction of 1,2-acetonide-3-oleoyl glycerol using Amberlyst-15 in methanol produced 1-monoolein with a yield of 72.8%. However, *Novozym-435* lipase enzyme catalyst is expensive and not economically recommended. Additionally, utilization of methanol in a deprotection reaction could produce a toxic 1-monoolein, and the esterification of oleic acid with 1,2-acetonide glycerol could give a low yield product because it is a reversible reaction. Jumina *et al.* [14] have successfully synthesized unsaturated 1-monolinolein via the protected glycerol as 1,2-*O*-isopropylidene glycerol (1,2-acetonide glycerol). Based on their studies, we proposed that the same method can also be applied to produce 1-monoolein from ethyl oleate.

In this work, 1-monoolein was firstly prepared through transesterification of ethyl oleate with 1,2-acetonide glycerol in the presence of sodium carbonate as a catalyst to produce 1,2-acetonide-3-oleoyl glycerol. The deprotection reaction of 1,2-acetonide-3-oleoyl glycerol was then conducted using Amberlyst-15 in ethanol to obtain 1-monoolein. Transesterification is preferred than esterification reaction because it is a direct reaction so that 1,2-acetonide-3-oleoyl glycerol could be afforded in a higher yield. Meanwhile, sodium carbonate catalyst is available abundantly, and utilization of ethanol as a solvent in a deprotection reaction enables to give a non-toxic 1-monoolein.

2. Experimental

2.1 Materials

Chemicals used in this experiment were ethyl oleate, 1,2-acetonide glycerol, sodium carbonate (Na_2CO_3), anhydrous sodium sulfate (Na_2SO_4), sodium hydrogen carbonate ($NaHCO_3$), Amberlyst-15, dichloromethane, n-hexane, chloroform, acetone, methanol, ethanol, and distilled water. All chemicals, except distilled water that was obtained from the Laboratory of Fundamental Chemistry Universitas Gadjah Mada, were purchased from E. Merck in analytical grade and were used without any further purification.

2.2 Instrumentation

Instruments used in this research were laboratory glassware and analytical mass balance (Mettler AT200). The ^{1}H– and ^{13}C–NMR spectra were recorded at 500 MHz (^{1}H) and 125 MHz (^{13}C) (JEOL JNM-MY spectrometer) using tetramethylsilane (TMS) as an internal standard. The chromatogram and mass spectra were measured on a Mariner Liquid Chromatography-Mass Spectrometry (LC-MS) using C-18 column (15 × 1 mm) and an injection volume of 2 μL. The mobile phase used was methanol with a flow rate of 0.1 mL/min.

2.3 Synthesis of 1,2-acetonide- 3-oleoyl glycerol

Amount of 0.005 mol ethyl oleate (1.55 g) and 0.03 mol of 1,2-acetonide glycerol (3.96 g) was placed into three-neck round-bottomed flask that equipped with thermometer and condenser. To the mixture was added 5% Na_2CO_3 (w/v from the total of reactants) and then refluxed with stirring at 140 °C for 24 hours. After completion of the reaction, the mixture was dissolved in n-hexane and washed with distilled water. The organic layer was then dried using anhydrous Na_2SO_4 and evaporated to remove the solvent. The product of 1,2-acetonide-3-oleoyl glycerol with the highest purity was analyzed using GC-MS (Gas Chromatography-Mass Spectra).

2.4 Synthesis of 1-monoolein

Amount of 1,2-acetonide-3-oleoyl glycerol (0.0029 mol; MW: 396 g/mol; 1.14 g) was mixed with 0.43 g of Amberlyst-15 in ethanol (15 mL) and stirred at room temperature for 24 hours. After completion of the reaction, the mixture was evaporated to remove the ethanol and extracted using dichloromethane. The product was then analyzed using Liquid Chromatography-Mass Spectrometry (LC-MS) to identify the formation of the target compound.

Purification of 1-monoolein was carried out by Preparative Thin Layer Chromatography (Prep. TLC) by dissolving 0.15 g of crude 1-monoolein in acetone and placed it on the TLC plate (20×20 cm). The separation was conducted using a mixed solvent of chloroform: acetone: methanol (9.5:0.45:0.05) as the eluent. The purification product was collected based on the visualization under a UV light lamp (265 nm). After it was dissolved in acetone, the product was evaporated and recrystallized from n-hexane to afford 1-monoolein. Characterizations of the purified 1-monoolein were conducted by ^{1}H– and ^{13}C–NMR Spectrometry.

3. Result and Discussion

3.1 Synthesis of 1,2-acetonide-3-oleoyl glycerol

Monoolein is a monoacylglycerol from oleic acid, which can be present as 1-monoolein and 2-monoolein. If the oleoyl group is attached to C number 2 (C2) from the glycerol framework, it will produce 2-monoolein, and if it is attached to C number 1 (C1) atoms, it will give 1-monoolein compound. Monoolein compounds are known to exhibit strong antioxidant activity as well as anti-arteriosclerotic. It also has a high nutritional function for humans.

There were some disadvantages in the synthesis of 1-monoolein from oleic acid that was performed by Wang *et al*. [13]. In their work, the *Novozym-435* enzyme was used in the esterification reaction of oleic acid and 1,2-acetonide glycerol, which was costly. Some efforts are still needed in order to find a less costly synthesis pathway of 1-monoolein. One of the reaction pathways that is expected to provide a solution to this problem is the utilization of transesterification reactions. The transesterification is a direct reaction compared with esterification, which is an equilibrium reaction. Unidirectional or irreversible reactions are easier to handle because they require simpler reaction systems and produce more products. Therefore, in this work, the synthesis of 1-monoolein will be conducted through transesterification reaction of ethyl oleate and 1,2-acetonide glycerol using Na_2CO_3 as a catalyst. The utilization of Na_2CO_3 base catalysts was proposed to reduce the cost in the synthesis of 1-monoolein compared with *Novozym-435* enzyme.

In this study, the transesterification reaction was carried out at 120 °C for 18 hours, with a 6:1 mol ratio between 1,2-acetonide glycerol and ethyl oleate. The color of the mixture was observed to turn to brownish when the reaction was stopped. Isolation of the 1,2-acetonide-3-oleoyl glycerol from the mixture was carried out in n-hexane and afforded the desired compound as a pale yellow liquid with a yield of 74%. The results of the analysis of 1,2-acetonide-3-oleoyl glycerol using GC-MS are presented in Figure 1 and Figure 2.

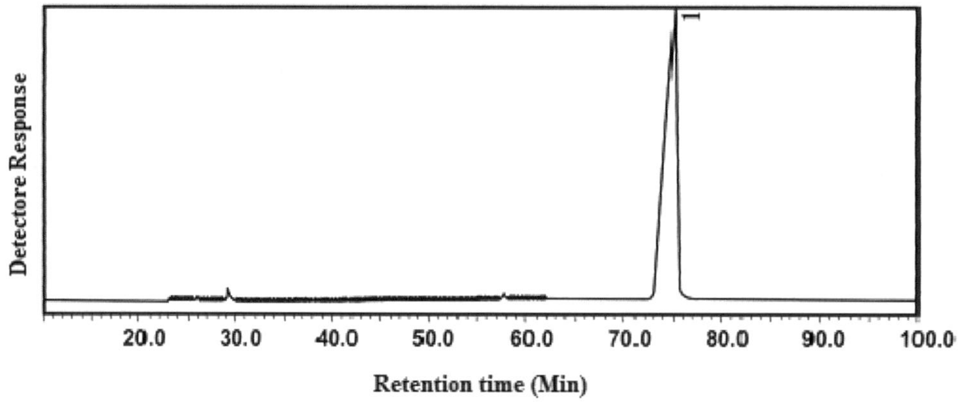

Figure 1. Chromatogram of 1,2-acetonide-3-oleoyl glycerol

Based on the chromatogram of the prepared 1,2-acetonide oleoyl glycerol in Figure 1, a single peak is present at a retention time of 75.44 minutes. The 1,2-acetonide-3-oleoyl glycerol compound relatively has more polar properties because of the double bond in the oleoyl chain. The mass spectrum of the compound was presented in Figure 2 and showed a fragment with m/z=395, which was formed due to the release of 1 proton from the 1,2-acetonide-3-oleoyl glycerol compound. The molecular ion of 1,2-acetonide-3-oleoyl glycerol compound with m/z=396 did not appear because it was less stable. Fragments with m/z=257 appeared due to fragment release with the molecular formula $C_{10}H_{19}$• (m/z=139). The base peak that appears at m/z=183 was formed from the release of $C_3H_6O_2$• fragments (m/z=74).

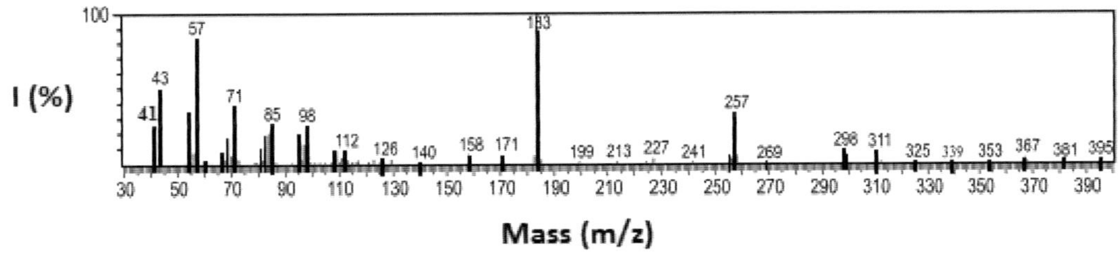

Figure 2. Mass spectra of 1,2-acetonide-3-oleoyl glycerol

3.2 Synthesis of 1-monoolein

The deprotection reaction by stirring of 1,2-acetonide-3-oleoyl glycerol using Amberlyst-15 in ethanol at room temperature for 18 hours could give 1-monoolein compound and some side products such as oleic acid and ethyl oleate. Amberlyst-15 is a heterogeneous catalyst that can be separated from the reaction product through the filtration process. The final reaction product was then dissolved in dichloromethane due to the pH of the product solution was 4. The dichloromethane phase was then washed with 5% $NaHCO_3$ solution (w/v) to reach neutral pH. After the evaporation, 1-monoolein was afforded as a soft white solid with a yield of 59%. The elucidation of the obtained 1-monoolein from the dichloromethane phase was conducted by LC-MS (Figure 3 and Figure 4).

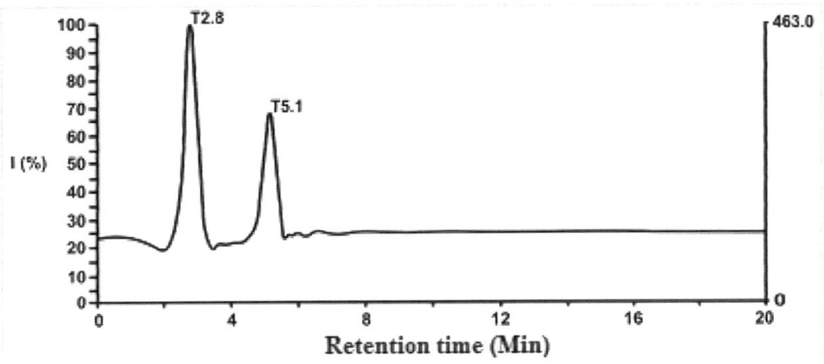

Figure 3. Chromatogram of 1-monoolein

The chromatogram in Figure 3 showed the presence of 2 peaks with different retention times (R_t). The stationary phase used in the liquid chromatography was nonpolar (reversed-phase), so the polar compounds will pass through the column and eluted first. Based on this fact, it was proposed that the peak with a retention time of 2.8 minutes was 1-monoolein compound and will be eluted through a column first. Meanwhile, the peaks at a retention time of 5.1 minutes were predicted to be impurities. The relative content of the compound with an R_t of 2.8 minutes was 62.83%.

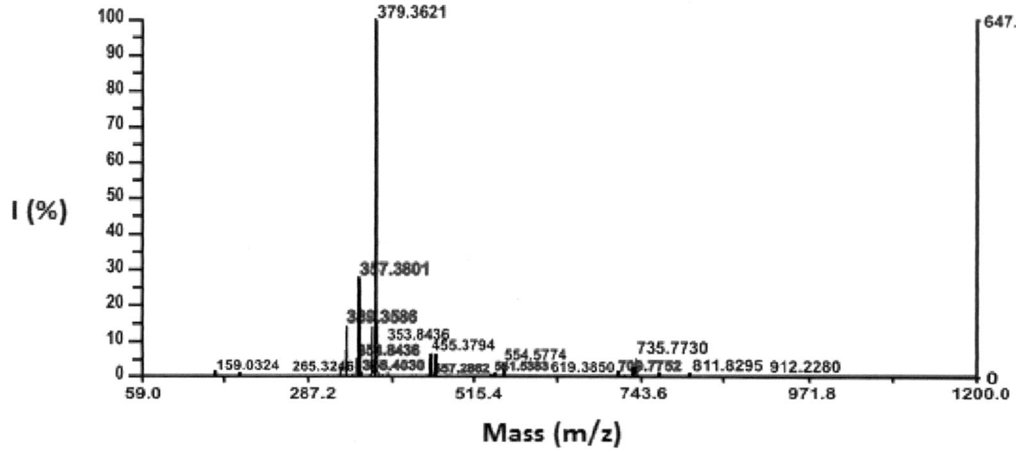

Figure 4. Mass spectra of a peak R_t = 2.8 min

The mass spectra in Figure 4 indicate the presence of 3 peaks of fragments (m/z) i.e. M + H = 357.58; M + Na = 379.36; and 2M + Na = 735.77. These fragments referred to 1-monoolein compound with a molecular weight of 256 g/mol.

The TLC analysis showed that the two compounds in the chromatogram (Figure 3) could be separated using a mixture eluent of chloroform-acetone-methanol (0.95: 0.45: 0.005). The same solvent ratio was eventually used for the purification of 1-monoolein product by preparative-TLC, which gave a soft white solid (unstable) with 59% of yield and melting point around 35-37 °C.

The proton ^1H–NMR (500 MHz, $CDCl_3$) spectrum of 1-monoolein was displayed in Figure 5. Characterization of each proton from the obtained compound was (δ ppm): 0.87 (3H, t, CH_3 of C_{18}), 1.23-1.29 (26H, m, CH_2 of $C_{4\text{-}8}$ and $C_{11\text{-}17}$), 1.58-1.62 (2H, m, CH_2 of C_3), 2.00 (2H, dd, CH_2 of -(OH)CH-C**H**$_2$-OCO-), 2.27 (2H, t, CH2 of -CO-C**H**$_2$-CH_2-), 2.61-2.75 (1H, m, -CH_2-(OH)C**H**-CH_2-), 4.11 (2H, dd, -(OH)CH-C**H**$_2$-OCO-), 5.33 (2H, s, (-C**H**=C**H**-)), 7.26 (2H, s, OH).

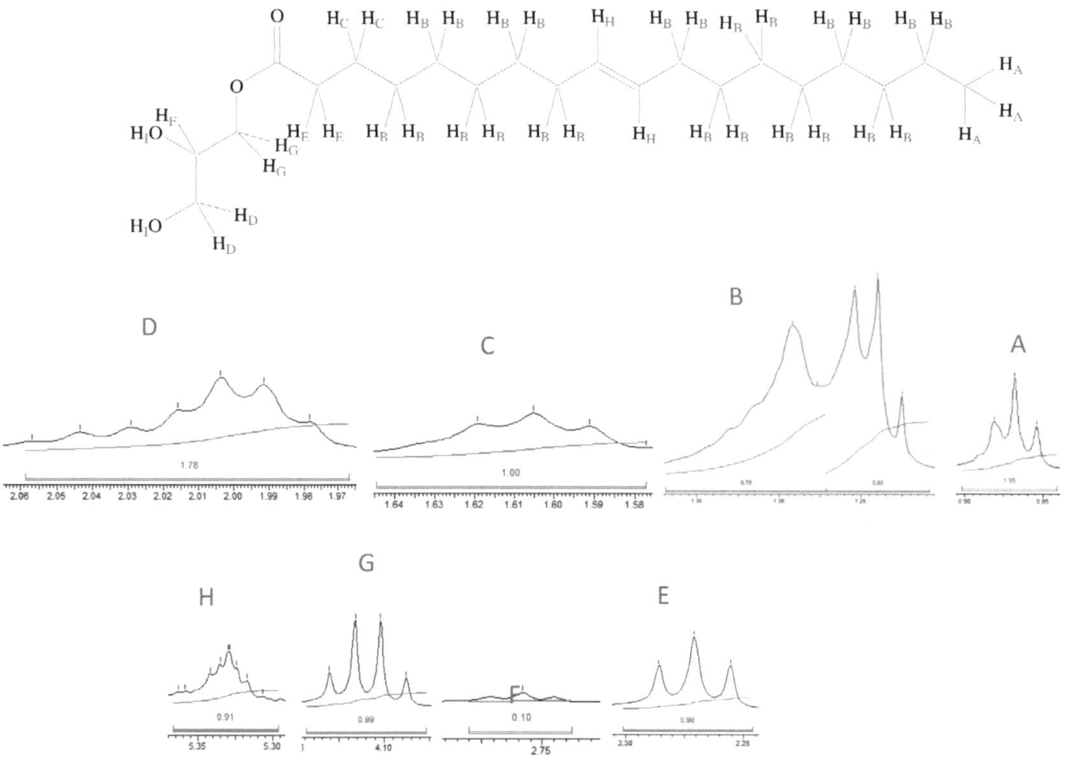

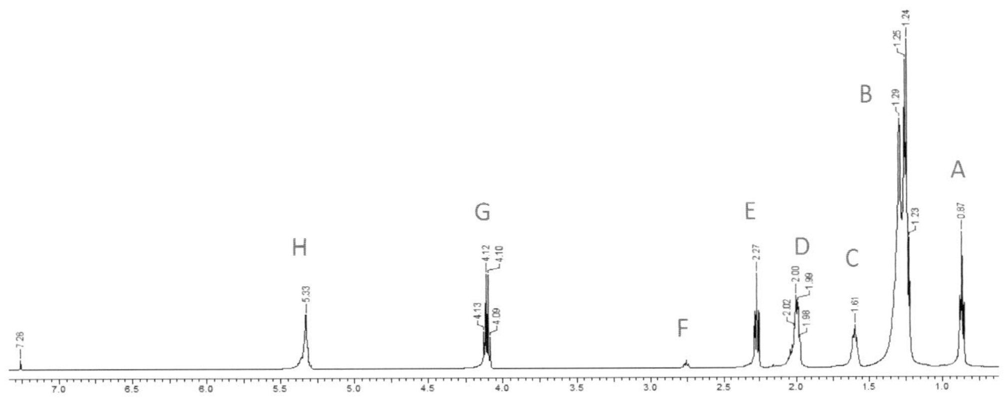

Figure 5. The ^{1}H–NMR spectra of 1-monoolein

Peak A (0.87 ppm, 3H, *triplet*) was ascribed as the protons of the methyl group (-**CH**$_3$**)** at C_{18}. Peak B (1.23-1.29 ppm, 26H, *multiplet*) appeared as the resonance of methylene protons (-CH_2-**CH**$_2$-) at C_4-C_8 and C_{11}-C_{17} atoms from the oleoyl groups. Meanwhile, peak C (1.58-1.62 ppm, 2H, *multiplet*) was proposed as the methylene protons of the C_3 atom of the oleoyl group.

Peak D (2.00 ppm, 2H, *dd*) was interpreted as the resonance of two methylene protons of -(OH)CH-**CH**$_2$-OCO-, while peak E (2.27 ppm, 2H, *triplet*) was the peak of methylene proton of -CO-**CH**$_2$-CH_2- group. The multiplet peak F (1H) appeared from the resonance of methine proton of -CH_2-(OH)C**H**-CH_2-. Peak G (4.11 ppm, 2H, *dd*) as a *doublet of doublet (dd)* signal assigned as two methylene protons from -(OH)CH-**CH**$_2$-OCO- group. A *singlet* peak H (5.33 ppm, 2H, *singlet*) and I (7.26 ppm, 2H, *singlet*) were interpreted as the proton signal of alkene (-C**H**=C**H**-) and hydroxy (–O**H)** group, respectively.

The ^{13}C–NMR (125 MHz, $CDCl_3$) spectra of 1-monoolein in Figure 6 showed 14 carbon atom types with a different chemical environment. Generally, the electronic environment of each carbon atom can be presented as (δ ppm): 14.25 (C18), 22.70 (C17), 24.98 (C16), 27.17 (C15), 29.11 (C_{14}, C_{13}, C_6, and C_5), 29.18 (C_7, C_8, C_{11}, and C_{12}), 29.34 (C4), 29.69 (C3), 29.77 (C2), 31.92 ((OH)CH_2-(OH)CH-CH_2-), 34.37 (-(OH)**C**H-CH_2-OCO-), 60.13 (-(OH)CH-**C**H_2-OCO-), 129.98 (-CH_2-**C**H=**C**H-CH_2-), 173.86 (C=O).

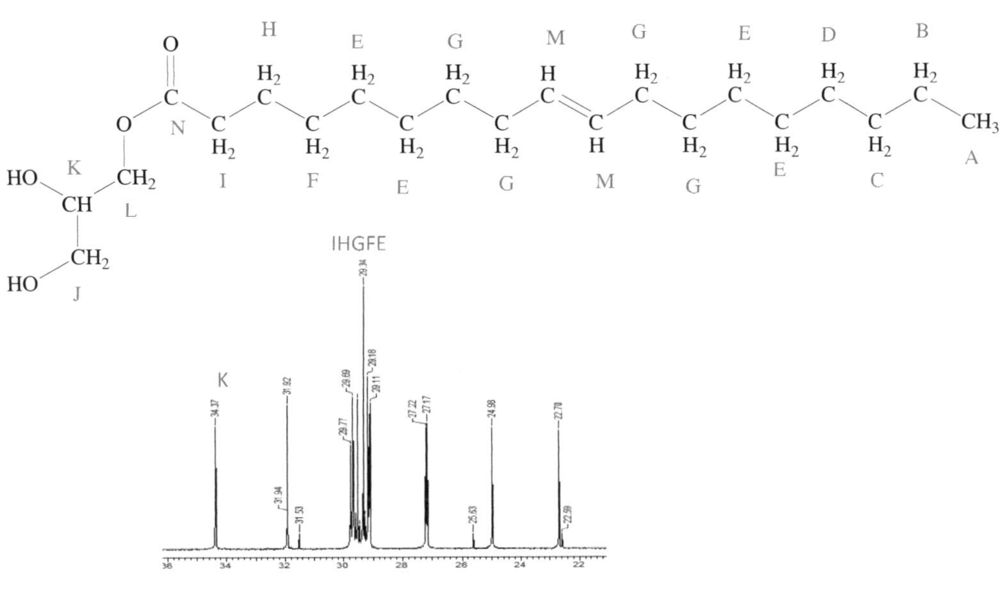

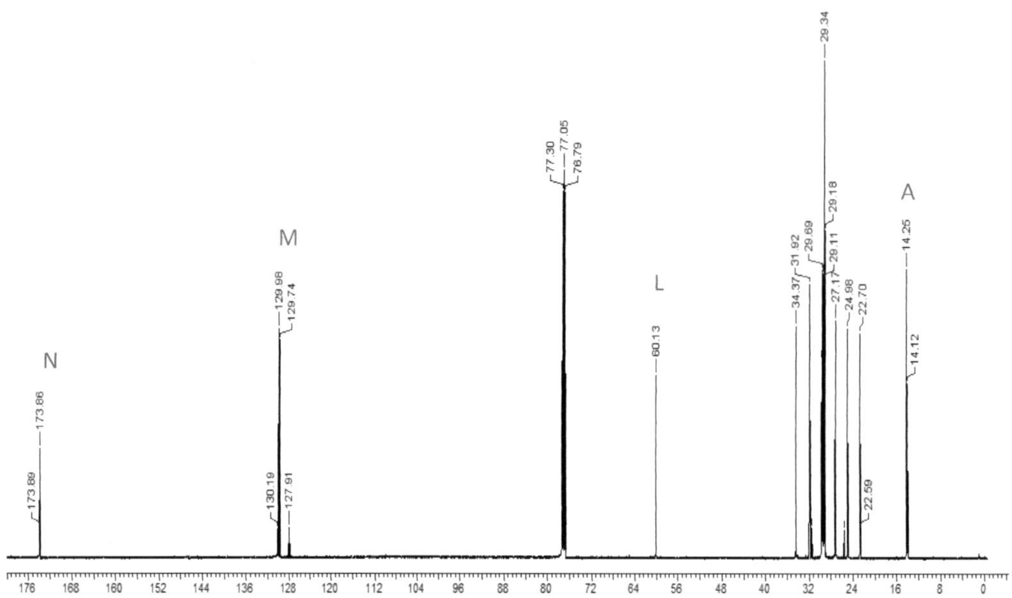

Figure 6. The ^{13}C NMR spectra of 1-monoolein

Peak A at a chemical shift (δ) of 14.25 ppm was due to the methyl group (-CH_2-$\mathbf{CH_3}$) at C_{18}, while peak B at 22.70 ppm was a signal carbon C_{17} of the oleoyl group. Peak C and D that appeared at 24.98 and 27.17 ppm were assigned as the signal of atom C_{16} and C_{15} of the oleoyl group. Peak E at 29.11 ppm was interpreted as the signal of carbon C_{14}, C_{13}, C_6, and C_5 from the oleoyl group, while peak F at 29.18 ppm corresponded with atom C_4 on the same group. The signal of atom C_7, C_8, C_{11}, and C_{12} (-$\mathbf{CH_2}$-$\mathbf{CH_2}$-CH=CH-$\mathbf{CH_2}$-$\mathbf{CH_2}$-) of the oleoyl group can be assigned as peak G at a chemical shift of 29.34 ppm. Peak H and I at 29.69 and 29.77 ppm respectively appeared as the signal of atom C_3 and C_2 of the oleoyl group.

Peak J and K appeared in a more downfield area (31.92 and 34.37 ppm) due to the signal of C atom at (OH)$\mathbf{C}H_2$-(OH)CH-CH_2 and -(OH)$\mathbf{C}$H-CH_2-OCO-, respectively. Peak L at 60.13 ppm was ascribed from C atom of -(OH)CH-$\mathbf{C}H_2$-OCO-, while peak M at 129.98 ppm was alkene carbon of -CH_2-$\mathbf{C}$H=$\mathbf{C}$H-CH_2-. Peak N at the most downfield environment (173.86 ppm) was inferred as the C atom of the carbonyl group (-$\mathbf{C}$O-).

This work showed that 1-monoolein could be successfully prepared through transesterification of 1,2-acetonide glycerol with ethyl oleate, followed by a deprotection reaction using Amberlyst-15. This reaction allows a low-cost and straightforward synthesis of 1-monoolein compared with the previously reported method by Wang *et al.* [13]. In this work, sodium carbonate was used as a catalyst in the transesterification reaction, instead of costly *Novozym-435* enzyme. Moreover, the transesterification reaction of ethyl oleate could give a higher yield of 1-monoolein than the esterification of oleic acid by Wang *et al.* [13] because transesterification is a unidirectional reaction.

4. Conclusion

1-Monoolein could be produced from the reaction of ethyl oleate as a raw material with 1,2-acetonide glycerol (1:6 ratio mole) in the presence of sodium carbonate as a catalyst to obtain intermediate 1,2-acetonide-oleoyl glycerol as a pale yellow liquid (74%). Deprotection reaction of 1,2-acetonide-oleoyl glycerol using an Amberlyst-15 catalyst in ethanol could afford 1-monoolein as a fine white solid with a yield of 59%.

Acknowledgment
We are thankful for the Organic Chemistry Laboratory in the Department of Chemistry, Universitas Gadjah Mada for the research facilities.

References

[1] Nitbani F O, Jumina, Siswanta D and Sholikhah E N 2015 Reaction path synthesis of monoacylglycerol from fat and oils *Int. J. Pharm. Sci. Rev. Res.* **35**(1) 126–36

[2] Zhong N, Li L, Xu X, Cheong L-Z, Li B, Hu S-Q and Zhao X 2009 An efficient binary solvent mixture for monoacylglycerol synthesis by enzymatic glycerolysis *J. Am. Oil. Chem. Soc.* **86** 783–9

[3] Nitbani F O, Jumina, Siswanta D, Sholikhah E N and Fitriastuti D 2018 Synthesis and antibacterial activity of 1-monolaurin *Orient. J. Chem.* **34**(2) 863–67

[4] Nitbani F O, Jumina, Siswanta D, Sholikhah E N and Fitriastuti D 2016 Synthesis and antibacterial activity of 2-monolaurin *Orient. J. Chem.* **32**(6) 3113–20

[5] Krislee A, Fadly C, Nugrahaningsih D, Nuryastuti T, Nitbani F O, Jumina and Sholikhah E N 2019 The 1-Monolaurin inhibit growth and eradicate the biofilm formed by clinical isolates of *Staphylococcus epidermidis BMC Proc.* **13** (Suppl 11) 19 doi: 10.1186/s12919-019-0174-9

[6] Seleem D, Chen E, Benso B, Pardi V and Murata R M 2016 In vitro evaluation of antifungal activity of monolaurin against *Candida albicans* Biofilms *Peer J.* **4** e2148 doi: 10.7717/peerj.2148

[7] Thormar H and Hilmarsson H 2007 The role of microbicidal lipids in host defense against pathogens and their potential as therapeutic agents *Chem. Phys. Lipids.* **150**(1) 1–11

[8] Nitbani F O, Jumina, Siswanta D and Sholikhah E N 2016 Synthesis and antibacterial activity test of 1-monocaprin *Int. J. Pharm. Sci. Rev. Res.* **39**(16) 74–80

[9] Ma M, Wen X, Xie Y, Guo Z, Zhao R, Yu P, Gong D, Deng S and Zeng Z 2018 Antifungal activity and mechanism of monocaprin against food spoilage fungi *Food Control.* **84** 561–8

[10] Jumina, Nurmala A, Fitria A, Pranowo D, Sholikhah E, Kurniawan Y and Kuswandi B 2018 Monomyristin and monopalmitin derivatives: Synthesis and evaluation as potential antibacterial and antifungal agents *Molecules.* **23**(12) 3141

[11] Cho K-H, Hong J-H and Lee K-T 2010 Monoacylglycerol (MAG)-Oleic acid has stronger antioxidant, anti-atherosclerotic, and protein glycation inhibitory activities than MAG-Palmitic Acid *J. Med. Food.* **13**(1) 99–107

[12] Feltes M M C, Villeneuve P, Baréa B, Barouh N, de Oliveira J V, de Oliveira D and Ninow J L 2012 Enzymatic production of Monoacylglycerols (MAG) and Diacylglycerols (DAG) from fish oil in a solvent-free system *J. Am. Oil. Chem. Soc.* **89**(6) 1057–65

[13] Wang X, Jin Q, Wang T, Huang J and Wang X 2013 An improved method for the synthesis of 1-monoolein *J. Mol. Catal. B. Enzym.* **97** 130–6

[14] Jumina J, Lavendi W, Singgih T, Triono S, Kurniawan Y S and Koketsu M 2019 Preparation of monoacylglycerol derivatives from Indonesian edible oil and their antimicrobial assay against *Staphylococcus aureus* and *Escherichia coli Sci. Rep.* **9**(1) 1–8

IOP Conf. Series: Materials Science and Engineering **823** (2020) 012005 doi:10.1088/1757-899X/823/1/012005

Effect of garlic extract (*Allium sativum*) administration of the total cholesterol level for white rats (*Rattus novergicus*) with hypercholesterol

Rr. Listyawati Nurina[1], Wilhelmina Hewen[2], I Nyoman Sasputra[3], Debora Shinta Liana[4], Anita Lidesna Shinta Amat[5]

[1]Department of Pharmacology and Pharmacotherapy, Faculty of Medicine, Universitas Nusa Cendana
[2]Faculty of Medicine, Universitas Nusa Cendana
[3]Department of Anatomical Pathology, Faculty of Medicine, Universitas Nusa Cendana
[4]Department of Pediatric, Faculty of Medicine, Universitas Nusa Cendana
[5]Department of Biochemistry, Faculty of Medicine, Universitas Nusa Cendana

E-mail: listyawati_nurina@staf.undana.ac.id

Abstract. Hypercholesterolemia is one of the non-contagious diseases which is the main cause of death globally. Accumulation of excessive cholesterol or hypercholesterolemia will be atherosclerosis which causes strokes and cardiovascular diseases such as coronary heart disease. Garlic has an active agent allicin which an alkaloid compound that was contained in garlic, can be used to reduce cholesterol levels, in addition: vitamin C and niacin. Knowing that effect of garlic extract (*Allium sativum*) to decrease cholesterol total levels in white rats (*Rattus norvegicus*) which are deliberately hypercholesterol.Experimental research through *Pre test-post test with control group design* approaching in laboratory. Research sample was 25 white rats selected randomly and divided into 5 groups are normal control group and treatment group. Data analysis by One Way ANOVA showed that total cholesterol levels on the 32, 33, 37 and 42 days was significance value of $p < 0.05$, its means the statistical test showed that given the garlic extract (*Allium sativum*) can reduce total cholesterol levels in white rats with hypercholesterolemia. There is the effect of garlic extract (*Allium sativum*) on decreasing cholesterol total levels in white rats (*Rattus novergicus*) with hypercholesterol and a significant value ($p < 0.05$).

1. Introduction

Hypercholesterolemia is one of the non-communicable diseases which is the main cause of death globally [1]. Excess cholesterol or what is called hypercholesterolemia will cause atherosclerosis which will cause strokes and cardiovascular diseases such as coronary heart disease [2,3,4,5]. Globally 2015 World Health Organization (WHO) data, this increase in cholesterol caused 2.6 million deaths (4.5% of all deaths) and 29.7 million Disability Adjusted Life Year (DALYs), or 20% of the total DALYS [6]. Data in Indonesia on residents aged> 15 years found abnormal total cholesterol of 35.9%, High-Density Lipoprotein (HDL) low at 22.9%, Low-Density Lipoprotein (LDL) not optimal with near optimal borderline combined categories high 60.3% and very high 15.9%, abnormal triglycerides with high borderline category 13.0% and very high category 11.9% [7].

Garlic has an active agent allicin which is an alkaloid compound that is contained in garlic which can be used to reduce cholesterol levels, besides that in garlic there are also: vitamin C, niacin, germanium and sulfur-binding compounds [2,6].

Based on the description of the data above, the researchers felt interested to find out more about the effect of giving garlic extract (Allium sativum) on the reduction of total cholesterol levels in Sprague dawley strain white rats that were hypercholesterol.

2. Method

Experimental laboratory study with pretest-posttest design with control group. The study sample was 25 randomly selected white rats divided into 5 groups, namely the control group and the treatment group. This research was conducted for 30 days. Research, maintenance and measurement of total cholesterol levels were carried out in the dry lab in the Faculty of Medicine, Universitas Nusa Cendana. The extraction process was carried out at the Bioscience Laboratory of the Universitas Nusa Cendana. The process of making garlic extract by maceration method. Giving hypercholesterol cholesterol feed consists of 30 grams of quail egg yolk PTU (Propiltiourasil) 0.1% and water up to 1000 mL. Making hypercholesterol feed is by using PTU 100 mg dissolved in 1000 mL of distilled water, quail egg yolk is dissolved with PTU solution that has been made. For 25 days this feed was given to mice with each volume of 0.5 cc per mouse by means of pressure [9] and followed by examination of cholesterol levels; after that, it was continued with the administration of the treatment in the form of simvastatin1.8 mg / 200 gram per day, dose I garlic extract: 0.108 grams, dose II: 0.144 grams, dose III: 0.18 grams. Examination of total cholesterol levels was carried out on the 1st, 5th and 10th day after administration of the test material.

3. Result

3.1. Extraction

Garlic used in this study came from the Kapan area, South Timor Tengah Regency, East Nusa Tenggara Province. Garlic is peeled and then weighed 1000 grams. Next, the garlic is washed and dried in an oven at 400C for 2 days. Then the garlic is smoothed by blending 800 grams. Garlic is then soaked in 2 liters of 70% ethanol for 3x24 hours. The next process is evaporation carried out in the Bioscience Laboratory Undana on the 4th day using a Vaccum rotatory evaporator for 14 hours; so that we get 144,282 grams of thick garlic extract.

3.2. Body-weight measurement

In general, the weight gain of the test animals occurred on the 0th day until the 32nd day; whereas on the 37th day and 42nd day there was a decrease in body weight in the test animals. Weight gain occurred on day 0 to day 32 because the test animals were induced hypercholesterol feed in the form of a mixture of egg yolks and propylthiouracil drugs are given for 25 days where this condition caused a state of hypercholesterolemia. Feed consumption of hypercholesterol causes an increase in cholesterol in the blood, cholesterol is a type of fat. Fat stored in adipose tissue is not directly used but stored in the form of triglycerides; until it is needed to produce fat energy, it will be hydrolyzed to free fatty acids and glycerol [10].

This administration aims to reduce the condition of hyperthyroidism in the condition of normal mice to be a hypothyroid state so that there is an increase in cholesterol levels in the blood; its manifestations are slow metabolism and becoming fat [11].

The weight loss that occurred on the 37th day and 42nd day was suspected because the treatment of the test material in the form of simvastatin and garlic extract was given to reduce the state of hypercholesterol.

IOP Conf. Series: Materials Science and Engineering **823** (2020) 012005 doi:10.1088/1757-899X/823/1/012005

3.3. Cholesterol total level measurement

There was an increase in total cholesterol levels on the 32nd day in the positive control group, treatment group I, treatment group II and treatment group III. This is because on the 32nd day hypercholesterol cholesterol feed has been induced for 25 days and then the total cholesterol level is checked; Whereas on the 33rd day, 37th day and 42nd day there was a decrease in total cholesterol levels in the positive control group, treatment group I, treatment group II and treatment group III because simvastatin and garlic extract were given treatment. Results of data analysis and discussion.

Table 1. The results of the One Way Anova test analysis of all groups of test animals

		Sig.	Nilai p
Cholesterol Level Total Day 0	Between Groups	,306	>0,05
	Within Groups		
	Total		
Cholesterol Level Total Day 32	Between Groups	,000	<0,05
	Within Groups		
	Total		
Cholesterol Level Total Day33	Between Groups	,022	<0,05
	Within Groups		
	Total		
Cholesterol Level Total Day37	Between Groups	,048	<0,05
	Within Groups		
	Total		
Cholesterol Level Total Day42	Between Groups	,001	<0,05
	Within Groups		
	Total		

Annotation: p<0,05 : significant

Table 1 shows that total cholesterol levels on day 32, day 33, day 37 and day 42 have a significance value of p <0.05, which means that this statistical test shows that administration of garlic extract (*Allium sativum*) can reduce total cholesterol levels in white rats with hypercholesterolemia; So that H0 is rejected and H1 is accepted. A follow-up test from One Way Anova carried out a post hoc test to compare cholesterol change data between groups.

Table 2. The results of the *Post hoc* test analysis of all groups of test animals

Dependent Variable	Group of test animals	Group of test animals	Sig.	Annotation
Cholesterol Level Total Day 32	Normal control group	Positive control group	,000	Significant
		Treatment group I	,001	Significant
		Treatment group II	,000	Significant
		Treatment group III	,001	Significant
	Positive control group	Normal control group	,000	Significant
		Treatment group I	,973	
		Treatment group II	1,000	
		Treatment group III	,998	
	Treatment group I	Normal control group	,001	Significant
		Positive control group	,973	
		Treatment group II	,975	
		Treatment group III	,998	
	Treatment group II	Normal control group	,000	Significant
		Positive control group	1,000	

		Treatment group I	,975	
		Treatment group III	,999	
		Normal control group	,001	Significant
	Treatment group III	Positive control group	,998	
		Treatment group I	,998	
		Treatment group II	,999	
		Positive control group	,836	
	Normal control group	Treatment group I	,633	
		Treatment group II	,822	
		Treatment group III	,349	
		Normal control group	,836	
	Positive control group	Treatment group I	,996	
		Treatment group II	1,000	
		Treatment group III	,056	
		Normal control group	,633	
Cholesterol Level Total Day 33	Treatment group I	Positive control group	,996	
		Treatment group II	,997	
		Treatment group III	,026	Significant
		Normal control group	,822	
	Treatment group II	Positive control group	1,000	
		Treatment group I	,997	
		Treatment group III	,053	
		Normal control group	,349	
	Treatment group III	Positive control group	,056	
		Treatment group I	,026	Significant
		Treatment group II	,053	
		Positive control group	,129	
	Normal control group	Treatment group I	,135	
		Treatment group II	,086	
		Treatment group III	,886	
		Normal control group	,129	
	Positive control group	Treatment group I	1,000	
		Treatment group II	,999	
		Treatment group III	,522	
		Normal control group	,135	
Cholesterol Level Total Day 37	Treatment group I	Positive control group	1,000	
		Treatment group II	,999	
		Treatment group III	,538	
		Normal control group	,086	
	Treatment group II	Positive control group	,999	
		Treatment group I	,999	
		Treatment group III	,401	
		Normal control group	,886	
	Treatment group III	Positive control group	,522	
		Treatment group I	,538	
		Treatment group II	,401	
Cholesterol Level Total Day 42	Normal control group	Positive control group	,025	Significant

	Treatment group I	,001	Significant
	Treatment group II	,004	Significant
	Treatment group III	,001	Significant
Positive control group	Normal control group	,025	Significant
	Treatment group I	,706	
	Treatment group II	,900	
	Treatment group III	,653	
Treatment group I	Normal control group	,001	Significant
	Positive control group	,706	
	Treatment group II	,994	
	Treatment group III	1,000	
Treatment group II	Normal control group	,004	Significant
	Positive control group	,900	
	Treatment group I	,994	
	Treatment group III	,988	
Treatment group III	Normal control group	,001	Significant
	Positive control group	,653	
	Treatment group I	1,000	
	Treatment group II	,988	

Table 2 shows the post hoc test to find out the strength of differences between treatment groups, data shows total cholesterol levels on day 32: normal control group towards positive control group, treatment group I, treatment group II, treatment group III, showing significant (significant) difference; positive control group, treatment group I, treatment group II, treatment group III for the normal control group showed significant differences due to the normal control group not given the test material treatment. Data on 33rd day: treatment group I against treatment group III; Treatment group III on treatment group I showed a significant difference in strength; this was due to multilevel dosing between treatment group I and treatment group III. The 37th day data did not have the strength of differences between groups. Data on 42 days: normal control group toward positive control group, treatment group I, treatment group II, treatment group III showed the strength of differences between groups; positive control group, treatment group I, treatment group II, treatment group III towards the normal control group showed there was a significant strength of difference between groups.

Table 3. Results of analysis of comparison of hypercholesterol levels before (day 32) and after administration of garlic extract (day 33, day 37 and day 42)

Group of test animals	No. Variable	Measurement results of hypercholesterol level (mg/dl)			
		Day 32	Day 33	Day 37	Day 42
Treatment group I	T3	277	204	172	144
	T25	238	131	130	113
	T13	241	140	131	138
	T28	212	124	114	109
	T24	238	136	134	104
p value			0.022	0.003	0.150
Treatment group II	T27	244	138	137	141
	T9	258	231	152	130
	T5	216	116	108	100
	T10	262	157	141	139
	T2	269	127	125	120

p value			0.507	0.232	0.267
Treatment groupIII	T8	217	194	160	118
	T26	232	186	118	106
	T19	241	208	159	125
	T14	237	196	145	100
	T18	302	283	216	154
p value			0.007	0.086	0.076

Annotation:
p value < 0,05 : Significant
p value > 0,05 : Not Significant

Table 3 shows the comparison of hypercholesterol levels before (day 32) and after administration of garlic extract (day 33, day 37 and day 42), among others: treatment group I hypercholesterolemia levels before administration of garlic extract (day 32nd) compared to after giving garlic extract (day 33 and day 37): has a value of p <0.05, which means that there are significant differences in the level of hypercholesterolemia before and after administration of garlic extract; While the comparison (day 32) with (day 42): has a value of p> 0.05, which means that there are no significant differences. Treatment group II hypercholesterol levels before administration of garlic extract (day 32) and after administration of garlic extract on (33rd day, 37th and 42nd): have a value of p> 0.05 which means that there is no significant differences. Treatment group III hypercholesterol level before giving garlic extract (day 32) compared to after giving garlic extract at (33rd day): has a value of p <0.05 which means that there are significant differences in the level of hypercholesterolemia before and after administration garlic extract; While the comparison (day 32) with (days 37 and 42): has a value of p> 0.05, which means that there are no significant differences.

The normal control group was the control group that was not given test material treatment. This group was used by researchers to determine normal total cholesterol levels in research rats that were not treated with hypercholesterol cholesterol.

Table 4. The measurement results of cholesterol total levels of test animals (mg/dl)

Group of test animals	No. Variabel	Day 0	Day 32	Day 33	Day 37	Day 42
Normal control group	T15	137	195	198	196	195
	T22	119	153	156	154	154
	T12	100	175	177	176	174
	T16	152	171	173	170	170
	T17	137	172	172	171	170
Positive control group	T7	169	253	156	120	141
	T23	155	236	144	141	139
	T11	177	282	164	152	139
	T20	167	245	144	104	108
	T21	153	234	164	162	152
Treatment group I	T3	151	277	204	172	144
	T25	115	238	131	130	113
	T13	157	241	140	131	138
	T28	165	212	124	114	109
	T24	158	238	136	134	104
Treatment group II	T27	112	244	138	137	141
	T9	137	258	231	152	130

	T5	188	216	116	108	100
	T10	189	262	157	141	139
	T2	126	269	127	125	120
Treatment group III	T8	194	217	194	160	118
	T26	103	232	186	118	106
	T19	186	241	208	159	125
	T14	159	237	196	145	100
	T18	151	302	283	216	154

The positive control group was the group given the treatment in the form of hipcholesterol cholesterol and given simvastatin drug test material. The dose of simvastatin given to mice is 1.8 mg / 200 g bb per day. In table 4, the average total cholesterol level can be seen a decrease in total cholesterol levels from the state of hypercholesterolemia on the 33rd day, while on the 37th day and 42nd day, a decrease in the same number. Based on the statistical results on table 2 post hoc tests on day 32: the positive control group towards the normal control group showed a significant difference in strength; 42nd day data: positive control group towards normal control group showed significant strength difference. There was a decrease in hypercholesterolemia levels through the statin drug mechanism inhibiting 3-hydroxy-3-methylglutaril coenzyme A reductase converting acetyl-CoA to mevalonic acid. This reaction is one step in the formation of cholesterol in the liver. Statins reduce LDL levels and increase HDL levels [12].

The decrease in total cholesterol levels by the drug simvastatin was in accordance with the research conducted by Nurmelis in 2015 about monitoring lipid-cholesterol profiles in normal mice and hypercholesterol cholesterol mice after the administration of cat whiskers (*Orthosiphon stamuneus*) extract in the positive control group given simvastatin drugs and research results showed a decrease in total cholesterol levels in the test group dose of 250,500,1000 mg/kg BB which was significantly different from normal controls in normal cholesterol mice (p <0,05) [13].

The treatment group I was the group given the garlic extract (*Allium sativum*) with a dose: 0.108 grams. In table 4 the average total cholesterol level decreased the level of hypercholesterolemia on the 33rd day, 37th and 42nd day. Based on table 2 the post hoc test on the 33rd day (one day after administration of garlic extract) the treatment group I to the treatment group III showed a significant difference in strength on the 42nd day (after 10 days of giving garlic extract) to the treatment group I to Normal controls show the power of significant differences. In table 3 Comparison of the results of the analysis of hypercholesterol levels before administration of garlic extract (day 32) with after administration of garlic extract (day 33, day 37 and day 42) showed results including treatment group I day 32nd with 33rd day having significant value; 32nd day with 37th day showing significant results; 32nd day with 42nd day addressing results is not significant. The above data shows that giving garlic extract at a dose of 0.108 grams can reduce the level of hypercholesterolemia. This decrease occurs because garlic contains many alkaloid compounds, namely allicin. The mechanism for reducing blood cholesterol by allicin is thought to be able to bind to the SH-group which is a functional part of coenzyme-A so that there is direct inhibition of the enzyme 3-hydroxy-3-methylglutaril coenzyme (HMG-CoA) reductase by allicin.

The treatment group II was the group given the garlic extract (*Allium sativum*) with a dose of 0.144 grams. In table 4 the average total cholesterol level in the treatment group II decreased the level of hypercholesterolemia. Based on table 2 the post hoc test on day 42: treatment group II for the normal control group showed a significant difference in strength. In table 3 Comparison of the results of the analysis of hypercholesterol levels before administration of garlic extract (day 32) with after administration of garlic extract (day 33, day 37 and day 42) showed results including treatment group II results statistics show before and after administration of garlic extract did not show a significant difference. There was no significant change in hypercholesterolemia levels before and after the administration of garlic extract at a dose of 0.144 grams, possibly not working effectively in inhibiting the HMG CoA reductase enzyme in the liver.

The treatment group III was the group given 0.18 grams of garlic extract. In table 4 the average total cholesterol level in the treatment group III showed a decrease in the level of hypercholesterolemia. Based on table 2 the post hoc test on the 32nd day: Treatment group III for the normal control group showed significant results; Data on the 33rd day: treatment group III on treatment group I showed a significant difference in strength, this was due to multilevel dosing between treatment group I and treatment group III. Data on 42 days: treatment group III on the normal control group showed a significant difference in strength between groups. Table 3 results of comparison of hypercholesterolemia levels before administration of garlic extract (day 32) with after administration of garlic extract (day 33, day 37 and day 42) show significant results between 32nd day and 33rd day. The dose of extract given to this group is a dose that is increased 2 times the dose that should be given to humans, therefore the effect directly decreases hypercholesterolemia levels after one day of administration. This decrease occurs because Garlic contains many alkaloid compounds, namely allicin.

The mechanism for reducing blood cholesterol by allicin is thought to be able to bind to the SH-group which is a functional part of coenzyme-A so that there is direct inhibition of the enzyme 3-hydroxy-3-methylglutaril coenzyme (HMG-CoA) reductase by allicin. In addition, garlic is also contained in vitamin C (ascorbic acid) in protecting the heart as follows: prevents endothelial damage that normally initiates inflammatory responses and lipoprotein adhesions, destroys existing plaques by binding to lipoproteins and expelling them from the body, building collagen and increasing vessel elasticity blood [8]. There is also nicotinic acid (niacin) can reduce cholesterol levels by suppressing the activity of the lipoprotein lipase enzyme through inhibition of free fatty acid flow from adipose tissue, thereby reducing VLDL production in the liver and can inhibit fat mobilization so that total cholesterol production and LDL cholesterol can decrease [8]. The results of this study indicate that administration of garlic extract (*Allium sativum*) can reduce total cholesterol levels in Sprague Dawley rats with hypercholesterolemia; this study is in accordance with the results of research conducted by Pramitasari, Riana and Bahrudin regarding the effect of giving garlic extract (*Allium sativum*) on the improvement of lipid profile in wistar hypercholesterolemic Rattus norvegicus strains at various doses: 0.05 g / head / day, 0.1 g / head / day and 0.2 g / head / day obtained from garlic resulted in a significant decrease in total cholesterol (KT), LDL, TG, LDL / HDL ratio, and the ratio of KT / HDL at a dose of 0, 2 g / tail can reduce cholesterol almost to normal [3]. Brajawikalpa, Kautama with the title of the effect of giving garlic ethanol extract to total cholesterol, LDL and HDL levels in hypercholesterol cholesterol rats with a dose: 3.6 mg, 7.2 mg, 10.8 mg and 18 mg. Conclusion: Garlic ethanol extract has an influence on levels of total cholesterol, LDL and HDL; a dose of 3.6 mg / 200 g BB can reduce total cholesterol, LDL and increase HDL. The higher the garlic ethanol extract the greater the decrease in total cholesterol, LDL and increased HDL [8].

4. Conclusion

There is an effect of garlic extract (*Allium sativum*) on decreasing total cholesterol levels in white rats (*Rattus novergicus*) with hypercholesterol with significant value ($p < 0.05$). There are significant differences in hypercholesterolemia levels before being given garlic extract and after administration of garlic extract, among others: treatment group I before giving garlic extract (day 32) with after giving garlic extract (day 33) significant value 0.022 and before administration of garlic extract (32nd day) with after administration of garlic extract (42nd day) significant value 0.003; Treatment group III before giving garlic extract (day 32) with after giving garlic extract (day 33) significant value 0.007 which means smaller than p-value <0.05.

For further research, it is necessary to quantitatively test phytochemicals so that it can determine the amount of each group of compounds contained in garlic ethanol extract (*Allium sativum*) specifically which can reduce cholesterol levels. The measurements of total cholesterol levels can use an examination method that provides a higher level of accuracy such as laboratory testing of CHOD-PAP (cholesterol oxidase phenol amino phenazone).

5. References

[1] Buletin-ptm.pdf [Internet]. Available from: www.depkes.go.id/download.file=download/pusdatin/buletin/buletin-ptm.pdf
[2] Suryanto I and Pradana M S 2017 Terapi hiperkolesterol pada mencit (Mus musculus) Strain BALB / C betina umur 2 bulan menggunakan sari bawang **3**(2) 71–5
[3] Mustika R P 2012 Pengaruh ekstrak bawang putih (Allium sativum L) terhadap perbaikan profil lipid pada (Rattus norvegicus) strain wistar hiperkolesterolemia

IOP Conf. Series: Materials Science and Engineering **823** (2020) 012005 doi:10.1088/1757-899X/823/1/012005

[4] Hartini S and Suryani M E 2016 Uji Kualitas Serum Simpanan Terhadap Kadar Kolesteroldalam Darah di Poltekkes Kemenkes Kaltim. J Ilm Manuntung **2**(1) 65–9
[5] Widada S T, Martsiningsik M A and Carolina S C 2016 Gambaran perbedaan kadar kolesterol total metode CHOD-PAP (Cholesterol Oxidase – Peroxsidase Aminoantypirin) sampel serum dan sampel plasma EDTA *J Teknol Lab.* **5**(1) 1–4
[6] Kedokteran cermin dunia 2015 Bawang putih untuk menurunkan kadar kolesterol **42**(3)
[7] Badan Penelitian dan Pengembangan Kesehatan 2013 Riset Kesehatan Dasar (RISKESDAS) 2013 *Lap Nas 2013* 1–384
[8] Brajawikalpa R S and Kautama M G Pengaruh pemberian ekstrak etanol bawang putih terhadap kadar kolesterol total , LDL dan HDL pada tikus putih hiperkolesterol
[9] Umami S R, Hapizah S S, Fitri R and Hakim A 2016 Uji penurunan kolesterol pada mencit putih (Mus musculus) secara in-vivo menggunakan ekstrak metanol umbi talas (Colocasia esculenta L) sebagai upaya pencegahan cardiovascular disease *J Pijar MIPA* **11**(2) 121–4
[10] Nurmawati T 2017 The correlation of weight and blood cholesterol levels of white rat (Rattus norvegicus) with high-fat die *J Ners dan Kebidanan (Journal Ners Midwifery)* **3**(3) 202–6
[11] Wicaksono D Ir 2013 Pengaruh ekstrak buah Garcinia atroviridis terhadap kadar LDL pada darah tikus strain wistar yang diberi asupan lemak berlebih Available from: http://www.lib.ui.ac.id/naskahringkas/2015
[12] Williams H 2008 Dislipidemia – terapi obat Available from: https://lyrawati.files.wordpress.com/2008/07/dislipidemia_obat_hosppharm1.pdf
[13] Nurmeilis 2015 Penentuan profil lipid-kolesterol setelah pemberian ekstrak herba kumis kucing 1–63

IOP Conf. Series: Materials Science and Engineering **823** (2020) 012006 doi:10.1088/1757-899X/823/1/012006

Effect of *Moringa oleifera* leaf powder in diets on laying hens performance, β-carotene, cholesterol, and minerals contents in egg yolk

IGNG Bidura[1], IBG Partama[1], IAP Utami[1], DPMA Candrawati[1], E. Puspani[1], IM Suasta[1], DA Warmadewi[1], IA Okarini[1], AAP Wibawa[1], IM Nuriyasa, and NW Siti[1]

[1]Faculty of Animal Science, Udayana University, Denpasar-Bali, Indonesia Jl. PB. Soedirman, Denpasar-Bali, Indonesia

E-mail: bidura_unud@yahoo.com

Abstract. The present study was designed to investigate the influence of *Moringa oleifera* on Laying Hens Performance, β-carotene, cholesterol, and minerals contents in egg yolk. Two hundred and forty 30 weeks of healthy laying hens with homogeneous body weight in a complete randomized design with four treatments and 6 replications. Laying hens were randomly divided into four groups: M0: diets without administration of *Moringa oleifera* leaves, M1: diets with 2% *Moringa oleifera* leaves; M2: diets with *Moringa oleifera* leaves 4%; and M3: diets with 6% *Moringa oleifera* leaves, respectively. Each treatment consisted of six replication cages with 10 birds randomly assigned to each cage. This study showed that administration of the *Moringa* leaves powder were increased significantly different on egg productions, egg mass, feed efficiencies, yolk color, shell thickness, Mg and Ca contents in eggshell, but not the efficiency of feed consumption. The administration of 2-6% *Moringa* leaves powder in diets results in significantly lower yolk cholesterol contents. It was concluded that supplementation of 4-6% *Moringa* leaves powder in diets, increased egg production, egg mass, feed efficiencies, yolk color, shell thickness, β-carotene, Mg and Ca contents in the yolk, but decreased yolk cholesterol contents in laying hens.

1. Introduction

High consumer demand for food safety, such as eggs with the dangers of antibiotic residues, has made researchers look for other ways to replace antibiotics by creating animal feed that not only provides for the nutritional needs of poultry (energy, amino acids, vitamins, and minerals) but also safe for consumers. The use of antibiotics or antimicrobials as additives in poultry feed has lasted more than 40 years. The antibiotic is used as a growth promoter in a relatively small amount but can improve feed efficiency and productivity of poultry so that with the use of these additives, farmers can benefit more.

In Indonesia, the use of Antibiotic Growth Promoter (AGP) has been banned since 2018, even in some developed countries since the past has banned the use of antibiotics as additives in animal feed. This is due to two main factors, namely (1) the possibility of residues from antibiotics that will be toxic to consumers, and (2) antibiotics can create micro-organisms that are resistant in the human body. With the negative impact of the use of AGP, experts began to look for replacements focused on natural ingredients, such as the efficacy of herbs. Herbs that have received much attention lately are *Moringa oleifera* leaves. *Moringa* leaves contain 7 times vitamin C orange, 4 times Calcium milk, 4 times carrot vitamins, 3 times Potassium banana, and 2 times Yogurt protein [18]. The leaves contain quality proteins so they can compete with soybeans because *Moringa* leaves contain all essential amino acids in the right proportions. Carbohydrates, vitamins, minerals, and fatty acids (essential and non-essential), are very high.

The use of *Moringa* leaves in feed laying hens is expected to be able to improve egg quality and feed efficiency. Eggs are a product of laying hens that can experience a rapid loss of quality during the period of collection and storage before consumption. Thus, increasing and extending the shelf life of eggs is very important for farmers. Various efforts have been made to increase the thickness of eggshells and reduce cholesterol levels in eggs. Because of the high cholesterol content in egg yolks is an inhibiting factor for consumers to consume eggs. Reducing cholesterol in egg yolk and increasing the yolk color will benefit the poultry industry and public health [38,40].

The use of *Moringa oleifera* leaves as a mineral source for eggshell thickness, and sources of provitamin A and natural beta-carotene to increase the color of egg yolks can be used as an alternative natural fortification feed to overcome vitamin A deficiency, and natural provitamin A and beta-carotene compounds safer to consume than synthetic vitamin A [1,9,7,16]. The phytochemical compounds contained in *Moringa* include flavonoids, saponins, tannins, and several other phenolic compounds that have antimicrobial activity [1,15]. Phytochemical compounds in herbal plants have antimicrobial and antioxidant properties, can be used in the treatment of diseases in poultry as a substitute for antibiotics [5,50,53,39]. The antimicrobial activity of phytochemical compounds in *Moringa* leaves has been investigated and is likely to use it to develop new antimicrobial drugs [19,21,26]. Several previous studies confirmed that extracts or compounds isolated from *M. oleifera* had antioxidant, anti-carcinogenic, anti-diabetic, anti-inflammatory, and anti-hypertensive properties, as well as the ability to protect against liver damage [10,18,19,21,27].

The use of *Moringa* can be an alternative to reduce fat and cholesterol levels in eggs and meat of poultry [32,33,42,20] and replace antibiotics, because of the rich and diverse phytochemical compounds, and efficacious as an antibacterial agent and can increase immunity [27,19,70]. It has been thought to control many diseases, as an endothelial vasodilator and inhibits the activity of HMG Co-A (3-hydroxy-3-methyl-glutaryl Co-A), thereby inhibiting lipase lipoprotein which is responsible for plasma lipid hydrolysis and has health benefits in connection with their ability to change the fat profile and carcass yield of broilers [3,20,45].

Previous research found that the total carotene in *Moringa* leaves was 24735 μg/100 g. The high content of beta-carotene is very good for increasing the color and content of β-carotene in egg yolks. Beta-carotene supplementation is effective in increasing serum beta-carotene concentrations with little effect on serum retinol concentrations, reducing serum oxidative stress, and increasing biological antioxidant capacity [51]. β-carotene is provitamin A, the highest carotenoid in food and human tissue. This provides a number of beneficial functions for livestock, including humans, because of their ability to produce vitamin A [61]. When added to feed, bioactive compounds along with other phytochemicals can improve egg quality and have a positive effect on chicken health and performance. *Moringa* leaves can be used as an effective feed supplement for poultry to encourage results in relation to total weight gain and feed efficiency, reduce abdominal fat and cholesterol in race chicken and laying hens [8,20,30]. Herbal extracts in drinking water increase egg production and can reduce cholesterol in serum and egg yolks in laying hens [14,33,38,69].

Based on this, the study was conducted to determine the effect of *Moringa* leaves in the diet on egg production, content of β-carotene, cholesterol, mineral in egg yolks of Lohmann Brown laying hens up to 30-weeks old.

2. Material and Methods

Animals, treatments, and experimental design: This study was a feeding trial using two hundred and forty 30 weeks of healthy laying hens with homogeneous body weight 1685 ± 18.37 grams obtained from commercial poultry farms in a complete randomized design with four treatments and 6 replications. Laying hens were randomly divided into four groups: M0: diets without administration of *Moringa oleifera* leaves, M1: diets with 2% *Moringa oleifera* leaves; M2: diets with 4% *Moringa oleifera* leaves; and M3: diets with 6% *Moringa oleifera* leaves, respectively. Each treatment consisted of six replication cages with 10 birds randomly assigned to each cage at 150×70×45 cm (length× width×height). Each experimental diet was in the form of mash and birds have free access to feed and water during the experiment.

All diets were isonitrogenous and isoenergetic. Diets are formulated to meet nutritional requirements for poultry [43] for 12 weeks of trials. The ingredients and chemical composition of the feed are shown in Table 1.

Table 1. The ingredient and calculated nutrient content of the feed of laying hens up to 30 weeks old

Basal Diets		Compositions			
Ingredients (%):		M0	M1	M2	M3
Yellow corn		61.00	60.00	59.60	58.10
Rice bran		15.40	14.40	13.50	13.50
Soybean		7.00	8.00	7.70	8.00
Fish meal		10.80	9.80	9.40	8.70
Coconut oil		0.32	0.20	0.10	0.14
Moringa oleifera		0.00	2.00	4.00	6.00
Mineral-B12*)		5.48	5.60	5.70	5.56
Total		100	100	100	100
*Chemical composition **):*					
Metabolizable energy	(kcal/kg)	2901.7	2900.9	2901.4	2901.5
Crude protein	(%)	16.02	16.07	16.04	16.08
Crude Fiber	(%)	3.53	3.65	3.74	3.94
Ether Extract	(%)	6.93	6.83	6.61	6.68
Calsium	(%)	3.56	3.65	3.76	3.74
Phosphor	(%)	1.30	1.28	1.28	1.24
Arginine	(%)	1.15	1.17	1.19	1.22
Histidine	(%)	0.41	0.42	0.42	0.43
Isoleucine	(%)	0.85	0.86	0.87	0.87
Leucine	(%)	1.58	1.59	1.61	1.62
Lysine	(%)	1.11	1.11	1.11	1.12
Methionine	(%)	0.38	0.38	0.38	0.37
Phenylalanine	(%)	0.82	0.83	0.85	0.86
Threonine	(%)	0.73	0.73	0.74	0.75
Tryptophan	(%)	0.18	0.19	0.19	0.20
Valine	(%)	0.87	0.87	0.88	0.89

*) The mineral-B12 composition per 10 kg contains: Calcium: 49%; Phosphor 14%; Iron: 40000 mg; Manganese: 27500 mg; Mg: 27.500 mg; Zinccum: 25 mg; Vit-B12: 4.50 mg and Vit D3: 500000 IU. (PT. Eka Farma. Deptan RI No. D 8109127 FTS)

**) Based on calculation according to Scott *et al.* (1982)

Process of making flour *Moringa* leaves: *Moringa oleifera* leaves were dark green, thinly sliced, and dried at room temperature for 1-2 days, then dried in an oven at 50°C for 24 hours. Then the *Moringa* leaves were pounded into fine powder.

Live performance. Continuous lighting and access to feeding and water were provided during the experiment. The hens were weighed at the start (age 30 weeks) and the end (age 42 weeks) of the experiment. Eggs were collected every day and egg production was expressed on a day-to-day basis (% of chicken days). Individual egg weights were recorded and then used to calculate the average egg weight for all trial periods. The total egg mass was calculated by multiplying the weight of the egg with egg production. Feed intake was measured based on cages (hens) every week. Daily feed intake per bird was calculated based on the total cage intake for the entire trial period and the number of days in all periods. Feed conversion ratio (gram feed/gram egg mass) for all periods was calculated based on the cage of egg production, egg weight, and feed consumption. Egg quality parameters were measured using a multi-egg tester.

Quality of eggs and yolk minerals. Eggs are collected and labeled every day at 08.00 and 14.00 hours during the trial period. The percentage of egg production was calculated. Examination of egg and eggshell quality (shell weight, eggshell thickness, egg yolk weight, and albumin, egg yolk color, albumen, and egg yolk height) were carried out at the end of the experiment. For this purpose, two eggs placed between 08.00 and 12.00 hours were taken randomly from each group on the 42nd day of the week (a total of 12 eggs per group during the experiment). Eggs were weighed individually and the specific gravity of the egg, as the index of the thickness of the shell, was measured. After the egg was broken on the EQM measurement stand, the albumen and yolk height was measured. The intensity of the yolk color was evaluated and recorded according to Roche's egg yolk fan method. Albumen's weight was calculated by reducing the weight of the yolk and shells of the overall egg weight. To measure the weight of the shell, the eggshell was cleaned from the albumen which attaches to the membrane was removed; Eggshells were then dried at room temperature and expressed as a percentage of all eggs. Evaluation of egg quality was carried out on individual eggs, similar to the weight of the eggs measured. The mineral concentration of egg yolk (Ca and Mg) was measured by the AAS method. Egg yolk samples were analyzed for dry matter content (DM) by drying the sample at 105^{0}C for 24 hours in a forced-air oven. Crude proteins were determined by the Kjeldahl method[11] and crude fat was determined by the Soxhlet method [11].

The cholesterol content of egg yolk is evaluated from a sample of egg yolk. These samples are mixed until smooth and homogeneous then stored for cholesterol analysis. Samples of egg yolks frozen in sample bottles were used to determine yolk total cholesterol levels with commercial kits [23]. Cholesterol levels were analyzed according to the Lieberman-Burchard method [34]. The cholesterol of egg yolk was calculated and expressed in milligram/gram of egg yolk. Measurement of Beta-carotene content: namely by entering as much as 0.10-0.50 into the centrifuge tube then adding 5 ml of acetone and 5 ml of pure petroleum ether (PE), then stirring evenly and centrifuging for 5 minutes at a speed of 3000 rpm. The supernatant was taken and stored in a test tube, while the sediment was added 5 ml of acetone and then centrifuged again until the supernatant was colorless (the supernatant becomes clear). The collected supernatant was then inserted into a separator tube and rinsed with 15 ml of distilled water and repeated three times. The rinsing water was then removed and the top of the tube (clear) was inserted into the test tube, then 1 g of $NaSO_4$ was added, then vortex. Then the clear part was taken and the PE solution was added until the volume becomes 10 ml and then read on the spectrometry absorbent (abs) at λ = 450 ηm. Total carotene (ug/100 g) = (total volumexabsx100)/(0.2 x sample weight).

Statistical analysis. All data were analyzed by ANOVA to determine the difference between treatments. If differences were found, further analysis is carried out with Duncan's multiple range test.

3. Results

The results are presented that the final body weight, feed consumption, egg production, and feed efficiency in groups fed the experimental diets are shown in Table 2. The treated laying hens exhibited higher significantly different ($p<.05$) on egg production, egg mass, egg numbers, and feed efficiencies (feed consumption/egg mass) than the control bird. No significant differences ($p>.05$) in the final body weight and feed consumption were observed among the dietary treated groups. Laying hens in M2 and M3 groups corresponded with higher egg production, egg mass, egg numbers ($p<.05$) and better feed efficiencies ($p<.05$) than those Group M0 and M1.

Table 2. Effect of *Moringa oleifera* leaves powder in diets on egg production and feed efficiencies in laying hens up to 30 weeks of ages

Variables	Groups[1]				SEM[2]
	M0	M1	M2	M3	
Initial body weight (g/hens)	1682.35a	1690.71a	1685.04a	1687.42a	18.631
Final body weight (g/hens)	1713.81a	1742.63a	1736.28a	1745.35a	20.062
Feed consumption (g/hen/d)	128.46a	127.35a	117.47a	115.60a	4.072
Egg production (%)	88.37a[3]	88.75a	90.92b	91.17b	0.306
Egg mass (g/hens)	55.61b	56.10b	57.30a	57.51a	0.285
Feed efficiency (feed consumption/egg mass)	2.31a	2.27a	2.05b	2.01b	0.017

[1] M0: The diet without *Moringa* powder (control); M1: The diet with 2% *Moringa* powder; M2: The diet with 4% *Moringa* powder; and M3: The diet with 6% *Moringa* powder, respectively.
[2] SEM: standard error of treatment means
[3] Means with different superscripts within raw values are significantly different ($P<0.05$)

The effects of dietary supplementation with *Moringa* leaves powder on external egg quality characteristics of laying hens are presented in Table 3. The treated laying hens exhibited higher significantly different ($p<.05$) on shell thickness, Mg and Ca in eggshell, yolk color and yolk cholesterol than the control of laying hens. No significant differences ($p>.05$) in the albumen, haugh unit, egg shape, and specific gravity, Mg and Ca in egg yolk were observed among the dietary treated groups. Laying hens in M2 and M3 groups corresponse with higher shell thickness, Mg and Ca in eggshell and yolk color, but lowering in yolk cholesterol ($p<.05$) than those Group M0 and M1.

In addition, the results show that an additional *Moringa* leaves powder (group M2 and M3) in diets resulted in a significant ($P<0.05$) increase in shell thickness of birds were: 22.28% and 26.08%, respectively higher than control (Table 3). The mineral content of Ca in the eggshell of the M2 and M3 groups were: 18.27% and 18.23% significantly different ($P<0.05$) higher than control. The mineral content of Ca in yolks group M2 was 21.15%, followed by group M3 was 22.89%, respectively significantly different ($P<0.05$) higher than group M0 (control).

Table 3. Effect of *Moringa oleifera* leaves powder in diets on external egg quality, yolk color and yolk cholesterol in laying hens

Variables	Groups[1]				SEM[2]
	M0	M1	M2	M3	
Shell thickness (mm)	0.395b[3]	0.408b	0.483a	0.498a	0.024
Magnesium in eggshell (ppm)	0.3925a	0.3901a	0.3953a	0.3948a	0.117
Magnesium in egg yolk (ppm)	0.1472a	0.1469a	0.1502a	0.1527	0.103
Calcium in eggshell (ppm)	27.821b	28.037b	32.905a	32.892a	1.073
Calcium in egg yolk (ppm)	12.138b	12.207b	14.705a	14.916b	0.509
Yolk colour (1-15)	6.35b	8.21b	8.35a	8.51a	0.513
β-carotene (μg/100 g)	129.35b	168.19a	179.37a	185.39a	10.502
Egg yolk cholesterol (mg/g yolk)	8.92a	8.65a	7.14b	7.29b	0.116

[1] M0: The diet without *Moringa* powder (control); M1: the diet with 2% *Moringa* powder; M2: the diet with 4% *Moringa* powder; and M3: the diet with 6% *Moringa* powder, respectively.
[2] SEM: standard error of treatment means
[3] Means with different superscripts within raw values are significantly different ($P<0.05$)

The content of β-carotene in egg yolk increased significantly ($P<0.05$) in the presence of *Moringa* leaf flour in feed at Group M1, M2, and M3 (Table 3). The β-carotene content in hens in Group M1, M2, and M3 was increased significantly different ($P<0.05$) than Group M0. Dietary *Moringa* leaf powder increased yellowness in yolk color (P<0.05) and decreased yolk cholesterol in egg-laying hens. The egg yolk color in M2 and M3 hen groups were increased by 31.50% and 34.02% than group M0. Yolk cholesterol content was tended to decrease by about 19.96% and 18.27% (P<0.05) in M2 and M3 groups rather than group M0, respectively.

4. Discussion

There was no significant difference in feed consumption between treatments. The same was reported by [57] and [20] which shows that *M.oleifera* does not have a toxic effect or contains factors that limit intake which is opposite to the absorption of nutrients. However, Ayssiwede *et a.l* [12] reported that the use of 24% *Moringa* leaves in the ration was a decrease in feed intake.

The use of *Moringa* in real feed increases feed efficiency in groups M2 and M3 rather than M0 and M1 groups. *Moringa* leaf extract can be useful to be used as an effective feed supplement in poultry to improve feed efficiency in poultry [8,57]. The main way of action of this active ingredient is the inhibition of microbial pathogens and endotoxins in the intestine and increased pancreatic activity, resulting in better metabolism and utilization of nutrients [68,28]. *Moringa oleifera* extract was found to be more effective in controlling gram-negative bacteria tested than gram-positive bacteria [19].

According to [26], the antimicrobial activity of plants is mainly caused by the presence of secondary metabolites. Plants are rich in various secondary metabolites, such as tannins, terpenoids, alkaloids, and flavonoids, which have been found *in vitro* to have antimicrobial properties. These active compounds in the digestive tract of poultry will be able to help absorb nutrients. As reported by [2], that herbal active compounds (Garlic) can increase villous height and crypto depth, and reduce the epithelial thickness and the number of villous cells in the duodenum, jejunum, and poultry ileum. Increased height of villi, as well as the thickness of epithelium in the duodenum, jejunum and ileum, will increase nutrient uptake. Results from[14] found that the administration of *Sauropus* leaf extract in drinking water can significantly improve feed efficiency in laying hens.

Increased egg production and egg weight in chickens given *Moringa* leaf is caused by the presence of phytochemical compounds on *Moringa* leaves, as reported by [53] that *Moringa* leaves are also a source of vitamin A, riboflavin, nicotinic acid, folic acid, pyridoxine, ascorbic acid, β-carotene, calcium, iron, and α-tocopherol. The same thing was reported by [8] that *Moringa oleifera* plants are also efficacious to strengthen the heart, eyes, brain, bile, and immune system. Supplementation of *Moringa* flour affects egg productions and egg mass [24] and egg yolk bioactive compounds [4]. Antioxidants, flavonoids, carotenoids, amino acids, proteins, and energy levels that result in a decrease in egg water content can be the reason for increasing nutrient density in egg yolk [45,44].

According to [37], *Moringa oleifera* seed flour inclusions can improve the quality of external eggs, and increase fatty acid profiles. Some of the results of research on the effect of herbal extracts in poultry were carried out by [8,30] that administration of *Moringa* leaf can significantly increase body weight and feed efficiency in broiler chickens. Siti *et al.* [59,60,41] reported that *Moringa* leaves can increase egg production and egg quality, but decreasing yolk cholesterol in laying hens. Bidura *et al.* [14] reported that administration of 5 cc/100 cc herbal extracts (*Sauropus* and *garlic* leaves) in drinking water increased egg production and total egg weight. The results of [69] reported that the addition of garlic powder at level 5 or 10 g/kg in feed, showed an increase in chicken egg production. The same was reported by [33] that laying hens given garlic (2-8%) showed higher egg productions.

Increased egg production and feed efficiency due to the role of *Moringa* in the digestive tract of chickens can increase feed digestibility. As reported by [31,45], that plant extract supplements can increase the digestibility of nutrients in the digestive tract of poultry. Herbal extracts (Garlic) can increase the activity of pancreatic enzymes and microenvironmental conditions for better utilization of nutrients in mice [54]. *Moringa oleifera* leaves are known to be very poor in anti-nutritional content and have been used in ruminant rations [62] and other poultry or monogastric. This result is contrary to that reported by [64] that the high use of *Moringa* leaves in feed can cause increased levels of saponin as an antinutrient which can reduce digestion and absorption of nutrients, especially lipids.

The yolk colors in Groups M1, M2, and M3 are significantly higher compared to the control. The increase in egg yolk color in this study showed that *Moringa* leaves are rich in vitamin A or carotenoid pigments which are efficiently absorbed and utilized by chickens. The yellow color increase can be attributed to the carotenoid content of *Moringa* leaf powder. In this study, the carotene of *Moringa* was: 24735 μg/100 g. Besides that, *Moringa* pods are enriched with carotenoids and flavonoids, which are powerful natural antioxidants that can modify the levels of β-carotene and quercetin egg yolk [24,35,29]. According to [6,55,56], β-carotene in *Moringa* pods ranges from 2.7 to 3.10 mg/100 g dried pods. When added to feed, this bioactive, along with phytochemicals, increases egg production and has a positive effect on chicken health. Carotenoids play an important role in the development of different color scores in egg yolk. Especially, lutein is an active yellow dye. According to [17] that increasing the amount of olive leaf powder in food results in a linear increase in the color of the yolk. This increase in the yolk color can be attributed to the carotenoid content of olive leaf powder [44]. Some researchers report that herbal extract supplements show the potential for an increase in the yolk color. *Moringa oleifera* seed inclusion increases egg yolk color [37], also mulberry leaves, ginkgo [36,72], *Allium sativum* and *Sauropus androgynus* [14], olive leaf powder [71], and carrots in feed [29]. This observation is supported by the findings of [48,49] that the use of *Moringa oleifera* leaves 10-20% in broiler feed or laying can significantly increase the yellow color of the skin and egg yolk.

The results showed that *Moringa* leaf supplementation at a level of 2-6% in feed could reduce cholesterol levels of egg yolk. According to [52] that reduction of cholesterol and triglycerides by alkaloids is partly due to a reduction in lipogenic enzyme activity and an increase in excretion of bile acids in feces. The ability of β-carotene to reduce cholesterol is related to the hydroxymethyl glutaril-CoA enzyme. This enzyme plays a role in mevalonic formation in cholesterol biosynthesis. The synthesis of cholesterol and β-carotene is together through the mevalonic pathway derived from acetyl CoA. If the consumption of β-carotene is greater than saturated fatty acids, the biosynthesis process by the HMG-CoA enzyme will be directed to the synthesis of β-carotene, so that saturated fatty acids are not converted to cholesterol [3,27,54]. Mevalonic is needed in the process of cholesterol synthesis by inhibiting enzymes, thus inhibiting cholesterol formation [63].

Several studies have shown that supplementation of *Moringa oleifera* leaves affects egg mass, serum biochemistry, and egg yolk bioactive compounds positively [4,39,20,67], as well as reducing cholesterol and triglycerides [22,32]. Some researchers report the use of herbal compounds can reduce cholesterol, as released by [18] that *Olea europaea* leaf powder can be used to reduce cholesterol levels of egg yolks, S*auropus* leaves [63], Garlic [66], *Tanacetum balsamita* [46].

Egg cholesterol levels are also influenced by antioxidants (flavonoids and carotenoids) in the feed. Antioxidants potentiate the production of bile salts, which produce fat emulsification and decrease fat absorption, thereby reducing cholesterol levels [58,46]. Sterols and antioxidants in plants, can reduce the absorption process and accelerate the disposal of cholesterol through feces [13]. In addition, a decrease in egg cholesterol levels can be caused due to fiber contents of *M. oleifera*, which plays an important role in binding and removing cholesterol [25,47,48]. On the other hand, Zangeneh and Torki [71] found no effect of herbs (*olive* leaves) on blood cholesterol levels in laying hens.

5. Conclusion

We conclude that supplementation in diets of 4-6% *Moringa oleifera* leave powders was increased egg production, egg mass, feed efficiencies, yolk color, shell thickness, Ca and β-carotene contents in the yolk, but decreasing yolk cholesterol contents in laying hens up to 30 weeks of ages.

6. Acknowledgments

The author would like to thank the staff of the Laboratory of Nutrition and Biomolecular Sciences, Faculty of Animal Husbandry, Udayana University, for their assistance in sample chemical analysis. The author also wishes to thank the Rector of Udayana University for his support during this research including research funding.

References

[1] Abbas T E 2013 The use of *Moringa oleifera* in poultry diets *Turk. J. Vet. Anim. Sci.* **37** 492-96

[2] Adibmoradi M, Navidshad B, Seifdavati J and Royan M 2006 Effect of dietary garlic meal on histological structure of small intestine in broiler chickens *J. Poult. Sci.* **43** 378-83

[3] Adriani L, Indrayati N, Rusmana D, Hernawan E and Rochana A 2017 Effect of Noni (*Morinda citrifolia*) fruit flour on antioxidant status and hematological indices of laying japanese quail *Int. J. Poult. Sci.* **16**(3) 93-97

[4] Ahmad S, Khalique A, Pasha T N, Mehmood S, Hussain K, Ahmad S, Shaheen M S, Naeem M and Shafiq M 2017 Effect of *Moringa oleifera* (Lam.) pods as feed additive on egg antioxidants, chemical composition and performance of commercial layers *SAJAS.* **47**(6) 864-74

[5] Akinmoladun A C, Ibukun E O, Afor E, Obuotor E M and Farombi E O 2007 Phytochemical constituent and antioxidant activity of extract from the leaves of *Ocimum gratissimum*. *Sci. Res. Essay* **2** 163-66

[6] Amaglo N K, Bennett R N, Curto R B, Rosa E, Turco V L, Giuffrida A, Curto A L, Crea F and Timpo G M 2010 Profiling selected phytochemicals and nutrients in different tissues of the multipurpose tree *Moringa oleifera* L, grown in Ghana *Food. Chem.* **15** 1047-54

[7] Andarwulan N, Kurniasih D, Apriady R A, Rahmat H, Roto A V and Bolling B W 2012 Polyphenols, carotenoids and ascorbic acid in underutilized medicinal vegetables *J. Fungsional Food* **4** 339-47

[8] Akhouri S, Prasad A and Ganguly S 2013 *Moringa oleifera* leaf extract imposes better feed utilization in broiler chicks *J. Biol. Chem. Research.* **30**(2) 447-50

[9] Alikwe P C N and Omotosho M S 2013 An evaluation of the proximate and phytochemical composition of *moringa oleifera* leaf meal as potential feedstuff for non-ruminant livestock *Agrosh.* **13** 17-27

[10] Ashok V, Gomashe P A G, Megha P J and Neeta A D 2014 Antimicrobial activity of indian medicinal plants: *Moringa oleifera* and *Saraca indica Int. J. Curr. Microbiol. App. Sci.* **3**(6) 161-69

[11] Association of Official Analytical Chemists (AOAC) 2005 *Official Methods of Analysis of AOAC International* 18th Ed. (Arlington, USA: Association of Official Analytical Chemists)

[12] Ayssiwede S B, Dieng A, Bello H, Chrysostome C A A M, Hanel M B, Mankor A, Dahouda M, Houinato M R, Hornick J L and Missohou A 2011 Effects of *Moringa oleifera* (Lam.) leaves meal incorporation in diets on growth performances, carcass characteristics and economics results of growing indigenous senegal chickens *Pakistan J. of Nutrition* **10**(12) 1132-45

[13] Benakmoum A, Larid R and Zidani S 2013 Enriching egg yolk with carotenoids and phenols *Int. J. Biolog. Biomole. Agri. Food and Biotechnol. Eng.* **7** 489-93

[14] Bidura I G N G, Partama I B G, Putri B R T and Watiniasih N L 2017 The effect of water extract of two leaves (*Allium sativum* and *Sauropus androgynus*) on the egg production and yolk cholesterol level in layer hens *Pakistan J. of Nutrition* **15**(1) 23-31

[15] Bukar A, Uba T I and Oyeyi 2010 Antimicrobical profile of *Moringa oleifera* ekstracts against some food-borne microorganism *Bayero J. of Pure and Applied Sci.* **3** 43-48

[16] Burham B 2017 Phytochemical, proximate composition and minerals contents of *Moringa oleifera Chemistry Research Journal* **2**(2) 78-83

[17] Cayan H and Erener G 2015 Effect of olive leaf (*Olea europaea*) powder on laying hens performance, egg quality and egg yolk cholesterol levels *Asian-Aust. J. Anim. Sci.* **28** 538-43

[18] Chukwuebuka E 2015 *Moringa oleifera* "the mother's best friend" *International J. of Nut. and Food Sci.* **4**(6) 624-30

[19] Dalukdeniya D A C K, DeSilva K L S R and Rathnayaka R M U S K 2016 Antimicrobial activity of different extracts of leaves bark and roots of *Moringa oleifera* (Lam) *Int. J. Curr. Microbiol. App. Sci.* **5**(7) 687-91 doi: http:// dx.doi.org/10.20546/ijcmas.2016.507.078

[20] Ekayuni A A, Bidura I G N G and Partama I B G 2017 The effect of water extract of two leaves (*Moringa oleivera* and *Sauropus androgynus*) on growth performance and meat cholesterol levels in broilers *J. Biol. Chem. Research.* **34**(1) 33-39

[21] Elangovan M, Dhanarajan M S, Rajalakshmi A, Jayachitra A, Mathi P and Bhogireddy N 2014 Analysis of phytochemicals, antibacterial and antioxidant activities of *Moringa oleifera* leaf extract an *in vitro* study *Int. J. Drug Dev. & Res.* **6**(4) 173-80

[22] El-Sheikh N I, El-Shazly E S, Abbas E A, Ghada I A and El-Gobary 2015 Effect of *Moringa* leaves on lipid content of table eggs in layer hens *Egypt. J. Chem. Enviro. Health.* **1** 291-302

[23] Ezema C and Didacus C Eze 2015 Probiotic effect of yeast (*Saccharomyces cerevisiae*) on hen-day egg performance, serum and egg cholesterol levels in laying chicken *Pakistan J. of Nutrition* **14**(1) 44-46

[24] Gakuya D W, Mbugua P N, Mwaniki S M, Kiama S G, Muchemi G M and Njuguna A 2014 Effect of supplementation of *Moringa oleifera* leaf meal in layer chicken feed *Int. J.Poult. Sci.* **13** 379-84

[25] Ghasi S, Nwobodo E and Ofili JO 2000 Hypo-cholesterolemic effects of crude extract of leaf of *Moringa oleifera* in high-fat diet fed Wistar rats *J. Ethno. Pharmacol.* **69** 21-25

[26] Goel A 2013 Anticancerous potential of plant extracts and phytochemicals *J. Biol. Chem. Research.* **30**(2) 537-58

[27] Godinez-Oviedo A, Guemes-Vera N and Acevedo-Sandoval O A 2016 Nutritional and phytochemical composition of *Moringa oleifera* lam and its potential use as nutraceutical plant: A Review *Pakistan J. of Nutrition* **15**(4) 397-405

[28] Grashorn M A 2010 Use of phytobiotics in broiler nutrition: an alternative to infeed antibiotics *J. Anim. Feed. Sci.* **19** 338-47

[29] Hammershoj M, Kidmose U and Steenfeldt S 2010 Deposition of carotenoids in egg yolk by short-term supplement of coloured carrot (*Daucus carota*) varieties as forage material for egg-laying hens *J. Sci. Food. Agric.* **90**(7) 1163-71

[30] Hassan K U, Khalique A, Pasha T N, Akram M, Mahmood S, Sahota A W, Imran M S and Saleem G 2017 Influence of *Moringa oleifera* decorticated seed meal on broiler performance and immunity *Pak. Vet. J.* **37** 47-50

[31] Hernandez F, Madrid J, Garcia V, Orengo J and Megías M D 2004 Influence of two plant extracts on broilers performance, digestibility, and digestive organ size *Poultry Sci.* **83** 169-74

[32] Irfan H M, Asmawi M Z, Khan N A K and Sadikun A 2016 Effect of ethanolic extract of *Moringa oleifera* lam leaves on body weight and hyperglycemia of diabetic rat *Pakistan J. of Nut.* **15**(2) 112-17

[33] Khan S H, Sardar R and Anjum M 2007 Effects of dietary garlic on performance and serum and egg yolk cholesterol concentration in laying hens *Asian J. Poult. Sci.* **1** 22-27

[34] Lieberman A and R Burchard 1980 Enzymatic method to determined cholesterol *Engl. J. Med.* **271** 915-24

[35] Liu H N, Liu Y, Hu L L, Suo Y L, Zhang L, Jin F and Li Y 2014 Effects of dietary supplementation of quercetin on performance, egg quality, cecal microflora populations, and antioxidant status in laying hens *Poult. Sci.* **93** 347-53

[36] Lokaewmanee K, Mompanuon S, Khumpeerawat P and Yamauchi K 2009 Effects of dietary mulberry leaves (*Morus alba L.*) on egg yolk color *J. Poult. Sci.* **46** 112-15

[37] Mabusela S P, Nkukwana T T, Mokoma M and Muchenje V 2018 Layer performance, fatty acid profile and the quality of eggs from hens supplemented with *Moringa oleifera* whole seed meal *South African Journal of Animal Sci.* **48**(2) 234-43

[38] Mahmoud K Z, Saad M, Gharaibeh, Hana A, Zakaria and Qatramiz A M 2010 Garlic (*Allium sativum*) supplementation: influence on egg production, quality, and yolk cholesterol level in layer hens *Asian-Aust. J. Anim. Sci.* **23**(11) 1503-09

[39] Mbikay M 2012 Therapeutic potential of *Moringa oleifera* leaves in chronic hyperglycemia and

dyslipidemia: a review *Front. Pharmacol.* **3**(24) 1-12

[40] Meliandasari D, Dwiloka B dan Suprijatna E 2015 Optimasi daun kayambang *(Salvinia molesta)* untuk penurunan kolesterol daging dan peningkatan kualitas asam lemak esensial *J. Ap. Tek. Pang.* **4**(1) 22-27

[41] Mohammed K A F, Sarmiento-Franco L, Santos-Recalde R and Solorio-Sanchez J F 2012 The nutritional effect of *Moringa oleifera* fresh leaves on Rhode Island Red hen egg production and quality *Trop. Anim. Health. Prod.* **44** 1035-40

[42] Moreki J C and Gabanakgosi K 2014 Potential use of *Moringa oleifera* in poultry diets *Global. J. Anim. Sci. Res.* **2**(2) 109-15

[43] National Research Council 1994 *Nutrient Requirements of Poultry* 9th rev. ed. (Washington, DC: National Academy Press)

[44] Nimalaratne C and Wu J 2015 Hen egg as an antioxidant food commodity: A Review *Nutrients.* **7** 8274-93

[45] Nkukwana T T, Muchenje, Pieterse E, Masika P J, Mabusela T P, Hoffman L C and Dzamab K 2014 Effect of *Moringa oleifera* leaf meal on growth performance, apparent digestibility, digestive organ size and carcass yield in broiler chickens *Livest. Sci.* **161** 139-46

[46] Nobakht A and Moghaddam M 2013 The effects of different levels of costmary (*tanacetum balsamita*) medicinal plant on performance, egg traits and blood biochemical parameters of laying hens *Iran. J. Appl. Anim. Sci.* **3**(2) 307-12

[47] Oladunjoye I O, Ojebiyi O and Amao O A 2010 Effect of feeding processed cassava (*Manihot esculenta crantz*) peel meal based diet on the performance characteristics, egg quality and blood profile of laying chicken *Agric. Trop. Subtrop.* **43** 119-26

[48] Olugbemi T S, Mutayoba S K and Lekule F P 2010a Effect of Moringa (*Moringa oleifera*) inclusion in cassava based diets fed to broiler chickens *Int. J. Poult. Sci.* **9** 363-67

[49] Olugbemi T S, Mutayoba S K and Lekule F P 2010b Evaluation of *Moringa oleifera* leaf meal inclusion in cassava chip based diets fed to laying birds *Livest. Res. Rural Dev.* **22**(6) URL address: [http://www.lrrd.org/lrrd22/6/olug22118.htm]

[50] Oriabi A G 2016 Moringa oleifera in vitro culture and its application as anti-diabetic in alloxan induced diabetic albino mice *Int. J. Curr. Microbiol. App. Sci.* **5**(2) 43-49

[51] Otomaru K, Ogawa R, Oishi S, Iwamoto Y, Hong H, Nagai K, Hyakutake K, Kubota C and Kaneshige T 2018 Effect of beta-carotene supplementation on the serum oxidative stress biomarker and antibody titer against live bovine respiratory syncytial virus vaccination in japanese black calves *Vet. Sci.* **5**(102) 2-9 doi:10.3390/vetsci5040102

[52] Patil R H, Prakash K and Maheshwari V L 2010 Hypolipidemic effect of *Celastrus paniculatus* in experimentally induced hypercholesterolemic Wistar rats *Indian. J. Clin. Biochem.* **25** 405-10

[53] Prasad A and Ganguly S 2012 Promising medicinal role of *Moringa oleifera*: a review *J. Immunol. Immunopathol.* **14**(1) 1-5 DOI: 10.5958/j.0972-0561.14.1.001

[54] Ramakrishna R R, Platel K and Srinivasan K 2003 *In vitro* influence of spices and spice-active principles on digestive enzymes of rat pancreas and small intestine *Nahrung* **47** 408-12

[55] Saini R K, Shetty N P, Prakash M and Giridhar P 2014a Effect of dehydration methods on retention of carotenoids, tocopherols, ascorbic acid and antioxidant activity in *Moringa oleifera* leaves and preparation of a RTE product *J. Food. Sci. Technol.* **51** 2176-82

[56] Saini R K, Shetty N P and Giridhar P 2014b Carotenoid content in vegetative and reproductive parts of commercially grown *Moringa oleifera* lam. cultivars from India by LC-APCI-MS *European. Food. Res. Tech.* **238** 971-78

[57] Sanchez N R, Sporndly E and Ledin I 2005 Effect of different levels of foliage of *Moringa oleifera* to creole dairy cows in intake, digestibility, milk production and composition *Livest. Sci.* **2810** 8

[58] Srinivasan K 2005 Spices as influencers of body metabolism: an overview of three decades of research *Food. Res. Int.* **38** 77-86

[59] Siti N W, Bidura I G N G and Utami I A P 2017 The effect of water extract of leaves *Moringa oleifera* on egg production and yolk cholesterol levels in egg laying hens *J. Biol. Chem. Res.* **34**(2) 657-65

[60] Suarjana I P, Siti N W dan Bidura I G N G 2018 The effect of *Morinda citrifolia* water extract on drinking water on physical quality of egg in Lohmann Brown layer hens aged 22-30 age *Journal of Tropical Animal Sci.* **6**(1) 129-39

[61] Shete V and Quadro L 2013 Mammalian metabolism of β-carotene: gaps in knowledge *Nutrients* **5** 4849-68 doi:10.3390/nu5124849

[62] Soliva C R, Kreuzer M, Foidl G, Machmüller A and Hess H D 2005 Feeding value of whole and extracted *Moringa oleifera* leaves for ruminants and their effects on ruminal fermentation in vitro *Anim. Feed Sci. Technol.* **118** 47-62

[63] Syahruddin E, Herawaty R and Ningrat R W S 2013 Effect of fermented katuk leaf (*Sauropus androgynus* l. merr.) in diets on cholesterol content of broiler chicken carcass *Pakistan Journal of Nutrition* **12**(11) 1013-18

[64] Teteh A, Lawson E, Tona K, Decuypere E and Gbeassor M 2013 *Moringa oleifera* leave: hydro-alcoholic extract and effects on growth performance of broilers *International Journal of Poultry Sci.* **12**(7) 401-05

[65] Tourniaire F, Gouranton E, vonLintig J, Keijer J, Bonet M L, Amengual J, Lietz G, Landrier J F 2009 Beta-carotene conversion products and their effects on adipose tissue *Genes Nutr.* **4** 179-87

[66] Wibawa A P P, Utami I A P and Bidura I G N G 2016 The effect of garlic (*Allium sativum*) extract on performance, abdominal fat and serum cholesterol of broiler *Journal of Biological and Chemical Research.* **33**(2) 404-11

[67] Widnyana I K, Bidura I G N G and Candrawati D P M A 2017 Pemberian ekstrak air daun katuk (*Sauropus androgynus*) dan daun kelor (*Moringa Oleifera* lam) melalui air minum terhadap distribusi lemak dan kolesterol darah broiler *E-Journal Peternakan Tropika* **5**(1) 64-77

[68] Windisch W, Schedle K, Plitzner C and Kroismayr A 2008 Use of phytogenic products as feed additives for swine and poultry *J. Anim. Sci.* **86** 140-48

[69] Yalcin S, Onbaslar I, Sehu A and Yalcin S 2007 The effect of dietary garlic powder on the performance, egg triats and blood serum cholesterol of laying quails *Asian-Aust. J. Anim. Sci.* **20** 944-50

[70] Yuniza A and Yuherman 2015 Content of phytochemical compound and antibacterial activity of cinnamon leaf (*Cinnamomum burmanii*) and Noni fruit and leaf (*Morinda citrifolia* l) mixture extract to replace antibiotics *Pakistan Journal of Nutrition* **14**(8) 492-97

[71] Zangeneh S and Torki M 2011 Effects of *b*-mannanase supplementing of olive pulp included diet on performance of laying hens, egg quality characteristics, humoral and cellular immune response and blood parameters *Global Vet.* **7** 391-98

[72] Zhao L, Zhang X, Cao F, Sun D, Wang T and Wang G 2013 Effect of dietary supplementation with fermented ginkgo-leaves on performance, egg quality, lipid metabolism and egg-yolk fatty acids composition in laying hens *Live. Sci.* **155** 77-85

IOP Conf. Series: Materials Science and Engineering **823** (2020) 012007 doi:10.1088/1757-899X/823/1/012007

Synthesis of mercaptoethyl ester of palm fatty acid distillate: *Effects of dehydration methods*

I Dewa Gede Arsa Putrawan[1], Dicka Ar Rahim[2], Andri Patuan Anggi Sitompul[3] and Salma Rifasari Wakana[3]

[1]Research Group on Chemical Engineering Product Design & Development, Faculty of Industrial Technology, Institut Teknologi Bandung, Bandung, Indonesia.
[2]Research Group on Chemical Engineering Energy & Process System, Faculty of Industrial Technology, Institut Teknologi Bandung, Bandung, Indonesia.
[3]Study Program of Chemical Engineering, Faculty of Industrial Technology, Institut Teknologi Bandung, Bandung, Indonesia.

E-mail: idewa@che.itb.ac.id

Abstract. This research was aimed to evaluate the synthesis of mercaptoethyl ester of palm fatty acid distillate (PFAD) where water as a byproduct was removed by using vacuum distillation and azeotropic distillation. The reactions were done in a batch reactor using para-toluene sulfonic acid (pTSA) as a catalyst. The performances of the synthesis were evaluated by measuring the sulfhydryl content and yield of the product. Compared to dehydration by vacuum distillation, dehydration by azeotropic distillation gave lower sulfhydryl content and yield but faster reaction completion. Azeotropic distillation gave sulfhydryl content and yield in the range of 6–7% and 57–74%, respectively. Vacuum distillation gave sulfhydryl content and yield in the range of 7–9% and 75–85%, respectively.

1. Introduction

Mercaptoethyl ester of fatty acid (MEFA) is a raw material for the manufacture of organotin based polyvinyl chloride (PVC) thermal stabilizers. Thermal stabilizers are essential for PVC to avoid degradation during the extrusion process [1]. PVC thermal degradation is also known as dehydrochlorination since hydrochloric acid is released during the process. As a result of thermal degradation, PVC changes its color from clear to yellow, red, brown, and finally black. Along with the severity of degradation, the mechanical properties of PVC changes, and in the worst case, PVC could not be formed into the desired final product.

MEFA is produced from free fatty acid (FFA) and mercaptoethanol (ME). The common source of FFA is Tall Oil Fatty Acid (TOFA). TOFA is produced as a byproduct in the production of pulp from pine wood. It mainly comprises of unsaturated fatty acid, up to 90% of its weight consists of oleic and linoleic acid [2]. Recently, Putrawan et al. [3] found that MEFA could be synthesized from Palm Fatty Acid Distillate (PFAD), a byproduct of palm oil refining processes. As with TOFA, PFAD is also mainly composed of FFA. The reaction between PFAD and ME can be described as follows [3]:

$$RCOOH + HSCH_2CH_2OH \rightleftharpoons RCOOCH_2CH_2SH + H_2O \quad (1)$$

The important parameter for MEFA quality is sulfhydryl (SH) content. Sulfhydryl group is used as a quality parameter since this functional group will be converted into sulfur bounds which substitute allylic chlorine atoms, the initiation sites of thermal degradation of PVC bone [4].

During the reaction, the system must be dehydrated to shift the reaction toward the product side. There are several dehydration methods, including distillation, extraction, and adsorption. Based on operating pressure, distillation can be divided into atmospheric and vacuum distillations. Distillation can also be divided into azeotropic and ordinary distillations, whether it involves entrainer or not, respectively. The separation task of dehydration, in this case, is to remove water as a by-product while keeping ME in the liquid phase. The peculiarities of the separation are the small amount of water to be removed (less than 5% of the total mixture), the presence of sulfur compounds, the reaction temperature is mild, and the product is sufficiently sensitive and may decompose or polymerize at its normal boiling point. Atmospheric distillation is not appropriate to apply as removing water at the end of the reaction will boil and decompose the product. Extraction is faced with the problem of the difficulty in finding solvents that can separate water and ME which are both polar. Adsorption could not be applied since MEFA synthesis is done at temperatures significantly above room temperature. Synthesis of MEFA from PFAD is best to work at 70–80°C [3] where the adsorption capacities of most desiccants are low. Besides, most water desiccants are poisoned by sulfur compounds. Thus, only vacuum distillation and azeotropic distillation can be explored.

Vacuum distillation mentioned here is ordinary distillation which operates below atmospheric pressure. The operation at low pressure is expected to enable water vaporization at lower temperatures to prevent thermal decomposition. Both temperature and pressure can be set independently. At a determined temperature, keeping the pressure at a certain value is not beneficial because it will create an equilibrium which means that water is left in the liquid phase. Lowering pressure, however, lowers the difference in boiling temperatures between the light key and the heavy key components. Moreover, water is formed gradually during the reaction. To remove water completely, the pressure must be reduced gradually in order not to vaporize ME.

Azeotropic distillation uses an entrainer as a separating agent [5]. An entrainer forms a low boiling point azeotrope with water but immiscible with water when condensed. Two popular entrainers for water removal are benzene and cyclohexane. Berenbaum et al. [6] performed azeotropic distillation using benzene as entrainer for the synthesis of copolymers at atmospheric pressure and temperatures up to 110°C within 1.5 hours. Azeotropic distillation has also been studied by Cane [7] at lower temperatures (60°C) to dehydrate ethanol using cyclohexane as entrainer. When an entrainer is added to the reaction mixtures of MEFA synthesis, it will bind the produced water during the reaction in the form of an azeotrope. Distilling the formed azeotrope will vaporize the azeotrope and, after condensation, separates into water and entrainer layers. The entrainer is then recycled and the dehydration cycle repeats.

Since the above mentioned two dehydration methods remove water in different ways while the effectiveness of water removal affects the reaction performance, both methods were thought to affect the reaction performance, i.e., the quality and yield of MEFA. It is the purpose of this research to study the effects of dehydration method on the quality and yield of product in the synthesis of MEFA from PFAD. The dehydration methods involved were vacuum distillation and azeotropic distillation. The synthesis was done in batch operations with para-toluene sulfonic acid (pTSA) as a catalyst. The quality of MEFA was measured as sulfhydryl content since this functional group takes an important role in stabilizing PVC.

2. Experiments

2.1. Materials

PFAD was obtained from a local palm oil refinery. Benzene (>99.7%) and cyclohexane (>99.5%) were purchased from Merck. pTSA with a purity of more than 99% was purchased from Fluka.

2.2. *Procedure*
For the reaction with vacuum distillation for removing water, the apparatus consisted of a jacketed beaker glass having a volume of 500 mL, a condenser, a condensate collector, and a vacuum system. The synthesis procedure referred to the previous work [3] that PFAD and ME were placed in the beaker and conditioned to the desired temperature by circulating hot water through the jacket, catalyst was added after the temperature achieved the desired value, the pressure was reduced gradually until full vacuum at 20 torr, the product was washed with water, and finally dried in vacuum. The reaction temperature was varied in the 60-90˚C range.

For the reaction with azeotropic distillation for removing water, a Dean-Stark apparatus was used. The apparatus consisted of a boiling glass still with a volume of 500 mL which was connected to a condenser with a water trap at the bottom. At first, 80 g of PFAD was placed in the still. ME, entrainer, and catalyst with determined ratios were added into the still. The still was then heated using a mantle heater. After boiling, the entrainer and water vaporized as an azeotrope. The azeotrope separated into two liquid phases, i.e., water and entrainer, after condensed. Water was trapped in the water trap, the entrainer was recycled back to the still and the cycle of water removal repeated. The reaction was stopped when the water level in the water trap became constant. The product was then washed and dried.

The dehydration method applied affects the management of reaction temperature. For the vacuum distillation, the temperature could be set independently by controlling the hot water circulating through the reactor jacket. For the azeotropic distillation, once the entrainer-to-PFAD ratio was set, the reaction temperature was fixed by the bubble point of the reaction mixture. Each run was done twice. Excess in ME and catalyst-to-PFAD mole ratio were fixed for all runs at 10% and 1%, respectively. Acid, saponification, and iodine values were determined according to AOCS methods Cd 3d-63, Cd 3c-91, and Cd 1d-92, respectively [8]. Sulfhydryl (SH) content was determined by iodine titration [9]. IR spectra were measured using spectrophotometer Shimadzu IR-Prestige-21.

3. Results and Discussion

3.1. *Characteristics of FFA and MEFA*

The PFAD used had acid, saponification, and iodine values of 181 mg KOH/g, 203 mg KOH/g, and 53 g I_2/100 g, respectively. High acid value ensured that PFAD is mostly composed of FFAs. Analogous to MET (mercaptoethyl tallate), a MEFA derived from TOFA, MEP (mercaptoethyl palmate) is used here to name MEFA synthesized from PFAD. Figure 1 shows the IR spectra of PFAD, ME, and MEP. Because most of PFAD components are FFAs, the characteristics of PFAD spectra are the same as those of carboxylate acids which characterized by the existence of a strong broad O–H stretching band in the 3500–2500 cm^{-1} range and the C=O stretching band at near 1700 cm^{-1}. Unsaturated FFAs have a typical peak at 3050 cm^{-1} showing the existence of =C–H stretching band (between 3100-3000 cm^{-1}). The =C-H peak in PFAD was not very sharp because half of PFAD components are saturated fatty acid. ME was characterized by the existence of a weak –SH stretching band near 2550 cm^{-1} and a strong broad O-H stretching band in the 3600–3000 cm^{-1} range. It is clearly shown that the spectra of MEPs obtained using the vacuum distillation and azeotropic distillation were similar. The broad O–H stretching band is significantly reduced in MEP spectra and an –SH stretching band was found at near 2550 cm^{-1}. The spectra of MEPs using both dehydration methods were identic to that of MET. This convinced that MEPs could be synthesized regardless of the methods used to remove water as a side product.

3.2. *Sulfhydryl content and yield of MEP with vacuum distillation as a dehydration method*
Figure 2 shows the sulfhydryl content and yield of MEP obtained at various reaction temperatures when vacuum distillation was used to remove water. Both sulfhydryl content and yield increased with temperature. However, the increase in both performance indicators declined as temperature increased. Higher temperature increased the reaction rate. However, it also increased the volatility of ME. Higher temperature increased the possibility for ME to leave from the liquid phase where the reaction occurred. At all temperatures, a perfect reaction with a 100% yield could not be achieved. The sulfhydryl content

was also still significantly lower than the theoretical value. The theoretical value of sulfhydryl content is 9.6 with was predicted based on palmitic and oleic acids. As mentioned previously, it is due to the existence of side reactions [3], among them are the formation of thioesters [11], co-polymerizes [6], and the attachment of double bonds by ME [10].

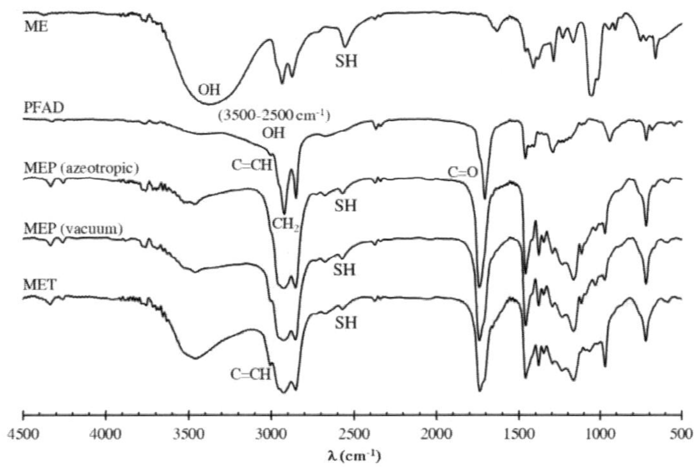

Figure 1. IR Spectra of FFA and MEFA.

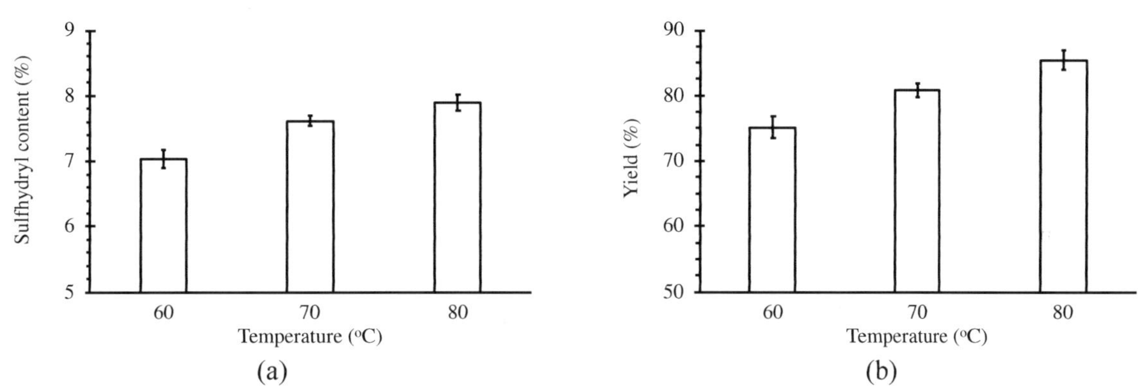

Figure 2. Sulfhydryl content (a) and yield (b) of MEP in vacuum distillation.

3.3. Sulfhydryl content and yield of MEP with azeotropic distillation as a dehydration method

At first, the performances of benzene and cyclohexane as entrainers were compared. The synthesis was done at an entrainer-to-PFAD weight ratio of 3.0. The sulfhydryl contents of MEPs obtained using benzene and cyclohexane as entrainers were (6.9±0.1)% and (6.7±0.2)%, respectively. Statistical tests showed that this difference in sulfhydryl content was not significant. As explained earlier, the role of entrainer is to dehydrate the reaction mixture by forming an azeotrope with water. The azeotrope composition of water-benzene and water-cyclohexane systems are almost the same, i.e., 9%-wt water, with azeotrope temperature in the 69–70°C range [14]. For the same entrainer-to-PFAD ratio, therefore, the capacities of the entrainers are almost the same and the extents of the reactions become the same. Since benzene is known to be carcinogenic, cyclohexane is considered to be more favorable.

The effects of entrainer-to-PFAD ratio were then studied using cyclohexane as entrainer. Figures 3a and 3b show the effects of entrainer-to-PFAD ratio on the sulfhydryl content and yield of MEP, respectively. When using vacuum distillation to remove water, the reaction temperature could be regulated independently. In contrast to the case, when using azeotropic distillation, the reaction temperature was fixed by the entrainer-to-PFAD ratio since entrainer-to-PFAD determines the bubble point of the reaction mixture. The reaction temperature observed for each entrainer-to-PFAD ratio is shown at the top of each bar in figure 3. Both sulfhydryl content and yield were low at the entrainer-to-PFAD ratio of 1. It is due to the high reaction temperature at this entrainer-to-PFAD ratio. The reaction between PFAD and ME occurs in the liquid phase. At the entrainer-to-PFAD ratio equals to 1.0, the reaction temperature was high (91°C). Although the normal boiling point of ME is higher than 90°C, i.e., 157°C [13], with the existence of cyclohexane, ME starts vaporizing at 90°C. Loss in one of reactant due to distillation made the extent of reaction low. A significant increase in sulfhydryl content was found when the entrainer-to-PFAD ratio increased from 1.0 to 2.0. Higher entrainer-to-PFAD ratio resulted in lower reaction temperature which could keep ME in the liquid phase. A further increase in the entrainer-to-PFAD ratio to 3.0 did not increase the sulfhydryl content and yield significantly. It is due to the long reaction time necessary at this entrainer-to-PFAD ratio. The reaction times at the entrainer-to-PFAD ratio of 3.0 were 4.0 hours and those at the entrainer-to-PFAD ratios of 1.0 and 2.0 were 3 hours. Longer reaction time increased the quantity of side products which lowered the sulfhydryl content and yield of MEP.

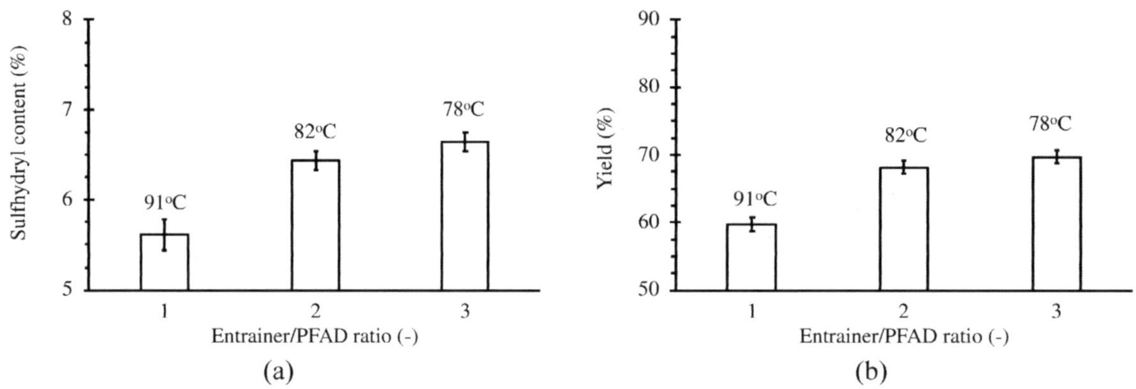

Figure 3. Sulfhydryl content (a) and yield (b) of MEP in azeotropic distillation.

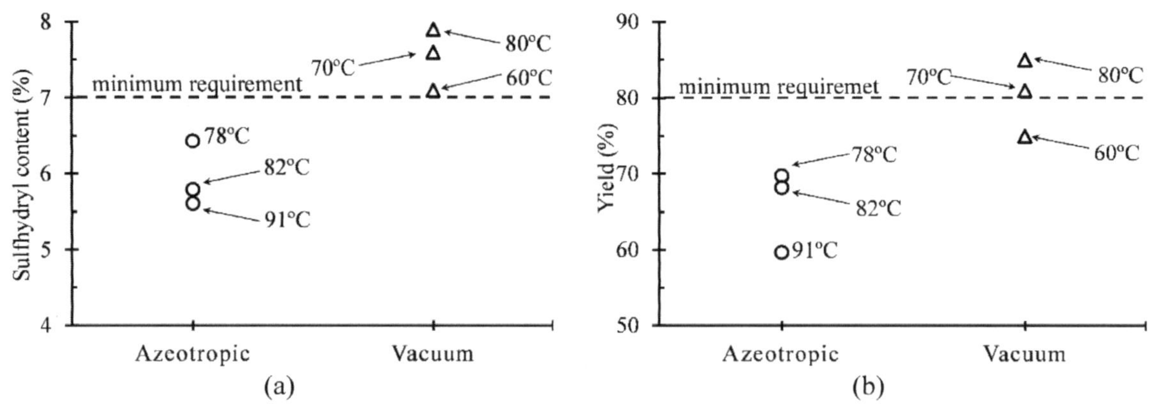

Figure 4. Comparison of dehydration methods in sulfhydryl content (a) and yield (b).

3.4. Effects of dehydration methods on reaction performance

The performances of reaction involving the two dehydration methods in the downstream were compared in figure 4. Each point in the graph is also provided with reaction temperature. Higher sulfhydryl content and yield were obtained when water was dehydrated using vacuum distillation. Using azeotropic distillation, there is a tendency that ME is also carried in the vapor phase since ME is soluble in benzene and cyclohexane. In other words, the azeotrope formed is, in fact, a ternary system. When the vapor phase is condensed, some ME is left in the condensate trap since it is miscible with water. This made the concentration of ME in the reaction mixture depleted, and less product is obtained. In vacuum distillation, reaction temperature could be set independently using a temperature controller. It is possible using vacuum distillation to control reaction mixture from 60 to 80°C. Using azeotropic distillation, however, the reaction mixture is fixed by the entrainer-to-FFA ratio. It is not possible to have reaction temperatures lower than the boiling point of the entrainer used. The lowest reaction temperature observed during the experiment was 78°C when using cyclohexane as an entrainer at an entrainer-to-FFA ratio of 3.0. It is a little bit lower than the normal boiling point of cyclohexane since the atmospheric pressure during the experiments was 60 to 80 mmHg lower than 1 atm. The advantage of azeotropic distillation is shorter reaction time, 3 to 4 hours. By using vacuum distillation, the reaction needed 6 hours for completion. The minimum sulfhydryl content technically required and the minimum yield of MEP to be economic as raw materials for PVC thermal stabilizer are also shown in figure 4. It could be seen that azeotropic distillation could not satisfy these requirements. On the other hand, the MEP by vacuum distillation at 70°C and 80°C could be accepted by both requirements.

4. Conclusions

Mercaptoethyl ester of palm fatty acid distillate has been synthesized using two dehydration methods: vacuum distillation and azeotropic distillation. Compared to vacuum distillation, dehydration by azeotropic distillation gave lower sulfhydryl content and yield but faster reaction completion. Azeotropic distillation gave sulfhydryl content and yield in the range of 6–7% and 57–74%, respectively, which are still lower than the minimum requirement. Vacuum distillation gave sulfhydryl content and yield in the range of 7–9% and 75–85%, respectively, and satisfied the minimum requirement.

Acknowledgment

This research was financially supported by the Indonesian Management Agency for Palm Crop Estate Fund under the Palm Research Grant 2018.

References

[1] Folarin O M and Sadiku E R 2011 Thermal stabilizers for poly(vinyl chloride): a review *Int. J. Phys. Sci.* **6** 4323–30
[2] Magee T V and Zinkel D F 1992 Composition of American distilled tall oils *J. Am. Oil Chem. Soc.* **69** 321–4
[3] Putrawan I D G A, Azharuddin A, Arum K R, Adityawarman D and Rahim D A 2018 Synthesis of mercaptoethyl ester of palm fatty acid distillate *MATEC Web Conf.* **156** 06010
[4] Wypych G 2015 *PVC Degradation and Stabilization* (Toronto: Chem Tec Publishing)
[5] Humphrey J L 1997 *Separation Process Technology* (McGraw-Hill)
[6] Berenbaum M B, Edward B and Christena R C 1967 Mercaptoethanol/mercaptopropanol copolymers and process *US3317486 A*
[7] Cane C 1979 Dehydration of alcohols, preferably ethanol, using a cyclohexane entrainer *EP0001681 A1*
[8] AOCS 1993 *Official Methods and Recommended Practices of the American Oil Chemists' Society* (Washington, DC: AOC Press)
[9] Jeffery G H, Basset J, Mendham J and Denney R C 1989 *Vogel's Quantitative Chemical Analysis* (New York: John Wiley & Sons, Inc.)
[10] Hayashi T, Kazlauciunas A and Thornton P D 2015 Dye conjugation to linseed oil by highly-effective thiol-ene coupling and subsequent esterification reactions *Dyes Pigments* **123** 304–16
[11] Sonnet P E and Moore G G 1989 Thiol esters of 2-mercaptoethanol and 3-mercapto-1,2-propanediol *Lipids* **24** 743–5

[12] Knezevic M, Katsikas L and Popovic I 2005 The synthesis and characterization of 2-mercaptoethyl methacrylate *Hem. Ind.* **59** 321–3
[13] Grogan C H, Rice L M and Reid E EMMET 1955 Dithiols and derivatives *J. Org. Chem.* **20** 50–9
[14] Horsley L H 1973 *Azeotropic Data—III* (Washington, DC: American Chemical Society)

IOP Conf. Series: Materials Science and Engineering **823** (2020) 012008 doi:10.1088/1757-899X/823/1/012008

The quality and mineral content of the community salt conventional in Kupang City and Kupang District East Nusa Tenggara Province

Umbu Paru Lowu Dawa[1], Dewi Setyowaty Gadi[1], Renya Rosari[2], Yunialdi Hapynes Teffu[1]

[1]Fisheries Product Technology Department-Marine Science and Fisheries Faculty, Artha Wacana Christian University
[2]Accounting Department of Economics Faculty, Artha Wacana Christian University

E-mail: umbupaki@yahoo.com

Abstract. The production of conventional salt produced by the community salt processing in the city of Kupang and Kupang district has quality at below the standards set. This study aims to analyze the quality of conventional salt and mineral contents. The sample is taken by *random sampling* method. Analysis of moisture content based on the AOAC method, NaCl content based on the method of wet dewatering and mineral content using *Inductively Coupled Plasma Optical Emission Spectrophotometer* (ICP OES). The quality of conventional salt is produced based on the criteria for the content of NaCl, water content and the content of Iodine is still below the Indonesian National Standard. The mineral content of magnesium, calcium and potassium is still in the range of Indonesian National Standard.

1. Introduction

An increase in population and industry has resulted in an increase in national salt demand from year to year. Industrial needs that require high-quality salt, namely with a NaCl content of more than 95%. Food and beverage industries, for example, which require an average salt per year of 509.6 thousand tons per year also desperately need government wisdom in seeing the national salvation situation that happened all this time. This is because salt is a raw material that is very essential in the production process of the food and beverage industry. Based on data from the Ministry of Maritime Affairs and Fisheries [1], national salt production in 2015 reached 2.84 million tons. A total of 2.5 million tons are produced by people's salt which is used to meet the needs of consumption salt and the remainder in the form of industrial salt produced by the salt company. The MMAF data [1] also shows that the quality of local salts produced, especially by salt farmers (community salt) is not uniform so that the sale of salt from pond farmers is also classified into several classes according to their quality. The first quality (KW1) is salt with a NaCl level of 95% -98%, the second quality (KW2) contains NaCl between 90% -95%, and the third quality (KW3) has a NaCl level of less than 90%. The production of salt in Indonesia generally through the solar evaporation method in small plot areas has resulted in the quality of the salt produced varying. The content of domestically produced salt NaCl only ranges from 81%-96%. There are many factors that are thought to contribute to the low productivity of salt in Indonesia. First, the production techniques and equipment used are still very traditional and the production of salt is very dependent on the weather and allows the production of salt in just 4 months [2]. This production period is much shorter when compared to Australia where the climate allows to produce salt for up to 8 months so as to produce far more salt with high quality. This prompted the government through the Indonesian Salt Users Industry Association (ISUIA) to set the ENT region as a potential location for national salt production because it has a summer (dry) that is longer than 7-8 months compared to other regions in Indonesia. East Nusa Tenggara (NTT) is one of the areas in Indonesia that has the potential for salt ponds estimated at around 60,000 hectares, but has not been managed optimally following national directions and policies to support salt self-sufficiency. Some areas that have potential for salt, including the areas of Kupang City, Kupang Region, the Middle North Timor Region, Ende, Manggarai, Nagekeo, Sumba, Alor and Sabu. Kupang Regency at Bipolo Village, has a potential production area of 318 hectares and utilization of up to 12 hectares in 2017 with production reaching 3.7 tons of salt. The potential of community salt land that exists if managed properly, ENT can fulfill its own needs. While some potential areas for industrial salt, can be focused on becoming a production pillar to meet national salt needs because it is supported by good weather conditions, vast and clean coastal and marine areas. The purpose of this study is to analyze the quality and content of traditionally produced cooking iodine salt and minerals.

2. Material and Method

The main materials in the study were cooked salt and krosok salt taken from the cooking salt production sites in Kupang City and Kupang Regency. Chemicals used include: aquades, potassium chromate solution (K_2CrO_4) 5%, silver nitrate solution ($AgNO_3$) 0.1 M, sodium thiosulfate with starch indicator, H_2SO_4 2 N, KI 10% solution, and $Na_2S_2O_3.5H_2O$ 0.005 N. The tools used include desiccator, oven, burette, Erlenmeyer, *Inductively Coupled Plasma Optical Emission Spectrophotometer* (ICP OES).

2.1. Research procedure

The sampling method uses random sampling, by determining the homogeneous population and then taken randomly into a representative sample. Quantitative methods are used for laboratory tests of water content, salt content (NaCl) and mineral (magnesium, potassium calcium) and iodine content. Samples are in the form of folk salt produced in Kupang City (Oesapa Village) and Kupang Regency (Oeteta Village and Pariti Village)

2.2. Water content analysis [3]

The analysis procedure is that the porcelain cup is washed and then dried in an oven at 105°C for one night. The cup is put into a desiccator which contains a silica gel to absorb the moisture that is still attached to the cup. The porcelain saucer was weighed using an analytical scale and weighed 1 g of sample and inserted into a porcelain dish. The porcelain cup is put into an oven with a temperature of 105°C for 5 to 6 hours or until the weight is constant. The porcelain cup was then put into the desiccator for 30 minutes then weighed the final weight. Salinity is calculated by the formula:

$$(\%)\ \text{Water Content} = \frac{B-C}{B-A} \times 100\% \quad (1)$$

Information:
A: The weight of the empty proselene cup the weight of the empty proselene cup (g)
B: Porcelain plate weight the proselene weight + initial sample weight initial sample weight (g)
C: Final porcelain cup weight Final weight of proselene cup (g)

2.3. Analysis of NaCl levels

NaCl analysis was carried out with wet fogging referring to Day and Underwood [4], namely salt weighed as much as 250 mg, washed with distilled water as much as 10 mL and transferred to Erlenmeyer 250 mL, potassium chromate solution (K_2CrO_4) 5% added 3 mL and titrated with silver nitrate solution ($AgNO_3$) 0.1 M. The endpoint of the titration is achieved when the first orange or orange color appears. NaCl levels can be calculated using the following formula:

$$Salt\ (\mathrm{NaCl})(\%) = \frac{(T\ x\ M\ x\ 58{,}4)}{\mathrm{W\ (mg)}} x100\% \tag{2}$$

Information:
T = Volume of the standard solution of AgNO3 0.1 M
M = Molarity of silver nitrite
W = sample weight

2.4. Mineral content analysis [5]

Analyze mineral magnesium, calcium and potassium carried out by using *Inductively Coupled Plasma Optical Emission Spectrophotometer (ICP OES).* The wavelength used for determining minerals is adjusted based on the wavelength of the type of mineral to be tested. The stages of testing the mineral content of magnesium, calcium and potassium are presented in Figure 1.

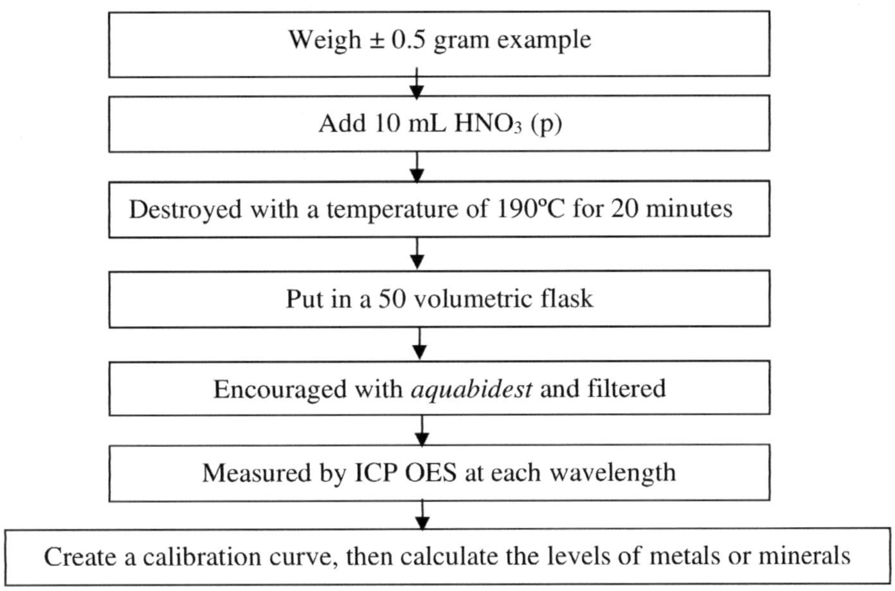

Figure 1. Flow chart of metal and mineral tests with ICP OES

Calculation:
Slope = Y (absorbance) / X (concentration, ppm) (3)
Metal and mineral concentrations, µg / mL of sample curves = absorbance / slope (4)
The metal and mineral content in the sample is calculated using the following formula:

$$\text{Metal / mineral content } (\mu g / g) = \frac{(\mu g \text{ metal or mineral / mL from the calibration curve) x V)}}{m} \tag{5}$$

Information:
V is the dissolution volume (mL)
M is the sample weight (gram)

2.5. Iodine level analysis (KIO_3)

Iodine level analysis was carried out with the principle of free iodine titration in the sample solution using sodium thiosulfate with starch indicator. The testing phase was carried out by: Samples were weighed as much as 25 g and put into Erlenmeyer 300 ml. Next, the sample was dissolved with 50 ml of distilled water. Samples were added with 2 ml H_2SO_4 2 N, 5 ml KI 10% solution, then placed in a dark place for 10 minutes to achieve optimal results. The samples were titrated using $Na_2S_2O_3.5H_2O$ 0.005 N until the yellow color disappeared, then added 2 ml of the starch indicator and continued the titration until there was a color change from dark blue to colorless (clear). Calculation of KIO_3 content of original material, namely:

$$\text{KIO3 content of original material (mg/kg)} = \frac{\text{V x 35,67 x N x 1.000}}{\text{W}}$$

Information:
V = the volume of $Na_2S_2O_3$ required for imaging, expressed in milliliters (ml);
N = the normality of $Na_2S_2O_3$;
W = the sample weight expressed in grams (g) 35.67 is the equivalent weight of KIO_3

Calculation of KIO_3 content of dry ingredients:

$$\text{KIO3 level on the basis of dry ingredients (mg/kg)} = \frac{100 \text{ x X}}{100 - \text{water content}}$$

Information:
X is KIO_3 origin material

3. Results and Discussion

3.1. Public salt in Kupang Regency

3.1.1. Making salt in Oeteta Village. Traditional folk salt produced in Oeteta Village, Sulamu District, Kupang Regency generally is pond salt produced traditionally by utilizing sunlight. The salt pond which is the research center belongs to Mr. Melkianus Y. Lae'iu with a total size of 10,000 m^2 (100 m long, 100 m wide). Visualization of the conditions of the salt ponds is presented in Figure 2. The production of pond salts takes around 30 days (4 weeks) after they can be harvested. Several stages of pond salt making are carried out, that is:

1. Extracting seawater from the mangrove forest to the first reservoir with the help of a water pump. In the first reservoir, seawater is stored for 1 week.
2. Seawater is forwarded to the second reservoir with the help of a water pump and stored for one week.
3. Seawater in the second tub, then proceed to the third reservoir and stored for one week. The time needed for hatching seawater before reaching the drying tub is for 3 weeks.
4. Seawater from the hatchery is forwarded to the drying table with the help of a water pump.
5. Seawater that has been moved to the first drying tub, then stored for 1 week and forwarded to the second drying tub and stored for one week.
6. Seawater from the second drying tub is forwarded to the third drying tub with the help of a water pump and stored for 1 minute, then continued to the fourth drying tub and stored for one week. In the fourth drying tub, the seawater has turned into old water.
7. From the fourth drying tub, old water is flowed to the fifth tub and stored for one week. In the fifth tank, seawater is ready to be channeled into the pond to be processed into people's salt.

The salt produced in Oeteta Village consists of 2 types, namely fine salt and coarse salt (krosok). Refined salt is sold at Rp 2,000 / kg, while coarse salt (krosok) is sold at Rp 500/kg. The salt production from one tank is 26 sacks (50 kg capacity/sack) and there are 5 tanks used to produce people's salt in Oeteta Village. The size of the drying tub has an area of 150 m^2 (length of 30 m^2 and width of 5 m^2). Harvesting carried out ones a week, so that within one month there are 4 (four) harvest times.

Figure 2. Condition of salt ponds in Oeteta Village, Sulamu District, Kupang Regency. (a) hatchery; (b) Drying tanks; (c) Drying and salt baths; (d) Pond salts

3.1.2. Making salt in Pariti Village (A2). Salt produced in Pariti Village is cooking salt sourced from filtered seawater from ash. The owner of the people's salt business is Ms. Lefina B. Tamoes. He worked as a salt farmer for 8 years, starting from 2000 until now. Visualization of the conditions of salt ponds is shown in Figure 3. The stages of making cooking salt from ash are as follows:

1. Taking ash. Ash is collected and in filtering containers and added with seawater as many as 12 buckets of other brands (capacity of 20 kg / bucket).
2. Filtering. Filtering seawater through ash is carried out for ± 2 hours. Filtered seawater is collected in a container measuring 185 cm^2 in length and 50 cm wide and then cooked for ± 12 hours.
3. The cooking process uses stoves and wood fuel. Cooking temperatures range from 85 to 103°C.
4. Salt that has been cooked for 12 hours, then removed using a spoon made of plate.
5. Salt is drained for 1 hour, after draining the salt, it is ready to be sold for Rp 10,000 / kg.

Figure 3. Condition of cooking salt in Pariti Village, Sulamu District, Kupang Regency. (a) Ash used; (b) Containers for taking ash; (c) Dissolving ash using sea water; (d) Salt water that is ready to be cooked; (e) Salt cooking stove; (f) Cooked salt produced

3.1.3. Quality of public salt. The results of mineral, iodine, water and heavy metals content in traditional salts produced by the people of Kupang Regency, namely traditional pond salt in Oeteta Village and traditional cooking salt from abu in Pariti Village are presented in Table 1. The results of mineral analysis on pond salts produced by the community in Oeteta higher content of Mg and K compared to cooked salt from ash produced in the Pariti. The presence of large amounts of magnesium minerals for salt producers can cause a decrease in the quality of the salt produced. The presence of large amounts of magnesium in the salt produced can reduce the NaCl content. This is evident from the results of the research conducted, the presence of large amounts of magnesium minerals, causing the value of the resulting NaCl content to be low.

Table 1. Mineral content, iodine, moisture content, and heavy metals in people's salt in Kupang regency

Parameters Test	Level	
	Salt Traditional Cooking	Cook Salt from ash
Mineral		
Magnesium (Mg)	1015,33 mg/100 g	492,93 mg/100 g
Potassium (K)	206,37 mg/100 g	124,74 mg/100 g
Calcium (Ca)	145,82 mg/100 g	487,86 mg/100 g
Sodium Chloride (NaCl)	71,79%	80,21%
Iodine (KIO_3)	0,77 mg/kg	1,28 mg/kg
Water (H_2O)	9,73%	5,87%
Heavy metal		
Lead (Pb)	Undetectable (mg/kg)	Undetectable (mg/kg)
Cadmium (Cd)	Undetectable (mg/kg)	Undetectable (mg/kg)

For buyers, the presence of high magnesium minerals as impurities can reduce the quality of salt. Salt needs for processing purposes, especially those related to the processing of salted fish can be a problem with the low quality of salted fish stored, if the presence of magnesium minerals is high. This is in contrast to the desired conditions in the pharmaceutical industry, which is in dire need of high amounts of magnesium minerals [6]. The calcium content in cooking salt is higher than that of pond salt. The high calcium content is very helpful in meeting the human body's need for calcium. According to Amandia and Sulistiani [7] that calcium requirements for children and adolescents will increase according to age, with several functions including helping in the formation and maintenance of bones and teeth, preventing osteoporosis, glycogen storage, accelerating the function of muscles, brain and nervous system. Murray et al. [8] explained that minerals Ca, Na, K and Mg are the main minerals that are needed in the human body because the intake of minerals recommended for consumption every day for each individual human body must reach> 100 mg per day. The content of NaCl produced by traditional pond salt samples and samples of cooked salt from ash is still below the requirements of SNI 3556-2016 for the category of iodized salt consumption, which is a minimum NaCl content of 94%. Purbani [9] classifies salts based on the content of NaCl as the main element in 3 categories, namely "very good" containing NaCl> 95%; "Good" contains NaCl 90-95%; "Medium" contains NaCl between 80-90%. Therefore, cooked salt from ash (A2) is included in the "moderate" category because it has a content of 80.21% NaCl but does not meet the salt requirements of raw materials for iodized consumption salt [10]. Sulistyaningsih et al. [11] explained that the constituents of the largest chemical compounds in salt were NaCl, and impurities were magnesium chloride, magnesium sulfate and calcium sulfate. Low NaCl content under SNI requirements is influenced by the mineral content of magnesium present in the salt produced. Furthermore, Zainuri et al. [6] explained that magnesium content in salt granules is one of the elements that can reduce NaCl levels from salt. The National Standard Agency [12] in the SNI 3556: 2016 document requires a minimum standard content of iodine-forming compounds (KIO_3) of 30 mg/kg. This condition is far different from the iodine content produced from pond salt (A1) in Oeteta Village and cooking salt (A2) in Pariti Village. Government regulation through Presidential Decree No. 69 of 1994, stipulates that all salts circulating in the community must contain iodine enriched with potassium iodate [13]. The iodine content produced is different from people's salt produced from several regions in East Nusa Tenggara. In some districts in East Nusa Tenggara that produce people's salt, iodine content is not found. This is interesting, because geographical conditions in a region can also determine the presence of certain minerals. The people's saltwater content that meets the requirements of SNI 3556: 2016 and SNI 4435: 2017 (maximum 7%) [10,12] is found in cooked salt from ash with a value of 5.87% while in salt ponds it does not meet the standard because it exceeds the standard set. The salt water level of the pond that is high than the standard is thought to be influenced by the time of draining and drying carried out and the storage conditions. Moist conditions also affect the pond's salt water content because of the hygroscopic nature of the salt, which allows an increase in the salt water content of the pond. The analysis of Pb and Cd Heavy metals in folk salt produced in Oeteta Village (A1) and traditional pond salts in Pariti Village was not detected. The requirements stipulated in SNI 3556-2016 and SNI 4435: 2017 stipulate the content of heavy metal lead (Pb) in iodized salt-forming compounds to a maximum of 10 mg / kg and cadmium (Cd) to a maximum of 0.5 mg/kg.

3.2. Profile of public salt in the City of Kupang

3.2.1. Making krosok cook salt in Oesapa Village (B1 and B2). Krosok salt obtained as a raw material is pond salt obtained from Oli'o Village. Cooked salt from Krosok salt produced in Oesapa Village, precisely at RT 002 / RW 001 is a business run by Mr. Marthen Fafoh since 2001 until now. The visualization of making cooking salt in Oesapa Village (B2) is presented in Figure 4. The salt making process is as follows:

1. Preparation of raw materials in the form of krosok salt taken from salt ponds in Oli'o Village, Kupang Regency.
2. Weighing 16 kg of krosok salt, then put into a filtering container and dissolved with as much as 100 liters of fresh bargain. The filtering process (filter) lasts for 2 hours.
3. The filtered water is put into a container (drum) cooking measuring 87 cm long, 58 cm wide, and 28 cm high. The cooking process lasts for 7-8 hours until salt crystals are formed.
4. Salt is removed from the cooking container using a spoon and placed in a drying container to melt for 1-2 hours. The amount of salt produced in one cooking is 25 kg.

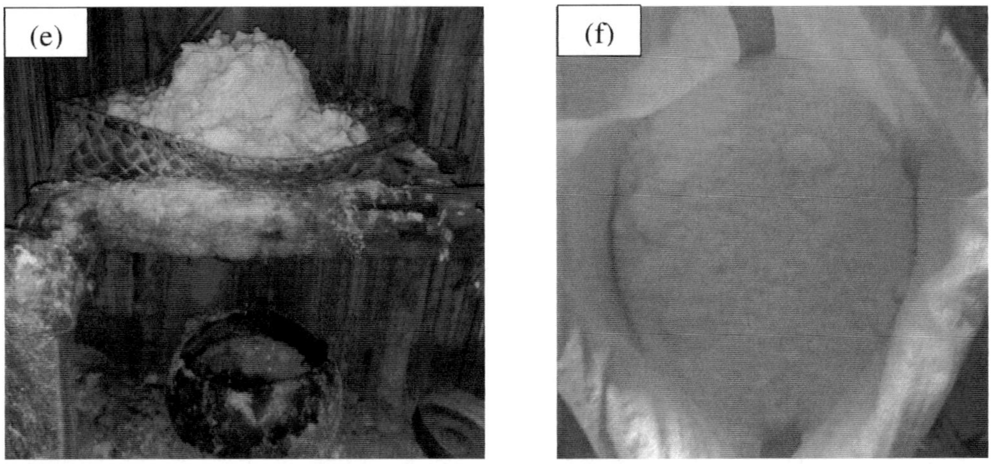

Figure 4. Condition of cooking salt from salt krosok in Oesapa Village (B2), Kupang City. (a) Krosok salt to be used; (b) Imitation of using fresh water (c) Salt water that is ready to be cooked; (d) Salt that has crystallized; (e) Thinning of cooking salt; (f) Cooked salt is ready for sale.

3.2.2. Making salt from ash in Oesapa Village (B3). Traditional cooking salt from ash is a work done by Mr. Philip Fallo from 2000 to the present, but it is located in RT 001 / RW 001 in Oesapa Barat Village, Kupang City. The source of the raw material used for ash comes from the soil obtained from the area of salt ponds and seawater used also sourced from the waters around the West Oesapa Village. Visualization of the condition of cooked salt from ash (B3) in Oesapa Barat Village is presented in Figure 5. The stages of the process are as follows:

1. Prepare as much as 8 baskets of ash (equivalent to 112 kg) into the filtering container which has been given sand. This functions as a filter.
2. Dissolution using seawater as many as 20 buckets (equivalent to 300 liters), then let the water drip into the storage container. The filtering time is 2 hours.
3. The filtered water is put into a boiling container which is 87 cm long, 58 cm wide, and 28 cm high.
4. Cooking is done in a furnace measuring 38 to 40 cm high. Cooking lasts for 10 hours, until salt crystals are formed.
5. Salting and drying of salt is carried out for 2 hours. From one boiling can get 25 kilos of salt to cook salt.

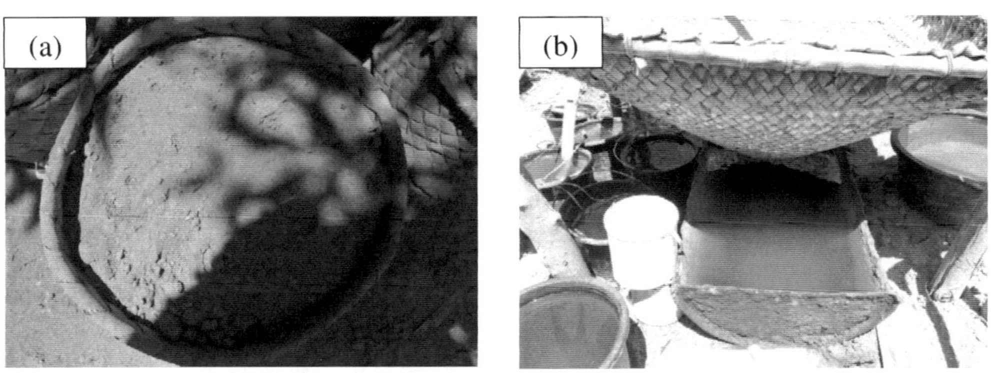

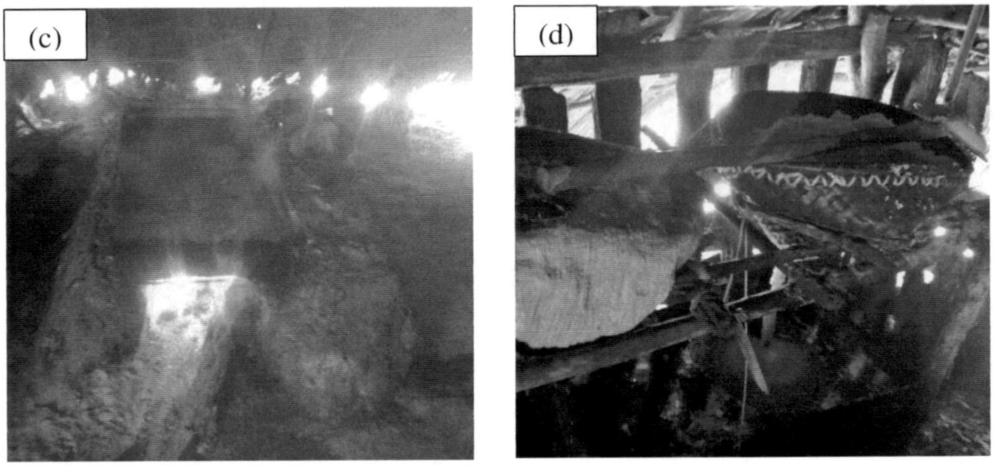

Figure 5. Condition of cooking salt from ash in Oesapa Barat Village (B3), Kupang City. (a) Ash used; (b Dissolving ash using sea water; (c) Salt water ready to be cooked on the stove; (d) cooking salt produced

3.2.3. Quality of public salt. The results of the analysis of several test parameters for traditional salts produced by the people of Kupang City, namely salt krosok from traditional salt ponds in Oli'o Village (B1); cooking salt from krosok salt (B2) in West Oesapa Village; and traditional cooking salt from ash (B3) in Oesapa Village is presented in Table 2.

Table 2. Mineral content, iodine, water and heavy metals in traditional salt in Kupang City

Test Parameters	Level		
	Krosok Salt	salt krosok from cooked salt	Salt cooked From ash
Mineral			
Magnesium (Mg)	387,16 mg/100 g	559,05 mg/100 g	222,33 mg/100 g
Potassium (K)	97,91 mg/100 g	168,55 mg/100 g	60,71 mg/100 g
Calcium (Ca)	152,96 mg/100 g	458,44 mg/100 g	193,74 mg/100 g
Sodium Chloride (NaCl)	80,61%	78,16%	74,45%
Iodine (KIO_3)	4,87 mg/kg	1,28 mg/kg	9,15 mg/kg
Water (H_2O)	5,33%	11,33%	7,00%
Heavy metal			
Lead (Pb)	Undetectable (mg/kg)	Undetectable (mg/kg)	Undetectable (mg/kg)
Cadmium (Cd)	Undetectable (mg/kg)	Undetectable (mg/kg)	0,25 mg/kg

Note: B1 is the sample code for krosok salt; B2 is the sample code for cooking salt from krosok; B3 is cooking salt from ash. Description: B1 is the code for the krosok salt sample; B2 is a code for samples of cooking salt from krosok; B3 is cooking salt from ash.

The results of mineral analysis on the salt of krosok from salt ponds in Oli'o Village, Kupang Regency (B1) have lower mineral content of Mg, K and Ca compared to cooking salt produced in Oesapa Village (B2) with values tending to increase. The presence of large amounts of magnesium minerals in B2 has implications for decreasing the NaCl content and consequently decreases the quality of the salt produced. Unlike the case with the mineral content of Mg and K produced by ash cooking salt (B3), it tends to be lower compared to B1 and B2, while for Ca minerals (193.74 mg / 100 g) it is higher than B1 (152.96 mg / 100 g). According to Amandia and Sulistiani [7] that calcium requirements for children and adolescents will increase according to age, with several functions including helping in the formation and maintenance of bones and teeth, preventing osteoporosis, glycogen storage, accelerating muscle, brain and nervous system functions. Murray et al. [8] stated that minerals Ca, Na, K and Mg are the main minerals that are very necessary in the human body because the intake of the recommended amount of minerals to be consumed every day for each individual human body must reach> 100 mg per day. The content of NaCl produced by sample B1 is higher than that of samples B2 and B3, but it is still below the requirements of SNI 3556-2016 for the category of iodized salt consumption, which is a minimum NaCl content of 94%. The content of NaCl in B1 salt (80.61%) is categorized as "moderate" according to Purbani [9] because it contains NaCl between 80-90%. Sulistyaningsih et al. [11] explained that the constituents of the largest chemical compounds in salt were NaCl, and impurities were magnesium chloride, magnesium sulfate and calcium sulfate. Low NaCl content under SNI requirements is influenced by the mineral content of magnesium present in the salt produced. Furthermore, Zainuri et al. [6] explained that magnesium content in salt granules is one of the elements that can reduce NaCl levels from salt. Iodine content (KIO_3) required by SNI 3556: 2016 [12] that the content of iodine salt-forming compounds, which is 30 mg / kg minimum. The iodine content in B1 salt is higher than B2 salt which is the result of cooking salt. This shows that iodine decreases due to the dissolution process with fresh water and cooking that occurs. It is different from the iodine content in B3 salt (9.15 mg / kg) higher than that of salt B1 and B2. The iodine content produced is quite high compared to people's salt produced from several other regions in East Nusa Tenggara. In parts of Kupang City that produce people's salt, iodine content is not found. This is interesting because geographical conditions in a region can also determine the presence of certain minerals. Government regulation through Presidential Decree No. 69 of 1994, stipulates that all salts circulating in the community must contain iodine enriched with potassium iodate [13]. Public salt water content (H_2O) that meets the requirements of SNI 3556: 2016 and SNI 4435: 2017 (maximum 7%) [10,12] is found in B1 salt (5.33%) and in B3 salt (7.00%), whereas in B2 salt it is not meet the standards. Salt water levels are thought to be influenced by the time of draining and drying carried out and storage in humid conditions due to the hygroscopic nature of the salt which allows for an increase in the amount of water content. The results of analysis of Pb and Cd metal contamination in salts B1, B2 and B3 were not detected, but the Cd metal in B3 salt was detected 0.25 mg / kg certainly did not exceed the required quality standards SNI 3556-2016 and SNI 4435: 2017, namely metal contamination cadmium (Cd) a maximum of 0.5 mg / kg.

4. Conclusion

The quality of cooking salt produced based on the Indonesian National Standard for kitchen salt, cooking salt produced based on NaCl addiction criteria is below the standard 94.6% and water content above 5%. Mineral content including iodine is within the range of the standard set.

References

[1] Ministry of Maritime Affairs and Fisheries 2015 *One Year Ministry of Maritime Affairs and Fisheries Performance Report (KKP)* Available from: roren.kkp.go.id/arsip/file/123/buku-laporan-kinerja-kkp-27112015.pdf. (referenced 2016 June 10)

[2] Ministry of Maritime Affairs and Fisheries 2014 *Marine and Fisheries in 2014 Figures* (Center for Statistics and Information Data, Marine and Fisheries Ministry)

[3] Association of Official Analytical Chemist (AOAC) 2005 *Official Methods of Analysis (18 ed) of the Association of Official Analytical Chemist Inc.* (Mayldan, USA)

[4] Day R A and Underwood A L 1994 *Analisa Kimia Kuantitatif* 4th edition (Jakarta: Erlangga)

[5] PT Saraswanti Indo Genetech 2013 *Metal and Mineral Test Work Instructions with Inductively Coupled Plasma Optical Emission Spectrophotometer (ICP OES)* Available from: http://www.saraswanti.com

[6] Zainuri M, Anam K and Susanti A P 2016 Relation of the Content of Sodium Chlorida (NaCl) and Magnesium (Mg) from People's Salt on Madura Island *Proceedings of the National Marine Seminar* pp 167-172 (Trunojoyo Madura University, July 27, 2016)

[7] Amandia D P S and Sulistiyani 2010 Effect of calcium on the growth of dental teeth in children *Stomatognatic (J. K. G. Unej)* **7**(3) 40-44
[8] Murray R K, Granner D K, Mayes P A and Rodwell V W 2003 *Biokimia Harper* 25th edition ed. Bani A P and Sikumbang T M N (Jakarta : EGC)
[9] Purbani D 2006 *Process of Forming Salt Crystals* (Marine and Non-Biological Research Centers, Marine and Fisheries Research Agency Ministry of Maritime Affairs and Fisheries)
[10] National Standardization Agency 2017 *Salt of Raw Materials for Iodized Consumption Salt* NSI 4435-2017 (Jakarta)
[11] Sulistiyaningsih, Sugiyono and Sedyawati 2010 Pemurnian garam dapur melalui metode kristalisasi air tua dengan bahan pengikat pengotor $Na_2C_2O_4$-$NaHCO_3$ dan $Na_2C_2O_4$-Na_2CO_3 *Jurnal Kimia* **1**(8) 26-33
[12] National Standardization Agency 2016 *Salt of Iodized Consumption* SNI 3556-2016 (Jakarta)
[13] Sugiani H, Previanti P, Sukrido and Pratomo U 2015 Determination of the effect of heating and storage time of iodized salt on potassium iodate *Chimica et Natura Acta.* **3**(2) 66-9

IOP Conf. Series: Materials Science and Engineering **823** (2020) 012009 doi:10.1088/1757-899X/823/1/012009

Identification of biomordant in Hundihopo Village, East Rote District, Rote Ndao Regency

N A Bako[1], A C Sabuna[1], and Y Daud[1]

[1] Department of Biology Education, Universitas Kristen ArthaWacana, Kupang 85227 Indonesia

E-mail: alan.sabuna@gmail.com

Abstract. Information of plant species, mordants, and the mordanting process of the yarn of ikat weaving in Hundihopo village are still minimal and not well documented. Therefore, the aim of this study was to identify biomordant in Hundihopo Village. Samples were collected by exploration methods. Parts of plants, mordanting process, and kinds of color were recorded based on observation, interview with craftsmen and documentation. A total of 12 species were traditionally used as biomordants in Hundihopo village namely *Sterculia foetida* L., *Erythrina varegata*, *Areca catechu* L., *Ceriops tagal*, *Datura metel*, *Calotropis gigantea*, *Abrus precatorius* L., *Symplocos sp.*, *Jatropha curcas* L., *Citrus aurantifolia*, beura and faliti. The plant organs used as a source of mordant are leaves, barks, fruits, and seeds. The method of processing was to be burned, crushed, boiled and soaked and the resulted color was red, black and yellow.

1. Introduction

Indonesia is one of the countries that are abundant in natural resources. Colored plants become one of the natural resources which have the potency to be used as the woven dye in Indonesia, especially in the development of a product that has nuances of the naturalist, imitative, cultural, and exclusive. Colored plants could also be the raw material for the weaving industry that has high economic value [1].

In the ancient time, the process of coloring ikat weaving in East Nusa Tenggara used dyes derived from natural ingredients. However, along with the discovery of synthetic dyes, the craftsmen began to shift their choice using synthetic dyes to dye the yarn, because synthetic dyes are considered easier to obtain, affordable prices, have a variety of colors, and the coloring process also does not require a long period time [2].

However, synthetic dyes often cause environmental problems because they contain heavy metals that are harmful to human health. Moreover, if pollutants are discharged into the water can pollute the environment [1]. Considering the impacts, one alternative to overcome them is to return to natural dyes, because the waste from the dyeing weaving is considered safer, environmentally friendly, less pollute, has a more natural color, shows the impression of cold, soft, and comfortable [3]. However, behind these advantages, natural dyes also have a weakness that is the color produced easily faded. Therefore, it is necessary to add mordant to increase the stickiness of various dyes on the fabric.

IOP Conf. Series: Materials Science and Engineering **823** (2020) 012009 doi:10.1088/1757-899X/823/1/012009

According to Sabuna et al. [4], the weavers at Hundihopo Village in East Rote Sub-district of Rote Ndao have used the mordant plants but have not been known clearly regarding the types of those plants. Therefore, we research the Identification of Biomordant in Hundihopo Village, Rote Timur District, RoteNdao Regency.

2. Materials and Methods

2.1 Study area

Our study was conducted in Hundihopo Village. Hundihopo is located in the western part of Rote Island. Administratively, Rote Ndao district is the southernmost district in the Republic of Indonesia. This regency has an area of 1280.10 km^2 which consists of 96 islands and is geographically located between 10 degrees 25 - 11 degrees South Latitude, and 121 degrees 49 - 123 degrees 26 East Longitude with clear boundaries, to the north bordering the Savu Sea, to the east is bordered by the Pukuafu Strait, to the west is the Savu Sea, and to the south is the Indian Ocean. The Rote Ndao regency is climatologically the same as the climate in other regions in NTT, namely a dry climate influenced by monsoons. The land surface is generally hilly and mountainous (32,625 Ha) and partly consists of lowlands (45,250 Ha) with an average slope of up to 45%. The contours of the island of Rote vary, at the coast, the altitude is 0-10 m above sea level while in the middle it reaches an altitude of 200 - 1500 m with a slope of 40 - 60% [5].

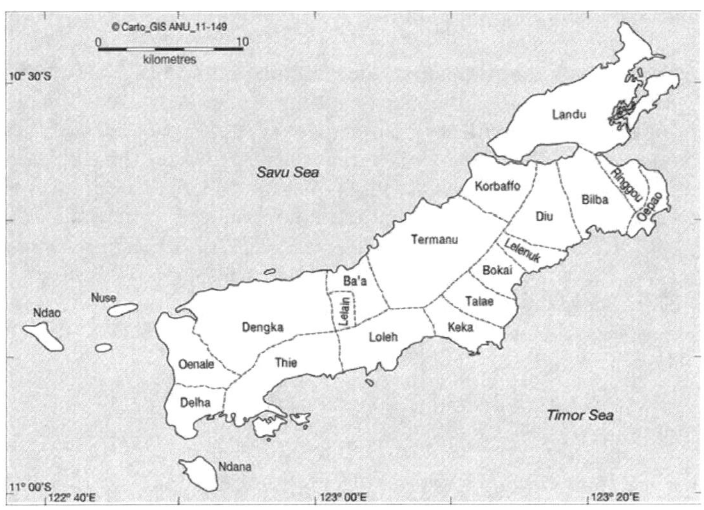

Figure 1. Map of Rote Ndao

2.2 Materials

The instruments used in this study arestationeries,camera, knife, plastic bag, mortar, hot plate, analytic scale, measuring cup, glass beaker, and spatula. The materials used are cotton thread, water, and the organ of coloring plants and mordanting plants.

2.3 Selection of participants

The participant selection technique used in this initial observation is the purposive sampling method [6]. The chosen figures to be interviewed are two local authorities and one traditional elder. From our preliminary observation, the data of prospective informants for the next stage are suitable to be interviewed. The selection of participants at this interview stage was carried out by the snowball sampling method. There are four key participants were selected purposively and systematically based on the recommendations of knowledgeable elders. The key participants are traditional experts who guard the indigenous knowledge of natural dye. Nowadays, most of the waivers at Hundihopo village use synthetic dye to color the yarn, particularly for the new generation.

2.4 Data collection

Our study used four data collection techniques, namely semi-structured interviews, field observations, documentation, and simple experiments. To interview the key informants, we ready a list of questions in Bahasa, such as does the coloring yarn use additional material to strengthen the color? What mordanting plants are used? Which mordanting plant organs are used and other questions? Every answer to the question is recorded. We then carried out field observations with the help of informants to collect data for morphology features and habitats of mordant plants species. Samples of mordanting plant organs were collected in the field, stored in a plastic bag, and delivered to the Biology laboratory, ArthaWacana Christian University, for characterizations. Characterizationswere performed by weighing natural dyes and mordanting plant organs and processed in such a way as to produce colors in the ikat yarn. The experiment was designed to visibly compare the effect of mordant on the color of the yarn, that is, yarn soaked using mordant and without mordant. The resulting color was documented using a camera.

2.4 Data analysis

All specimens of mordanting plants were classified and identified using Flora's book [7]. Once identified, the collected data were analyzed descriptively and presented in the form of local names, scientific names, families, the used organ, processing methods, and the resulted color.

3. Results and Discussion

3.1. The types of mordant plants found at Hundihopo village

The mordant plants used by weavers at Hundihopo Village were 12 species and 9 families, namely Malvaceae, Rhizoporaceae, Solanaceae, Asclepiadaceae, Fabaceae, Euphorbiaceae, Rutaceae, Arecaceae, and Symplocaceae (Table 1).

Table 1. The types of mordant plants found atHundihopo Village

No.	General Name	Local Name	Scientific Name	Family
1	Kelumpang	Nitas	*Sterculia foetida* L	Malvaceae
2	Dadap	Deras	*Erythrina variegata*	Fabaceae
3	Areca Nut	Pua	*Areca catechu* L	Aracaceae
4	Soga tingi	Tene	*Ceriops tagal*	Rhizoporaceae
5	Kecubung	Loloa	*Datura metel*	Solanaceae
6	Biduri	Koleng susu	*Calotropis gigantea*	Asclepiadaceae
7	Saga	Memea	*Abrus precatorius* L	Fabaceae
8	Loba	Loba	*Symplocos sp*	Symplocaceae
9	Castor Oil Plant	Jarak pagar	*Jatropha curcas* L	Euphorbiaceae
10	Lime	Jeruk nipis	*Citrus aurantifolia*	Rutacea
11	-	Faliti	Unidentified	Unidentified
12	-	Beura	Unidentified	Unidentified

Source: The 2019 result of study

To compare with the previous study, which conducted in Kaliuda village, East Sumba, 3 types of plants used as mordant such as Loba (*Symplocossp*), Candlenut (*Aleuritesmoluccana* L.) and Dadap (*Erythrinavariegata* L.) [8]. While Oktiarni [9] found that star fruit used as mordant in natural dyes with Guava leaf (*Psidium guava*), Dewandaru leaf (*Eugenia uniflora*), and Rosella leaf (*Hibiscus sabdariffa* L.) as natural plant dyes. Furthermore, Martalinda [10] used coconut water as a mordant on the extract of Curcuma tuber (*Curcuma xanthorriza roxb*) to produce the variations of yellow color. Suppose there are lots of mordants have been used a long time ago and every place has their resources for coloring the ikat yarn.

3.2. Plant parts used as mordant

Weavers of the study area harvest different plant parts for the preparation of coloring woven yarn such as barks, leaves, fruits, and seeds, which listed in the following table.

Table 2. Plant parts used as mordant

No	Plant name	Scientific Name	Organ Used
1	Kalumpang	*Sterculia foetida* L	Fruit shell and Seed
2	Dadap	*Erythrina variegata*	Bark
3	Loba	*Symplocos sp*	Bark
4	Soga tingi	*Ceriops tagal*	Bark
5	Kecubung	*Datura metel*	Leaf
6	Biduri	*Calotropis gigantea*	Leaf
7	Saga	*Abrus precatorius* L	Leaf
8	Areca Nut	*Areca catechu* L	Seed
9	Castor oil plant	*Jatropha curcas* L	Bark
10	Lime	*Citrus aurantifolia*	Fruit
11	Beura	Unidentified	Leaf
12	Faliti	Unidentified	Leaf

Source: The 2019 result of study

Based on Table 2, it could be seen that the organ of Kelumpang used as the mordant is fruit shells and dried seeds, which are commonly found in the house yard of weavers, while the fruit shells of Kelumpang plant have thick skin and its edge is tapered shaped. The level of maturity depends on the individual and the factor of the growing place.

Inside the shell of Kelumpang fruit, there is a shiny black seed containing much oil which it can be used to dulcify the thread. There is also another benefit of kepuh seed, which as industrial material in making shampoo, soap, paint, and plastic [11].

The organ of Dadap (*Erythrinavariegata*) used is the bark of stem. On the stem bark of Dadap contains saponins, flavonoids, and polyphenols, which roles in fabric coloration [12].

Loba is a shrub plant to the tree. The inside of the stalk (stem) contains some amount of aluminum (reaching 50% of the ash), which is one type of mordant substance that functions as a binding agent in the coloration of thread [12]. Similar to the sogatingi plant, the organs used are dried bark of stem. The bark stem of sogatingi also contains tannins that are good to be used in the process of threads coloration and also as tanners and color substance for the paint [13].

Kecubung includes shrub species. Its root, stem, leaf, and seed contain an alkaloid, steroid, flavonoid, phenol, and tannin [14]. Biduri is a large shrub plant with a height of 4 meters to 10 meters. This plant has an oval-shaped pale green leaf containing tannin, saponin, and calcium oxalate, which can treat itching, trachoma, coughing, constipation, and scabies [15].

Saga is a shrub plant, has compound leaves, oval-shaped, and also small in size. Saga leaf contains the chemical substance of flavonoid, tannin, triterpenoid, and steroid [16], while areca nut contains 15% of tannins [17]. Unripe areca nut mashed and mixed with alkali can produce a wine-carmine color. Therefore in India, they have used areca nuts to dye cloth. Limealso contains citric acid, hesperidin, and aurantiamarin, which could be used as mordant [18].

Based on the review above, the reason for the weavers at Hundihopo Village used these organs as mordant because the ancestors had already used them, and they believed that on those organs containing substances of mordant.

3.3. Processing method of mordant plant
The weavers at Hundihopo Village generally process the mordant plants in various traditional ways, which are burned, pounded, boiled, and then soaked together with coloring plants. The longest treatment process happens in the soaking process because it takes three days. Before proceeding to theyarn coloration, weavers do the initial soaking of the thread with the dried kalumpangfruit and then burnt it,the ash of the combustion of the kalumpangfruit is mixed with sufficient water and soak it for three days after that separate the solution from the result of the soaking of kalumpangash.Next step, add the kalumpang seeds, saga leaves, amethyst leaves, biduri leaves, and finely crushed dadap skin into the kalumpang fruit bath. Prepare the yarn, put it in the marinade, and kneaded until completely mixed, then soak for two days. After that, yarn is removed and dried. Based on the interview, the purpose of this initial immersion is to strengthen the basic color, so it does not fade easily. We found that the purchased yarn was not used immediately,instead of through initial treatment because the distinctive color of the East Rote woven fabric was grayish-white.

In coloration process, there is known 3 types of colors that will be formed which are black, red, and yellow.

3.3.1. Red color. The needed plant is the root skin of mengkudu, Loba bark, and water. Each organ pounded until smooth, and then adds water sufficiently while stirred until the bubbles appeared on the surface. Insert the thread into the extract, and cook it but avoid boiling. Afterward, the thread soaked in the container for 2 days. Lastly, the thread is removed and dries it until completely dried.

3.3.2. Black color. The initial step is to collect tarum leaves, dried areca nut, Jatropha bark, and sogatingi bark. Pound all ingredients until smooth, except for sogatingi bark. Mix all ingredients with sufficient water while stirring, then put the yarn, cook but avoid boiling, stir, then soak for two days until the black color accumulates. Afterward, removed and dried. Boil the sogatingi bark with sufficient water, and add the yarn. Soak it for five minutes. Lastly, remove and dry it.

3.3.3. Yellow color. The materials used to produce yellow color are turmeric, lime, and water. Turmeric rhizome is pounded until smooth, add lime juice sufficiently, then put in the screw, cook until boiling, and then soak for 2 days. Afterward, the yarn is removed and dried until it is completely dry.

4. The Resulted Color
Based on our research in the village of Hundihopo, several things were found, namely the mordant plant used by the weavers, and the steps in the coloring process,however the craftsmen made the colors do not use exact measurements. The results of this study intend to find a definite measure of each material used by craftsmen so it will provide the understanding and possibly reciprocity of the entire process made. The results of our simple research are as follows:

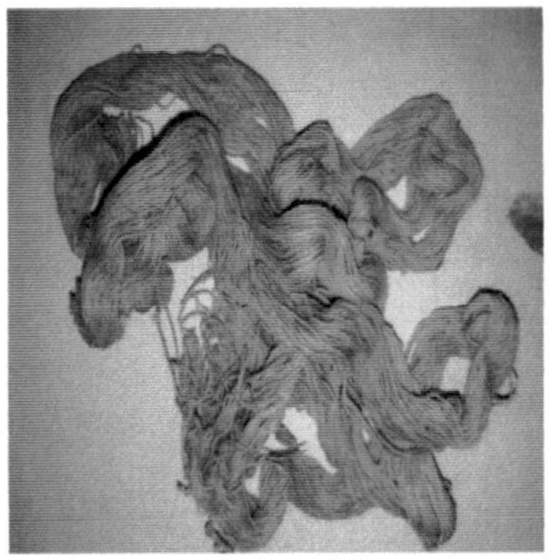

Figure 2. The initial soaking of the thread

Based on Figure 2, before the yarn is colored, the yarn is soaked first in the mordant solution. The method of making mordant begins by burning the kelumpang. The formed ashes are taken 3.600 grams, then add a half-liter of water and soak for three days. Furthermore, the amethyst leaves, biduri leaves, saga leaves, dadap bark and kelumpang seeds each pounded, then each ingredient weighed as much as 15 grams of Saga leaves, 45 grams of amethyst leaves, 45 grams of biduri leaves, 38 grams of kelumpang seeds, and 30 grams of dadap bark.After all the ingredients have been weighed, the next step is to separate the solution from the soaking ash. All of the weighed ingredients were placed together with the yarn in the ashes solution, while the thread is kneaded, then soaked for 2 days. Afterward, the thread is removed and dried. This process is called pre-mordanting. Before the yarn is dyeing, the yarn is soaked with mordanting plants to strengthen the basic color, so the thread does not easily fade.

Table 3. The resulted colors through themordanting process

No	Type of Plants		Processing Method		The Resulted Color
	Mordant Plant	Color Plant	Mordant Plant	Color Plant	
1	Loba Bark	Mengkudu Root	Pounded	Pounded	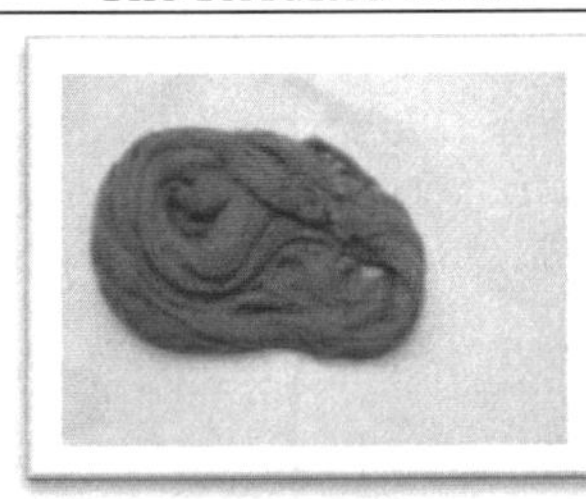 Wine Color
2	[1]Dry Areca Nut [2]SogaTingi Bark [3]Skin of Castor Oil Plant	Tarum Leaf	Pounded Pounded Pounded	Pounded	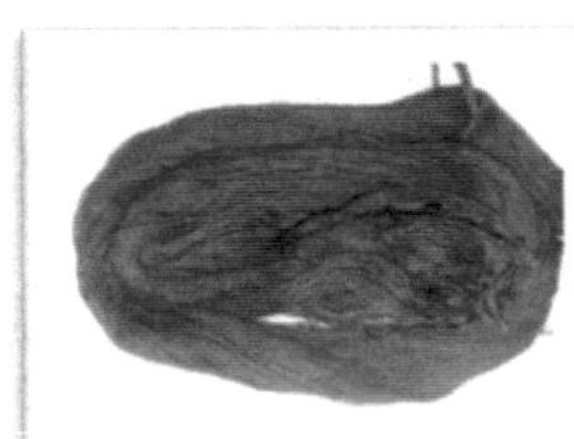 Black

No	Type of Plants: Mordant Plant	Type of Plants: Color Plant	Processing Method: Mordant Plant	Processing Method: Color Plant	The Resulted Color
3	Lime Juice	Turmeric	Squeezed and juice taken	Pounded	 Yellow

Source: The 2019 result of study

Based on Table 3 for the red color, the needed plants are the root bark of mengkudu (*Morinda citrifolia L*), Loba skin (*Symplocos sp.*), and water. Previously each material is cleaned and then pounded until smooth, then weighed 6 grams of Loba bark, 46 grams of mengkudu root skin, and 350 ml of water. All materials are cooked but avoid boiling, then put yarn inside, while stirring until evenly distributed, then the thread is soaked for two days. After that, the thread is removed and dries it until completely dry.

For black, the plants needed are 70 grams of tarum, 5 grams of dried areca nut, 20 grams of Jatrophabark, sogatingi skin, and 350 ml of water. All ingredients are ground until smooth, except for the sogatingi skin. Mix all crushed ingredients and add water, then the yarn is inserted. Cook the solution but keep it from boiling, then soak the yarn for two days, until the black color accumulates. Afterward, remove and dry it. Next step, boil the sogatingi skin with 350 ml of water, add the yarn, and soak for about 5 minutes. In the last stage, remove the thread and dry it in the sun.

For yellow color, the waivers used turmeric, lime, and water. Turmeric weighed 45.0 grams smoothed, 30 ml of lime juice, and 350 ml of water. Mix and cook the materials until boiling, put into the thread while stirring until the yellow color accumulated. After that, soak the yarn for two days. The last step is the yarn removed and dries it until dry.

Table 4. The resulted color with and without the addition of mordant

No	Coloring Plant	Without mordant	Mordant
1	Tarum (*Indigofera tinctoria*L)	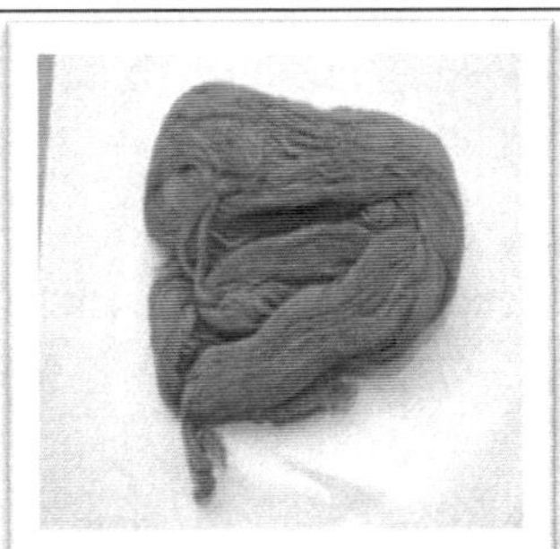 Pale Black	 Thick Black

IOP Conf. Series: Materials Science and Engineering **823** (2020) 012009 doi:10.1088/1757-899X/823/1/012009

No	Coloring Plant	Without mordant	Mordant
2	Mengkudu *(Morinda citrifolia* L*)*	 Pale Red	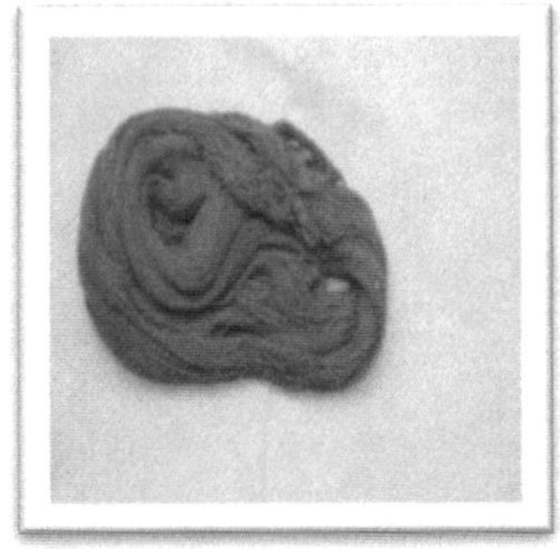 Wine Colored
3	Turmeric*(Curcuma longa* L*)*	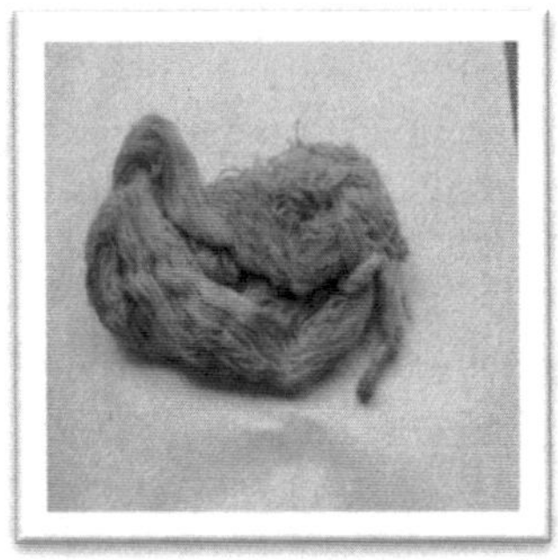 Brownish Yellow	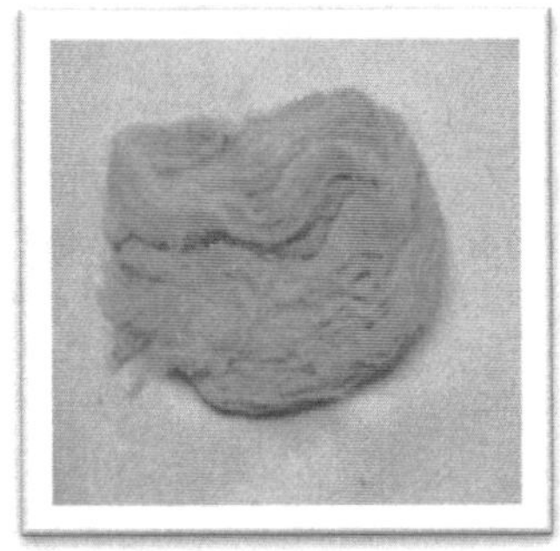 Bright Yellow

Source: The 2019 result of study

Based on Table 4, our results showed that the cotton yarn dyeing using tarum leaves extract without mordant produce a pale black color while in the coloring process with the addition of mordant produces a thick black color.

Likewise, the mengkudu root bark extract without adding mordant produces a pale red color and uneven color on the yarn, whereas in the coloring process using mordant, a color change occurs, namely the red heart color (wine). And the last color produced from turmeric extract without mordant produces a brownish yellow color and uneven color on the yarn. While the coloring process using mordant, produces bright yellow, and color evenly distributed.

Based on the description above, it can be concluded that the addition of mordan to the yarn coloring process can affect the color. This is supported by Sulistiami and Fathonah [19] which state that mordant is a color enhancer. The use of mordant can increase the stickiness of various dyes on the fabric. Mordant can affect the final color of a dye; therefore, the use of different mordants will produce a variety of colors. In addition, according to Prabhu and Bhute [20], mordant can improve the fixation of natural colors in fibers by forming complexes with dyes.

5. Conclusion

It was concluded that the type of mordanting plants used in the process of threads coloration by weavers at Hundihopo village are 12 plant species, namely *Sterculia foetida* L., *Erythrina variegata, Symplocos sp, Ceriops tagal, Datura metel, Calotropis gigantea, Abrus precatorius* L., *Areca catechu* L., *Jatropha curcas* L., *Citrus aurantifolia,* Beura, and Faliti. Plant organs used are leaf, bark, fruit, and seed. The processing methods are burned, pounded, boiled, soaked, and the resulted colour is black, red and yellow. In this study, there were 2 species of plants that have not been found the scientific names, faliti, and beura. Therefore, it is expected for the next researchers to identify the plants.

References

[1] Rosyida A and Zulfiya A 2013 Textile coloring by using jackfruit wood extract and coloring techniques to obtain optimal results *Rekayasa Proses Journal* **7**(2) 51-57

[2] Alvina, Adriani and Novrita S Z 2016 The Difference between Mordant Starfruit Sweet (*Avverhoa carambola*) and Starfruit (*Avverhoa bilimbi*) toward Immersion Results in Silk Material using Kesumba Seed Extract (*Bixa orellana* L.) *Journal* (Padang: Padang State University)

[3] Kwartiningsih E, Setyawardhani D A, Wiyatno A and Triyono A 2009 Natural textile coloring agent from mangosteen rind *Ekuilibrium Journal* **8**(1) 41-47

[4] Sabuna A Ch, Daud Y and Ngginak J 2017 Ethnobotany study of natural dyes by women in hundihopo village, east rote district, rote ndao regency *Proceedings of 1st International Conference on Climate Change, Biodiversity, Food Security, and Local Knowledge* (Kupang, 3-4 September 2018) pp 238-48

[5] Office of Communication, Information, Statistics and Encoding Rote Ndao Regency 2019 *Profile of Rote Ndao District* Available from: https://rotendaokab.go.id/profil-wilayah.php

[6] Martin G J 1995 Ethnobotany: A Method Manual, A "People and Plants" Conservation Manual (Londok, UK: Champman and Hall)

[7] Steenis V 2005 *Flora* (Jakarta: PT. Pradnya Paramita)

[8] Seran W and Hana Y W 2018 Identification of weaving dye plants types in kaliuda village, pahunga lodu district, east sumba regency *Agribisnis Perikanan Journal* **11**(2) 1-8

[9] Oktiarni D 2011 Utilization of Guava Leaf (*Psidium guava*), Dewandaru Leaf (*Eugenia uniflora*), and Rosella Leaf (*Hibiscus sabdariffa* l.) as Textile Natural Dyes in Cotton Fabric with Mordant Starfruit Wuluh *Thesis* (Bengkulu: Bengkulu University)

[10] Martalinda S 2013 Effect of mordant coconut water on dyeing of cotton using temulawak bulbs extract (*Curcuma xanthorriza* Roxb) *E-Journal Home Economic and Tourism* **2**(1)

[11] Herdiana N 2005 The potential of kepuh cultures (*Sterculia foetida* L.) *Seminar Proceedings, Plantation Forest Research Results* (Baturaja, 17 Desember 2017)

[12] Murniati and Takandjanji M 2015 The level of utilization of color producing plants in ikat weaving business in East Sumba regency *Penelitian Hutan Tanaman Journal* **12**(3) 223-37

[13] Jansen P C M and Cardon D 2005 *Plant Resources of Tropical Eropa, 3. Dyes and tannins* (Wageningen, Netherland: Prota Foundation)

[14] Handayani W, Yulita N and Nintya S 2012 Response of growth and production of alkaloids in callus rooted *Datura metel* L. to increase micronutrients from MS medium *Anatomi dan Fisiologi Bulletin* **20**(1) 29-36

[15] Dalimartha S 2006 Traditional ingredients for the treatment of cancer (Jakarta: Penebar swadaya)

[16] Nainggolan M 2011 Chemical content and activation of antioxidant extracts and saga leaf fractions (Adenanthera bicolor Moon) *Proceedings of Seminar Nasional Biologi* (Medan: Sumatra Utara University, January 22, 2011) pp 262-70

[17] Purwadaksi R 2007 *Utilizing The Yard for Family Medicinal Plants* (Jakarta: PT Agro Media Pustaka)

[18] Santosa E K and Kusumastuti A 2014 Utilization of tobacco leaves for staining silk fabric with mordant lime *Teknobuga Journal* **1**(1)

[19] Sulistiami and Fathonah 2013 The use of mordant and guava leaves amplifiers toward results of dyeing technique in cotton fabric *Buana Pendidikan Journal* **9**(16)

[20] Prabhu K H and Bhute A S 2012 Plant based natural dye and mordants: a review *J. Nat. Prod. Plant Resour.* **2**(6) 649-64

Synthesis and characterization of modified γ-Alumina-NaA and γ-Alumina-NaX zeolite composites as methanol dehydration catalysts in synthesis Dimethyl Ether (DME)

M Ulfah[1], S Octavia[1], H Suherman[2], M Laniwati[3], IGBN Makerthiharta[3] and Subagjo[3]

[1]Department of Chemical Engineering, University Bung Hatta, Padang, Indonesia
[2]Department of Mechanical Engineering, University Bung Hatta, Padang, Indonesia
[3]Department of Chemical Engineering, Institut Teknologi Bandung, Indonesia

E-mail: ulfahnad71@gmail.com

Abstract. Dimethyl ether (DME) is of great interest due to its wide application. It is currently being demonstrated for use as an alternative diesel fuel or substituting LPG for household. DME production via methanol dehydration reaction is affected by textural properties and catalyst surface acidity. The experimental study of textural properties in addition to surface acidity of both γ–Al_2O_3 prepared in this study and modified γ–Al_2O_3 over zeolite A and X mixing were conducted. The catalysts were characterized by N_2 adsorption-desorption, X-Ray diffraction (XRD), and temperature programmed desorption of ammonia (NH_3-TPD). Referring to N_2 adsorption/desorption isotherm using the Brunauer-Emmerr-Teller (BET) equation, the catalysts were observed possessing high specific surface area (170–279 $m^2\ g^{-1}$) and pore diameters of 5.6 – 8.8 nm (mesoporous). The XRD patterns of all catalysts still reveals clear peaks of γ–Al_2O_3 components. The strength of the acid sites of alumina modified with zeolite A is categorized as medium acid sites, whereas by modified by zeolite X varies.

1. Introduction

Alumina (Al_2O_3) has been used widely as a catalyst and heterogeneous catalyst support owing to its lower cost, thermal stability, high specific areas, surface acidity and interaction with transition-metals deposited onto the surface. It exists in α, γ, κ, ρ, η, and θ phases, each exhibits a unique properties [1]. The γ-alumina phase is commonly used in methanol dehydration reaction into dimethyl ether (DME).

Dimethyl ether, also known as methoxymethane (CH_3OCH_3), is the smallest aliphatic ether compound in the absence of a C-C bonds that benefits from being non-toxic, non-carcinogenic and non-corrosive. Due to its high cetane number (55-60), DME can be used in a diesel engine that produce low levels of CO and NOx, thus DME can be counted as an alternative fuels [2]. Furthermore, considering the similarity of its physical properties with those of liquefied petroleum gas (LPG), DME can easily add or replace LPG for home cooking, power generation, and LPG derived engines [3]. DME also used as a CFC aerosol-propellant replacement, raw material for chemical products, and solvent. The rising demand for DME in the near future is attributed to its widespread applications as

IOP Conf. Series: Materials Science and Engineering **823** (2020) 012010 doi:10.1088/1757-899X/823/1/012010

precedently reported. Moreover, the well-developed LPG infrastructures can be adapted for DME, thus makes DME formidable for practical use [4].

Methanol dehydration to DME reaction (MTD) carried out over solid acid catalysts is economically and thermodynamically favored [2]. Solid acid catalysts, such as alumina and zeolite, being the most common acid catalysts employed in the dehydration of methanol to produce DME commercially. Alumina and zeolite strong acid sites tend to lead to coke formation or other undesired by-products such as hydrocarbons, thus decreasing DME selectivity. The production of DME through MTD reaction is mainly related to sites with weak and medium acidity [2-10]. Accordingly, the surface acidity of solid acid catalysts, such as alumina and zeolite, is adjusted and modified to increase DME selectivity and catalytic stability. Modifying the catalyst by adding alkaline, alkaline earth, transition, rare earth metals, or other composite oxides is the strategy to adjust the surface acidity [7].

Adjusting the surface acidity of zeolite can be obtained over introduction of Na^+ on zeolite structures. Modified H-form zeolite over Na^+ exhibits higher MTD performance, high activity, selectivity and stability at temperatures of 230-340 °C [8, 9]. The catalyst surface acid sites display moderate acid strength which is responsible to inhibit hydrocarbon and coke formation.

Wang et al. (2013) observed that the key to MTD reaction is related to sites with weak acids and bases (bi-functional catalysts). Weak acidic sites facilitating the activation of O in methanol, while weak basidic sites activating H of hydroxyl group in surrogate methanol molecules [10]. The investigation prepared the bi-functional catalysts over aluminium acid and aminoacetic with triblock copolymer pluronic F127 as a template via and evaporation induced organic-organic assembly method.

Zhou et al. (2016) employed a sandwich FAU-LTA zeolite with dual-layer membranes as a catalytic membrane reactor of dimethyl ether synthesis (DME). Methanol dehydration into DME conducted on the upper layer of H-FAU zeolite with moderate acidity, and water as the by-product of MTD reaction was removed in situ on Na-LTA zeolite layer that possessed hydrophilic properties [11]. Na-LTA zeolite layer was located between porous alumina supporter and H-FAU. The combination provide 90.9% methanol conversion at 310 °C and DME selectivity of 100%. The dual-layer membrane of FAU-LTA zeolite was prepared on a porous α-Al_2O_3 tube over 3-aminopropyltriethoxysilane (APTES) as a covalent linker between the zeolite LTA layer and the α-Al_2O_3 support.

This research aims to develop and characterize of the γ-Al_2O_3-Zeolite NaA (Al-Zeo-Ao and Al-ZeoA-33) and γ-Al_2O_3-Zeolite NaX (Al-Zeo-Xo and Al-ZeoX-33) catalysts. The catalysts were prepared by mixing A or X zeolite with boehmite at ratio of 1:3, dried and calcinated at 570 °C for 3 hours. The effect of zeolite type and Na content on catalysts properties was investigated and compared with self-prepared γ-Al_2O_3 derived from boehmite. The catalysts produced in this studied were characterized by BET, XRD and NH_3-TPD.

2. Method

2.1. Catalyst preparation

2.1.1 γ-Al_2O_3 catalyst. Gamma alumina catalysts were prepared by using sol-gel method; the boehmite (AlOOH) was weighed and added to water, nitric acid 3 wt. % and sodium hydroxide 25 wt. %. After mixing, the mixture was extruded and stored at room temperature for 48 hours. The mixture was dried at 120 °C for 3 hours and calcinated at 570 °C for 2 hours [12].

2.1.2 Al-ZeoAo catalyst. Zeolite A (4Å) was synthesized as described in US Patents Document 2882243 [13]. The reactant mixture has a composition expressed in terms of molar ratio of:

Na_2O/SiO_2	SiO_2/Al_2O_3	H_2O/Na_2O
1.44	2.2	55

Al-ZeoA catalyst was prepared by adding zeolite 4Å into boehmite (weight ratio of 1:3), water, nitric acid 3 wt. % and sodium hydroxide 25 wt. %. The mixture was extruded and stored at room temperature for 24 hours, then dried at 120 °C for 2 hours and calcinated at 570 °C for 3 hours.

2.1.3 Al-ZeoXo catalyst. Zeolite X was synthesized as described in US Patents Document 2882244 [14]. The mixture of reagents were prepared with the Na_2O/SiO_2, SiO_2/Al_2O_3, and H_2O/Na_2O molar ratio of 1.3, 3.4 and 38, respectively. Al-ZeoX catalyst was prepared by adding zeolite X into boehmite (weight ratio of 1:3), water, nitric acid 3 wt. % and sodium hydroxide 25 wt. %. The paste was formed into shapes and stored at room temperature for 24 hours. The shaped paste then dried at 120 °C for 2 hours and calcinated at 570 °C for 3 hours.

2.1.4 Al-ZeoA33 catalyst. ZeoA-33 catalyst was prepared by homogenizing zeolite 4Å and metakaolin (from Lampung, Indonesia) with weight ratio of 4:3. Sodium hydroxide 33 wt. % was added into the mixture and homogenized to form paste. The shaped paste was inserted to sealed HDPE container and stored at room temperature for 4 days. After 4 days, the paste was crystallized at 100 °C for 17 hours in an oil batch. ZeoA33 pellets were washed then dried and calcinated at 570 °C for 2 hours. The pellets were crushed and added into boehmite, water, nitric acid 3 wt. % and sodium hydroxide 25 wt. %. The paste was shaped into pellets and stored at room temperatures for 24 hours. After 24 hours, the pellets were dried and calcinated.

2.1.5. Al-ZeoX33 catalyst. ZeoX-33 catalyst was prepared by homogenizing zeolite NaX and metakaolin with the ratio weight of 4:1. Sodium hydroxide 33 wt. % was added into the mixture and homogenized to form paste. The paste was shaped into pellets, inserted into sealed HDPE container and stored at room temperature for 4 days then crystallized. ZeoX33 pellets were washed then dried and calcinated at 570 °C for 2 hours. The next procedure is similar with ZeoA33 preparation as described previously.

2.2. Characterization catalysts

The BET surface area (SBET), total pore volume (Vp), and pore diameter (Dp) were measured using a NOVA 2200 instrument (Quantachrome, USA). The XRD patterns of all the calcined samples were recorded on a Bruker dB advance (40 kV, 30 mA) X-Ray diffractometer, using a Cu-Ka radiation source (λ=1.5406 Å) and a nickel filter in the 2θ range of 0–80°. The acidity of the samples was measured via NH_3-TPD, using a Quantachrome TPR Win v4.0 instrument, which included on-line thermal conductivity detection (TCD).

3. Results and Discussion

3.1. Textural properties

Table 1 summarizes the textural properties of modified γ-Al2O3 catalyst over two types of zeolite (A and X). The textural properties were derived attributed to nitrogen adsorption-desorption data using Brunauer, Emmett and Teller (BET) method to elaborate the specific surface area (S_{BET}), total pore volume (Pv, cc/g) and average diameter pore (Dp, nm). The crystallite size of alumina was determined over XRD results using Scherrer equation.

Methanol dehydration into DME is strongly attribute to the textural properties, degree of crystallinity and total amount of Al_2O_3 acid sites [15]. The study suggested that materials with higher porosity (larger than 1 cm^3 g^{-1}) or smaller crystallite size (lesser than 3.5 nm) are less active, most probably attributed to its highly amorphous nature. Furthermore, the investigation reported that higher conversion of methanol was presented by catalyst with average cylindrical pores of about 9-10 nm, total pore volume of 0.60-0.80 cm^3 g^{-1} and crystallite size of 5-5.5 nm. In particular, such materials possess specific surface area of about 195-240 m^2 g^{-1}.

The textural, acidic properties and catalytic activity of γ-alumina are greatly influenced by the preparation method. Nano crystalline γ-alumina prepared by sol-gel method via sucrose as template shows the highest catalytic activity. Concerning the NH_3-TPD analysis, Keshavarz et al. (2011) confirmed that smaller crystallite size and moderate acidic sites showed higher catalytic activity. Such catalyst have specific surface area of 294 $m^2\ g^{-1}$, pore volume of 0.54 $cm^3\ g^{-1}$, cylindrical pores of 7.4 nm and crystallite size of 4.2 nm [6].

Considering the extensive number of nanostructured gamma alumina application in different areas, the preparation of such materials with interesting morphology and high specific surface area over simpler methods has gained significant importance [16]. Potter et al. developed a nano sized gamma alumina via room temperature precipitation/digestion method and subsequent calcination of boehmite at 550 °C for 5 hours. The study investigated that the product surface area and average pore volume were 133.90 $m^2\ g^{-1}$ and 1.02 $cm^3\ g^{-1}$, respectively [17]. Prida et al. (2009) have prepared nano gamma alumina using sodium hydrogen carbonate, sodium carbonate and ammonium bicarbonate. The spherical nano sized particle with a surface area of 190 $m^2\ g^{-1}$ and pore volume of nearly 0.467 $cm^3\ g^{-1}$ was obtained over ammonium carbonate [18].

In the present study, it is focused on gamma alumina preparation based on sol-gel method and calcination at 570 °C with boehmite as raw material. The surface area of self prepared gamma alumina was about (249.209 m^2/g) while pore volume was about 0.421 $cm^3/\ g$. This value of surface area is greater than 125 m^2/g and it indicates that synthetic nano gamma alumina is very stable [17]. The physicochemical parameters of developed catalysts were expressed in Table 1.

Table 1. Physicochemical parameters of developed catalysts

	S_{BET} (m^2/g)	Pv cm^3/g	Dp nm	Crist. size Nm
γ-Al_2O_3	249.209	0.421	6.860	15.888
Al-ZeoA0	278.877	0.395	5.660	11.935
Al-ZeoA33	239.371	0.375	6.264	24.043
Al-ZeoX0	215.439	0.458	8.497	19.160
Al-ZeoX33	170.365	0.372	8.743	30.470

Alumina features a range of advantages including excellent thermal stability, relatively high surface area, high mechanical strength and appropriate acidity concerning MTD reaction exhibited by higher methanol conversion and 100% of selectivity towards DME [7]. However, the ability to absorb water (hydrophilic properties) better than methanol gave a disadvantages to alumina application. Zhou et al. (2016) developed alumina combined with NaA zeolite (LTA) and H-Y zeolite to enhance the reduced activity by water presence on methanol dehydration [11].

NaA zeolite is one of the foremost zeolite in chemical industry due to its hydrophilic properties. Nano-sized and uniform pores of NaA zeolite received great attention that has been reported to be very attractive catalyst in dehydration process [19]. With regard to increase the MTD catalytic activity, self-developed alumina over NaA zeolite were investigated in this study. NaA zeolite preparation was in accordance with US Patents Document 2882244 [14] that was successfully produce zeolite A with pores of 0.4 nm (4Å). Al-ZeoAo catalyst textural properties prepared by mixing boehmite and zeolite 4Å crystalline powder with ratio of 3:1 followed by calcination at 570 °C is shown in Table 1. Zeolite 4Å added will increase alumina specific surface area. On the contrary, pore volume, pore diameter and crystallite size of Al-ZeoAo is smaller than non-modified γ-alumina.

Aliphatic alcohol compound with carbon atoms of C1, C2 and C3 can be adsorbed onto zeolite 4Å [14], but not on zeolite 3Å. In this study, zeolite 3Å was also prepared with preparation method refers to US Patents Documents 3679604 [20]. Firtsly prepared zeolite 3Å was assorted with boehmite, water, HNO_3 (3 wt. %) and NH_4OH (25 wt. %) followed by calcination at 570 °C (Al-ZeoA33). Al-

IOP Conf. Series: Materials Science and Engineering **823** (2020) 012010 doi:10.1088/1757-899X/823/1/012010

ZeoA33 specific surface area was smaller than γ-alumina. However, pore volume, pore diameter and crystallite size is greater. Physicochemical properties of Al-ZeoA33 catalyst is shown in Table 1.

Another zeolite type that received great attention in chemical industry as catalysts or adsorbent is faujasite. Zeolite X and Y is classified as faujasite zeolite. Both were distinguished based on Si/Al molar ratio; zeolite X have Si/Al molar ratio of 1-3 and zeolite Y of 4-6. This molar ratio is a deciding factor of surface acidity properties, higher Si/Al ratio resulted in higher surface acidity. In this study, zeolite NaX was developed as reported in US Patents Document 2,882,244 [13]. Al-ZeoX0 catalyst was prepared by mixing zeolite NaX crystalline powder with boehmite, homogenized, introduced to water, diluted nitric acid and concentrated ammonium hydroxide. Al-ZeoX0 catalyst developed in this study possessed physicochemical properties as displayed in Table 1. Surface area, total pore volume and average pore diameter of γ-Al_2O_3 is decrease with zeolite NaX powder mixtured, however the crystallite size is enhanced. Similar trends were also displayed by AlZeoX-33 catalyst; surface area, total pore volume and average pore diameter of γ-Al_2O_3 is decrease with zeolite X-33 pellet added.

According to International Union of Pure and Applied Chemistry (IUPAC) notation, microporous materials have pore diameters of less than 2 nm and macroporous materials have pore diameters of greater than 50 nm; the mesoporous category thus lies in the middle [21]. Pore diameters of catalysts developed in this study were in a range of 2-50 nm, thus categorized as mesoporous. Highly ordered mesoporous Al_2O_3 has excellent thermal stability, high mechanical resistance, as well as high activity and DME selectivity as a result of its weak and moderate acid strength [22].

3.2. Analysis XRD

Figure 1 displays the XRD pattern of catalysts prepared in this study. All the peaks belonging to γ-Al_2O_3 at $2\theta = 37.6°$; 45.5° and 67.0° can be discriminated from the XRD pattern. Simultaneously, the new diffraction peak attributed to zeolite A appeared at $2\theta = 24°$; 27.2°; 30°; and 34.3° on Al-ZeoAo and Al-ZeoA33 catalysts as also reported by Milton and Buffalo [13]. The diffraction attributed to zeolite X peak appeared at $2\theta = 23.3°$; 26.7°; 31° and 33.6° on the XRD pattern of Al-ZeoXo and Al-ZeoX33 catalysts [14]. However, the peaks intensities of γ- Al_2O_3 were not significantly enhanced.

Figure 1 displays the XRD pattern of Al-ZeoA33 and Al-ZeoX33, giving higher zeolite ZeoA0 intensities as compared with ZeoA33, leading to better ZeoA0 crystallinity. Zeolite ZeoA33 prepared by mixing zeolite NaA powder with metakaolin and NaOH solution 33 %-wt. Kaolin was added as a binder to form crystalline powder of NaA, then NaOH solution was added [23]. According to XRD patterns displayed in Figure 1, the specific diffraction peaks of metakaolin at $2\theta = 21°$ and 27° was not observed, suggesting metakaolin was completely converted into zeolite, however low intensities. The aforemention result was demonstrated by ZeoX33 prepared through resembling method, resulting in lower intensity peaks than ZeoXo.

Degree of crystallinity was in accordance with crystallite size that generally correlates with MTD reaction catalytic activity [4, 7, 15]. The catalyst crystallite size was calculated using equation 1.

$$B = \frac{K\lambda}{\beta \cos\theta} \qquad (1)$$

Akarmazyan et al. (2014) reported that reducing crystallite size can further enhance methanol adsorption site, thus increasing methanol conversion [15]. However, further reduction will give opposite results due to excessively small crystallite size will turn into inactive amorphous phase. Crystallite size of developed catalysts were displayed in Table 1. Al-ZeoA33 crystallite was observed bigger than Al-ZeoA0, and so does mixture of γ- Al_2O_3 and zeolite type X.

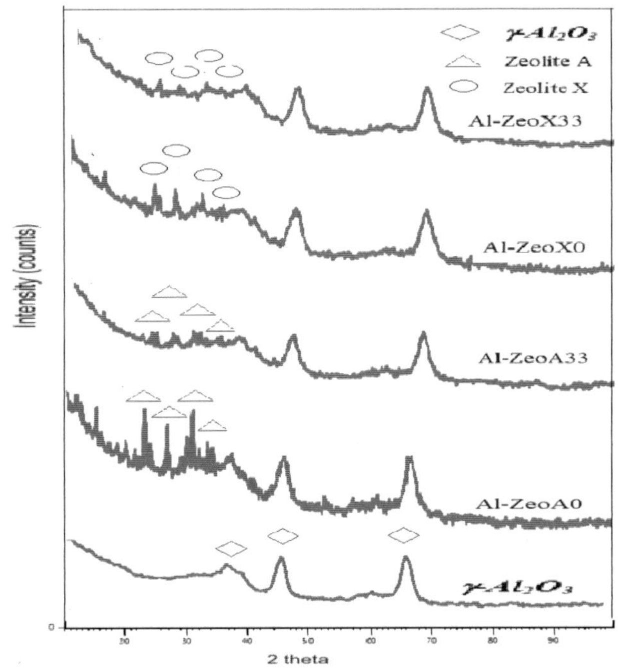

Figure 1. XRD patterns of self-prepared catalysts

3.3. Analysis ammonia temperature programmed desorption (NH3-TPD)
The acid site concentration and strength were determined by ammonia temperature programmed desorption (Table 2). The total volume of ammonia desorbed, represented by the area in Table 2, equal the total acid site concentration. The temperature at which desorption takes place, relates to the strength of the acid site. The stronger the acid site, the stronger ammonia is bound and the higher the temperature that is required to desorb the ammonia from the acid site.

Table 2. Acid site concentration and acid strength distribution determined by ammonia temperature programmed desorption

	Area (signal* time)			Peak NH_3 desorption temperature, °C		
	Peak 1	Peak 2	Peak 3	weak	medium	strong
γ-Al_2O_3	23.527	999.319	34.370	148.795	252.980	472.096
Al-ZeoA0	3079.302				324.876	
Al-ZeoA33	1889.928				306.397	
Al-ZeoX0	1458.184	27.837			308.212	518.832
Al-ZeoX33	559.140				259.638	

Some researchers have reported that MTD reaction requires high concentration of acidic sites with weak or medium acid strength [2-8].

4. Conclusion

The textural properties and acidity of modified alumina catalyst over zeolite A and X has been investigated. Referring to N_2 adsorption/desorption isotherm by using the Brunauer-Emmerr-Teller (BET) equation, the catalysts were observed possessing specific surface area of 170–279 m^2 g^{-1}, pore volumes of 0.35-0.46 cm^3 g^{-1}, and mesoporous pore diameters of 5.6 – 8.8 nm. The crystallite size of developed catalyst determined by X-ray diffraction technique was found to be in a range of 11.5-30.5 nm. Modifying alumina over zeolite A and X were not significantly enhancing the specific diffraction intensities with respect to XRD pattern. Self-developed gamma alumina catalyst expressed by the temperature of ammonia desorption possessed weak, medium and strong acid strength. Acid strength distribution of Al-ZeoAo, Al-ZeoA33 and Al-ZeoX33 catalysts was only at medium temperature of 259-325 °C. Al-ZeoXo catalyst possessed medium and strong acid strength; ammonia desorption was at temperature of 308.212 °C and 518.832 °C.

5. Acknowledgments

The financial supports from the Ministry of Technology Research and Higher Education Indonesia (011/L10/Ak-04/KONTRAK-PENELITIAN-J/2019), University Bung Hatta (Nomor: 024-08/LPPM-Penelitian/Hatta/IV-2019).

6. References

[1] Osman Ahmed A, Abu-Dahrieh J K, Rooney D W, Halawy S A, Mohamed A M, Abdelkader A and Abu Dahrieh J 2012 *J. Appl. Cat. B: Env.* **127** 307
[2] Yaripour F, Baghaei F, Schmidt I and Perregaard J 2005 *J. Catal. Commun.* **6** 542
[3] Tokay K C, Dogu T and Dogu G 2012 *J. Chem. Eng.* **184** 278
[4] Saravanan K, Ham H, Tsubaki N and Bae J W 2017 *J. Appl. Cat. B: Env.* **127** 307
[5] Fu Y, Hong T, Chen J, Auroux A and Shen J 2005 *J. Thermochima Acta* **434**(1) 22
[6] Keshavarz A R, Rezaei M and Yaripour F 2011 *J. Natural Gas Chemistry* **20** 334
[7] Sun J, Yang G, Yoneyama Y and Tsubaki N 2014 *J. ACS Catal.* **4**(10) 3346
[8] Hassanpour S, Taghizadeh M and Yaripour F 2010 *J. Ind. Eng. Chem. Res.* **49** 4063
[9] Vishwanathan V, Jun K W, Kim J W and Roh H S 2004 *J. Appl. Catal. A* **276** 251
[10] Wang B, Wen Y and Huang W 2013 *J. Energy Sources, Part A* **35** 1590
[11] Zhou C, Wang N, Qian Y, Liu X, Caro L J and Huang A 2016 *J. Angew. Chem. Int. Ed.* **55** 1
[12] Ulfah M and Subagjo 2012 *J. Reaktor* **14** 151
[13] Milton R M and Buffalo N Y 1959 *US Patents 2,882,243*
[14] Milton R M and Buffalo N Y 1959 *US Patents 2,882,244*
[15] Akarmazyan S S, Panagiotopoulou P, Kambolis A, Papadopoulou C and Kondarides D M 2014 *J. App. Catal. B: Env.* **145** 136
[16] Eliassi A and Ranjbar M 2014 *Int. J. Nanosci. Nanotechnol.* **10** 13
[17] Potdar H S, Jun K W, Bae J W, Kim S M and Lee Y 2007 *J. Appl. Catal. Gen.* **321** 109
[18] Parida K M, Prdhan A C, Das N J and Sahu N 2009 *Mater. Chem. Phys.* **113** 244
[19] Mirfendereski M and Mohammadi T 2016 *Proceedings of the World Congress on Recent Advances in Nanotechnology* (Prague, Czech Republic) ICNNFC 113
[20] Rosback D H and Elmhurst 1976 *US patents 3,969,276*
[21] Sing, K W S 1998 *J. Adv. Colloid Interface Sci.* **76** 3
[22] Luan C, Zhang A, Huang W and Yin L 2015 *J. Energy Sources, Part A: Rec., Util. and Env. Effects* **37** 2285
[23] Pryor J N, Catonsville and Chi C W *US Patents 4424144*

IOP Conf. Series: Materials Science and Engineering **823** (2020) 012011 doi:10.1088/1757-899X/823/1/012011

Central composite design based statistical modelling of curcuminoids extraction from *Curcuma zedoaria* using choline chloride based Natural Deep Eutectic Solvent (NADES)

O Rachmaniah, Istiqomah, Fakhrinanda, and M Rachimoellah

Department of Chemical Engineering, Faculty of Industrial Technology, Institut Teknologi Sepuluh Nopember, Surabaya, 60111, Indonesia

E-mail: orchidea@chem-eng.its.ac.id

Abstract. Ionic type of Natural Deep Eutectic Solvents (NADES) consisted of choline chloride and organic acids, i.e., choline chloride-malic acid (CCMA) and choline chloride-citric acid (CCCA), known as the most suitable liquefied solvent for curcuminoids extraction from *Curcuma zedoaria*. A laboratory-scale extraction of curcuminoids from *C. zedoaria* successfully conducted; indicated that solvent per feed ratio (S/F) and extraction time are the two important variables for yielding a high extraction rate of curcuminoids. Therefore, a central composite based design is used to optimize the extraction parameters such as S/F (3/10, 5/10, 7/10), time extraction (18, 20, and 22 h), and stirring speed (200, 250, and 300 rpm) at 500 mL capacity. Stirring speed is the most influential parameter among the studied parameters. However, due to the low bulk density of *C. zedoaria* powder, a homogenous mixture of *C. zedoaria* in the extraction tank was barely achieved; hence a low yield of curcuminoids is obtained.

1. Introduction

The World Health Organization (WHO) and the International Cancer Control Union (UICC) predict that in 2030, there will be an increase in cancer patients by 300% worldwide. Almost 20% of them will be in developing countries such as Indonesia. Thus, The Ministry of Health of the Republic of Indonesia declares that cancer is one of the main causes of death.

Curcumin, a kind of curcuminoids, is commonly found in the Zingiberaceae family such *Curcuma zedoaria*. It is known as an anticancer agent [1], antioxidant, etc. However, its efficacy and safety are good, curcumin currently not yet approved as an anticancer therapeutic agent. The limited supply of curcumin extract for medicinal purposes is one of the reasons [2]. Since human blood contains approximately 90% of water [3], low solubility of curcumin in the water (pH = 7.3), 4 ppb (4 μg / L), also a constraint factor. Therefore, the carrier material of curcumin that can maximally release curcumin needs to be studied to facilitate the oral delivery of curcumin in the human body maximally.

Various extraction processes of curcumin have been conducted for obtaining a high yield of curcumin. Organic solvents (such as ethanol, methanol, ethanol-water, etc.) are commonly used as solvents. However, organic solvents leave residues in the crude extract. They potentially deposited in the body. Therefore, an environmentally friendly solvent, which is health and save, is necessary to use in the extraction process especially in pharmaceutical industries.

The next generation of ionic liquids (IL), called eutectic solvents (DES), consisted of low toxicity chemical compounds in the form of cations and anions [4]. The main advantages of DES are easy to make. Just by melting two chemical compounds at high purity with moderate heat. The suitability of DES for biological applications can be seen at dissolving enzymes into DES [5]. There is no change in the enzyme conformation. The enzyme activity can be controlled by water addition. Enzyme deactivation can be made by minimizing the water content by means freeze-drying the enzyme and reactivate it back by diluting with water [5]. When DES consisted of primary metabolites such as sugar, sugar alcohol, polyalcohol, organic bases, organic acids, and amino acids, they referred to as Natural Deep Eutectic Solvent (NADES) [6]. Various macro- and micro-molecules are well-dissolved NADES such as gluten and DNA [7]. A micro-molecules such routine, quercetin, cinnamic acid, taxol, carthamin, ginkgolide B provide good solubility in some types of NADES [7]. Some of them showed a higher solubility in NADES compare in water and methanol. NADES proline-malic acid (PMH) and lactic-glucose (LGH) are both used to extract phenolic, cartormin, and carthamin from safflower (*Flos Carthami*) and corolla from *Carthamus tinctorius* L. [8]. These safflower natural dyes showed high solubility in PMH and LGH. Sugar-based NADES reported to be stable to store natural phenolic pigments, cyanidin, both at higher temperatures and at longer storage times [8]. Moreover, Zullaikah et al. [9] showed suitability of NADES for extracting curcuminoids. Neutral type of NADES (FS-H_2O), ionic type of NADES (CCMA-H_2O) and acidic type of NADES (CAS-H_2O and MAS-H_2O) are better-extracted curcuminoids from *Curcuma mangga* compared to other kinds of NADES. The highest yield of curcuminoids given by CCMA-H_2O (1: 1: 2) of 0.355 mg curcuminoids/g dry powder of *C. mangga*.

The extraction process with a high yield of curcuminoids requires optimum conditions of the extraction parameters such as particles size, type of solvents/NADES, porosity and diffusivity, pH, extraction time, agitation speed, temperature, solvent to feed (S/F) ratio, and mode of the extraction process. Therefore, to increase the extractability of curcuminoids from Curcuma powder by NADES the optimum condition of the extraction parameters need to be studied. This study focused on optimizing the extraction parameters of *Curcuma zedoaria*, i.e., extraction time, agitation speed, and solvent to feed (S/F) ratio, using Response Surface Methodology (RSM).

2. Materials and Methodology

2.1. Reagents and chemicals

Curcuminoids standard purchased from Merck (Darmstadt, Germany) as well as solvents such as acetonitrile, acetic acid (HOAc), and methanol (MeOH). All the solvents are in High Pure Analytical grade. While an individual compound of NADES such as choline chloride and malic acid purchased from Sigma Aldrich (St. Louis, MO, USA). While citric acid in food-grade purchased at Surabaya local market (Gajah, Jakarta, Indonesia). Buffer solution (pH = 2.0) from Mediss (India). Moreover, fine powder of *Curcuma zedoaria* kindly donated from Sari Herbal (Sukun, Malang, Indonesia).

2.2. Natural Deep Eutectic Solvents (NADES) preparation

All NADES compounds used in this study were prepared based on the method by Rachmaniah et al. [10], which modification method is described earlier [7].

2.3. Extraction of curcuminoids using Natural Deep Eutectic Solvents (NADES)

Curcuma zedoaria powder was accurately weighed based on the S/F variable as determined. It was firstly placed in the extractor (amber glass extractor), equipped with a thermometer that set at 40°C. Subsequently, NADES, either CCMA-H_2O (1: 1: 18) or CCCA-H_2O (1: 1: 18), was added. Each experiment conducted using 200 mL of NADES as a solvent. The extraction was conducted with IKA RW-20 digital rotor (IKA, Staufen, Germany) included its 4-blades propeller IKA R-1342 (IKA, Staufen, Germany). After the extraction is complete, 1 mL sampled and stored at freeze for further step of analysis.

2.4. Analysis of curcuminoids using spectrophotometer

Extracted curcumin was determined by spectrophotometric at λ_{max} = 425 nm. As the spectrophotometric method cannot differentiate curcumin, desmethoxycurcumin, and bisdemethoxycurcumin, hence it expressed as total curcuminoids. A 400 μL was accurately pipetted and diluted by adding a mixture of methanol-3% v of HOAc to 2 mL, pH adjusted to pH = 1.8-2.0. Subsequently, the diluted sample was analyzed using a UV-Vis Spectrophotometer T-60 (PG Instruments, Leicestershire, UK). The extracted yield of curcuminoids expressed as a weight of extracted curcuminoids (mg) per dry weight of *Curcuma zedoaria* powder (g).

2.5. Data analysis using Response Surface Methodology (RSM)

The optimum condition for extraction time, S/F ratio and agitation speed were simulated by responses surface methodology (RSM) with Minitab program.

3. Result and Discussion

Table 1. Central Composite Matrix Design

Run Order[a]	Standard Order[b]	Optimized variables[c]			Yield (mg curcuminoidsss/g)	
		Agitation Speed (rpm)	S/F (mL/mg)	Time (hour)	CCCA-H_2O (1:1:18)	CCMA-H_2O (1:1:18)
1	17	250 (0)	5/10 (0)	20 (0)	6.086	7.246
2	15	250 (0)	5/10 (0)	20 (0)	6.086	7.206
3	13	250 (0)	1/6 (-1,68)	20 (0)	2.004	1.820
4	12	250 (0)	5/10 (0)	23.36(+1.68)	5.069	7.066
5	9	200 (-1,68)	5/10 (0)	20 (0)	1.534	3.829
6	10	300 (+1.68)	5/10 (0)	20 (0)	1.534	3.469
7	18	250 (0)	5/10 (0)	20 (0)	6.086	7.266
8	3	220 (-1)	3/10 (-1)	22 (+1)	2.653	2.645
9	5	220 (-1)	7/10 (+1)	18 (-1)	5.738	7.039
10	20	250 (0)	5/10 (0)	20 (0)	6.086	7.226
11	14	250 (0)	5/6 (+1.68)	20 (0)	5,424	6.168
12	8	280 (+1)	7/10 (+1)	22 (+1)	5.835	6.787
13	4	280 (+1)	3/10 (-1)	22 (+1)	3.818	2.597
14	1	220 (-1)	3/10 (-1)	18 (-1)	2.501	2.609
15	16	250 (0)	5/10 (0)	20 (0)	6.086	7.326
16	11	250 (0)	5/10 (0)	16.64(-1.68)	4.653	3.489
17	7	220 (-1)	7/10 (0)	22 (+1)	5.997	7.123
18	19	250 (0)	5/10 (0)	20 (0)	6.086	7.246
19	6	280 (+1)	7/10 (+1)	18 (-1)	5.771	4.969
20	2	280 (+1)	3/10 (-1)	18 (-1)	4.012	2.681

[a] not randomized order; [b] randomized order; [c] code on experimental design: -1 low level; 0 mid level, +1 high level

Three variables, i.e., extraction time (h), solvent to feed ratio (S/F), and agitation speed, optimized in these experiments. There are two methods, i.e., Box-Behnken or Central Composite Design, for optimizing two or more variables. However, using Box-Behnken (BB), 15 experiments are required. Whereas 20 experiments are needed when using Central Composite Design (CCD). The CCD method designed 5 levels (-1.68, -1, 0, +1, +1.68) while only 3 levels (-1, 0, +1) for BB. The two additional levels (-1.68 and +1.68) are the extreme conditions (far from the central level, 0). CCD

IOP Conf. Series: Materials Science and Engineering **823** (2020) 012011 doi:10.1088/1757-899X/823/1/012011

well applied in continuous experiments such as this study, in which the previous extraction conditions (extraction time, solvent ratio to feed (S/F), and agitation speed) applied as the mid-level (0). The matrix design of the experiments both for CCMA-H_2O (1: 1: 18) and CCCA-H_2O (1: 1: 18) is displayed in Table 1 as well as yield curcuminoids.

Curcuma zedoaria contains 1.96% of curcuminoids [9]. Applying S/F ratio of 5/10 with 200 mL of NADES as solvent, 400 mg of *C. zedoaria* powder will be used. Those, maximally 7.840 mg of curcuminoids, will be extracted, by assuming all the curcuminoids extracted. NADES CCMA-H_2O (1: 1: 18) successfully recovers 14-37% of curcuminoids from *C. zeodaria* while CCCA-H_2O (1: 1: 18) only 8-31% (Table 1). Even though the recovery is less than 50%; the obtained yield was higher compared to Zullaikah et al. [9]. It only yielded 0.355 mg curcuminoids/g with 4 hours longer of the extraction time (S/F = 3 mL/20 mg, 40 °C, 24 h of extraction). The obtained yield is comparable with microwave-assisted extraction (MAE) of *C. longa* with ethanol yielded 3.03 mg/g (S/F = 60 mL/0.3 g, 800 W, 2450 MHz, 7 min) [11]. Pulsed ultrasonic-assisted extraction (PUAE) of *C. longa* with ethanol yielded 10.30 mg/g (S/F = 20 mL/0.1 g, 3/1 s/s pulsed duration/interval time) [11]. Since *C. longa* contains higher curcuminoids, 4.4% [9], hence only 7% and 23% of the recovery, respectively for MAE and PUAE. In the case of extraction, instead of yield, the selectivity and type of solvent, as well as duration of extraction, were also necessary to be considered to obtain the optimum extraction condition. Mostly, the conventional extraction process i.e. Soxhlet extraction with volatile organic solvents (VOS) such acetone, ethyl acetate, ethanol, methanol, gives a high yield of curcuminoids [12], 24 mg curcuminoids/g of *C. longa* powder. However, this type of extraction method is claimed not to be classified as a green process due to the utilization of VOS, time-consuming, usage of huge amounts of hazardous solvent as well as the toxicity of VOS to humans and environment [13].

Table 2. ANOVA Analysis of Central Composite Design

NADES CCCA-H_2O		NADES CCMA-H_2O	
Parameters	*p*-value	Parameters	*p*-value
Linear	0.002	Linear	0.000
Speed	0.406	*Speed*	0.268
Time	0.745	Time	0.011
S/F	0.000	S/F	0.000
Square	0.001	Square	0.000
*Speed***Speed*	0.000	*Speed***Speed*	0.000
Time*Time	0.487	Time*Time	0.007
S/F*S/F	0.024	S/F*S/F	0.000
2-Way Interaction	0.661	2-Way Interaction	0.401
*Speed**Time	0.815	*Speed**Time	0.427
*Speed**S/F	0.241	*Speed**S/F	0.241
Time*S/F	0.874	Time*S/F	0.341

Quadratic polynomial regression model from ANOVA analysis (Table 2) shown significant *p*-values, both for CCCA-H_2O and CCMA-H_2O. Those had lower values than the Confidence Interval (CI), ca. 5% significances. Furthermore, the equations for the optimum condition are the following, where A is agitation speed (rpm), B is extraction time (h), and C is S/F ratio (mL/mg).

NADES CCCA-H_2O (1:1:18):

$$\text{Yield (mg curcuminoids/g } C.\ zedoaria) = -121.8 + 0.795*A + 1.77*B + 32.1*C - 0.001475*A^2 - 0.0378*B^2 - 13.92*C^2 - 0.00113*A*B - 0.0584*A*C + 0.114*B*C \quad (1)$$

NADES CCMA-H_2O (1:1:18):

Yield (mg curcuminoids/g *C. zedoaria)* = -133.9 + 0.608*A + 5.22*B + 35.3*C - 0.001315*A^2 - 0.1520*B^2 - 26.55*C^2 + 0.00336*A*B - 0.0506*A*C + 0.609*B*C (2)

Figure 1 is the response surface contour for studied variables versus the yield of curcuminoids (mg/g). Figure 1B approximately representing the ideal surface contour, a capsized bowl-shaped. Both the agitation speed and S/F ratio both have p-values <0.05 in the square parameter (Table 2), indicating that both parameters significantly affect the yield in the quadratic model. In other words, a slight change of both variables will result in a significant change in the yield of curcuminoids. Moreover, the *p*-value (Table 2) also supported this fact. Fernandez et al. [14] also observed that S/F ratio and Reynold numbers (related to agitation speed) were highly affected by the extraction process in the pilot-scale extraction of Mango leaves. In addition, the agitation speed significantly affects the yield due to the observed vortex formed when extraction conducted at >250 rpm, though at 250 rpm it is still in the transition flow range.

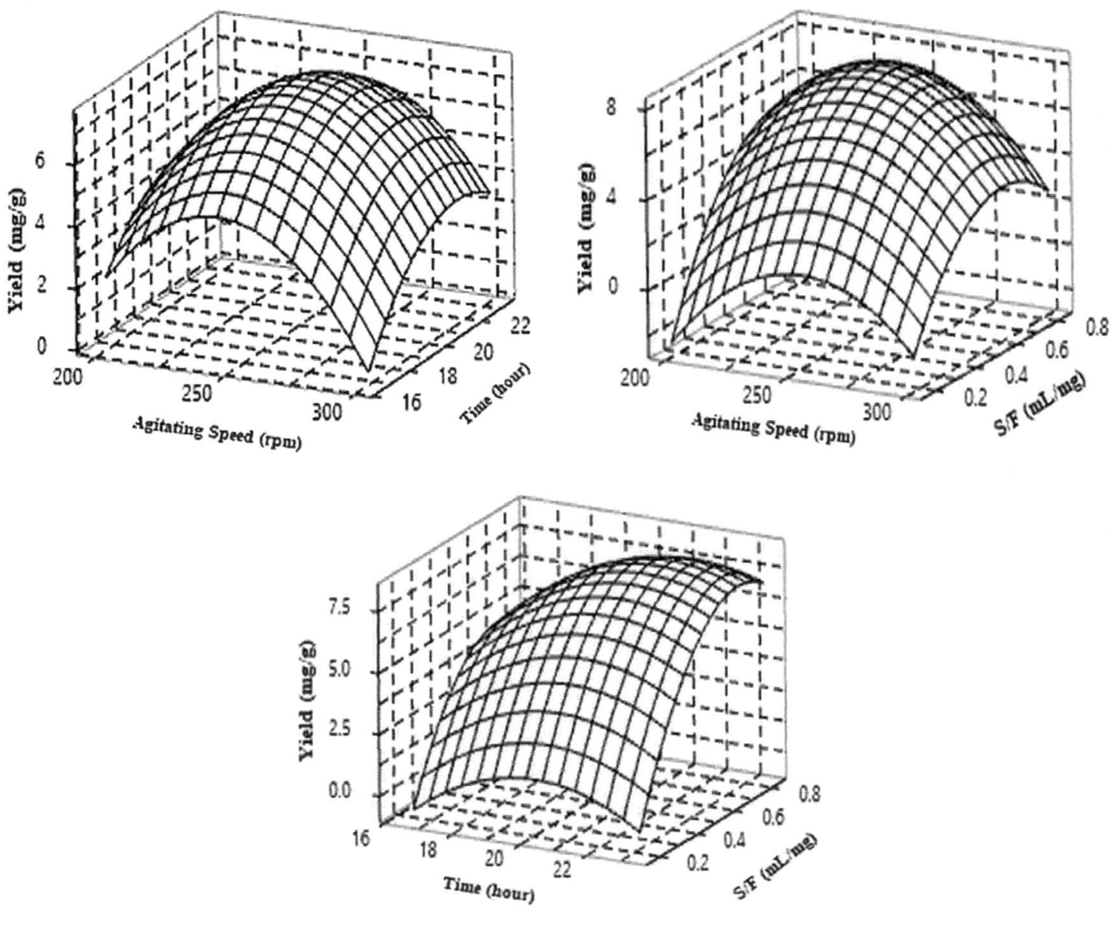

Figure 1. Response surface contour of the yield of curcuminoids (mg/g) with CCMA-H_2O as a NADES solvent: (**A**) agitation speed (rpm) vs extraction time (hours); (**B**) agitation speed (rpm) vs S/F ratio (mL/mg); and (**C**) extraction time (hours) vs S/F ratio (mL/mg).

Meanwhile, Figure 1A and 1C show that the extraction time is less significant to the yield. A doldrums surface contour was reflecting this relation. Extraction time is slightly affecting the yield.

Subsequently, the surface plot looks like a saddle shape. Hence, the S/F ratio variable is the most influential, both in the linear and quadratic models.

3.1. Effect of the extraction time on the yield of curcuminoids

The designed variables of the extraction time are 18 h (-1), 20 h (0), and 22 h (+1) both for CCMA-H_2O (1: 1: 18) and CCCA-H_2O (1: 1:18). The center level of the extraction time, ca. 20 hours, was selected due to it was yielding the highest yield of curcuminoids at our previous study (data not shown). Longer extraction time will increase the obtained yield. It is possibly due to the longer contacting time between the raw material of *C. zedoaria* powder and NADES solvent. Hence, the mass transfer of curcuminoids from the materials to the bulk of the liquid solvent will go maximally until reaching a saturation condition of the solvent [14]. Meaning, no mass transfer of curcuminoids from the materials to the bulk of liquid solvent after the saturation condition reached. Since the driving force of curcuminoids concentration approximately zero, the saturation condition observed happens beyond 20 hours of the extraction time. Those, the yield of curcuminoids remain constant and tend to decrease.

3.2. Effect of solvent/feed (S/F) ratio on the yield of curcuminoids

The S/F ratio = 5/10 (5 mL NADES/10 mg Curcuma powder) was chosen. Taking constant a volume of NADES ca. 200 mL, an excessive amount of powder (smaller value of S/F) decreased the contacted surface of powder with the NADES. By having a higher S/F ratio, a lower amount of powder applied. Higher S/F value may give a higher concentration gradient between the curcuminoids inside and at the surface of the solid (powder), which may result in a higher extraction rate and finally increase the obtained yield of curcuminoids. Hence, a lower yield value of curcuminoids resulted and vice versa. The resulted optimum condition of S/F is 7/10.

3.3. Effect of agitation speed on the yield of curcuminoids

The agitation conducted by applying a 4-blades propeller type mixer. It was chosen since it is suitable for the viscosity of NADES and produces an axial agitation pattern for homogenizing the *C. zedoaria* powder throughout the solvent [15]. The agitation speed significantly affects the obtained yield. A significant correlation between an agitation speed and the yield of curcuminoids shown in the ANOVA analysis (Table 2) for both types of NADES. The *p*-value of the agitation speed as a quadratic correlation is lower values than CI ca. 5% significances.

A significant increase in the yield of curcuminoids obtained at 200-250 rpm though a dispersed of *C. zedoaria* powder observed when agitated at 200-250 rpm. However, a vortex is present when agitated >250 rpm, resulting in a poor homogenization [15]. Thus, lower yield of curcuminoids is obtained.

3.4. Optimum conditions of the extraction

The optimum conditions of the extraction optimized using Minitab. The optimum value of time extraction, S/F ratio, and agitation speed shown in Table 3. Both types of NADES, more or less, have similar optimum conditions. However, CCMA-H_2O yielded higher curcuminoids. It is possibly due to the different characteristics of both NADES, such as density, viscosity, and acidity [7, 16].

Both CCCA-H_2O (1: 1: 18) and CCMA-H_2O (1: 1: 18), have lower viscosity compare to sugar-based on NADES [4]. A lower viscosity solvent with a higher diffusion coefficient increases the mass transfer, enhances the solubility and finally the extraction rate of curcuminoids [16]. Moreover, a high molar ratio of water, ca. 18 molar ratio, resulting in a density approximately to water. They are 1.1164 and 1.1431 g/mL, respectively, for CCCA-H_2O (1: 1: 18) and CCMA-H_2O (1: 1: 18). NADES CCCA-H_2O (1: 1: 18) and CCMA-H_2O (1: 1: 18) have viscosity 8,914 cP and 10,885 cP, respectively, resulting in a higher yield of curcuminoids in CCMA-H_2O than CCCA-H_2O (1: 1: 18).

In the case of polarity, both of the selected NADES have more or less a similar polarity, 47.8 kcal/mol and 47.91 kcal/mol (polarity is expressed as E_{NR}), respectively for CCMA-H_2O (1: 1: 18) and CCCA-H_2O (1: 1:18). Therefore, the polarity does not majorly affect the curcuminoids yield.

Table 3. The simulated optimum extraction conditions by CCD

NADES	Variabel			Predicted yield of curcuminoids (mg/g)	R^2
	Agitation Speed (rpm)	S/F (mL/mg)	Time (h)		
CCCA-H_2O	247.45	0.721	20.85	6.701	0.885
CCMA-H_2O	245.41	0.673	21.26	8.163	0.944

Based on the optimal variables obtained from the Minitab application (Table 3), verification experiments carried out in triplicates. Respectively, 12.953±0.135 and 15.239±0.121 mg/g were obtained for CCCA-H_2O (1:1:18) and CCMA-H_2O (1:1:18). Unfortunately, the obtained yield of curcuminoids at the optimum conditions was significantly different from the models. CCD models of RSM are well-optimized the condition of the curcuminoids extraction from *C. zedoaria*. However, the extraction process itself was not well-modeled by the CCD.

4. Conclusion

Extraction time (h), solvent to feed (S/F) ratio, and agitation speed (rpm) of curcuminoids extraction from *Curcuma zedoaria* optimized by means central composite design (CCD) with response surface methodology. The optimum condition for CCCA-H_2O (1:1:18) were S/F = 0.721 (mL/mg), 20.85 h extraction time and 247.45 rpm. While S/F = 0.673 (mL/mg), 21.26 h extraction time and 245.41 rpm for CCMA-H_2O (1:1:18). However, the verification experiments at optimum condition yielded 12.953±0.135 and 15.239±0.121 mg/g, respectively, for CCCA-H_2O (1:1:18) and CCMA-H_2O (1:1:18), of curcuminoids. CCD models of RSM can well optimize the condition of the curcuminoids extraction from *C. zedoaria* (R^2 = 0.885-0.944). Meanwhile, it does not well model the extraction of curcuminoids from *C. zedoaria*.

References

[1] Lopes M S, de Andrade Sena C F, Silva B L, de Souza C M, Ramos J P, Cassali G D, de Souza-Fagundes E M, Alves R J, de Oliveira M C and de Oliveira R B 2015 *Anticancer Agents Med. Chem.* **15** 206

[2] Bar-Sela G, Epelbaum R and Schaffer M 2010 *Curr. Med. Chem.* **17** 190

[3] Patra D and Sleem F 2013 *Anal. Chim. Acta* **795** 60

[4] De Maria D P and Maugeri Z 2011 *Curr. Opinion in Chem. Biol.* **2** 220

[5] Gorke J, Srienc F and Kazlauskas R 2010 *Biotechnol. Biopro. Eng.* **15** 40

[6] Choi Y H, van Spronsen J, Dai Y, Verberne M, Hollmann F, Arends I W C E, Witkamp G J and Verpoorte R 2011 *Plant Physiol.* **156** 1701

[7] Dai Y, van Spronsen J, Witksamp G J, Verpoorte R and Choi Y H 2013 *Anal. Chim. Acta* **766** 61

[8] Dai Y, Verpoote R and Choi Y H 2014 *Food Chem.* **159** 116

[9] Zullaikah S, Rachmaniah O, Utomo A T, Niawati H and Ju Y H 2018 *Green Separation of Bioactive Natural Products using Liquefied Mixture of Solids (Green Chemistry)* (InTechOpen) Chapter 2 p 27

[10] Rachmaniah O, Fazriyah L J, Seftiyani N H and Rachimoellah M 2018 *MATEC Web Conf.* **156** 01011

[11] Ming L, Ngadi M O and Ma Y 2014 *Food Chem.* **165** 29

[12] Popuri A K and Pagala B 2013 *Int. J Inn. Res. Studies* **2** 5

[13] Azmir J, Zaidul I S M, Rahman M M, Sharif K M, Mohamed A, Sahena F, Jahurul M H A, Ghafoor K, Norulaini N A N and Omar A K M 2013 *J Food Eng.* **117** 426

[14] Fernandez P T, Parjikolaei B R, Lari H N, Casas L, Mantell C and de la Ossa E J M 2016 *Chem.*

Eng. J **299** 420
[15] Geankoplis C J 2003 *Transport Processes and Unit Operations 2nd Edition* (London: Allyn and Bacon, Inc) p 433
[16] Mulia K, Krisanti E, Terahadi F and Putri S 2015 *Int. J. of Technol.* **7** 1211

IOP Conf. Series: Materials Science and Engineering **823** (2020) 012012 doi:10.1088/1757-899X/823/1/012012

Surface cover method to reduce evaporation rate of crude oil

Veronico[1], Livia Kristin[1], Yansen Hartanto[1], Antonius Indarto[2]

[1]Department of Chemical Engineering, Universitas Katolik Parahyangan, Bandung, Indonesia
[2]Department of Chemical Engineering, Institut Teknologi Bandung, ITB Campus, Indonesia

E-mail: antonius.indarto@che.itb.ac.id

Abstract. One of the problems that commonly found in petroleum industry is related to the evaporation of crude oil during storage. Hence the purpose of this study is to find lower-cost method to reduce crude oil evaporation. The method has the same principle as floating roof tank, but instead of surface covered by metal, low-density chemical substances, i.e., polyurethane foam and ethylene vinyl acetate foam (EVA mat), was used to cover the surface of crude oil. In comparison with EVA mat, polyurethane foam gave lower percentage of reduced crude oil evaporation. Increasing the thickness of the additive to 3 cm was found to be the best parameter to reduce the evaporation of crude oil. At temperatures of 40, 45, and 50°C, 3 cm thickness of polyurethane foam was able to reduce the evaporation of crude oil by 81.6 - 91.7%, 75.9 - 87.3%, and 74.2% - 86.6%, respectively. At temperatures of 40, 45, and 50°C, 3 cm thickness of EVA mat can reduce the evaporation of crude oil by 68.6 - 78.1%, 56.8 - 60.9%, and 42.4 - 49.6%, respectively.

1. Introduction

Crude oil is a non-renewable liquid fossil fuel that consists of volatiles hydrocarbons. The occurrence of the evaporation of volatile compounds of crude oil can cause economic losses. In Russia, the annual loss of evaporation from crude oil is estimated to be 500 - 600 thousand tons. In addition, the amount of evaporation that occurs in crude oil storage tanks on Khark Island, Iran, reaches 13,200 barrels per year. The amount of crude oil evaporated will be relatively much greater in countries closer to the equator like Indonesia [1]. Crude oil production in Indonesia reaches 777,000 barrels/day. By assuming 1% crude oil evaporation, the total economic loss reaches $488,267/day.

In addition to losses to the economy, crude oil evaporation can also have a negative impact on the environment. Crude oil can cause air pollution due to light hydrocarbons such as benzene, xylene, toluene and ethylbenzene [2-4]. Therefore, efforts to optimize the usage of crude oil are required. One of the apparent efforts is to reduce the evaporation of crude oil. Evaporation of crude oil mostly occurs in crude oil storage tank [1]. Inside the storage tanks, evaporation of crude oil occurs in the empty space (head space) above the surface of crude oil. At any time, the vapor form of crude oil will be released out of the tank through a breather valve that will open whenever crude oil is fed into or out of the storage tank causing the change in pressure. A number of approaches have either been applied or considered by experts in the attempt to reduced evaporation losses. One of them is to install a floating roof tank. However, this method is considered expensive because it requires reconstruction of the tank [5]. Therefore, in this study, a simple method that does not require tank reconstruction is proposed. Floating covers or mat or spheres on the surface of the crude oil could be suitable for this case. The aim of this study is to find the best additive to reduce crude oil evaporation and to study the effect of additives thickness, temperature and types of crude oil.

2. Experimental Setup

Crude oils used in this experiment were classified into light and medium crude, differentiated by the density. Crude oil A and C that have density of 0.812 g/mL and 0.793 g/mL were categorized as light crude while crude oil B (density of 0.892 g/mL) was medium crude. During the experiment runs, 50 mL of crude was put to the glass. To cover the crude surface, foam additives used in this experiment were potassium oleate, potassium palmitate, sodium oleate, sodium palmitate, polyurethane foam, and ethylene vinyl acetate (EVA) mat. The total duration of evaporation experiment was 180 minutes with weight interval measurement of 10 minutes for the first 30 minutes, 15 minutes for the next 60 minutes, and 30 minutes for the last 90 minutes. The thickness of the covered surface was varied from 1 cm to 3 cm. The temperature of crude oil evaporation was maintained at 40, 45, and 50°C using a thermostat. Data analysis was conducted by calculating the differences in weights of crude oil before and after the experiment which indicates the percentage of evaporation losses.

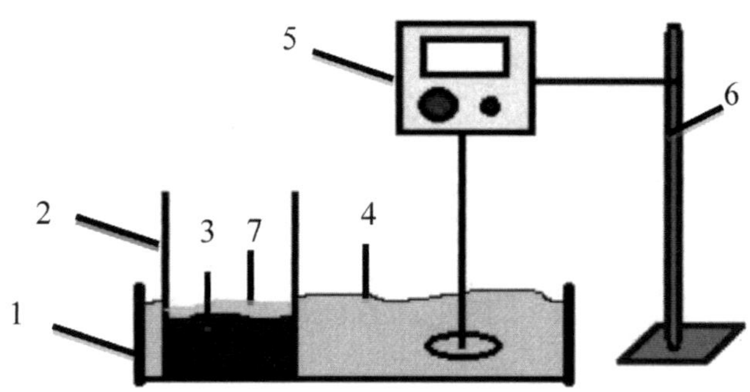

Components:
1. Water container
2. Beaker glass
3. Crude Oil
4. Water
5. Thermostat
6. Statif and clamp
7. Additives

Figure 1. Crude oil evaporation apparatus.

3. Result and Discussions

3.1. Effect of different types of crude oil and temperature on the evaporation of crude oil

Figure 2 shows that at higher density of crude oil, the mass of the evaporated oil was smaller. This occurs because higher crude oil density indicates that the crude oil consists mainly of long-chain hydrocarbons which have stronger intermolecular forces in comparison with short-chain hydrocarbons. Therefore, in order to evaporate long-chain hydrocarbon, greater kinetic energy is needed to overcome the intermolecular forces compared to short-chain hydrocarbons. In conclusion, long-chain hydrocarbons tend to be more difficult to evaporate [6].

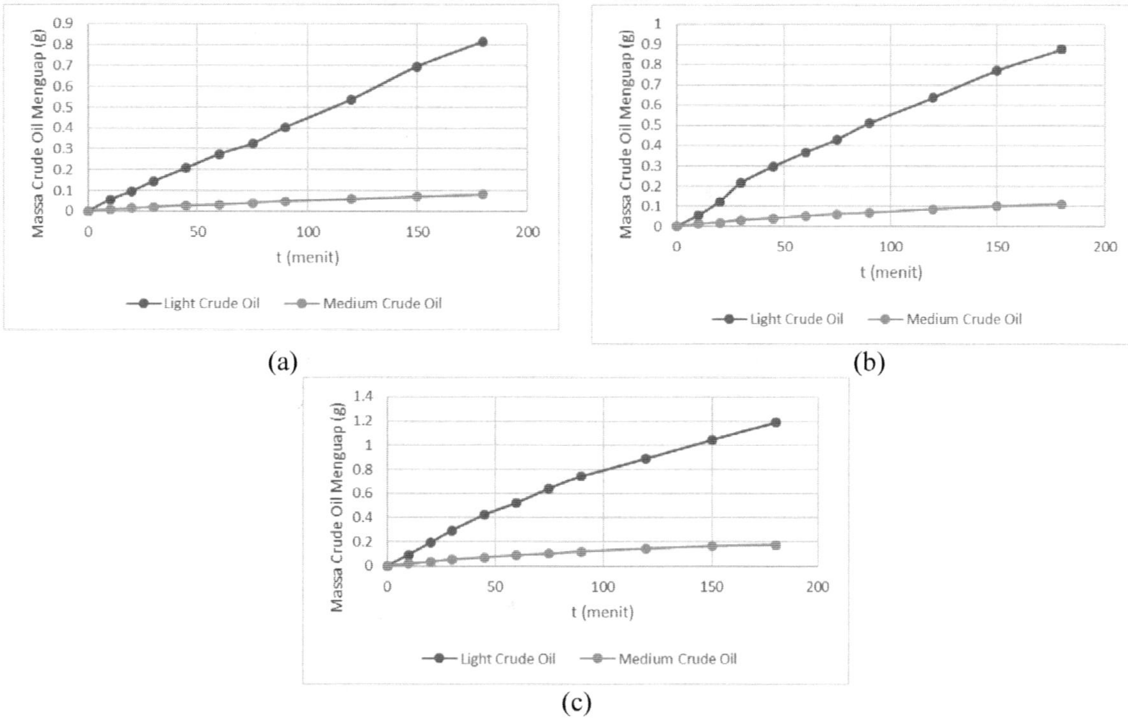

Figure 2. Evaporation profile of crude oil at (a) 40°C (b) 45°C; and (c) 50°C.

3.2. Effect of different types of additives on the evaporation of crude oil

In this experiment, several types of additives to cover the oil surface were used, *i.e.*, sodium palmitate foam, sodium oleate foam, potassium palmitate foam, potassium oleate foam, polyurethane foam, and EVA mats. All additives shall have a lower density than crude oil so that additive could stay above the oil surface.

Sodium palmitate, sodium oleate, potassium palmitate, and potassium oleate was dissolved in a certain amount of water to obtain solution with concentration of 7%-w/w. The solution is then shaken in a sprayer to produce foam. However, all types of foam have poor stability because the lifetime of the foam was too short to cover the surface of crude oil. At 40°C, potassium oleate foam, sodium oleate foam, potassium palmitate foam, and sodium palmitate foam was stable only for 4 minutes 12 seconds, 5 minutes, 13 minutes 45 seconds, and 5 minutes 33 seconds respectively on the surface of the oil (see Figure 3). The foam was easily rupture resulting from the vapor pressure of crude oil that put pressure on the lamellar layer of the foam causing it to be thinned and eventually ruptured [7]. Other than that, foam stability was influenced by the viscosity of each foam-forming solution. The higher viscosity of the solution causes the solution in the foam lamellar layer to be more resistant to flow towards the plateau border. According to Rivai et al. [8], a solution of potassium palmitate with a concentration of 7% has a greater viscosity (*ca.* 1.87 cP) when compared with a solution of potassium oleate, sodium oleate, and sodium palmitate.

Figure 3. Rupturing process of potassium oleate foam.

Figure 4. Stable EVA mat (left) and polyurethane foam (right).

Polyurethane foam sprayed on the surface of crude oil will float, cover, and isolate crude oil surface from the surrounding air, which will reduce crude oil evaporation. One of the reasons why polyurethane foam is capable to reduce the evaporation of crude oil is related to the process of forming polyurethane foams, where silicon surfactant contained in the foam plays its role to reduce the surface tension and causes the foam formed after the blow reaction to stabilize. In addition, the urethane polymer formed from gelation reaction has a solid phase; eventually polyurethane foam is a two-phase material consist of solids urethane polymer and CO_2 gas [9]. Furthermore, another reason is related to the low density of polyurethane foam and their resistance to most hydrocarbons including crude oil. While EVA mat additives in solid-state also have high resistance to chemicals so there will be no reaction between EVA mat and crude oil.

Table 1. Percentage of reduced crude oil evaporation using polyurethane foam

Type of Crude Oil	Additive thickness (cm)	Temperature (°C)	Evaporation Reduced (%)
Light	1	40	45.95
		45	42.85
		50	38.02

Type of Crude Oil	Additive thickness (cm)	Temperature (°C)	Evaporation Reduced (%)
		40	78.36
	2	45	70.32
		50	66.78
		40	91.01
	3	45	87.31
		50	85.96
		40	46.35
	1	45	39.94
		50	34.17
		40	70.44
Medium	2	45	65.75
		50	56.13
		40	81.63
	3	45	75.94
		50	74.27

Table 2. Percentage of reduced crude oil evaporation using EVA Mat

Type of Crude Oil	Additive Thickness (cm)	Temperature (°C)	Evaporation Reduced (%)
		40	43.46
	1	45	33.33
		50	28.39
		40	60.29
Light	2	45	39.90
		50	33.52
		40	73.42
	3	45	60.64
		50	45.02
		40	43.07
	1	45	36.73
		50	33.83
		40	58.03
Medium	2	45	42.33
		50	43.41
		40	77.01
	3	45	56.84
		50	49.69

Table 1 and Table 2 show the tendency that polyurethane foam additive is better at reducing evaporation of each type of crude oil. This is because the density of polyurethane foam is greater than that of EVA mat. The greater the density of the additives indicate the closer the distance between the polyurethane foam molecules so that the permeability decreases [10]. Two other main factors affecting polymer permeability to gas is the solubility and crystallinity of polymers. The solubility of gas in polymers is affected by the polarity of the polymer and the gas. Same polarity of polymer and gas results in high permeability of gas in the polymer. In this case, polyurethane foam consists of mainly non-polar molecules, whereas EVA mat consists of mainly polar molecules. From the solubility aspect, EVA mat is supposed to be better at reducing crude oil evaporation. However, from the experiment result, it is found that polyurethane foam is better in reducing crude oil evaporation. This is due to the crystallinity of the additives. Crystallinity refers to the degree of structural order in a solid. In polymers, a copolymer is considered more amorphous than a homopolymer; therefore, copolymer has lower crystallinity compared to homopolymer [11]. EVA mat is a copolymer of ethylene and vinyl acetate, whereas polyurethane foam is a homopolymer. Hence polyurethane has higher crystallinity when compared to EVA mat and better in reducing crude oil evaporation.

3.3. Effect of different additives thickness on the evaporation of crude oil

Table 1 and Table 2 also show that at thicker additive covers the amount of each type of crude oil the evaporated gets smaller. The tendency of these data is in accordance with the theory that the thicker the additives means the greater the pressure exerted to the surface of crude oil causing an increase in the resultant pressure acting towards the surface and finally the molecules of crude oil become less easy to evaporate. Polyurethane foam with thickness of 3 cm at 50°C could reduce the evaporation of light crude oil and medium crude oil up to 85.9% and 74.2%, respectively. Whereas, EVA mat with thickness of 3 cm at 50°C could reduce the evaporation of light crude oil and medium crude oil up to 45.1% and 49.6%, respectively.

4. Conclusions

An effort to reduce the crude oil evaporation by covering the liquid surface has been conducted. Some conclusions that could be made from the experiments are compared to EVA mat, polyurethane foams were better at reducing evaporation of crude oil; potassium oleate, potassium palmitate, sodium oleate, and sodium palmitate foam shows an unsatisfactory result in reducing crude oil evaporation due to stability reason; at temperatures of 40, 45, and 50°C, polyurethane foam with a thickness of 3 cm could reduce the evaporation of crude oil by 81.6% - 91.1%, 75.9% - 87.3%, and 74.2% - 85.9% respectively. While EVA mat with same thickness could reduce the evaporation of crude oil by 68.6% - 78.1%, 56.8% - 60.9%, and 42.4% - 49.6% respectively.

References

[1] Farhan M M, Al-Jumialy M M, Al-Muhammadi A D and Ismail A S 2017 Development of a new method for reducing the loss of light hydrocarbons at breather valve of oil tanks *Energy Procedia* **141** 471-78

[2] Indarto A 2012 Heterogeneous reaction of HONO formation from NO_2 and HNO_3: A review *Research on Chemical Intermediates* **38**(3-5) 1029-41

[3] Indarto A 2006 Treatment of CCl_4 and $CHCl_3$ emission in a gliding-arc plasma *Plasma Devices and Operations* **14**(1) 1-14

[4] Indarto A 2012 Decomposition of dichlorobenzene in a dielectric barrier discharge *Environmental Technology* **33**(6) 663-6

[5] Magaril E 2014 Reducing gasoline loss from evaporation by the introduction of a surface-active fuel additives *WIT Press* 233

[6] Chinenyeze M A and Ekene U R 2015 Physical and chemical properties of crude oils and their geologic significances *International Journal of Science and Research* **6**(6) 1514-21

[7] Exerowa D and Kruglyakov P M 1998 *Foam and Foam Films: Theory, Experiment, Application* (Netherlands: Elsevier)

[8] Rivai M, Hambali E and Suryani A 2017 Synthesis of palm oil fatty acid as foaming agent for firefighting application *IOP Conference Series Earth and Environmental Sciences* **65**(1) 012047

[9] Zhao Y 2015 *Modelling and Experimental Study of Polyurethane Foaming Reactions* (Columbia: University of Missouri)

[10] Ostrogorsky A G, Glicksman L R and Reitz D W 1986 Aging of polyurethane foams *International Journal of Heat and Mass Transfer* **29** 1169-76

[11] Kamath P M and Wakefield R W 1965 Crystallinity of ethylene-vinyl acetate copolymers *Journal of Applied Polymer Science* **9** 3153-60

IOP Conf. Series: Materials Science and Engineering **823** (2020) 012013 doi:10.1088/1757-899X/823/1/012013

Quality of eggs from fermented dragon fruit (*Hylocereus polyrhizus*) flour fed Isa Brown

Gusti Ayu Mayani Kristina Dewi[1], Made Wirapartha[1], I K. A. Wiyana[1], and N.P. M. Suartiningsih[1]

[1]Faculty of Animal Science, Udayana University, Denpasar, Bali, Indonesia

E-mail: kristinadewi@unud.ac.id

Abstract. The objective of this study was to study the quality of eggs coming from Isa Brown chickens nourished with fermented dragon fruit peel (*Hylozereus polyrhizus*) dietary for 4 weeks. The experimental design used was Completely Randomized Design (CRD) with 3 treatments for 5 replications in which each consisted of 10 chickens; so that the total number of chickens was 150 heads aged 85 weeks. The given treatments were R0: feed without fermented dragon fruit peel flour in their diet, R1: feed with 5% fermented dragon fruit peel flour in their diet, and R2: feed with fermented dragon fruit peel flour + 1% Calcium in their diet. Variable observed: egg production, egg weight, exterior and interior eggshell weight, egg thickness, HU. The results showed that treatment of R0, R1, and R2 are not significantly different (P>0.05) for yolk colors, pH, Index but % egg production, egg weight, HU, egg thickness R1 and R2 is significantly different (P>0.05) than R0. Conclude this research that quality of egg Isa Brown gave ration fermentation four skin dragon fruit (*Hylocereus polyrhizus*) 5% (R1) and with fermentation dragon fruit peel flour + 1% Calcium (R2) increase the egg production, egg weight, HU, egg thickness.

1. Introduction

Isa Brown chicken is one of meat and egg-producing poultries that is increasingly reared by Balinese community. Balance diet containing either macro and micronutrients is one factor that influences and has important role in laying period. Many factors interact affecting to optimal production of However, the maintenance of Issa Brown chickens faced with a variety of problems such as the increasing feed prices are enough sharply because the feed is a primary need at the cost of approximately 60-70%. The high price of feed indirectly requires that farmers are looking for alternative feed ingredients so it can lower the feed costs and maximize revenues.

According to Dewi et al. [1] and Ahmad [2], *Saccharomyces cerevisiae* yeast can increase fibrous fiber digestibility and can act as a probiotic in poultry. Another benefit of fermentation products is to suppress the enzyme activity of 3-hydroxy-3-methylglutarylCo-A reductase that serves to synthesize cholesterol in the liver [3] and can decrease the amount of broiler fat [4]. Application of feed technology is absolute must be applied in the optimization of waste utilization. Application of supplementation technology utilizing superior *Saccharomyces cerevisiae* origin of yeast is very potential developed.

According to Mustika et al. [5], dragon fruit peel is an agricultural waste that has not been widely used by the community, especially in Indonesia. According Citramukti [6] part of dragon fruit, 30-35% is peel and still rarely or even not been fully utilized, although some studies have reported peel dragon fruit contains high antioxidant and contents phenolics in the dragon fruit peel amounted 28.16 mg/100 g, in addition to having antioxidant also contain anthocyanins [7].

Research on dragon fruit peel for livestock feed is still rarely done according to Mustika et al. [5] dragon fruit peel can be given up to the level of 1% and Rosa et al. [8] can be given up to the level of 4%, without have negative effects on the body of livestock. Dewi et al. [1] used 5, 7% dragon fruit peel fermentation by *Saccharomyces cerevisiae* to increase performance broiler chickens at 5 weeks and Dewi et al. 9] gave significantly to increased productivity Lohman brown 23 aged. But shell quality a little bit thinner than commercial ration. While for From the description above, the researcher using dragon fruit peel meal without and fermented plush calcium as a feed ingredient in diets for egg quality for Issa Brown.

2. Materials and Methods

2.1. Animal, ration and feeding treatment

This research conducted over weeks, located in Teaching Farm, Bukit Campus Faculty of Animal Science, Udayana University. A total of 85 weeks with average body 1710 ± 15 g were kept in individual cages of 30 x 30 x 35 cm.

2.2. Diets

Diets used in this research were independently prepared by recommendation Scott et al. [10] which consists of yellow corn, fish meal, soybean meal, rice bran, dragon fruit peel meal, dragon fruit peel meal fermented, coconut oil, premix and $CaCO_3$. Diets given is iso energy (2900 Kcal/kg) and iso protein (20%), and commercial ration.

2.3. Instrument

Instruments used in this research were a diet and drinking water, torch lighting cage, machine grinding feed, knife, bowl, spoons stirrer, scissors, paper labels, markers, plastic bags, oven, stove, pans, trays, thermometer, wood, bamboo, wire, sprayer and digital scales.

2.4. Research methods

In this research, there are two stages making process meal dragon fruit peel, first making of dragon fruit peel meal is fresh dragon fruit peel chopped small, then dried and ground up into flour. Second process namely the making of dragon fruit peel meal fermented with *Saccharomyces Sp.* In the process of fermentation, the solution is ready for use. Fermentation process dragon fruit peel chopped small, be dried, inserted in a plastic, then moistened with solution fermentation, closed tightly (3-5 days), after it is dried, ground into flour and ready for use.

2.5. Research design

The design used was Completely Randomized Design (CRD) with 3 treatments, 5 replications in which each replication consisted of 10 chickens so that the total chicken used was 150 heads 85 weak aged. The treatments given were R0: ration without fermentation dragon fruit peel flour, R1: ration with 5% fermentation dragon fruit peel flour, and R2: ration with fermentation dragon fruit peel flour + 1% Calcium.

2.6. Observed variables

Variables observed are egg production, egg weight, exterior and interior eggshell weight, egg thickness, HU.

3. Result and Discussion

The results showed that treatment of R0, R1, and R2 are not significantly different (P>0.05) for yolk colors, pH, Index but % egg production, egg weight, HU, egg thickness R1 and R2 is significantly different (P>0.05) than R0.

According to McDonald et al. [11], energy and protein ration can be used the productivity. According to Weiss and Hogan [12], the material having the antioxidant content of livestock can reduce the effects of free radicals such as increasing feed consumption. According to Mustika et al. [5], it is because free radicals can cause oxidative stress in livestock resulting in lower feed consumption. Oxidative stress is a state of imbalance between the amount of free radicals and antioxidants in the body that can trigger the occurrence of cell damage and lowered immune system [7].

Table 1. The effect of treatment for egg quality of Isa Brown

Variable	Treatment [1)]			
	R0	R1	R2	SEM[3)]
Egg Production (%)	68,00[b]	70,00[a]	71 ,00[a]	3,6
Egg Weight (g)	60[b]	62[a]	63[a]	0,9
Egg Index	81[a]	82[a]	81[a]	0,60
Egg shell weight (g)	6,1[a]	6,3[a]	6,4[a]	0,102
Egg shell thickness (mm)	0,379[b]	0,389[a]	0,399[a]	0.06
Egg yolk colour	8,7[a]	8,9[a]	9[a]	0.02
HU	93[a]	96[a]	98[a]	1.2

Note: 1) R0: ration without fermentation dragon fruit peel flour, R1: ration with 5% fermentation dragon fruit peel flour, and R2: ration with fermentation dragon fruit peel flour + 1% Calcium 2) Means with different Superskrip in the same row differ significantly (P<0.05), 3) Standard Error of Mean

One of the candidates is dragon fruit peel which is not generally utilized. The dragon fruit peel is potential as a natural colorant [13] an antibacterial against nine bacteria of food pathogen [14] and natural antioxidant [15]. The average egg production for 85 weeks is shown in Table 1. Diet without flour peel dragon fruit produced significantly lower (P<0,05) egg production (HD) as compared to both treatments R1 and R2. This is presumably due to the same nutrients supply, such as protein, energy, and fat, derived from a similar amount of feed consumption. The amount of energy, protein, and fat consumptions can affect egg production. Energy and nutrient needs for the formation of eggs are obtained from daily feed intake [16]. High egg production is usually positively correlated with increasing egg mass because egg mass is the result of egg production multiplied by egg weight [17] and further [18] explained that the egg mass is influenced by egg weight and egg production.

The results of studies on eggshells weight showed non-significant different (P>0,05) between R0, R1, and R2. But in eggshell thickness showed a significant difference (P<0,05) and treatment R2 diet resulted in higher (P<0,05) eggshell thickness compared to R1 and R0. This is due to the reduced eggshell thickness and eggshell weight with the increased egg size. The average of eggshell thickness was 0,36 – 0,389 Mm. The eggshell thickness had the same relative value because the treatment diets have relatively similar Ca dan P contents. The quality of eggshell thickness can be influenced by many factors including mineral, calcium, magnesium and phosphorus are the major inorganic constituents of the eggshell [19]. According to Kebreab et al. [20] the higher calcium intake can improve the quality of the eggshell. According to Mine and D'Siva [21], standard yolk composition of bird's egg ranging from 32% -35%, egg whites 52%- 58% and eggshell 9% -14%

The results showed that the Haugh unit ranged between 93-98, with the R2 treatment showing higher results than the other treatments. Haugh unit value was not different because each egg was stored in the same place and time. Factors affecting the value of Haugh units were time and place of egg storage, age, the strain of livestock, nutrients, disease, and supplementation (Vit C or E) [22]. Further results by Omara [23] reported that the dietary combination of protein, methionine and choline did not significantly affect the Haugh unit.

4. Conclusion

This research has justified that the addition of fermented dragon (*Hylocereus polyrhizus*) fruit peel flour in their diet has successfully increased the egg production (egg weight, hu, and thickness) of the Isa Brown chickens.

References

[1] Dewi G A M K, Wiyana I K A and Wirapartha I M 2018 Optimalisasi peningkatan produksi ternak unggas dengan pemanfaatan limbah kulit buah naga (*Hylocereus sp*) terfermentasi *Laporan Kemajuan*

[2] Ahmad R Z 2005 Pemanfaatan kamir *Saccharomyces cerevisiae* untuk ternak *Wartazoa* **15**(1) 45-55

[3] Tanaka K, Youn B S, Santoso U, Otan S and Sakaida M 1992 Effect of fermented feed products from Chub Mackerel extract on growth and carcass composition, hepatic lipogenesis ando n various lipid praction in the liver and thigh muscle of broiler *Anim. Sci. Technol* **63** 32-37

[4] Katarin P P, Sinurat A P, Sainudin D, Purwadarta T and Kompiang I P 1999 Bungkil inti sawit dan produk fermentasinya sebagai pakan ayam pedaging *Jurnal Ilmu Ternak dan Veteriner* **4**(2) 107-12

[5] Mustika A A, Andriyanto, Wientarsih and Margarita M L 2014 The ability of various avian egg albumen as a chelator to overcome lead heavy metals toxicity *Veteriner Journal* **15**(3) 406-10

[6] Citramukti I 2008 Ekstraksi dan Uji Kualitas Pigmen Antosianin pada Kulit Buah Naga Merah (*Hylocereuscostaricensis*), Kajian Masa Simpan Buah dan Penggunaan Jenis Pelarut *Skripsi* (Malang: Universitas Muhammadiyah Malang)

[7] Nurliyana R, Syed Zahir I, Suleiman K M, Aisyah M R and Kamarul Rahim K 2010 Antioxidant study of pulps and peels of dragon fruit: A comparative study *International Food Research Journal* **17** 307-75

[8] Rosa R A, Malik M A, Prakoso I G, Djati R W and Purnamawati Y 2013 Suplemen Pakan Berbasis Limbah Kulit Buah Naga (*Hylocereus undatus*) Guna Menghasilkan Telur Puyuh yang Kaya Vitamin A dan Rendah Kolesterol (Bogor: Institut Pertanian Bogor)

[9] Dewi G A M K, Nuriyasa M and Wijana I W 2017 Effect of diet containing dragon fruit peel meal fermentation for productivity of kampung chickens *The 2nd International Conference on Animal Nutrition and Environment (ANI-NUE)* ISBN 978-616-438-084-4 Vol. II (Khon Kaen, Thailand)

[10] Scott M L, Nesheim M C and Young R J 1982 *Nutrition of the Chickens* 2nd edition (New York: M L Scott and Associates)

[11] Mc Donald P, Edwards R A, Greenhalgh J F D and Morgan C A 2002 *Animal Nutrition* 5th edition (New York: Longman Scientific and Technical)

[12] Weiss W P and Hogan J S 2007 Effects of dietary vitamin c on neutrophil function and responses to intramammary infusion of lipopolysaccharide in periparturient dairy cows *Journal of Dairy Science* **90**(2) 731-39

[13] Harivaindaran K, Rebecca O and Chandrana S 2008 Study of optimal temperature, pH and stability of dragon fruit (*Hylocereus polyrhizus*) *Pakistan J. Biol. Sci.* **11** 22592263 doi: 10.3923pjbs.2008.2259.2263

[14] Nurmahani M M, Osman A, Hamid A A, Ghazali F M and Dek M P 2012 Antibacterial property of *Hylocereus polyrhizus* and *Hylocereus undatus* peel extracts *Int. Food Res. J.* **17** 469-76

[15] Luo H, Cai Y, Peng Z, Liu T and Yang S 2014 Chemical composition and in vitro evaluation of the cytotoxinc and antioxidant activities of supercritical carbon dioxide extracts of pitaya (dragon fruit) peel *Chem. Central J.* **8** 1-7 doi: 10.1186/1752-153x-8-1

[16] Ghazvinian K, Irani M, Jamshidi R, Mirzaei-Aghsaghali A, Siadati S A and Javaheri-Vaighan A 2011 The effect of energy to protein ratio on production performance and characteristics of Japanese quail eggs *Annals. Bio. Res.* **2** 122-8

[17] Sh R, El.-Ghar A, Debes A, Ghanem H and Ali O 2013 Prediction of breeding values and selection for egg mass in a closed flock of white leghorn *Int. J. livest. Prod.* **1** 16-25

[18] Vercese F, Garcia E A, Sartori J R, Silva A de P, Faitarone A B G, Berto D A, Molino A de B and Pelicia K 2012 Performance and egg quality of Japanese quails submitted to cyclic heat stress *Braz. J. Poult. Sci.* **14** 37-41 doi: 10.1590/s1516-635x2012000100007

[19] King'ori A M 2011 A review of the uses of poultry eggshells and shell membranes *Int. J. Poult. Sci.* **10** 908-12 doi: 10.3923/ijps.2011.908.912

[20] Kebreab E, France J, Kwakkel R P, Leeson S, Kuhi H D and Dijkstra J 2009 Development and evaluation of a dynamic model of calcium and phosphorus flows in layer *Poult. Sci.* **88** 680-68 doi: 10.3382/ps.2008-00157

[21] Mine Y and D'Siva I 2008 *Egg Biosscience and Biotechnology* (New Jersey, US : John Wiley and Sons) pp 141-84

[22] Ahmadi F and Rahimi F 2011 Factors affecting quality and quantity of egg production in laying hens: A review *World Applied Sciences Journal* **12** 372-84

[23] Omara I I 2012 Protein-sparing effect of DL-Methionine and choline supplementation of low-protein diets on productivity in laying hens *Egyptian. J. Anim. Prod.* **49** 219-31

Application of rhamnolipid biosurfactant for bio-detergent formulation

Q Helmy[1,2], S Gustiani[3] and AT Mustikawati[3]

[1]Water and Wastewater Engineering Research Group, Faculty of Civil and Environmental Engineering, Institute of Technology Bandung
[2]Biotechnology and Bioscience Research Center, Institute of Technology Bandung
[3]Center for Textile, Indonesian Ministry of Industry, Ahmad yani 390 Bandung, West Java, Indonesia

E-mail: helmy@tl.itb.ac.id

Abstract. Biosurfactants, also known as biological surface-active agents, are a group of surface-active agents, which are produced by a variety of microorganisms. Biosurfactants possess the characteristic property of lowering surface tension, increase the solubility of poorly soluble compound, low toxicity, non-allergenic, and biodegradable. Growing public awareness about the environmentally friendly health care and associated product has stimulated the search for eco-friendly compounds in laundry detergents. In this research, the application of biosurfactants in the formulation of a washing detergent was investigated. Variation of biosurfactants and its mixture with sodium tripolyphosphate as a builder and sodium sulfate as filler was applied to wash cotton fabric which was contaminated with known amounts of stain (chocolate milk). The effects of various bio-detergent formulations, including the ratio of biosurfactants and builder, was examined. The formulation presented in this study was also compared with some standard detergent for the stains removal efficiency. The results showed that the rhamnolipid biosurfactants have a promising as a substitute for its synthetic counterpart. Based on the colorfastness to wash, color strength (K/S), and color difference (ΔE) value, rhamnolipid based bio-detergent have similar washing effectiveness compared to the synthetic detergent. Research result also found that Na-bicarbonate and Na-citrate have the potential as a substitute for phosphate-based detergent builder.

1. Introduction

Nowadays, laundry detergent is becoming more popular due to the increasing demand and use in washing machines, where it provides rapid stain removal, softness, resiliency to fabric, can be dispersed in water, and is delicate for our skin. There are many different brands or types of detergents, and most of them claim some special qualities regarding their products. Concerns over the persistence of detergent chemicals in the environment and possible contamination of groundwater, other surface water sources also their subsequent health-related issues have raised speculation over biodegradability. Detergents on the market are generally products with active ingredients in the form of LAS surfactants (Linear Alkylbenzene Sulfonate) that derived exclusively from petroleum derivatives: benzene and linear paraffin. Problems arising from the use of LAS surfactants are biodegradability problem. During its use, detergent is difficult to be degraded by bacteria in water, so detergent waste remains in the water; therefore the accumulation of the amount of detergent in water occurs. The accumulation of detergent in water can be a source of pollution in the water stream [1, 2].

The shift to an era of more environmentally friendly products has encouraged the industry to create environmentally friendly detergents. The solution to this problem can be overcome by using surfactants made from oleochemicals and other bio-based products. One example of a surfactant made from oleochemicals is Methyl Ester Sulfonate (MES) and Biosurfactants that produced from microbial origin. Biosurfactants are a group of surface-active agents, which are produced by a variety of microorganisms (bacteria, fungi). These materials are safer than their synthetic counterparts regarding environmental concerns [3]. Rhamnolipid is a biosurfactant produced by microorganisms. This compound has a hydrophilic group and a hydrophobic group so that rhamnolipid has diverse properties. In addition, rhamnolipid is much more easily degraded in nature (environmentally friendly) compared to synthetic surfactants. One of the advantages of rhamnolipid biosurfactants is that they are able to reduce surface tension and are able to form microemulsions between two different material phases. Besides that, the potential utilization of rhamnolipid is very broad i.e. as an emulsifier [4]; cosmetic mix ingredients [5]; controlling crop pests [6], and is used for bioremediation of land contaminated with hydrocarbon compounds [7].

This study aims to answer environmental problems caused by detergents, with detergent formulations using Rhamnolipid Biosurfactants which is more easily biodegraded, expected to create detergents that are environmentally friendly. In this research work, rhamnolipids biosurfactants were applied in the formulation of laundry detergent as a substitute for its synthetic counterpart (LAS) and their effectiveness in removing chocolate milk stains from cotton fabrics was examined.

2. Materials and Methods

2.1. Materials

Rhamnolipid biosurfactants with a purity greater than 90% were purchased from Pioneer Biotech, China. Sodium tripolyphosphate (STPP), Sodium bicarbonate, Sodium citrate, and Sodium sulfate obtained from Merck used as builder and filler, respectively. Sodium carboxymethyl cellulose from Bratachem and Lipase (64 LU/g) from DuPont.

2.2. Surface activity properties

Surface tension of aqueous solutions of Rhamnolipid was measured with a tensiometer (Fisher Sci. 21) by the duNouy ring method. Tests were performed at room temperature (25 °C) and pH 7.

2.3. *Bio-detergent formulation*

The making of bio-detergent in this study uses rhamnolipid as a substitute for its synthetic counterpart. The builders tested in the formulation were STPP, Na-bicarbonate, and Na-citrate. Na-sulphate as filler, Na- carboxymethyl cellulose as an anti-redeposition agent, and Lipase as enzyme additive. Water hardness before and after the washing test is not analyzed in this study. Composition of the bio-detergent formulations is shown in Table 1. Blank uses only water to wash the fabric, while control in this experiment uses only Rhamnolipid without the addition of other additives in order to know the strength of the Rhamnolipid alone in the washing performance.

2.4. Fabric washing test

Textile fabric washing efficiency was evaluated according to published guidelines [8] using the reference soap standard. Several washing baths containing rhamnolipid biosurfactants were investigated; STPP, Na-citrate, Na-bicarbonate; Na-sulphate; Na-carboxymethyl cellulose; Lipase also their mixtures with and without one or more of the other above ingredients. The white color fabric cotton samples were cut into 4x10 cm rectangle and each of them was stained with 2 ml of chocolate milk and dried overnight at room temperature. The stained fabric samples then treated in a Launder-

Ometer for 20 min at 90 °C with 5:1 v/w liquor to fabric ratio. Afterward, the fabric samples were removed from the washing bath, immersed in 100 ml deionized water at 30 °C for 20 min, removed, and allowed to dry at room temperature. Indicators of washing efficiency were the color fastness, color strength (K/S) and the color difference (ΔE). Colorfastness to wash was measured using greyscale for staining method as per SNI ISO 105-C06:2010 standard.

The stained fabric samples were assessed for the depth of color strength (K/S) and color difference value using a spectrophotometer. The K/S is a function of color depth and calculated by the Kubelka-Munk equation [9, 10].

$$\frac{K}{S} = \frac{(1-R)^2}{2R} \tag{1}$$

Where R is the reflectance at complete opacity, K is the absorption coefficient, and S is the scattering coefficient.

3. Results and Discussion

3.1. Bio-detergent formulation

Detergents are associated with cleaning solid objects, removing unwanted objects from the surface. This cleaning process can be done by various methods, including simple mechanical separation (e.g., rubbing and dipping the cloth into water), separating with solvents (e.g., the addition of organic solvents), and separation by adding water and chemicals such as surfactants. Washing system with detergent consists of objects to be cleaned called substrate, dirt and stains that will be cleaned through a washing process, and liquid bath (liquid containing water and surfactant to clean impurities). The results of washing will depend on the interaction of the elements and the washing conditions used, such as temperature, time, given mechanical energy, and hardness of water used. Detergents have a formula for cleaning dirty substrates under various washing conditions. Some detergents, such as toilet soap, only consist of one component. Some other detergents have more than one component. In general, detergent formulas contained more than one material consist of surfactants, builders, fillers and additives. Complementary materials such as water softeners/chelating agents, anti-redeposition agents, enzymes, bleaching agents, dispersing agents, perfumes, and other minor additives are also included in the commercial formulations [11, 12].

In this study, several washing baths containing surfactant (rhamnolipid biosurfactants); builders (STPP, Na-citrate, Na-bicarbonate); Na-sulphate (filler); Na-carboxymethyl cellulose (anti-redeposition); Lipase (additive) also their mixtures were investigated (Table 1). Knowing the relationship between the increase in phosphorus input into water bodies which has caused the phenomenon of eutrophication of water bodies has further raised public concern over the last few decades. This has led to actions by several countries, especially developed countries such as Germany, the Netherlands, Italy, the United States, and Japan to reduce the phosphorus content, especially those from urban and industrial sources [13]. An attempt to replace the role of STPP as a builder was tried by reducing and replacing it with several compounds, namely Na-bicarbonate and Na-citrate as well as a combination of both.

Table 1. Composition of bio-detergent formulations

Materials	Unit	Blank	Control	Var-1	Var-2	Var-3	Var-4
Rhamnolipid	%	0	20	20	20	20	20
STPP	%	0	0	30	10	0	0
Na-bicarbonate	%	0	0	0	10	15	15
Na-citrate	%	0	0	0	10	15	15
Na-sulphate	%	0	0	48	48	48	46
Na-carboxymethyl cellulose	%	0	0	2	2	2	2
Lipase	%	0	0	0	0	0	2
Total				100	100	100	100

Biosurfactants, derived from microbial extracellular are comparable with its synthetic counterpart in terms of reducing the surface tension of liquids. These amphipathic molecules have a hydrophilic head which is normally a simple sugar or peptide, and a hydrophobic tail which is normally a lipid. Rhamnolipids are a group of biosurfactants which are generally produced by *Pseudomonas aeruginosa* strains of bacteria. The critical micelles concentration (CMC) value of Rhamnolipids biosurfactants used in this study are shown in Table 2. CMC value was determined by measuring the surface tension at different concentrations [14, 15]. Below the CMC, the surface tension decreases with increasing biosurfactants concentration as the number of the active agent at the interface increases. While above the CMC, the surface tension of the solution is constant because the interfacial biosurfactant concentration does not change anymore. Compared with other surfactants, rhamnolipids have a much lower CMC value of 20 mg/l, as shown in Table 3.

Table 2. Surface tension value of various concentration of rhamnolipid biosurfactants and CMC value measurement (inset graph)

Rh (mg/l)	ST (mN/m)
0	71.42 ±0.52
2	49.85 ±0.21
4	46.20 ±0.26
6	43.60 ±0.10
8	40.03 ±0.22
10	38.23 ±0.17
20	35.20 ±0.14
30	34.03 ±0.22
40	33.43 ±0.21
50	33.00 ±0.10
60	32.58 ±0.17
70	31.95 ±0.07
80	31.45 ±0.13
90	30.80 ±0.10
100	30.17 ±0.25
110	29.57 ±0.15
120	29.00 ± 0.10

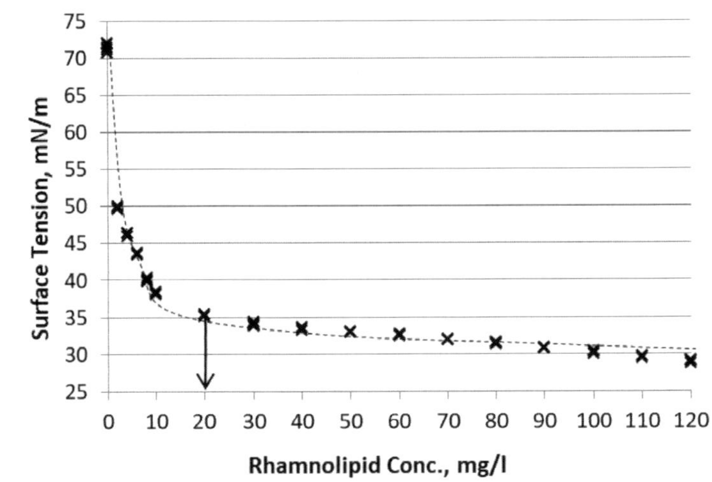

Rh: Rhamnolipid; ST: Surface tension

Table 3. Nature and properties of various surfactants

Surfactant	Nature	CMC value, mg/l	Ref.
Sodium dodecyl sulphate	Anionic	2380	[16]
Tween 20	Non-ionic	140	[16]
Linear alkylbenzene sulfonate	Anionic	120	[17]
Rhamnolipid	Anionic	20	This study

3.2. Washing performance

Detergency is the process of cleaning the surface from undesirable foreign objects by using washing / soaking liquid in the form of a solution containing surfactants. The detergency process occurs through the formation of micelles by surfactants that are able to form globules of impurities through a decrease in interface tension and with the help of electrostatic interactions between charges [12]. The result of the formulated bio-detergent washing test in removing the stain, visually can be seen in Figure 1.

Surfactants or surface-active agents are organic compounds that act as active ingredients in detergent formulations. Surfactants can reduce surface tension thereby allowing particles to stick to the fabric to be washed free from fibers and float or dissolve in water. The presence of active ingredients in detergent formulations is often associated with the ability of its detergency. Detergency is a specific property possessed by surfactants to clean a surface from impurities [12]. However, surface-active compounds cannot completely clean dirt from the surface without the presence of other additional compounds as support, namely builders, anti-redeposition, enzymes, and other additives. The test results in this study confirm this, where the washing results in controls that contain only 20% active ingredient in the form of rhamnolipid biosurfactant are visually not good enough to clean stains (mild stain can be seen visually on the fabric). The use of Na-bicarbonate and Na-citrate is intended to replace the role of STPP as a phosphate-containing builder, where the use of phosphate compounds in detergents began to be limited and avoided due to eutrophication issues [13,18,19]. Washing result indicates that both Na-bicarbonate and Na-citrate are suitable as a builder in replacing STPP. The color difference (ΔE) using greyscale for staining method shows the performance of each bio-detergent formulation used in this study (Table 4).

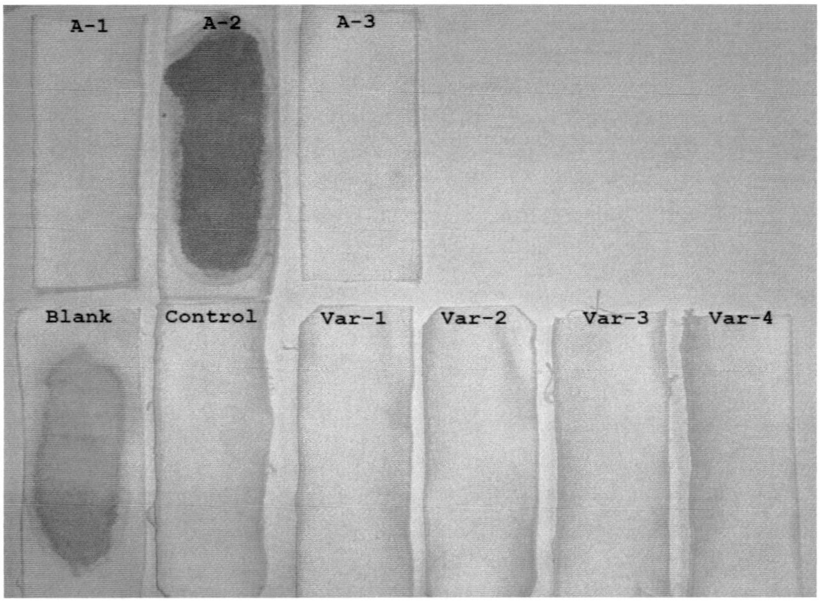

Figure 1. A-1: initial white cotton fabric; A-2: stained fabric with chocolate milk; A-3: washing result of standard detergent. Washing performance of various bio-detergent formulation: Blank (only water); Control; Var-1; Var-2; Var-3; and Var-4.

Table 4. Color fastness to wash, color strenght (K/S), and difference (ΔE) value of cotton fabric

Variation	Color Fastness	(K/S)	(ΔE)	Remark
Initial fabric	1-2	0.068	-	Very good
Stained fabric	5	2.088	32.75	Very bad
Wash with standard detergent	1-2	0.076	1.36	Very good
Blank (wash with water only)	4-5	0.321	12.94	Bad
Control (wash with Rhamnolipid)	4	0.085	1.78	Fairly bad
Var-1	3-4	0.066	0.47	Fairly good
Var-2	3-4	0.066	0.45	Fairly good
Var-3	3-4	0.061	0.49	Fairly good
Var-4	3-4	0.086	1.46	Fairly good

In detergent industry, enzymes are used enormously in terms of quantity and economic value. Lipase enzyme catalyzes the hydrolysis of triglycerides into diglycerides and monoglycerides by releasing fatty acid. Lipase (Triacil glycerol acyl hydrolase) is a group of enzymes that catalyze the hydrolysis of long chains of triglycerides. Lipase is a lipolytic enzyme that dissolves in water and can work in oil-in-water emulsions. This enzyme catalyzes the hydrolysis of fats and oils into glycerol and fatty acids in the presence of water. During hydrolysis, lipase takes the acyl group from the glyceride to form the acyl lipase complex; then the acyl group is transferred to the OH group from water. Commercial lipase enzymes generally are extracellular lipases produced from microbes [20]. This is because the production of enzymes using microbes has several advantages, i.e., microbes can be cultured quickly in a small space to produce enzymes in large quantities and a short time, the enzyme content produced is more easily controlled and predicted, the composition of media and other components can be arranged and costs of its production is cheaper compared to synthesizing enzymes using sources from plants or animals.

4. Conclusion

The results of the bio-detergent washing test with lipase enzyme additive did not show a significant difference compared to the performance of other bio-detergent formulations. This is because the types of stains tested in this study did not contain significant amounts of fat, so the role of the lipase enzyme was not seen. The use of lipase as a detergent additive is beneficial in terms of economy and environment. Normally fat or oil stains are difficult to remove in low-temperature conditions if using conventional detergents. But by using lipase as detergent, fat and oil stains are easily removed. This is a saving in energy use. In addition, the use of detergent lipases reduces the impact of chemical pollution on the environment because lipases can be decomposed naturally in the environment, are non-toxic, and do not leave harmful residues [20].

References

[1] Mousavi S A and Khodadoost F 2019 Effects of detergents on natural ecosystems and wastewater treatment processes: a review *Environ Sci Pollut Res Int.* doi: 10.1007/s11356-019-05802-x
[2] Anwar A F 2011 Effect of greywater irrigation on soil characteristics *2nd Int. Conf. on Env. Sci. and Dev.* IPCBEE pp 15–18
[3] Helmy Q, Kardena E, Funamizu N and Wisjnuprapto 2011 Strategies toward commercial scale of biosurfactant production as potential substitute for it's chemically counterparts *Int. J. Biotechnology* **12** 66-86 doi: 10.1504/IJBT.2011.042682
[4] Muthusamy K, Gopalakrishnan S, Ravi T K and Sivachidambaram P 2008 Biosurfactants: Properties, commercial production and application *Current Science* **94** 736-47
[5] Lourith N and Kanlayavattanaku M 2009 Natural surfactants used in cosmetics: glycolipids *International Journal of Cosmetic Science* **31** 255–61
[6] Vatsa P, Sanchez L, Clement C, Baillieul F and Dorey S 2010 Rhamnolipid biosurfactants as new players in animal and plant defense against microbes *International Journal of Molecular Science* **11** 5095-108
[7] Kaskatepe B and Yildiz S 2016 Rhamnolipid biosurfactants produced by Pseudomonas species *Braz. arch. biol. technol.* **59** 1-16 doi: 10.1590/1678-4324-2016160786
[8] SNI ISO 105-C06 2010 *Test for Colour Fastness-Part C06: Colour Fastness to Domestic and Commercial Laundering*
[9] Kubelka P and Munk F 1931 Ein beitrag zur optik der farbanstriche *Z. Techn. Phys.* **12** 593-601
[10] Hossain Md F, Asaduzzaman Md, Kamruzzaman Md, Talukder Md E, Miah M R and Quan H 2016 Application of new synthetic fifth generation thickeners for printing cotton fabric with reactive dyes *American Journal of Polymer Science & Engineering* **4** 123-32
[11] Smulders E 2002 *Laundry Detergents* (Weinheim: Wiley-VCH Verlag)
[12] Lynn J L 2009 Detergency and detergents *Kirk-Othmer Encyclopedia of Chemical Technology* (New York: John Wiley & Sons)
[13] Glennie E B, Littlejohn C, Gendebien A, Hayes A, Palfrey R, Sivil D and Wright K 2002 Phosphates and alternative detergent builders *EU Environment Directorate Report No. UC 4011* WRc plc Swindon

[14] Arsyah D M, Kardena E and Helmy Q 2018 Characterization of biosurfactant produced by petrofilic bacteria isolated from hydrocarbon impacted soil and its potential application in bioremediation *IOP Conf. Ser.: Earth Environ. Sci.* **106** 012101 doi: 10.1088/1755-1315/106/1/012101

[15] Effendi A J, Kardena E and Helmy Q 2018 *Biosurfactant-Enhanced Petroleum Oil Bioremediation* In: Kumar V, Kumar M, Prasad R (eds) Microbial Action on Hydrocarbons (Singapore: Springer) doi: 10.1007/978-981-13-1840-5_7

[16] Singh R P, Rawat J P and Kumar R 2000 Effect of cationic, non-ionic and anionic surfactants on the adsorption of carbofuran on three different types of indian soil *Adsorption Science & Technology* **18** 333-46

[17] Smith D L 1997 Impact of composition on the performance of sodium linear alkylbenzenesulfonate (NaLAS) *JAOCS* **74** 837-45

[18] Pattusamy V, Nandini N and Bheemappa K 2013 Detergent and sewage phosphates entering into lake ecosystem and its impact on aquatic environment *International Journal of Advanced Research* **1** 129-33

[19] Kundu S, Coumar M V, Rajendiran S, Rao A and Rao A S 2015 Phosphates from detergents and eutrophication of surface water ecosystem in India *Current Science* **108** 1320-5

[20] Amara A A, Salem S R and Shabeb M S A 2009 The possibility to use bacterial protease and lipase as biodetergent *Global Journal of Biotechnology & Biochemistry* **4** 104-14

Producing drop-in biofuel from feedstock grown on marginal land in Sumba island, Indonesia: a systems dynamics analysis

M Lamria[1,2] and R E H Sims[1]

[1] Massey University, Palmerston North 4442, New Zealand
[2] The Ministry of Energy and Mineral Resources, Jakarta 10320, Indonesia

E-mail: maslanl@yahoo.com

Abstract. To improve sustainability, liquid biofuel can be produced by growing biomass feedstock on marginal land and using innovative and appropriate conversion technology under development. Throughout the Indonesia Archipelago, oil fuel imports are high, and marginal land is abundant. This study assessed future potential for drop-in biofuel (DBF) production from *Pongamia pinnata* oilseed crops as driven by various policy and technical parameters. A systems dynamics approach was applied and a model developed to simulate the dynamics of developing DBF technology at the national level. The potential use of marginal land in Sumba Island was taken as a case study to assess the projected DBF production as well as any increase in the gross regional domestic product (GRDP) by 2045. The model outputs showed there is an inter-relationship between biofuel development and political elements, especially the level of sense of urgency (SU) by the country's President. In enhancing DBF production, and hence increasing GRDP, SU can be increased and sustained by empowering a future vision for the nation. Given a maximum future vision over time, 100% of liquid fuel self-sufficiency could be reached in Sumba by 2033. Then the GRDP increased from DBF-related revenues would exceed the current total GRDP.

1. Introduction

Indonesia is a large archipelago that is highly dependent on oil imports. On the other hand, abundant marginal land is available for growing energy crops. In addition, a conversion technology for drop-in biofuel (DBF) production has been under development. Improving liquid fuel self-sufficiency through sustainable biofuel production and utilization could be feasible both nationally as well as locally, especially on small islands. In Indonesia where fuel distribution is a challenge, developing DBF using feedstock grown on marginal land would benefit energy security.

In analyzing marginal land issues, Sumba, the iconic island for developing renewable energy systems was chosen as a case study given that petroleum oil resources are absent. Analysis of DBF technology readiness was conducted at the national level. Political elements drive the progress rate for marginal land development as well as DBF technology innovation.

Delayed actions in degraded land restoration, as well as energy technology innovation, can lead to acceleration of negative impacts on the economy, society, and environment in longer-term [1,2]

This study aims to better understand how policy implementation and liquid biofuel implementation could affect to one another. Considering time factor, it assessed political dimensions which are lacking in existing assessments of bioenergy deployment and sustainability.

For assessing impacts on liquid fuel self-sufficiency and gross regional domestic product (GRDP) systems that contain inter-relationships, a systems dynamics approach was applied [3,4].

In this paper, Section 2 describes the strategy of DBF production from marginal land-based feedstock; Section 3 provides a literature review on assessment of the strategy; Section 4 explains the methodology which includes the model development; Section 5 presents results and discussion and the conclusions are given in Section 6.

2. DBF Production from Feedstock Grown on Marginal Land

2.1. Drivers for liquid biofuel implementation

Liquid fuel demand will remain significant in the long term, especially for heavy-duty vehicles, marine ships, planes, and light-duty vehicles. In 2017, liquid fuel consumption in Indonesia reached around 70 Gl of which around 30 Gl was imported [5]. The high rate of oil imports has threatened the national economy in relation to the balance of trade (BOT).

On the other hand, the use of liquid biofuel in Indonesia has fluctuated and been unsatisfactory. It has progressed significantly only when BOT is in trouble, which drives the sense of urgency for accelerating biofuel utilization by the President who has the utmost cross-sectoral authority.

To reduce oil imports, Indonesia could produce and use DBF from biomass feedstock grown on marginal land as an integrated strategy. As the progress is influenced by the government's urgency to increase liquid fuel self-sufficiency, it is necessary to increase and maintain the urgency at high level. For example, by applying a driver that acts anticipatively rather than responsively [6] as does the pressure from BOT.

The anticipative driver could be a future vision that can be adopted from The Constitution 1945 preamble, paragraph four, which states the nation's idea to be a sovereign country. Moreover, the 7th President wrote, "The Indonesian Dreams 2085" [7] which included for Indonesia to become an independent country. This has been put into a shorter-term vision, "Golden Indonesia 2045" marking the 100th Independence Day as an important development milestone [7].

2.2. DBF technology development

Conventional liquid biofuels such as bioethanol and biodiesel have limitations in high-level blends with petroleum fuels. DBFs allow more flexible utilization in the whole existing infrastructure thanks to their equivalent characteristics with petroleum fuels.

This study assessed the indigenous DBF conversion technology being developed at the Institute of Technology of Bandung, a tertiary educational institution in Indonesia. "Hydrodeoxygenation" route requires hydrogen in the production process and has progressed closer to commercial scale, whereas "decarboxylation of metal soap" route can be undertaken without hydrogen and has better economic feasibility at a lower production scale [8].

Financial support from the government is very important for the success of degraded land utilization [9] as well as in new technology development for energy production [10]. SU level determines how effective the support would be provided.

The role of urgency in accelerating a technology innovation was shown in the development of the "Houdry Process", a primary technology applied in oil refineries up to the present. It took only three years from the initial concept to commercial production in 1942 [11], urged by the Allied Forces commander in overcoming fuel shortages.

2.3. Marginal land potential

Sumba is an iconic island for renewable energy in Indonesia that has around 900,000 population on an area of 1.1 Mha including around 475,000 ha of marginal land distributed across East Sumba and Central

Sumba regencies [12,13,14,15,16]. It has no oil resources but the marginal land is around 2% of around 25 Mha categorized as "critical" and "very critical" in Indonesia [16]. Much of this land has the potential for growing crops [17,18].

Pongamia pinnata was chosen as a preferred crop on marginal land-based on several criteria including high energy wood, nitrogen-fixing, fast-growing and able to grow on high-salinity soil. Sumba climate conditions and a soil test result indicated that Pongamia can be cultivated in Sumba. Therefore, having good availability of data and information, it was used as the model input. However, determination of the most suitable on-site crop should be based on further assessments including field trials.

2.4. Approach for the assessment

DBF production using marginal land comprise several key components: national supporting policy; conversion technology readiness; marginal land preparedness; DBF commercial production, and liquid fuel import demand. The components have interdependent relationships so that a systems dynamics approach was used for the assessment using Sumba as the case study.

A delay in any of these components leads to a delay in all components and sectors, thus, time is a crucial factor. Therefore, the political aspect of biofuel sustainability that is currently lacking was included in the assessment.

Systems dynamics modeling commonly assesses multi-disciplinary systems that involve various sectors and thus are capable of assisting stakeholders in communicating and playing a role in implementing and evaluating the strategy based on scenarios. In developing countries, a systems dynamics model could be a user-friendly tool for policymakers and non-experts, thanks to their transparency in generating results, flexibility in using input data and information, and ease of revision.

3. Assessment of The Strategy

Out of existing studies on liquid biofuel assessments, no study was found that assessed marginal land use and conversion technology innovation in an integrated fashion.

In Indonesia, before 2018, two studies that integrated conventional feedstock and technology were carried out by Rahmadi, Aye [19] and Jupesta [20], but neither applied a systems dynamics approach.

Assessments that used system dynamics approach have been conducted in Columbia [21], Iran [22], Latvia [23], Malaysia [24], USA [25], and South Africa [26]. However, none of them covered the inter-relationship between political sustainability and liquid biofuel sustainability.

Among bioenergy sustainability dimensions [27, 28 in [29]] as studied by Bautista, Enjolras [30], politics and technology are the least assessed in the literature. Unlike existing studies, this study explicitly shows political and technological sustainability which interrelate to one another allowed by treating policy parameter as an endogenous variable.

4. Methodology

Systems dynamics methodology was used by developing a model that has the capability of handling feedback loops. The main steps in the methodology comprise conceptualization, problem definition, model structuring, model validation, and scenario analysis.

The problem was defined using a causal loop diagram that shows inter-relationships between the liquid fuel import demand that indicates liquid fuel self-sufficiency, the proposed strategy, and the policy (Figure 1). The liquid biofuel development and the policy affect one another. An increase in policy increases biofuel production which affects the consumption, which in turn affects the policy itself.

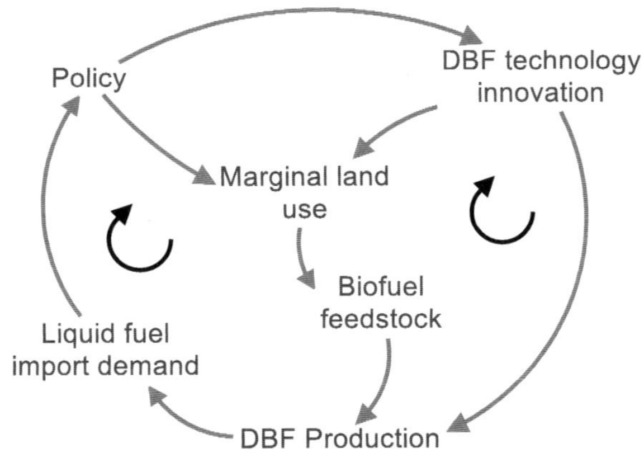

Figure 1. Problem definition in a causal loop diagram showing a feedback loop between policy and liquid biofuel development.

4.1. The model structure

The systems dynamics model was developed by building a stock and flow diagram to perform quantitative simulations. Determination of variables and their relationships refer to general theory or knowledge including from literature analysis, work experience, interviews, and personal communications.

The model was developed using Stella® Architect software applying the time horizon 2018-2045. The model consists of seven sectors representing policy, technology readiness, marginal land preparedness and liquid fuel import demand from a national perspective and feedstock production DBF production and GRDP on Sumba. Quantifying national-level indicators relating to the cultivation area was calculated using the multiplier factor from the Sumba area.

Methods used for data collection consisted of literature analysis, semi-structured interviews, focus groups, and personal communications as follows.

- Sumba marginal land development.
  This stage identified support and barriers that might exist in the program implementation. It involved landowners and policymakers in East Sumba and Central Sumba.
- Soil test.
  This work was conducted to analyze the suitability of the soil for growing the Pongamia energy crop on Sumba marginal land. Soil samples were taken from six locations across the two regencies and tests were conducted at an accredited soil laboratory in Indonesia.
- Focus group of DBF conversion technology
  This activity aimed to identify factors that could influence success in technology development. Funding continuity up to the commercialization was the greatest challenge.
- Focus group of cross-sectoral policy
  This focus group aimed to capture any cross-sectoral policies at the national level that might have not been considered in the study.

Important variables in the model are shown in a simplified stock and flow diagram (Figure 2).

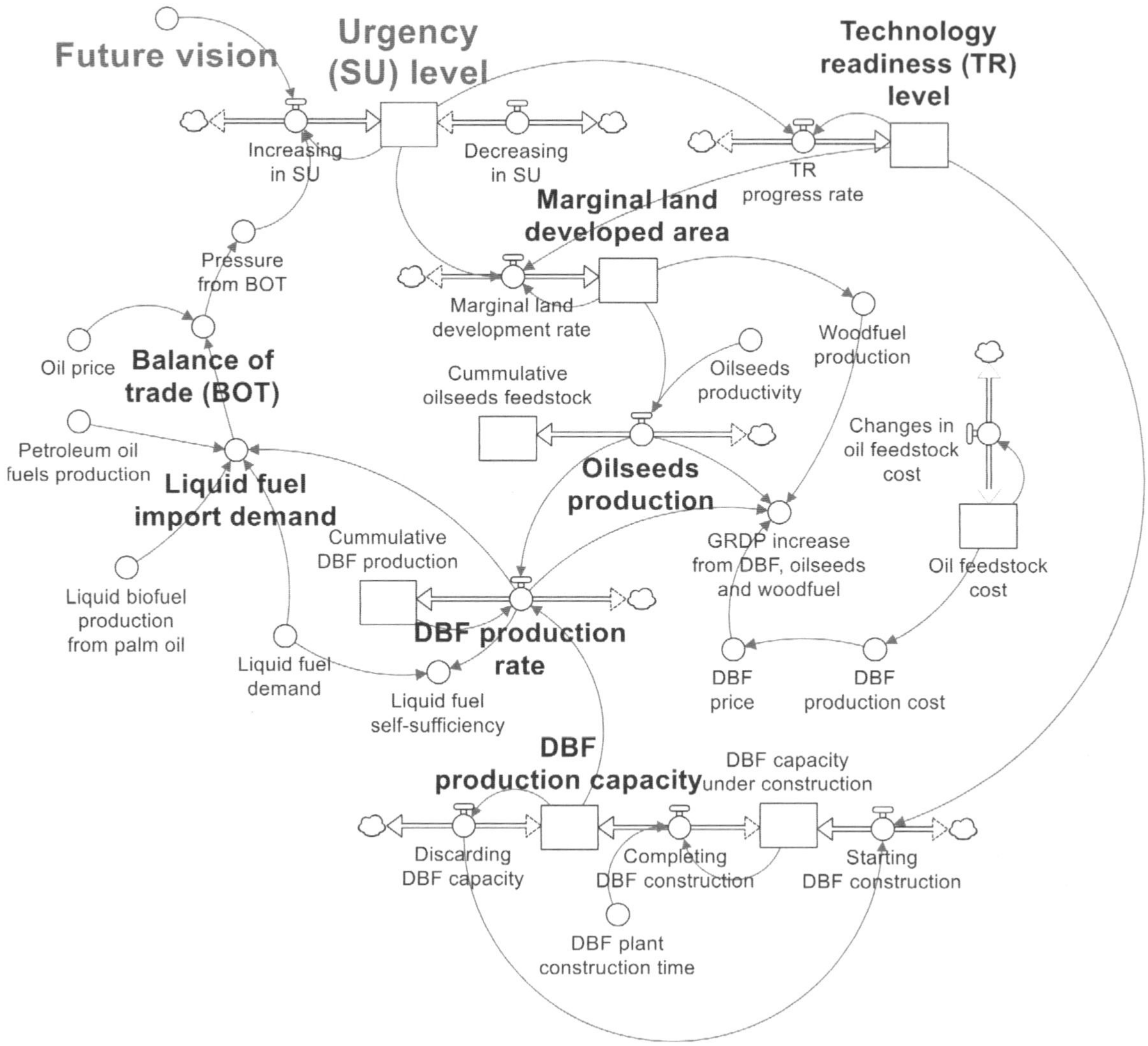

Figure 2. Simplified stock and flow diagram showing main variables and their relationships

4.1.1. Policy sector. The policy sector was modelled using variables that influence policy parameters, especially the sense of urgency (SU) by the President, which drives the liquid fuel self-sufficiency through DBF production.

SU can be maintained high and stable by the realization of a future vision for the nation. SU (dimensionless unit) ranges from zero to one where the initial value of SU is set 1 and the increasing in SU (r_{IU}) is calculated as:

$$r_{IU} = (PBOT * (1 - WVS) + FVS * WVS)/t_{SU} \quad (1)$$

where PBOT (dimensionless unit) is pressure from a balance of trade (section 2.1), WVS (dimensionless unit) is weight to vision, FVS (dimensionless unit) is future vision power, and t_{SU} (years) is time to change urgency.

The national balance of trade (BOT) is influenced by liquid fuel imports as determined by liquid fuel consumption volume and price. To make the model simpler, this study assumed that liquid biofuel consumption is equal to liquid biofuel production from both Pongamia DBF and palm biodiesel. When

DBF technology is commercially ready but marginal land feedstock is still unavailable, DBF will be initially produced from existing palm oil feedstock.

Thus, national liquid fuel import demand (NLI, in l/yr) is calculated as:

$$NLI = NLD - (NOP + NBP) \tag{2}$$

where NLD is national liquid fuel demand (kL/yr); NOP is national oil fuel production (kL/yr), and NBP is national liquid biofuel production (kL/yr).

4.1.2. Technology readiness sector. The technology readiness (TR) sub-model is changed by the provision of support for the investment by the sectors in-charge. In this study, support increases TR through accumulated investment driven by the SU as an input from the policy sub-model. This sub-model simulates the estimation of the future time when the technology will become ready.

4.1.3. Sumba marginal land preparedness sector. This sub-model shows the dynamics of conversion from marginal land available for energy crops into Sumba's developed marginal land area, through marginal land development rate.

Based on the interviews with landowners and local government officials, factors that influence marginal land development rate in Sumba comprise infrastructure readiness, local government commitment, landowners willing to cultivate, land status clarity, and pressure from TR to land development. As in the technology readiness progress rate, all those factors are driven by SU.

This sub-model determines delay in planting the crop which affects delay in DBF production. Planting delay reflects the difference between the actual and desirable year of planting start.

$$PLD = AYP - DYP \tag{3}$$

where PLD is planting delay (years), AYP is actual year of planting start (when the feedstock becomes available for running the first DBF production plant), and DYP is desirable year of planting start (the marginal land developed area can provide sufficient feedstock to start a DBF production plant) when future vision is maximum.

4.1.4. Sumba feedstock production sector. This sub-model calculates biomass feedstock grown on marginal land for DBF production at the island level based on the crop growth characteristics. Parameter values that relate to Pongamia crop's growth properties can be estimated [31]. Crop growth was simulated by applying a cohort structure to show the pattern of oilseed production by estimating expected yields based on the expected growth factors of each cohort. The establishment of a blocking system linked to the crop rotation cycle is useful in reducing investment risk.

The sub-model is divided into 15 age cohorts, each with a tree stock, an oilseed yield flow, a growing rate flow, and a planting and growing flow that represents the associated year as well as the planted block area.

4.1.5. Sumba DBF production sector. This sub-model calculates DBF production from Sumba marginal land feedstock, which is also used for the estimation at the national level.

Sumba DBF production (r_{DP}, in kL/yr) is constrained by production capacity and feedstock availability, and calculated as:

$$r_{DP} = MIN(DP, OFD * OFL) \tag{4}$$

where DP (kL/yr) is DBF production capacity stock, OFD (kL/yr) is oil feedstock for DBF production, and OFL (dimensionless unit) is oil feedstock conversion rate.

4.1.6. Sumba liquid fuel supply and demand sector. This sub-model calculates liquid fuel self-sufficiency through DBF production and use, which is a very important indicator since Sumba Island has no crude oil resources to meet the liquid fuel demand.

Sumba liquid fuel self-sufficiency (SLF, in dimensionless unit) is calculated as:

$$SLF = \frac{SLD-SLI}{SLD} \quad \backslash \tag{5}$$

$$SLI = SLD - r_{DP} \tag{6}$$

where SLD is Sumba liquid fuel demand (kL/yr) and SLI is Sumba liquid fuel import demand (kL/yr).

4.1.7. Economic sector. This sub-model calculates the potential economic benefits from DBF implementation in Sumba using the indicators of the increase in the gross regional domestic product (GRDP) and the national foreign exchange savings (FES).

Sumba's GRDP increase arises from DBF, oilseeds, and woodfuel (SGI, in USD/yr) calculated as:

$$SGI = r_{DC} * DBP * l/kl + SOH * OSP + \frac{SWH}{WCT} * WFP \tag{7}$$

where r_{DC} (kl/yr) is DBF production, DBP (USD/l) is DBF price, SOH (t/yr) is oilseeds for harvest, OSP (t/yr) is oilseeds price, SWH (USD/t) is woodfuel for harvest, WCT (years) is woodfuel consumption time, and WFP (USD/t) is woodfuel price. As a comparison, combined total GRDP for East Sumba and Central Sumba in 2013 was around USD 200 million (BPSKSTM [13]; [12]).

The national foreign exchange saving (FES, in USD/yr) from DBF implementation using feedstock grown on marginal land is calculated as:

$$FES = NDN * DBP \tag{8}$$

where NDN is national DBF consumption (kL/yr) which is assumed equal to national DBF production, and DBP is DBF price (USD/kL).

DBF price is calculated as Pongamia oil feedstock cost plus DBF conversion cost and DBF profit margin, subtracted by by-products revenues. The key variable in this sub-model is the cost of Pongamia oil feedstock which is interdependent with the rate of changes in Pongamia oil cost.

The values of parameters in the involved sectors are shown in Table 1.

Table 1. Parameters' values applied in the model.

Sector	Parameter	Type	Value	Unit	Notes/Source
Policy	Initial value (IV) for SU	Constant	1	dimensionless	Estimation
	Gas export volume	Time series		billion cubic feet/yr	[32]; Appendix
	Crude oil export volume	Time series		bbl/yr	[32] ; Appendix
	Non-oil & gas export value	Time series		USD/yr	assumed 5% growth of historical data [33]; Appendix

Sector	Parameter	Type	Value	Unit	Notes/Source
	Non-oil & gas import value	Time series		USD/yr	assumed 5% growth of historical data [33]; Appendix
	LPG import demand volume	Time series		Mt/yr	[32]; Appendix
	Crude oil import volume	Time series		bbl/yr	[32]; Appendix
	Gas import volume	Time series		billion cubic feet/yr	[32]; Appendix
	Oil price	Time series		USD/barrel	[34]; Appendix
	Gas price	Time series		USD/MMBtu	[35]; Appendix
	Palm oil price	Time series		USD/t	Extrapolating historical data of [36]; Appendix
DBF technology readiness	IV technical readiness (TR)	Constant	0.5	dimensionless	Estimation
	Desired TR	Constant	1	dimensionless	Max limit
	Expected time to progress TR	Constant	5	years	[37]; [38]
	Expected time of post-technical readiness	Constant	3	years	[39]
Marginal land preparedness	IV marginal land available for bioenergy	Constant	200,000	ha	Required minimum area for meeting Sumba liquid fuel demand in 2045 (342k kl/yr)
	Expected time to develop land	Constant	3	years	Estimation
	IV infrastructure readiness	Constant	0.2	dimensionless	Site visit and interview
	IV local government commitment	Constant	0.6	dimensionless	Site visit and interview
	IV land certification	Constant	0.2	dimensionless	Site visit and interview
	IV landowners respect	Constant	0.9	dimensionless	Site visit and interview
	IV landowners understanding	Constant	0.5	dimensionless	Site visit and interview
Oil feedstock production	Trees per ha	Constant	350	tree/ha	[31]
	Crop rotation cycle	Constant	15	years	Considering the peak period of Pongamia oilseeds

IOP Conf. Series: Materials Science and Engineering **823** (2020) 012015 doi:10.1088/1757-899X/823/1/012015

Sector	Parameter	Type	Value	Unit	Notes/Source
					harvest at around 8th year.
	Cultivation time before first harvest	Constant	3	years	[31]
	Growth year 1-15	Constant	0.5e-3; 2.5e-3; 17e-3; 25e-3; 30e-3; 20e-3; 35e-3; 25e-3; 0; 0; 0; 0; 0; 0; 0.	t/ree	[31]
	Oilseeds yield 1-15	Constant	0.00; 0.00; 0.00; 0.01; 0.01; 0.013; 0.014; 0.025; 0.025; 0.025; 0.025; 0.025; 0.025; 0.025; 0.025.	t/ree	[31]
	Oil content in seeds	Constant	0.4	dimensionless	[31]
	Woodfuel yield per tree	Constant	0.01	t/tree	Estimation
DBF production	DBF plant construction time	Constant	2	years	Common practice of similar oleochemical plants
	DBF capacity life time	Constant	20	years	Common practice
	Oil feedstock conversion	Constant	0.76	dimensionless	[40]
	Desired DBF production per plant	Constant	50,000	kl/yr	[8]
Liquid fuel import demand	Sumba liquid fuel demand	Time series		kl/yr	2018-2025: [41]; 2025-2045: assumed 5% growth; Appendix
	National liquid fuel demand	Time series		kl/yr	[32]; Appendix

Sector	Parameter	Type	Value	Unit	Notes/Source
	National oil fuels production	Time series		kl/yr	[32]; Appendix
	National DBF export quota	Constant	1 million	kl/yr	Assumption
	National to Sumba marginal land area multiplier	Constant	50	dimensionless	Estimation
	Oil palm plantation area	Constant	14	Mha	[42]
	Palm oil productivity target	Constant	5	t/ha	[42]
	Fraction palm oil consumption for food & oleo excluding biofuel	Constant	0.1	dimensionless	Current trend
	Fraction palm oil for export	Constant	0.1	dimensionless	Assumption
	National biodiesel export quota	Constant	1 million	kl/yr	Assumption
GRDP	Pongamia oil feedstock cost growth rate	Constant	0.001	dimensionless	Assumption (well controlled by government)
	Oilseeds price	Constant	15	USD/t	[31]
	Woodfuel price	Constant	50	USD/t	Assumption
	DBF profit margin	Constant	0.1	dimensionless	Common practice in oleochemical industries
	DBF conversion cost	Constant	0.36	USD/l	Assumption based on existing study [43]
	IV pongamia oil feedstock cost	Constant	5,000	IDR/l	[31]
	By-products revenues	Constant	1,000	IDR/l	Assumption for seed cake revenue

4.2. Scenarios design

The model simulations applied scenarios for the different magnitude of future vision to see how the systems respond to a change in future vision, and hence Sumba liquid fuel self-sufficiency (SLF), Sumba GRDP increases from DBF-related revenues (SGI), and national foreign exchange savings (FES). SU is a variable of the political aspect, while SLF, SGI, and FES are related to economic performance. Besides for SLF and SGI, the dynamics of some related variables were simulated: the sense of urgency (SU), technology technical readiness (TR), Sumba developed marginal land area (MD), Sumba oilseeds feedstock, and Sumba DBF production (r_{DC}).

The reference mode is the condition of "full pressure" where future vision is absent. Three vision scenarios were set:

- "full vision" (FV) which is a contrast to reference mode,
- "medium vision" (MV), and
- "low vision" (LV).

Future vision value is a combination of weight to vision (WVS) and future vision power (FVS) (Table 2).

Table 2. Design of vision scenarios.

Scenario	Weight to vision (WVS)	Future vision power (FVS)
Reference Mode / Full Pressure (FP)	0	0
Full Vision (FV)	1	1
Medium Vision (MV)	0.5	0.7
Low Vision (LV)	0.5	0.1

5. Results and Discussion

The reference mode defines the problem where SU is fully driven by pressure. Resulting from reference mode simulations (Figure 3), the SU profile drives TR to allow it to be completed by 2028 with commercialization realized in 2031. The SU profile drives marginal land development in 2027 so that the sufficient feedstock volume to run a DBF plant can first be harvested in 2031. Thus, DBF production is expected to commence in 2034 and then bring 100% liquid fuel self-sufficiency to Sumba by 2039 when SGI achieves USD 198 million. The SGI then peaks at USD 289 million in 2043.

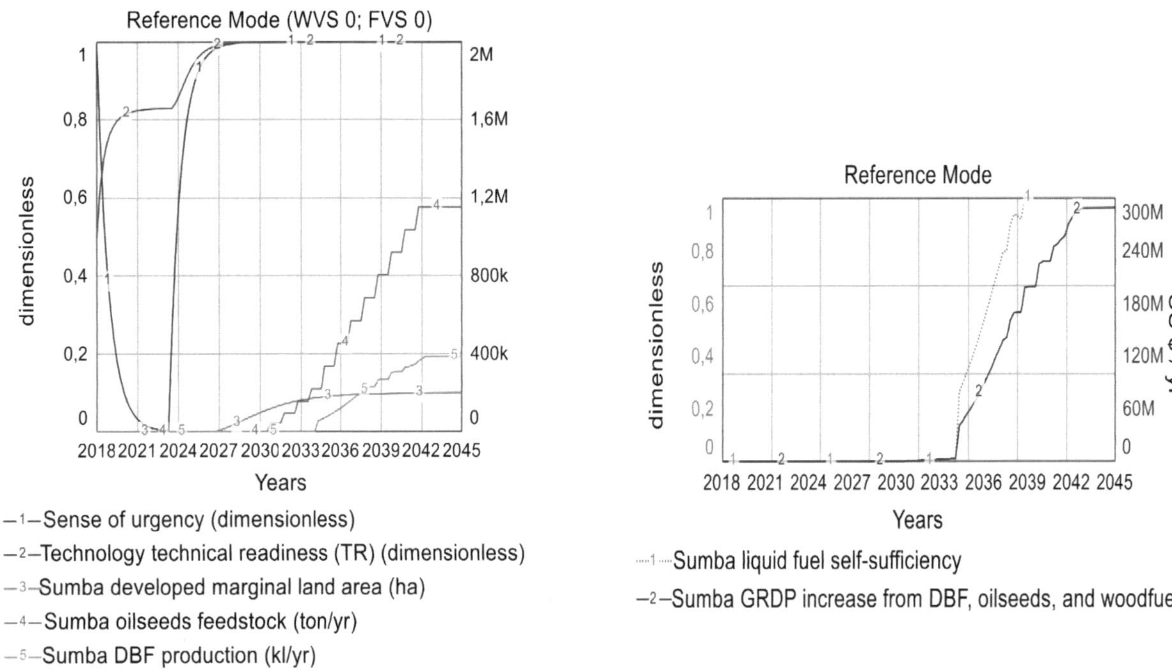

Figure 3. Dynamics of liquid fuel self-sufficiency, GRDP increase from DBF-related revenues, and relevant indicators for Reference Mode.

5.1. Implication across vision scenarios

The scenarios were assessed by applying planting delay, liquid fuel self-sufficiency, Sumba GRDP, and national FES. These indicators are determined by the dynamics of each stage driven by SU.

The modeling results (Figure 4) show that in reducing liquid fuel demand through the proposed strategy, SU is the key leverage point which is significantly affected by the future vision.

At the reference mode where future vision is absent, SU fluctuates the most. An increase in future vision at the same WVS increases SU which in turn improves the progress of technology readiness, marginal land development, oilseeds feedstock production, and DBF production.

Thus, in increasing liquid fuel self-sufficiency and hence GRDP of Sumba, the pressure should be minimized, while the future vision should be maximized and maintained high.

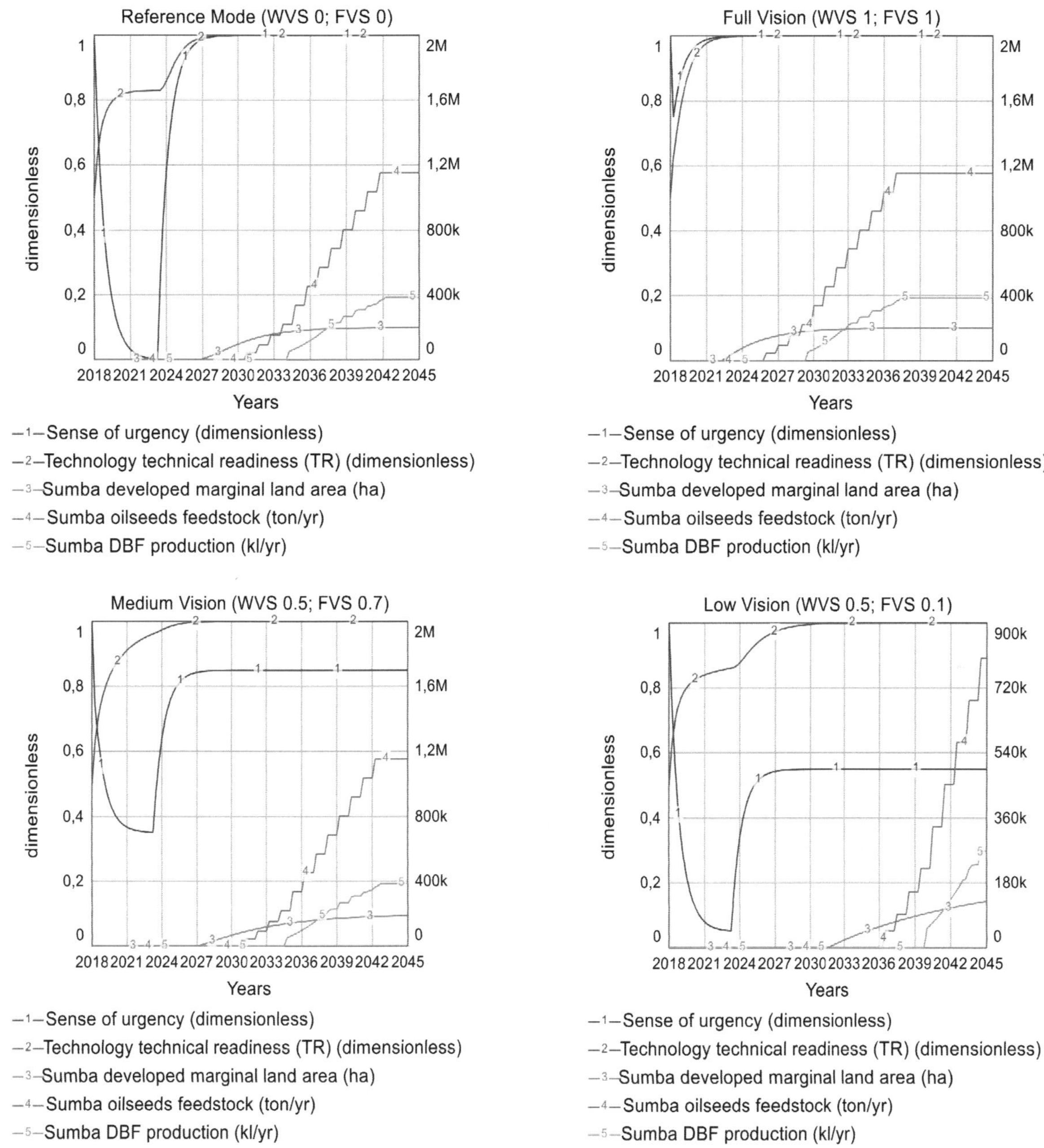

Figure 4. Dynamics of Sumba DBF production across the full (FV), medium (MV), and low (LV) vision scenarios.

IOP Conf. Series: Materials Science and Engineering **823** (2020) 012015 doi:10.1088/1757-899X/823/1/012015

Table 3 shows implications of future vision in planting delay. Planting delay at full vision (FV) scenario is three years, while at medium (MV) and low (LV) vision scenarios it is eight and fourteen years respectively which are significantly longer.

It is interesting that FP and MV result in a similar planting delay. This is resulted by given assumptions that generate SU through pressure from BOT, such as oil price and liquid fuel supply and demand volumes.

Table 3. Implications across the full (FV), medium (MV), and low (LV) vision scenarios for a range of variables.

Parameters	Reference Mode / Full Pressure (FP) Scenario	Full vision (FV) scenario	Medium vision (MV) scenario	Low vision (LV) scenario
Weight to vision (WVS)	0	1	0.5	0.5
Future vision power (FVS)	0	1	0.7	0.1
Year when conversion technology ready (TR)	2028	2023	2026	2030
Desirable year of planting (year when TR of full vision completed – cultivation time before first harvest = A)	2020	2020	2020	2020
Year of starting marginal land area development	2027	2023	2027	2032
Year of first harvest of oilseed feedstock	2031	2026	2031	2037
Actual year of planting (year of feedstock first harvest - cultivation time before first harvest = B)	2028	2023	2028	2034
Year of starting DBF production from feedstock grown on marginal land	2034	2030	2035	2040
Planting delay (towards expected time to TR at full vision = B – A)	8 yrs	3 yrs	8 yrs	14 yrs

Figure 5 shows the dynamics of SU, SLF, GRDP increases, and FES across scenarios. The reference mode produces the most fluctuated SU, which becomes more stable by applying a higher future vision. At full vision (FV) scenario, SU profile allows the first harvest of feedstock in 2026 and then DBF production in 2030, four years earlier than the reference mode. This rate can bring full SLF by 2033 or six years faster than the reference mode. At MV, full SLF is reached by 2039 or six years later than FV, while at LV it did not become full until 2045.

At FV, SU profile allows GRDP to increase from DBF-related revenues reaches USD 289 million by 2041 which exceeds current GRDP. At MV, it peaks at USD 289 million by 2043 or two years later than FV, while at LV it does not even peak by 2045.

SU profile at FV allows FES to reach USD 13.46 billion by 2041. At MV, it peaks at USD 13.46 billion by 2043 or two years later than FV, while at LV it does not even peak by 2045.

An increase in future vision power (FVS) improves liquid fuel self-sufficiency, GRDP increase, and national FES. It is interesting that the performance at reference mode is between medium (MV) and full

(FV) vision scenarios. This results from the SU profile generated by the pressure from BOT. The same urgency level will result in higher performance when the future vision is higher.

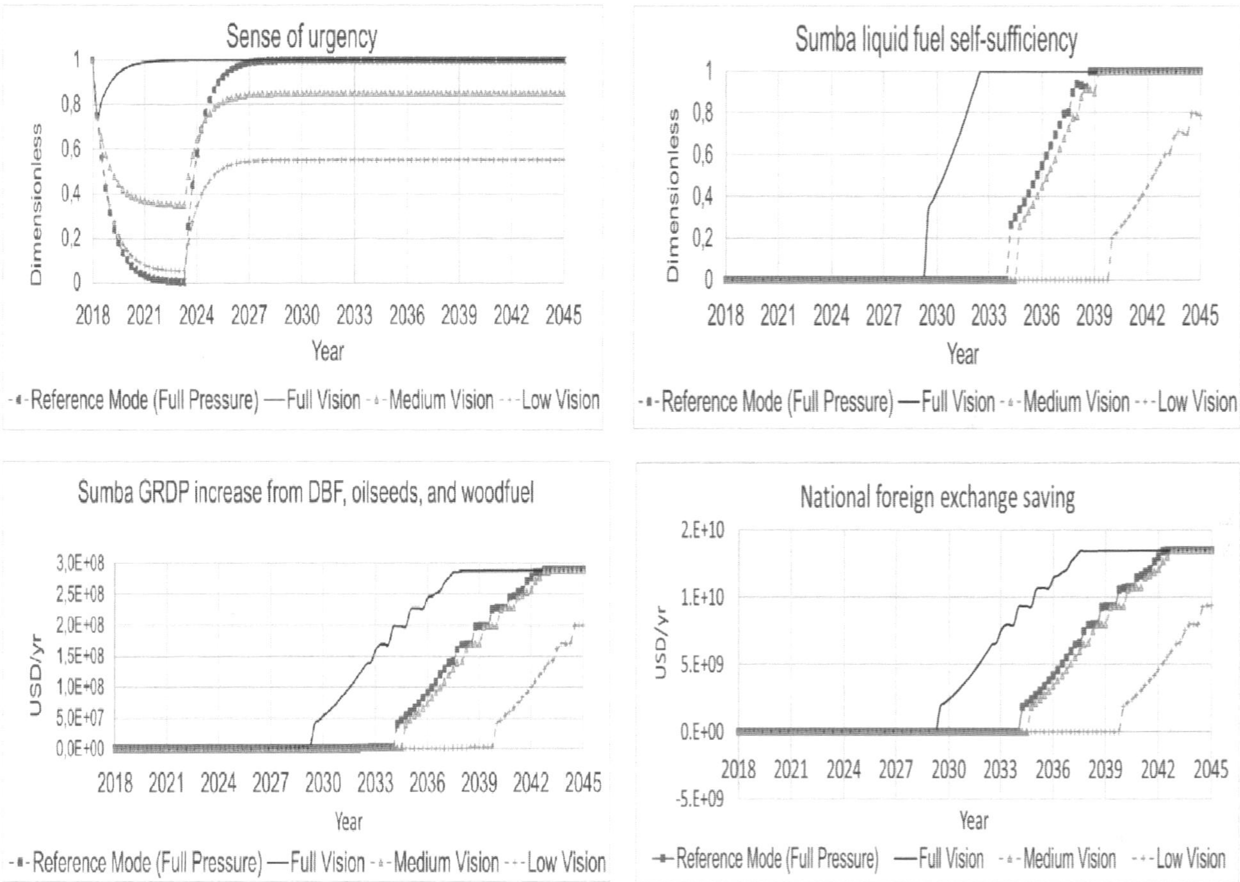

Figure 5. Selected indicators across the full (FV), medium (MV), and low (LV) vision scenarios.

5.2. Limitations and recommendation for further research

The model has been developed to meet the research aim, yet limitations still exist mainly due to time constraints and participants' availability. This provides opportunities for model improvement through further works such as:

- exploring further around SU structure, for example, investigating factors other than BOT that drives SU;
- calculating BOT that affects SU profile more representatively;
- performing an analysis of sensitivity to oil price which is the main factor of oil imports value, although this is not critical as BOT is inherently fluctuating and highly uncertain that a future vision is needed to improve SU;
- describing dynamics of TR more accurately by dividing TR into several phases;
- building the function of effects to differentiate between the influence of SU to TR progress rate and SU to marginal land development rate;
- calculating national parameters that consider marginal land area by disaggregating islands that include islands outside of Sumba;

- considering marginal land feedstock management that would affect feedstock productivity and price;
- improving the accuracy of DBF production model by disaggregating several DBF plant units;
- determining DBF production cost more representatively, such as by applying a wide range of by-product revenues and inclusion of transportation cost;
- improving GRDP model by considering multiplier effects on the local economy;
- presenting the dynamics of liquid fuel supply and demand more representatively, by modeling the transition from oxygenated biofuel to drop-in biofuel (DBF) production, also useful in planning for investments;
- simulating the effect of SU on biofuel consumption, to understand better the biofuel demand that affects BOT;
- improving validation, such as by:
  - conducting further direct structure tests that involve experts from a wider area; and
  - improving variable quantifications, especially in modeling the marginal land preparedness.

6. Conclusions

This study has provided a better understanding of how the implementation of marginal land use and biofuel technology innovation needs to be implemented as an integrated strategy to increase liquid fuel self-sufficiency.

The modeling showed that there is an inter-relationship between biofuel development and a sense of urgency (SU) by the highest-level political authority in executing such cross-sectoral strategy. It showed that SU can be increased and sustained by empowering a future vision for the nation. This is important knowledge in improving biofuel development by empowering its political sustainability dimension.

The simulation results show that given a maximum future vision over time, through DBF production from marginal land feedstock, Sumba can achieve full liquid fuel self-sufficiency by 2033. This would make the GRDP increased from DBF-related revenues to exceed the current total GRDP.

This study provided policy recommendations in producing DBF from marginal land-based feedstock by:

- applying maximum future vision all the time in maintaining a high level of SU to minimize the delay and hence maximizing liquid fuel self-sufficiency, Sumba GRDP, and national FES.
- actualizing the future vision in setting apart investment in DBF technology and giving early instruction in planting energy crops on marginal land.
- providing specific supports which are SU-driven, such as:
  - in DBF utilization: DBF pricing through direct subsidy.
  - for marginal land development: supports on infrastructure readiness, local government commitment, landowners willing to cultivate, and land status clarity.

Appendix

Timeseries values for the variables in Table 1

Gas_export_volume [billion cubic feet/yr]:
(2018.00, 900), (2019.00, 800), (2020.00, 750), (2021.00, 600), (2022.00, 600), (2023.00, 400), (2024.00, 300), (2025.00, 300), (2026.00, 250), (2027.00, 225), (2028.00, 200), (2029.00, 175), (2030.00, 150), (2031.00, 125), (2032.00, 100), (2033.00, 75), (2034.00, 50), (2035.00, 25), (2036.00, 0), (2037.00, 0), (2038.00, 0), (2039.00, 0), (2040.00, 0), (2041.00, 0), (2042.00, 0), (2043.00, 0), (2044.00, 0), (2045.00, 0)

Crude_oil_export_volume [bbl/yr]:
(2018.00, 80000000), (2019.00, 76000000), (2020.00, 72000000), (2021.00, 68000000), (2022.00, 64000000), (2023.00, 60000000), (2024.00, 56000000), (2025.00, 52000000), (2026.00, 48000000), (2027.00, 43000000), (2028.00, 39000000), (2029.00, 35000000), (2030.00, 31000000), (2031.00, 27000000), (2032.00, 23000000), (2033.00, 19000000), (2034.00, 15000000), (2035.00, 5000000),

(2036.00, 0), (2037.00, 0), (2038.00, 0), (2039.00, 0), (2040.00, 0), (2041.00, 0), (2042.00, 0), (2043.00, 0), (2044.00, 0), (2045.00, 0)

Non_oil_&_gas_export_value [USD/yr]:
(2018.00, 160738000000), (2019.00, 168775000000), (2020.00, 177214000000), (2021.00, 186074000000), (2022.00, 195378000000), (2023.00, 205147000000), (2024.00, 215404000000), (2025.00, 226174000000), (2026.00, 237483000000), (2027.00, 249357000000), (2028.00, 261825000000), (2029.00, 274917000000), (2030.00, 288662000000), (2031.00, 303095000000), (2032.00, 318250000000), (2033.00, 334163000000), (2034.00, 350871000000), (2035.00, 368414000000), (2036.00, 386835000000), (2037.00, 406177000000), (2038.00, 426486000000), (2039.00, 447810000000), (2040.00, 470201000000), (2041.00, 493711000000), (2042.00, 518396000000), (2043.00, 544316000000), (2044.00, 571532000000), (2045.00, 600108000000)

Non_oil_&_gas_import_value [USD/yr]:
(2018.00, 139303000000), (2019.00, 146268000000), (2020.00, 153581000000), (2021.00, 161260000000), (2022.00, 169323000000), (2023.00, 177790000000), (2024.00, 186679000000), (2025.00, 196013000000), (2026.00, 205814000000), (2027.00, 216104000000), (2028.00, 226910000000), (2029.00, 238255000000), (2030.00, 250168000000), (2031.00, 262676000000), (2032.00, 275810000000), (2033.00, 289600000000), (2034.00, 304080000000), (2035.00, 319284000000), (2036.00, 335249000000), (2037.00, 352011000000), (2038.00, 369612000000), (2039.00, 388092000000), (2040.00, 407497000000), (2041.00, 427872000000), (2042.00, 449265000000), (2043.00, 471729000000), (2044.00, 495315000000), (2045.00, 520081000000)

LPG_import_demand_volume [Mt/yr]:
(2018.00, 4.6), (2019.00, 4.9), (2020.00, 5.2), (2021.00, 5.2), (2022.00, 5.2), (2023.00, 5.5), (2024.00, 5.8), (2025.00, 5.4), (2026.00, 5), (2027.00, 5.5), (2028.00, 6), (2029.00, 6.4), (2030.00, 6.8), (2031.00, 6.9), (2032.00, 7), (2033.00, 7.3), (2034.00, 7.6), (2035.00, 7.6), (2036.00, 7.6), (2037.00, 8.1), (2038.00, 8.6), (2039.00, 8.8), (2040.00, 9), (2041.00, 9.3), (2042.00, 9.6), (2043.00, 10.1), (2044.00, 10.5), (2045.00, 10.5)

Crude_oil_import_volume [bbl/yr]:
(2018.00, 175000000), (2019.00, 1.9e+08), (2020.00, 2e+08), (2021.00, 2.2e+08), (2022.00, 2.5e+08), (2023.00, 3.5e+08), (2024.00, 4e+08), (2025.00, 5e+08), (2026.00, 5e+08), (2027.00, 5e+08), (2028.00, 5e+08), (2029.00, 5.8e+08), (2030.00, 6.4e+08), (2031.00, 6.4e+08), (2032.00, 6.4e+08), (2033.00, 6.4e+08), (2034.00, 6.4e+08), (2035.00, 7e+08), (2036.00, 7.5e+08), (2037.00, 7.5e+08), (2038.00, 7.5e+08), (2039.00, 8e+08), (2040.00, 8.5e+08), (2041.00, 8.5e+08), (2042.00, 8.5e+08), (2043.00, 8.5e+08), (2044.00, 8.5e+08), (2045.00, 9.5e+08)

Gas_import_volume [billion cubic feet/yr]:
(2018.00, 0), (2019.00, 0), (2020.00, 0), (2021.00, 0), (2022.00, 0), (2023.00, 0), (2024.00, 25), (2025.00, 50), (2026.00, 100), (2027.00, 200), (2028.00, 300), (2029.00, 500), (2030.00, 700), (2031.00, 800), (2032.00, 900), (2033.00, 900), (2034.00, 1000), (2035.00, 1100), (2036.00, 1300), (2037.00, 1700), (2038.00, 1900), (2039.00, 2000), (2040.00, 2200), (2041.00, 2300), (2042.00, 2350), (2043.00, 3000), (2044.00, 3500), (2045.00, 3800)

Oil_price [USD/barrel]:
(2018.00, 65.0), (2019.00, 66.7), (2020.00, 66.4), (2021.00, 67.1), (2022.00, 67.9), (2023.00, 68.6), (2024.00, 69.3), (2025.00, 70.0), (2026.00, 70.0), (2027.00, 70.0), (2028.00, 70.0), (2029.00, 70.0), (2030.00, 70.0), (2031.00, 69.0), (2032.00, 68.0), (2033.00, 67.0), (2034.00, 66.0), (2035.00, 65.0), (2036.00, 65.0), (2037.00, 65.0), (2038.00, 65.0), (2039.00, 65.0), (2040.00, 65.0), (2041.00, 65.0), (2042.00, 65.0), (2043.00, 65.0), (2044.00, 65.0), (2045.00, 65.0)

Gas_price [USD/MMBtu]:
(2018.00, 8.80), (2019.00, 8.90), (2020.00, 9.10), (2021.00, 9.30), (2022.00, 9.40), (2023.00, 9.60), (2024.00, 9.70), (2025.00, 9.90), (2026.00, 9.90), (2027.00, 9.90), (2028.00, 10.00), (2029.00, 10.00), (2030.00, 10.00), (2031.00, 9.90), (2032.00, 9.80), (2033.00, 9.70), (2034.00, 9.70), (2035.00, 9.60),

(2036.00, 9.50), (2037.00, 9.40), (2038.00, 9.30), (2039.00, 9.20), (2040.00, 9.10), (2041.00, 9.00), (2042.00, 9.00), (2043.00, 8.90), (2044.00, 8.80), (2045.00, 8.70)

Palm oil price [USD/t]:
(2018.00, 490), (2019.00, 500), (2020.00, 500), (2021.00, 510), (2022.00, 510), (2023.00, 520), (2024.00, 520), (2025.00, 530), (2026.00, 550), (2027.00, 560), (2028.00, 560), (2029.00, 560), (2030.00, 560), (2031.00, 560), (2032.00, 560), (2033.00, 570), (2034.00, 580), (2035.00, 590), (2036.00, 590), (2037.00, 600), (2038.00, 610), (2039.00, 620), (2040.00, 620), (2041.00, 620), (2042.00, 620), (2043.00, 620), (2044.00, 640), (2045.00, 640)

Sumba_liquid_fuel_demand [kl/yr]:
(2018.00, 91595), (2019.00, 96175), (2020.00, 100983), (2021.00, 106033), (2022.00, 111334), (2023.00, 116901), (2024.00, 122746), (2025.00, 128883), (2026.00, 135327), (2027.00, 142094), (2028.00, 149198), (2029.00, 156658), (2030.00, 164491), (2031.00, 172716), (2032.00, 181352), (2033.00, 190419), (2034.00, 199940), (2035.00, 209937), (2036.00, 220434), (2037.00, 231456), (2038.00, 243029), (2039.00, 255180), (2040.00, 267939), (2041.00, 281336), (2042.00, 295403), (2043.00, 310173), (2044.00, 325681), (2045.00, 341966)

National_liquid_fuel_demand [kl/yr]:
(2018.00, 7.5e+07), (2019.00, 7.5e+07), (2020.00, 8e+07), (2021.00, 8.6e+07), (2022.00, 9.2e+07), (2023.00, 9.7e+07), (2024.00, 1.03e+08), (2025.00, 1.09e+08), (2026.00, 1.15e+08), (2027.00, 1.21e+08), (2028.00, 1.27e+08), (2029.00, 1.33e+08), (2030.00, 1.38e+08), (2031.00, 1.44e+08), (2032.00, 1.5e+08), (2033.00, 1.58e+08), (2034.00, 1.67e+08), (2035.00, 1.75e+08), (2036.00, 1.84e+08), (2037.00, 1.92e+08), (2038.00, 2.01e+08), (2039.00, 2.09e+08), (2040.00, 2.2e+08), (2041.00, 2.28e+08), (2042.00, 2.36e+08), (2043.00, 2.44e+08), (2044.00, 2.52e+08), (2045.00, 2.6e+08)

National_oil_fuels_production [kl/yr]:
(2018.00, 4.5e+07), (2019.00, 4.5e+07), (2020.00, 4.5e+07), (2021.00, 5e+07), (2022.00, 5e+07), (2023.00, 5.5e+07), (2024.00, 6e+07), (2025.00, 6.5e+07), (2026.00, 8e+07), (2027.00, 8.5e+07), (2028.00, 9e+07), (2029.00, 9.5e+07), (2030.00, 1e+08), (2031.00, 1e+08), (2032.00, 1e+08), (2033.00, 1.05e+08), (2034.00, 1.05e+08), (2035.00, 1.1e+08), (2036.00, 1.1e+08), (2037.00, 1.15e+08), (2038.00, 1.2e+08), (2039.00, 1.2e+08), (2040.00, 1.25e+08), (2041.00, 1.25e+08), (2042.00, 1.25e+08), (2043.00, 1.3e+08), (2044.00, 1.3e+08), (2045.00, 1.35e+08)

References

[1] Arneth A *et al* 2019 *Climate Change and Land: An IPCC Special Report on Climate Change, Desertification, Land Degradation, Sustainable Land Management, Food Security, and Greenhouse Gas Fluxes in Terrestrial Ecosystems Summary for Policymakers* (IPCC: IPCC International Panel on Climate Change)

[2] Anadon L D and Holdren J P 2008 Policy for energy-technology innovation *Acting in Time on Energy Policy Conf.* (Cambridge, USA: Harvard Kennedy School)

[3] Sterman J D 2000 *Business Dynamics: Systems Thinking and Modeling for a Complex World* (USA: McGraw-Hill Higher Education)

[4] Maani K E and Cavana R Y 2007 *Systems Thinking and Modelling: Understanding Change and Complexity* (Auckland, New Zealand: Pearson Education New Zealand)

[5] MEMR 2018 *Handbook of Energy & Economic Statistics of Indonesia 2018* (Jakarta, Indonesia: MEMR)

[6] Sasmojo S 2018 in *Workshop of Re-Knitting the Country's Energy ("Merajut Kembali Energi Negeri") (Part 2)* ed ITB (Bandung, Indonesia: Institut Teknologi Bandung (ITB)) Available from: https://www.youtube.com/watch?v=aEh3LSd9UDU

[7] MNDP 2017 The Indonesian vision 2045 ("Visi Indonesia 2045") *Scientific Speech in The Faculty of Economy and Business - University of Indonesia* (Jakarta, Indonesia: MNDP)

[8] Soerawidjaja T H 2 November 2016 *Development of DBF Production Technology* ed M Lamria (Bandung, Indonesia)
[9] Sims R E H, Mabee W, Saddler J N and Taylor M 2010 An overview of second generation biofuel technologies *Bioresour. Technol.* **101** pp 1570-80
[10] Sims R E H, Taylor M, Saddler J and Mabee W 2008 *From 1st- to 2nd- Generation Biofuel Technologies. An Overview of Current Industry and RD&D Activities* (Paris, France: International Energy Agency (IEA))
[11] Hook M 1996 *The Houdry Process for The Catalytic Conversion of Crude Petroleum To High-Octane Gasoline: A National Historic Chemical Landmark* (Pennsylvania, USA: American Chemical Society)
[12] BPSKSTG 2014 *Central Sumba in Figures 2014* (Central Sumba, Indonesia: BPS Statistics of Central Sumba Regency)
[13] BPSKSTM 2014 *East Sumba in Figures 2014* (East Sumba, Indonesia: BPS Statistics of East Sumba Regency)
[14] BPSKSBD 2014 *South West Sumba in Figures 2014* (South West Sumba, Indonesia: BPS Statistics of South West Sumba Regency)
[15] BPSKSBR 2014 *West Sumba in Figures 2014* (West Sumba, Indonesia: BPS Statistics of West Sumba Regency)
[16] MOF 2013 *Book of Spatial Forestry Database 2013 ("Buku Basis Data Spasial Kehutanan 2013")* (Jakarta, Indonesia: The Ministry of Forestry)
[17] EastSumbaLandAgency 30th May 2016 *Sumba Supports and Barriers* ed M Lamria (East Sumba, Indonesia)
[18] EastSumbaForestryAgency 30th May 2016 *Sumba Supports and Barriers* ed M Lamria (East Sumba, Indonesia)
[19] Rahmadi A, Aye L and Moore G 2013 The feasibility and implications for conventional liquid fossil fuel of the Indonesian biofuel target in 2025 *Energy Policy* **61** pp 12–21
[20] Jupesta J 2012 Modeling technological changes in the biofuel production system in Indonesia *Appl. Energy* **90** pp 211-17
[21] Espinoza A, Bautista S, Narvaez P C, Alfaro M and Camargo M 2017 Sustainability assessment to support governmental biodiesel policy in Colombia: A system dynamics model *J. Clean. Prod.* **141** pp 1145-63
[22] Azadeh A and Arani H V 2016 Biodiesel supply chain optimization via a hybrid system dynamics mathematical programming approach *J. Renew.* **93** pp 383-403
[23] Barisa A, Romagnoli F, Blumberga A and Blumberga D 2015 Future biodiesel policy designs and consumption patterns in Latvia: a system dynamics model *J. Clean. Prod.* **88** pp 71-82
[24] Applanaidu S D, Abidin N Z, Sapiri H and Zabid M F M 2015 Impact of Various Palm-Based Biodiesel Blend Mandates on Malaysian Crude Palm Oil Stock and Price: A System Dynamics Approach *Asian Soc. Sci.* **11** pp 190-203
[25] Jeffers R F, Jacobson J J and Searcy E M 2013 Dynamic analysis of policy drivers for bioenergy commodity markets *Energy Policy* **52** pp 249–63
[26] Jonker W D, Brent A C and Musango J K 2015 Modelling the production of biofuel within the Western Cape Province, South Africa *24th Int. Conf. for Management of Technology* (Cape Town, South Africa: IAMOT)
[27] MIT 2015 5 April 2015 *System Dynamics, Sustainability, and Earth Day* Available from: https://executive.mit.edu/blog/system-dynamics-sustainability-and-earth-day
[28] Sachs I 1999 *Social sustainability and whole development: exploring the dimensions of sustainable development*
[29] Musango J K 2012 Technology Assessment of Renewable Energy Sustainability in South Africa *Dissertation* (South Africa: Stellenbosch University)

[30] Bautista S, Enjolras M, Narvaez P, Camargo M and Morel L 2016 Biodiesel-triple bottom line (TBL): A new hierarchical sustainability assessment framework of principles criteria & indicators (PC&I) for biodiesel production. Part II-validation *Ecol. Indic.* **69** pp 803–17

[31] Murphy H T, O'Connell D A, Seaton G, Raison R J, Rodriguez L C, Braid A L and Kriticos D J 2012 *A Common View of the Opportunities, Challenges and Research Actions for Pongamia in Australia* (Australia: BioEnergy Research)

[32] BPPT 2018 *Indonesia Energy Outlook 2018 Sustainable Energy for Land Transportation* ed Yudiartono *et al* (Jakarta, Indonesia: Agency for The Assessment and Application of Technology) [cited 2019] Available from: www.bppt.go.id

[33] BPS 2018 *Non-Oil & Gas Exports and Imports Value* (Statistics Indonesia) [cited 2018] Available from: www.bps.go.id

[34] IEA 2017 *World Energy Outlook 2017* (IEA: International Energy Agency) [cited 2018] Available from: iea.org/weo/

[35] WorldBank 2018 *Commodity Markets Outlook Oil Exporters: Policies and Challenges* (Washington, DC: World Bank Group)

[36] IndexMundi 2019 *Palm oil Monthly Price* (Index Mundi) [cited 2019] Available from: https://www.indexmundi.com/commodities/?commodity=palm-oil&months=300

[37] DBF-TechnologyGroup 2016 *Focus Group on DBF Technology* ed M Lamria (Bandung, Indonesia)

[38] Soerawidjaja T H 2018 *Progress of DBF Technology Development Project at ITB* ed M Lamria (Jakarta, Indonesia)

[39] Saparita R 2017 *Model Face Validation* ed M Lamria (Bandung, Indonesia)

[40] Soerawidjaja T H 2018 *Progress of Biofuel "Distillation Pyrolysis" Development Project at ITB* ed M Lamria (Bandung, Indonesia)

[41] TSIID 2012 *Blueprint and Roadmap of Development Program for Sumba as Iconic Island of Renewable Energy 2012-2025 ("Blueprint dan Roadmap Program Pengembangan Pulau Sumba sebagai Pulau Ikonik energi Terbarukan 2012-2025")* (Jakarta, Indonesia: Team for Sumba Iconic Island Development (TSSID))

[42] GAPKI 2018 *Recent Progress of Indonesian Palm Oil Industry ("Perkembangan Mutakhir Industri Minyak Sawit Indonesia")* [cited 15 October 2018] Available from: https://gapki.id/news/3971/perkembangan-mutakhir-industri-minyak-sawit-indonesia#more-3971

[43] Pearlson M, Wollersheim C and Hileman J 2013 A techno-economic review of hydroprocessed renewable esters and fatty acids for jet fuel production *BIOFUEL BIOPROD BIOR* **7** pp 89-96

Acknowledgment
This paper is associated with the Ph.D. study of Maslan Lamria at Massey University New Zealand, supported financially by NZ-ASEAN Scholars Awards, Massey University Doctoral Scholarship and Massey SF&AT resources.

IOP Conf. Series: Materials Science and Engineering **823** (2020) 012016 doi:10.1088/1757-899X/823/1/012016

Study of electrode modification and microbial concentration for microbial fuel cell effectivity from molasses waste and reduction of heavy metal Cr (VI) by continue dual chamber reactor

I F Nuryana[1], R Puspitasari[1], S. R Juliastuti[1]

[1] Chemical Engineering Department, Institut Teknologi Sepuluh Nopember, Surabaya 60111, Indonesia

E-mail: julizainul@gmail.com

Abstract. The rapid growth of textile industry has resulted in increasing amounts of heavy metal chromium waste. Metal waste production reaches 80% of the total amount of water used in the batik process. Microbial Fuel Cell (MFC) is a promising alternative because of its ability to reduce metal waste Cr^{6+} while processing organic waste and producing bioelectric. In addition, molasses still contain 50-60% sugar which has the potential to be degraded. *Shewanella oneidensis MR-1* was used in this research due to its high productivity. MFC used dual chamber continue to optimize the process. The objective of this research is to analyze the effect of carbon modification and bacterial concentration on MFC electricity production, reduction of Cr^{6+} and reduction of BOD. Carbon electrode was modified with rice husks and strong acid-base to improve MFC performance. The higher concentration of bacteria and modification of electrodes could produce greater electricity and can reduce more hexavalent chromium metal Cr^{6+} also able to reduce the most optimum BOD. Variable of 10^{11} cells/mL bacteria with electrode modification produced the best performance by producing electricity that was relatively more stable and the most optimum with total power density of 690.9mW/m^2, reduced heavy metals Cr^{6+} to Cr^{3+} up to 92.78% and BOD removal of 33.11%.

1. Introduction

Industry in Indonesia is rapidly growing, one of which is the textile and textile products (TTP) industry. Exported TTP increased by 3% in 2017 from the previous period. With the rapid growth of the industry, the amount of waste produced has also increased. The production process produces liquid metal waste which amounts to 80% of the total amount of water used in the batik process [1]. Chromium, one of metal waste, is one of the most dangerous waste for the environment because it cannot decompose. Generally, waste is disposed of in the water, so that it causes a negative impact on the ecosystem and food chain of living things in the water. Chromium metal will contaminate aquatic organisms (plants, fish, etc.) which will be consumed by humans. So, even though exposure to contamination in small amounts and indirectly, it can still endanger human health [2]. Moreover, the toxicity of this waste depends on the form of chromium, its oxidation number, and pH. Thus, acid pH and higher temperatures will increase the toxicity of Cr^{6+} to aquatic organisms [2].

There are several conventional methods for treating Cr^{6+} waste, such as reduction with ferro sulfate, binding of chromium compounds with resin (ion exchange) and thickening [3]. However, these three

IOP Conf. Series: Materials Science and Engineering **823** (2020) 012016 doi:10.1088/1757-899X/823/1/012016

methods require other chemical compounds which are less stable so that they require additional processing and costs. Microbial Fuel Cell (MFC) is a promising alternative because of the ability to reduce Cr^{6+} metal waste while at the same time processing organic waste and producing bioelectric (double track research). Microbial Fuel Cell (MFC) is a biology-based fuel cell that converts chemical energy into electrical energy by catalytic reactions of microorganisms [4]. MFC by utilizing industrial waste is able to become a promising alternative energy source. In addition, molasses as organic waste becomes particular concern because of its high production of 1.56 million tons in 2010 [5]. While more than 30% of molasses waste is still not optimally utilized. Molasses still contain 50-60% sugar which has the potential to be degraded and utilized further as a substrate agent of Microbial Fuel Cell (MFC) with higher productivity than other substrates [5].

The use of both wastes, heavy metal chromium and molasses waste in MFC system will be able to obtain 3 benefits at once, such as heavy metal processing, molasses waste treatment, and bioenergy production. So far, mostly developed-MFC [6] still use a batch system. The use of a batch system has several disadvantages on an industrial scale, which requires a large volume and space, other than that the bacteria used are counted. Thus, Microbial Fuel Cell (MFC) is needed to develop alternative energy from molasses waste and reduce Cr (VI) in heavy metal waste with dual batch systems that use continuous systems which require small volume and space.

Shewanella oneidensis MR-1 is a good bacterial agent to degrade molasses substrate and capable to produce high energy densities of 2 W/m^2 [7]. This bacterium will be used in this research. In addition, electrodes are also a very influential component in the process of reduction and oxidation. The anode surface area affects bacterial activity and electron transfer. In this study, the anode in the form of a carbon rod will be modified with the addition of an area. Rice bran contains high silica [8], so it can be used to increase the anode contact area. Silica will stick to the carbon rod so that additional particles form. This study analyse the effect of bacterial concentration and the effect of electrode modification on the effectiveness of MFC dual batch using a continuous recirculation system, where it is expected that MFC can degrade molasses and reduce chromium (reduce toxicity) while producing the optimum bioelectric. The objectives of this research are to analyze the effect of carbon modification and bacterial concentration on MFC electricity production, reduction of Cr^{6+}, and reduction of BOD.

2. Method

The system was conducted in at atmospheric condition with temperature of 30°C and pressure 1 atm. Molasse waste was used as feed in anode chamber and chromium solution as representative of contaminated water was used as feed in cathode chamber. The anode and cathode reactor was separated with salt bridge. The bacteria used in this research were *Shewanella oneidesis MR-1* with 2 variable concentrations of 10^9 cell/mL and 10^{11} cell/mL.

The experiment was conducted using a dual-chamber reactor with recirculated continue system, as illustrated in figure 1. Due to continue process, the fresh feed would be injected with flowrate 0,4 L/hour for both anode and cathode chamber. The experiment was running during 96 hours, so fresh feed should be prepared every 24 hours to maintain continue system.

The molasse fresh feed was poured into chamber number 11 and chromium solution into chamber number 12. In the first 25 hours, the fresh feed was used to fulfill the reactor anode and cathode first; then after 25 hours, valve number 10 (to chamber) was opened to fill chambers number 7 and 8. After 25 hours, the bottom valve number 10 (bottom product) was opened and at the same time, the pump was turned on to return the bottom chamber feed into reactor again. Recycle ratio was 0,85 or in another sentence, the pump was operated with flowrate 0,34 L/hour. The system was maintained until 96 hours. The analysis was conducted every 12 hours such as BOD (Biological Oxygen Demand), Chromium Hexavalent concentration, bacterial concentration by counting chamber and acidity. However, power or electricity production was analyzed every 6 hours.

2.1. Substrate preparation for anode chamber

500 mL of molasse and distilled water were sterilized using an autoclave to ensure there is no other microorganisms will be involved in the process. To prepare starter of *Shewanella oneidensis MR-1*, 1.5 L nutrient broth was prepared and sterilized. Then, inoculated bacteria were displaced into nutrient broth and incubated during 13 hours in temperature 30°C. After 13 hours, the bacteria would be counted if it was qualified for the concentration variable, starter was ready to use. Starter, molasse and buffer phosphate were mixed together then distilled water was added up to 10 L of total volume. In the continue MFC system, substrate for anode chamber should be prepared every 25 hours.

2.2. Cathode solution preparation

Chromium hexavalent 8 ppm solution was prepared by diluting 0,226-gram $K_2Cr_2O_7$ in distilled water up to 10 L. Buffer citrate pH 3 was added into solution in order to maintain the acidity in pH 3.

2.3. Electrode modification

Modification was the variable of this research in order to analyze the impact of anode modification. Chemical-base modification was used to activate the anode by soaking the electrode in HCl 1 M for 24 hours then rinse the anode by distilled water. After that, the anode was soaked again in KOH 1 M for 24 hours then rinsed by distilled water [9]. In order to increase the surface area, modification using rice bran was conducted. Rice bran contained high amount of silica which semiconductor and able to stick around the electrode's surface [8]. 320 grams of rice bran with homogenous size of 50 mesh was added into 2000 mL of distilled water then stirred around an hour. Then, rice bran was precipitated gravitationally. 1500 mL of solution was displaced and mixed with 100 grams of activated carbon powder.

2.4. Salt bridge preparation

Salt bridge was used to transfer the proton produced by the system from anode to cathode chamber. In this research, KCl-agar salt bridge was used. 30 grams of solid agar was diluted in 1000mL hot distilled water. Then, 40 grams of potassium chloride was added into agar solution. The salt agar solution was poured into containment and cooled until it turned into solid.

2.5. Experiment using dual chamber reactor

The experiment was conducted using dual-chamber reactor with recirculated continue system. Due to continue process, the fresh feed would be injected with flowrate 0,4 L/hour for both anode and cathode chamber. The experiment was running for 96 hours, so fresh feed should be prepared every 24 hours to maintain continue system.

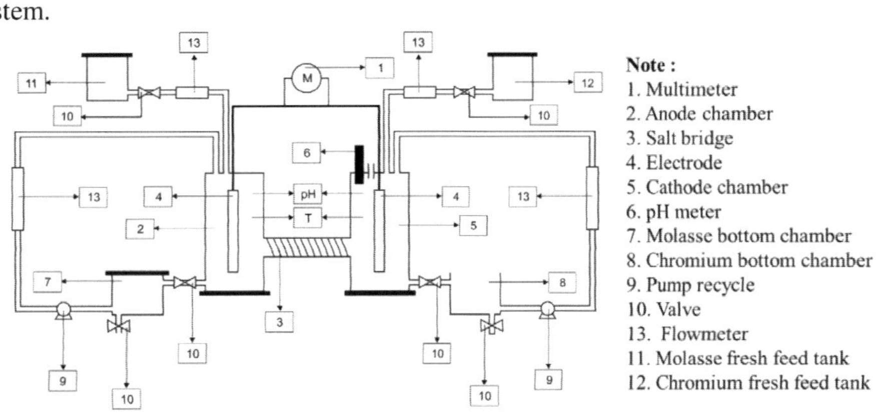

Figure 1. Dual chamber recirculated continue system

3. Experimental Result and Analysis

3.1. Microbial fuel cell – dual chamber continue

In this study, Microbial Fuel Cell (MFC) used a dual-chamber, which was distinguished between the anode chamber and the cathode chamber. The anode compartment contained organic sugar mill waste (molasses) with a concentration of 5% as a substrate. However, the cathode compartment contained chromium hexavalent solution as the representative of wastewater with Cr (VI) concentration of 8 ppm. *Shewanella oneidensis MR-1* carried out metabolism that involves molasses solution as a substrate which caused degradation of organic compounds in the anode compartment. This metabolism took place anaerobically which produces electrons (e-) and protons (H +) with reactions as follows:

$$C_6H_{12}O_6 + 6H_2O \longrightarrow 6CO_2 + 24H^+ + 24^{e-} \qquad E^0 = 0{,}01\ V \qquad (1)$$

electrons (e^-) and protons (H^+) that have been produced would be transferred to the cathode compartment by following the working principle of the voltaic cell. The electrons formed from the reaction accumulate on the electrode (graphite) in the anode compartment. The potential difference between the anode chamber and the cathode caused the flow of electrons to produce power density. The flow of electrons streamed through the electrode circuit and copper wire, while the protons (H^+) pass the Proton Exchange Membrane (PEM) from the anode compartment to the cathode compartment. Then, electrons (e^-) and protons (H^+) reacted with dichromate ions ($Cr_2O_7^{2-}$) in the cathode compartment, these ions were formed by the following reaction:

$$K_2Cr_2O_7 \longrightarrow 2K^+ + Cr_2O_7^{2-} \qquad (2)$$

electrons (e^-) and protons (H^+) reacted with dichromate ions ($Cr_2O_7^{2-}$) to form Cr^{3+} ions or commonly known as reduction reactions as follows:

$$Cr_2O_7^{2-} + 14H^+ + 6^{e-} \longrightarrow 2Cr^{3+} + 7H_2O \qquad E^0 = 1{,}33\ V \qquad (3)$$

As a result of the transfer of electrons from the anode to the cathode, the charge on each chamber became unbalanced. Therefore, the load balancing process was occurred by utilizing a Proton Exchange Membrane (PEM) containing K^+ and Cl^- ions. In order to stabilize anode compartment due to electron loos, the Cl^- ion of the PEM would move towards the anode compartment. Whereas in the cathode compartment that received electrons has a negative charge, so it requires K^+ ions from PEM to stabilize the charge. K^+ enters the cathode compartment to balance dichromate ion or reduced permanganate levels. Then the H^+ ion from the anode compartment moves to the cathode compartment to replace the K^+ ion.

3.2. The profile and effect of acidity to Shewanella oneidensis MR-1 growth

Bacteria played an important role in the process of Microbial Fuel Cells (MFC), which was to oxidize organic substrates. In this study, the bacteria used was *Shewanella oneidensis MR-1*. Bacteria were injected into molasses solution in the anode compartment when in the log phase with the amount corresponding to the concentration variable. In the log phase, bacteria experienced a rapid growth. In addition, the need for energy for bacteria in the log phase was higher compared to other growth phases. So, it was expected that many protons and electrons were produced from the metabolic processes of these bacteria [10]. In order to analyze the growth profile of *Shewanella oneidensis MR-1, the* experiment was conducted. The result was compared with the research of Jeong, 2006.

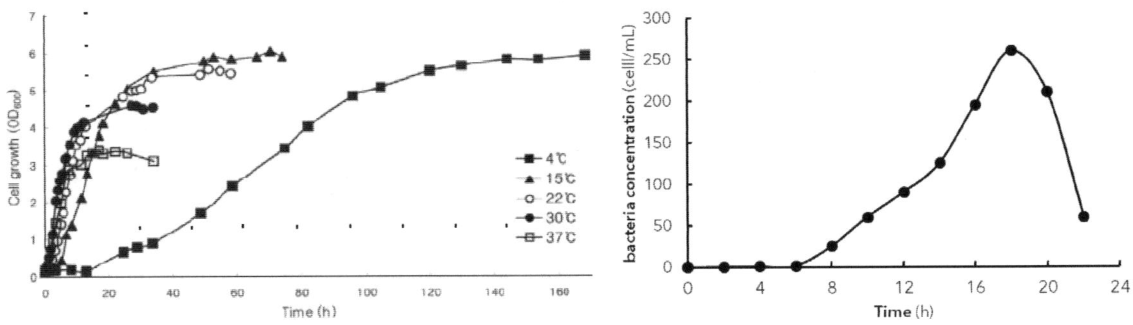

(a) (b)

Figure 2. Growth profile of *Shewanella oneidensis MR-1* from (a) the research of Jeong [11]; (b) in this research

Based on Figure 2 (a), this *Shewanella oneidensis MR-1* was able to live at extreme temperatures close to 0°C even though the optimum temperature growth ranges from 30-35°C [11]. In lower temperatures, *Shewanella oneidensis MR-1* experienced a lower growth rate and vice versa. The profile of the microbial growth of the experiment in Figure 2 (b) had a tendency similar to the previous one. The log phase was reached in 16-18 hours then undergoes a phase of stagnation or stationary phase before then experiencing a death phase. Therefore, for a continuous recycled fresh feed MFC process, bacteria were inoculated into the media and incubated for 13 hours to reach the initial log phase before mixing with molasse as the substrate. 13 hours were chosen considering the log phase.

Shewanella oneidensis MR-1 was optimally working in acidity pH 6-7 to produce MFC with optimum electricity production [12]. Whereas indirectly, electricity production was influenced by population and bacterial activity. Therefore, the main factor in this study was the pH control of the anode compartment to maintain the stability of population conditions and bacterial activity. the pH of anode chamber was monitored every 12 hours. Before KOH addition, the pH of the anode compartment always decreased for all variables (curve (b)) in Figure 3 due to the metabolic reaction of the *Shewanella oneidensis MR-1* with molasses that produces protons or H^+. The protons produced from this reaction should be transferred to the cathode chamber, but unfortunately, the protons did not completely be transferred. As a result, it caused a decrease in pH. Low pH (acid) in the anode compartment would deactivate microorganisms and reduce MFC performance [13].

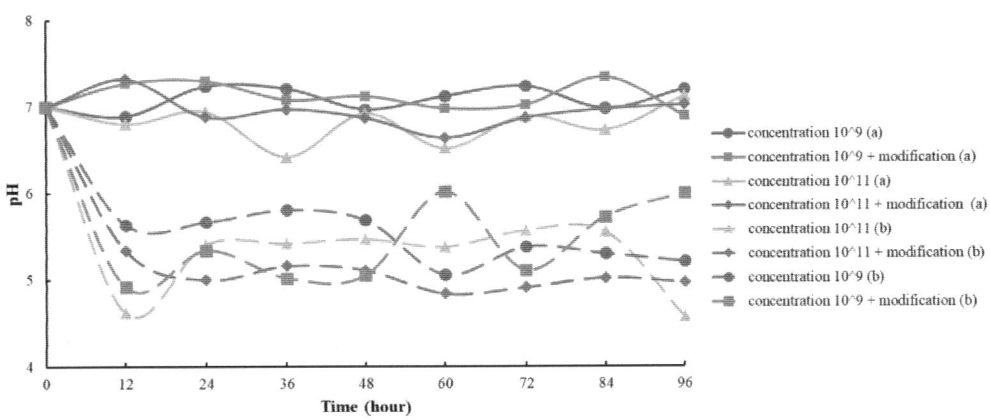

Figure 3. pH of anode chamber before (b) and after (a) pH control using KOH

The use of buffer phosphate pH 7 was unable to anticipate pH reduction due to the production of H^+. Salt bridge did not work optimally to balance the anion and cations. The H^+ proton rate towards the cathode compartment was much lower than the rate of H^+ production in the anode compartment. So, every 6 hours a strong base solution or KOH was added to the reactor to neutralize the reactor pH and maintain the bacterial population. So, it can be seen in the curve (a) of Figure 3, that after KOH addition, the pH rises to around 6-7.

3.3. The effect of bacteria concentration and electrode modification into power production

Systematically, the continuous system processed feeds and produced products continuously. In this process, the substrate fresh feed (anodes) and chromium solution (cathode) were injected continuously. Likewise, electricity should be produced stably. Electricity production was influenced by several factors such as the rate of substrate conversion, PEM (Proton Exchange Membrane), internal and external barriers in cells and the potential difference between the anode and cathode [14]. Biofilm formation around the electrode could also be a barrier to the maximum absorption of electrons from bacteria [15].

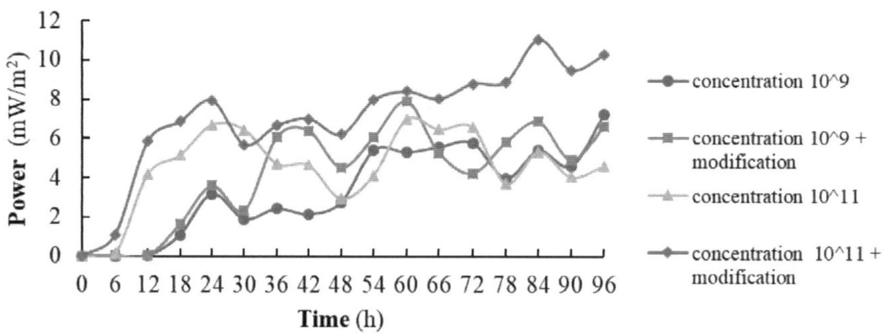

Figure 4. Power production of MFC dual chamber

Based on Figure 4, the increasing concentration of bacteria caused higher production of electricity. The concentration of bacteria 10^{11} cells/mL either without or with modification was able to produce a higher power than the concentration of 10^9 cells/mL. The higher the concentration of bacteria, the more substrates were degraded by bacteria to produce electrons.

Electricity was generated every time unit by the system; therefore, an intergalactic calculation was carried out to find out all the electricity produced during the process. This refers to the area under each curve for each variable. For the variable concentration of 10^{11} cells/mL without modification, it produced a total of 446.06 mW/m^2 or 39.73% higher than the concentration of 10^9 cells/mL without modification which was 319.23 mW/m^2. Likewise, for the variable with electrode modification, where the concentration of 10^{11} cells/mL was able to produce a total of 690.9 mW/m^2 or 66.95% higher than the concentration of 10^9 cells/mL which was 413.84 mW/m^2.

Another factor affecting power production was anode modification. The electrodes used in the manufacture of MFC must have good electrical conductivity, a broad surface, low resistivity, non-corrosive, biocompatible, chemically and mechanically stable to obtain results that can be produced continuously [16]. Anode modification with strong acid-base and treatment of rice bran aimed to increase the surface area of the electrode. Silicon inside rice bran was expected to be able to attach to carbon electrodes so that it helps increase the contact area of electron transfer from bacteria. In this study, electrode modification was able to increase the profile of electricity production. For total electricity produced, the concentration of 10^{11} cells/mL bacteria with modification produced 54.89% more electricity than without modification. While the variable concentration of 10^9 cells/mL produced 29.64% higher than without modification.

3.4. The effect of bacteria concentration and electrode modification into BOD removal

BOD or (Biological Oxygen Demand) was a measure of the amount of oxygen used by microbial populations contained in waters in response to the entry of decomposed organic matter [17]. BOD removal in this study was shown in the Figure 5.

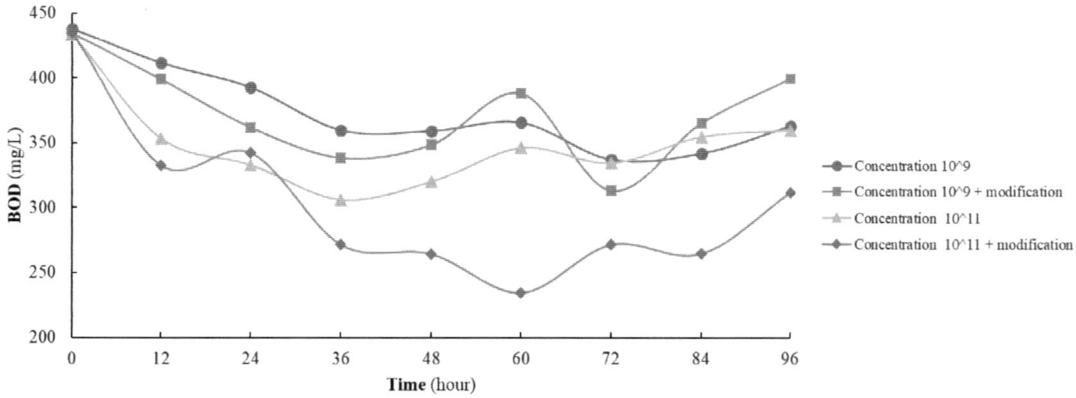

Figure 5. BOD removal of molasse substrate in MFC dual chamber

BOD removal was strongly influenced by bacterial activity, while bacterial activity was influenced by bacterial concentration, bacterial population, and pH of the anode space. Based on bacterial concentration, 10^{11} cells/mL either without or with modification was able to produce BOD removal higher than the concentration of 10^9 cells/mL.

3.5. The effect of bacteria concentration and electrode modification into Cr^{6+} removal

Cr^{6+} was one of the compounds that can put environment and water in danger. Therefore, the decrease in hexavalent chromium levels in this study was highly preferred. Hexavalent chromium metal reduction occurs by utilizing the metabolic results of *Shewanella oneidensis MR-1* in the form of protons and electrons in the anode chamber which moved to the cathode compartment due to potential differences and then reduces Cr^{6+} to Cr^{3+} in the cathode compartment.

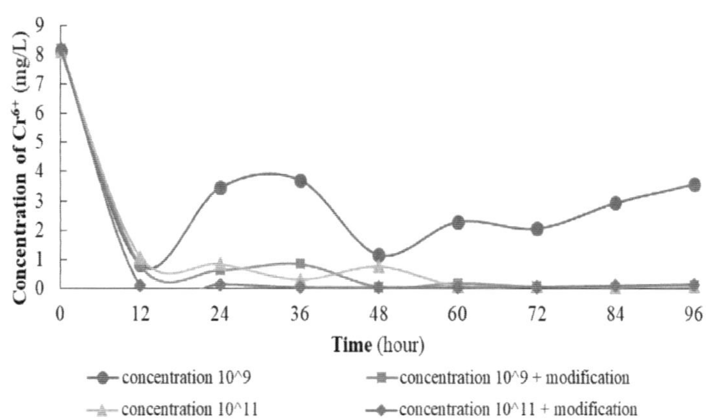

(a)

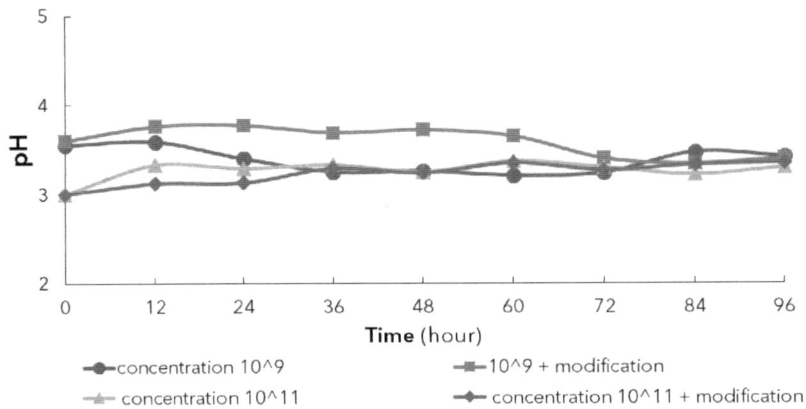

(b)

Figure 6. (a) Cr^{6+} removal in MFC dual chamber; (b) pH in cathode chamber

Based on Figure 6, Cr^{6+} concentration levels drastically decreased from the concentration of 8 ppm to almost touching the 0-ppm point in the first 12 hours. Then at the 24 to 96 hours, the levels were relatively stable between 0 and 1 ppm. This fluctuation of hexavalent chromium levels occurred because of the fresh feed entering the main chamber, this results in an increase in the amount of hexavalent chromium every time along with the presence of electric currents which caused the reduction of Cr^{6+} to Cr^{3+}.

Based on these data, it showed that the higher the concentration or bacterial population the greater the chromium reduction that occurs. The large population of bacteria involved in anaerobic metabolic reactions affects the number of electrons formed. The higher the concentration of bacteria, the more bacteria that metabolize molasses so that more electrons were formed. Then, electrons would be transferred to the cathode chamber so that it can reduce Cr^{6+}. The modification of the electrode also affected the results of hexavalent chromium metal reduction. Electrode modification allowed higher removal than without modification due to the modification by using strong acids, strong bases and rice bran capable to increase the surface area of carbon rod electrodes. As a result, more electrons were transferred to the cathode chamber. So, the reduction of hexavalent chromium was higher.

The pH of the cathode compartment was maintained at pH 3 with the help of a citrate buffer. The optimum pH for Cr^{6+} removal was in the range of pH 3-4 [18]. This was in line with the reduction of hexavalent chrome that carried out at acidic conditions.

Figure 6 (b) was the pH profile of the cathode compartment which has no significant fluctuation. Even though protons could not well transfer to the cathode, the available protons still able to reduce Cr^{6+}. The acid condition was chosen for the cathode chamber because Cr^{6+} can be easily or spontaneously reduced to Cr^{3+} [18]. So, it was easier for hexavalent chrome to be reduced in an acid solution than neutral or alkaline.

4. Conclusion

This research showed that dual-chamber microbial fuel cell (MFC) continuous recirculation system was capable to reduce metal Cr (VI), produce electrical energy and reduce BOD. The higher concentration of bacteria and modification of electrodes produce greater electricity and more efficient to reduce hexavalent chromium metal Cr^{6+}. Variable concentration of 10^{11} cell / mL bacteria with electrode modification produced more stable and the most optimum electricity with a total power density of 690.9 mW / m^2. It also reduced heavy metals Cr^{6+} with a percent removal of 92.78% and BOD removal of 33.11%.

IOP Conf. Series: Materials Science and Engineering **823** (2020) 012016 doi:10.1088/1757-899X/823/1/012016

5. Acknowledgment

We would like to show our gratitude to Mr. Dimas Eko Prasetyo as Director of PT Energi Agro Nusantara for supplying molasses as the substrate of this research. We are also immensely grateful to Chemical Engineering Department, Institut Teknologi Sepuluh Nopember specially to Waste Water Treatment Laboratory for all chemical supplies and facilities to support the research.

6. References

[1] Watini 2009 Pengaruh waktu kontak eceng gondok (Eichornia crassipes) terhadap penurunan kadar Cd dan Cr pada air limbah industri batik (Home Industry Batik di Desa Sokaraja Lor) Kota Purwokerto *Skripsi* (Purwokerto: Universitas Jenderal Soedirman)

[2] Jain A 2010 Removal of Copper(II) from aqueous solution using spent tea leaves (STL) as potensial sorbent *Water SA* **36**(3) ISSN 1816-7950

[3] Idaman N 2010 Metoda penghilangan logam berat (As, Cd, Cr, Ag, Cu, Pb, Ni Dan Zn) di dalam air limbah *Industri Jurnal Air Indonesia* **6**(2) 1

[4] Li Z, Xingwang Z and Lecheng L 2008 Electricity production during the treatment of real electriplating containing Cr^{6+} using microbial fuel cell *Process Biochemistry* **43** 1352

[5] Yakinudin A 2010 *Bioetanol Singkong sebagai Sumber Bahan Bakar Terbaharukan dan Solusi untuk Meningkatkan Penghasilan Petani Singkong* (IPB : Bogor)

[6] Mohan, Lens S V and Nancharaiah Y V 2015 Metal removal and recovery in bioelectrochemical system: A review *Bioresource Technol* **195**(1) 96

[7] Ringeisen B, Ray R and Little B 2007 A miniature microbial fuel cell operating with an aerobic anode chamber *Journal of Power Sources* **165** 591

[8] Agung F 2013 Ekstraksi silika dari abu sekam padi dengan pelarut KOH *Jurnal Konversi* **2**(1)

[9] Novitasari D 2011 *Optimalisasi Kinerja Microbial Fuel Cell (MFC) untuk Produksi Energi Listrik Menggunakan Bakteri Lactobacillus bulgaricus* (Jakarta : Departemen Teknik Kimia FT Universitas Indonesia)

[10] Setyati, Wilis A, Erni M, Triyanto, Subagiyo and Zainuddin M 2015 Kinetika pertumbuhan dan aktivitas protease isolat 36k dari sedimen ekosistem mangrove, Karimunjawa, Jepara *Ilmu Kelautan* **20**(3) 163

[11] Jeong Y, Song S, Lee S and Hur B 2006 The growth and EPA synthesis of Shewanella oneidensis MR-1 and expectation of EPA biosynthetic pathway *Biotechnology and Bioprocess Engineering* **11** 127

[12] Biffinger J C, Pietron J, Bretschger O, Nadeau L J, Johnson G R, Williams C C, Nealson K H and Ringeisen B R 2008 The influence of acidity on microbial fuel cells containing *Shewanella oneidensis Biosensors & bioelectronics* **24** 906-11 doi: 10 1016/j bios 2008 07 034

[13] Li Z, Xingwang Z and Lecheng L 2008 Electricity production during the treatment of real electriplating containing Cr^{6+} using microbial fuel cell Process *Biochemistry* **43** 1352

[14] Jadhav G S, More T T and Ghangrekar M M 2008 *Microbial Fuel Cell: Application in Wastewater Treatment* (India: Indian Institute of Technology)

[15] Sevda S and Sreekrishnan 2012 Effect of salt concentration and mediator in salt bridge microbial fuel cell for electricity generation from synthetic wastewater *Journal of Environmental Science and Health* **47**(6) 878

[16] Akbar N, Kirom M R and Iskandar R 2017 Analysis of the effect of metals as an electrode in microbial fuel cell to the electrical energy *Production Proceeding of Engineering* **4**(2) ISSN : 2355 – 9365

[17] Erwin 2014 Tingkat pencemaran pada saat pasang dan surut di perairan pantai Kota Makassar *Skripsi* (Makassar: Universitas Hasanuddin)

[18] Silva B, Figueiredo H, Neves I C and Tavares T 2009 The role of pH on Cr (VI) reduction and removal by *Arthrobacter Viscosus International Journal of Chemical. and Biological Enginering* **2**(2) 100

Biomass gasification of candlenut shell and coconut coir with the updraft gasifier method being alternative fuel

Makaborang M [1], Kabeko S S [1], Dethan J J S [1], and Lano M L [1]

[1]Technology Agriculture Faculty, Universitas Kristen Artha Wacana, Kupang, NTT

E-mail: marthenmakaborang@yahoo.com

Abstract. Lately, the use of fossil fuels has caused many environmental problems such as global warming as a result of greenhouse gas emissions, which has led to increased fuel prices and erratic availability. These encouraged the interest of researchers to find alternative fossil fuels sourced from renewable raw materials. Candlenut shells and coconut fiber are high-potential plantation and agricultural commodities in East Nusa Tenggara. Statistical data shows that candlenut production reached 1,262.38 tons, while coconut reached 68,496 tons in 2016. The by-products of shells and coir can be used as a good raw material for the gasification process into alternative fuels. One of the established gasification processes in the industry is updraft gasifier. This research has varied the use of candlenut shells and coconut coir in the range of 0-100% to study the characteristics of its gasification products. The purpose of this study was to determine the effect of variations of candlenut shells and coconut coir on syngas ignition time; the effective time of the gasification process, remaining charcoal and ash, the efficiency of the gasification process and to find out the best variation between candlenut shells and coconut fiber on gas products (syngas) updraft gasifier. The results obtained indicate that the process of gasification of candlenut shells and coconut coir using updraft gasifier was very influential at the time of ignition of syngas, and the effective time of gasification. The highest syngas ignition time was obtained in a variation of 50% candlenut and 50% coconut husk which was 43.14 minutes and the lowest was a variation of 0% shells and 100% coconut husk which was 39.75 minutes. While the highest effective time of the gasification process was obtained by variations of 100% candlenut and 0% coconut husk which was 33.13 minutes and the lowest in variations of 75% shells and 25% coconut husk which was 37.40 minutes. The best gasification process was shown in a variation of 50% candlenut shells and 50% coconut husk.

1. Introduction

The use of fossil fuels is increasing causing world oil reserves to run low, so that fuel prices continue to rise. The issue of using fossil fuels also causes global warming and causes environmental damage. To minimize the worst possible impact of the use of fossil fuels, the development of renewable energy sources is an alternative to fossil fuels.

Candlenut and coconut fiber are forestry commodities that are growing rapidly in Indonesia, especially East Nusa Tenggara. In 2013 the area of candlenut plantations in East Nusa Tenggara reached 79,354 ha, and produced a total fruit production of 24,301 tons, while the production of coconut plants in East Nusa Tenggara province in 2016 reached 68,496 tons, with candlenut production reaching 1,262.38 tons [1].

Candlenut consists of four layers which include the skin of the fruit, mesocarp, fruit flesh, shell, and seed meat. Between the shell and the flesh of the seeds is the epidermis. The percentage of each part of the whole hazelnut is 20% outer skin, middle layer 5%, and 75% shell and seeds. Comparison between shell and seed weight is 7: 3 (Dali and Ginting, 1981 in [2]) so that from 24,301 tons of candlenut in NTT has the potential to produce shells of 13,098.28 tons/year.

Coconut fruit consists of several parts, namely, epicarp (outer skin whose surface is slippery, rather hard and less than 1/7 mm thick); mesocarp (middle skin called coir which consists of hard fibers 3 - 5 cm thick); endocarp (hard shell, 3-6 mm thick, the inner part is attached to the outer shell of the seeds or endosperm); institute or endospermic pistils 8-10 mm thick [3]. Coconut fruit consists of 35% coconut fiber, 15% coconut shell, 29% fruit flesh and 21% coconut water [3]. Thus, from 68,496 tons of heads in NTT has the potential to produce coconut husk of 23,973.6 tons/year.

Today the use of candlenut shells and coconut husks is limited to smoking and making charcoal. This type of candlenut shells and coconut husk waste in very large quantities has not been utilized properly. And if the waste is not used, there will be the possibility of environmental pollution. In general, people just throw the candlenut shell and coconut coir away, although some people use it as fuel for smoking and also as ordinary charcoal or for other cooking activities. However, smoking method carried out by the community can result in environmental pollution. On the other hand, the waste or pecan shell endangers pedestrians, especially those who do not use footwear because the pecan shell is usually hard and sharp. In the utilization of biomass waste converted into heat energy, one of the ways that is often used is the combustion process with limited air (partial oxidation). The gas produced can be used for domestic purposes and can be used as a source of heat energy in the process of drying, roasting agricultural material.

The results of the proximate analysis of the candlenut shell showed that the pecan shell has a volatile value of 81.9 %, fixed carbon 12.60 %, ash 3.10 %, and moisture 2.40 %. The ultimate analysis of the candlenut shell showed that the candlenut shell contained carbon (C) 53.78%, hydrogen (H) 4.37, oxygen (O) 41.50 %, nitrogen (N) 0.37% and calorific value of 14.25 KJ kg^{-1}. While the proximate analysis of coconut fiber has a volatile value of 74.59 %, fixed carbon 21.6 %, ash 1.34 %, and moisture 2.45 %. The ultimate analysis of coconut fiber contained with carbon (C) 47.6 %, hydrogen (H) 5.7 %, Oxygen (O) 45.6 %, nitrogen (N) 0.2 % and calorific value 3942.751 cal/kg [4]. Based on the proximate and ultimate analysis, the biomass of candlenut shells and coconut fiber can be used for gas production through gasification.

Gasification is the process of converting biomass material into gas fuel and is usually carried out on a reactor called a gasifier. Gasifier is a chemical reactor where complex chemical and physical processes occur such as drying, heating, pyrolysis, oxidation, and reduction.

Based on the background, the authors are interested in conducting research with the topic "Biomass Biomass Candlenut and Coconut Fiber Biomass by Using the Updraft Gasifier Method into Alternative Fuels". The aim of the study is to find out the effect of variations in the treatment of candlenut shells and coconut coir on syngas ignition time, effective time gasification process, remaining charcoal and ash, the efficiency of the gasification process and to find out the best variation between candlenut shells and coconut coir in gas product (syngas) updraft gasifier.

2. Material and Method

This research has been carried out at the Agricultural Mechanization Engineering Laboratory, Faculty of Agricultural Technology Universitas Kristen Artha Wacana Kupang from September to October 2016. This research was used as the main tool gasifies type updraft batch capacity of 46 liters, which was equipped with a blower (combustion air supply), temperature sensor (thermocouple), stove or burner. Research support tools such as hotwire anemometer to measure airflow and flow rates, digital thermometers to record temperature performance, stopwatches, roller meters, and scales. While the main ingredients to be used in this study are candlenut shells and coconut fiber and supporting materials include charcoal and kerosene for initial ignition.

This study used a basic design of RAL (Randomized Complete Design), 5 treatments and 3 replications to produce 15 experimental units. The treatment in this study was to examine the percentage of candlenut shell material (TK) and coconut fiber (SK) consisting of:

Treatment A is 100% hazelnut and 0% coconut fiber
Treatment B is 75% hazelnut and 25% coconut fiber
Treatment C is 50% hazelnut and 50% coconut fiber
Treatment D with 25% hazelnut and 75% coconut fiber
Treatment E with 0% hazelnut and 100% coconut fiber

Observation Variables:

1. Syngas Ignition Time.
   Syngas ignition time is calculated based on the time interval at which syngas is turned on stably without interruption.
2. Effective time of gasification.
   The effective time of gasification is calculated based on the time needed for the conversion of all biomass in the gasifier and each kilogram of material.
3. Charcoal and Ashes Remaining
   The mass of ash and charcoal residual gasification results are collected and dried until the water content reaches 10%, then weighed.
4. Efficiency of Gasification Process
   The efficiency of the gasification process is calculated using the following formula:

$$Eff = \frac{Mbb - Mab}{Mbb} \ x \ 100\% \quad (1)$$

Where : Eff = Efficiency of Gasification Process
M_{bb} = The Mass of Material Used
M_{ab} = The Mass of Charcoal and Ashes from Gasification

2.1. Data retrieval

Gasifier was filled with charcoal feed (0.2 kg) and ignited evenly, candlenut shells and coconut coir were added (3 kg), ignite feeders (charcoal) with kerosene, then if the charcoal has been burned fairly evenly then data collection the temperature in the gasifier, the temperature of the ignition for the gas. Data retrieval was carried out for 108 minutes with observation time every 3 minutes when the resulting gas was able to ignite until the gas was able to run out. After the experiment, the gasifier closes and the blower was shut down.

2.2. Data analysis

Data were analyzed using graphical analysis and analysis of variance (ANOVA) to see the effect of treatment on parameters. If it had a significant effect then we will continue the post hoc Duncan's Multiple Distance Test (GD) and to see the best treatment combination at 1% test level.

The mathematical model for the factorial RAL (complete random design) experiment with the percentage TK factor (3 levels) and the SK percentage factor (3 levels) can be written as follows [5]:

$$Y_{ijk} = \mu + \alpha_i + \beta_j + (\alpha\beta)ij + \varepsilon_{ijk} \quad (2)$$

i = 1,2,…,a
j = 1,2,…,b
k = 1,2,…,n

For variations in the percentage of TK and SK repeated 3 times, then the mathematical equation can be written as follows:

IOP Conf. Series: Materials Science and Engineering **823** (2020) 012017 doi:10.1088/1757-899X/823/1/012017

$$Y_{ijk} = \mu + TK_i + SK_j + (TKSK)ij + \varepsilon_{ijk} \quad (3)$$

i = 1,2,3
j = 1,2,3
k = 1,2,3

Where:

Y_{ijk}	=	The observation of TK factor level-i, SK factor level-j and on the k-replication
μ	=	General median value
TK_i	=	The influence of TK factor on level-i
SK_j	=	The influence of SK factor on level-j
(TKSK)ij	=	the effect of the interaction of the TKSP factor on the i-level (from the TK factor), and the j-level (from the SK factor
ε_{ijk}	=	random influence (random error) on i level (TK factor), j-level (SK factor), TKSK interactions on i and j and in k-replications

3. Result and Discussion

3.1. Syngas ignition time

The ignition time of syngas or gas producer was calculated based on the time interval where syngas starts to burn stably without interruption in the process of gasification of the candlenut shell and coconut husk. The rate of ignition of syngas on the burner is affected by the air flow rate and material size. The longest ignition time occurred in treatment C (TK50% VS SK50%) with a material size of 1.1-3.0 cm with an effectiveness value of 43.14 minutes. The average time of ignition of syngas in each treatment can be seen in (Figure 1).

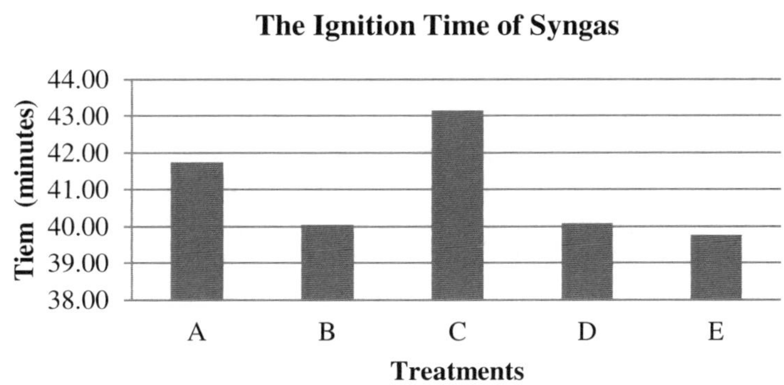

Figure 1. Average time to start syngas at each treatment

The duration of ignition of syngas depends on several factors, including how fast the fuel can be decomposed or decomposed into gas. If the fuel can be decomposed in a short time, the ignition time for syngas will be short. But conversely if the decomposition of fuel into gas lasts longer, then the ignition of syngas can take place in a longer time. Another factor causing fast or slow ignition time of syngas is very dependent on the temperature of the gasification reactor where if the oxidation process takes place under limited air conditions and the right material composition, the time of ignition of syngas will last longer. But on the contrary if the combustion process takes place perfectly, the syngas ignition time will take place faster [2, 6].

IOP Conf. Series: Materials Science and Engineering **823** (2020) 012017 doi:10.1088/1757-899X/823/1/012017

3.2. Effective time of the gasification process

The effective time of the gasification process is the time required to carry out thermal conversion of all biomass in the gasification reactor. The effective time of the gasification process is calculated based on the time needed for the conversion of all biomass in the updraft gasifier reactor per kilogram of material. The updraft gasifier capacity scale used in this study was 3 kg while the feeder (charcoal) was 0.2 kg so that the total material used in each treatment was 3.2 kg.

The analysis showed that the effective time of the treatment is 35.17 minutes, which is the average of all treatments. But for the treatment with the best effective time occurs in treatment A (Candlenut Shell 100% VS Coconut Fiber 0%). The average effective time of the gasification process at each treatment is shown in (Figure 2). The analysis results in the figure show that the effective time of the gasification process tends to be stable and does not experience a decrease and a significant increase in the range of 33-37 minutes.

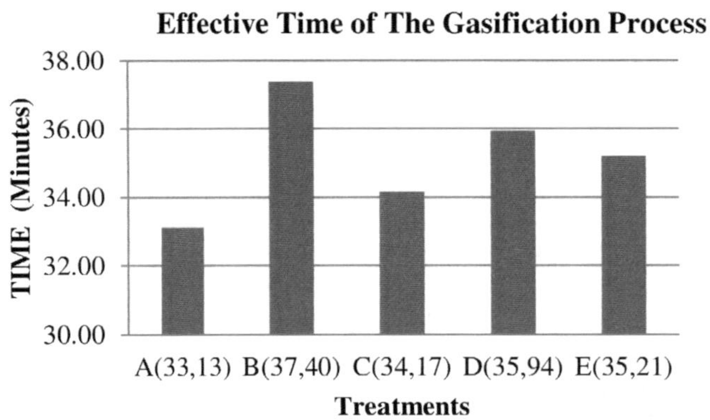

Figure 2. Average effective time of the gasification process

3.3. Mass of charcoal and ashes remaining

Measurement of the rest of the process of gasification of the candlenut shell and coconut fiber in the form of ash and charcoal was carried out to determine the residual combustion process. Ash and charcoal measurements were carried out after drying with a moisture content <10%. From the results, it is known that the higher the percentage of candlenut shell, the remaining combustion in the form of ash and charcoal the higher as seen in Figure 3.

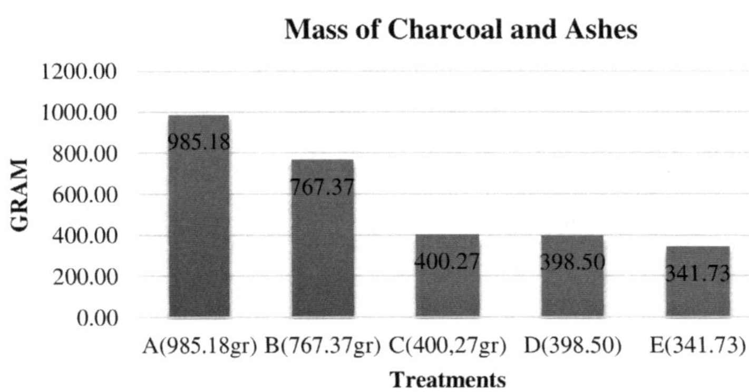

Figure 3. Mass of charcoal and ashes

From the graph above shows that the greater the percentage of candlenut shell in the treatment of the gasification process, the greater the mass of ash and charcoal.

3.4. *Efficiency of gasification process*

The efficiency of the gasification process is the percentage of the gasification process from the Candlenut Shell and Coconut Fiber which is changed to syngas or the ratio between the syngas formed with fuel consumption. The results of the analysis show that the higher the amount of coconut fiber, the better the combustion efficiency of the gasification process, as seen in Figure 4.

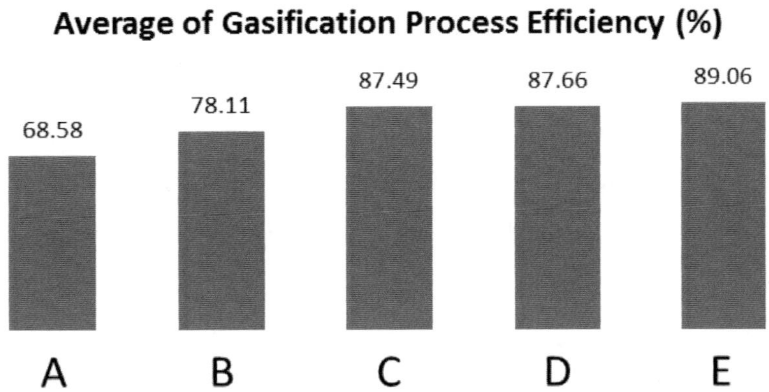

Figure 4. The efficiency of gasification process

The results show that the amount of residual combustion gets smaller if the percentage of coconut coir in the treatment is higher than the percentage of the candlenut shell.

4. Conclusion

Provision of treatment in the process of gasification of the candlenut shell and coconut fiber using an updraft gasifier is very influential in the ignition of syngas, and the time of gasification effectiveness. Syngas ignition time was 43.14 minutes (C), and the gasification process time was 35.17 minutes (A). The most remaining ash and charcoal in treatment A was 985.18 grams and the lowest in treatment E was 341.73 gr, and the efficiency of the gasification process was 68.58%. The gasification process of candlenut shell and coconut fiber, the best combination occurred in the TK50% VS SK50% treatment, where the syngas flame reaches 43.14 minutes, the effective time for the gasification process is 34.17 minutes. The remaining combustion was 400.27 grams and the efficiency of the gasification process was 87.49%.

References

[1] Anonim 2018 *Badan Pusat Statistik Propinsi NTT* (Kupang)
[2] Abineno J and Ibrahim H 2013 Pemanfaatan limbah tempurung kemiri sebagai energi alternatif melalui pada reaktor updraft gasifier *Laporan Penelitian Hibah Bersaing*
[3] Setyamidjaja D 2008 *Bertanam Kelapa* (Yogyakarta: Kanisius)
[4] Abineno J and Purwantana B 2012 Gasifikasi tandan kosong kelapa sawit menggunakan updraft gasifier *Prosiding Seminar Nasional Perhimpunan Teknik Pertanian Indonesia* (Bandung)
[5] Gomez K A and Gomez A A 1995 *Prosedur Statistik untuk Penelitian Pertanian* 2nd edition E Syamsudin dan J S Baharsjah Eds. (Jakarta : UI – Press) pp 13 – 16
[6] Abineno J C and Koylal J A 2018 Gasifikasi limbah tempurung kemiri sebagai energi alternatif menggunakan updraft gasifier pada laju aliran udara berbeda *Jurnal Teknik Pertanian Lampung* 7(3) 175-80

Enzymatic pretreatment effect on product composition in biohydrocarbon synthesis via metal soap decarboxylation

A N Istyami[1], M Pratiwi[1], T Prakoso[1], and R Purwadi[2]

[1]Department of Bioenergy Engineering and Chemurgy
Institut Teknologi Bandung, Kabupaten Sumedang 45363 Indonesia
[2]Department of Food Engineering
Institut Teknologi Bandung, Kabupaten Sumedang 45363 Indonesia

E-mail: anistyami@che.itb.ac.id

Abstract. The recent demand of green energy sources has driven technology development into production of renewable fuels. Among the types of renewable fuels, biohydrocarbon is one of the most attractive forms. It can be applied onto current automotive machine without blending with fossil fuel. Biohydrocarbon can be produced from plant oils via decarboxylation of metal soap. Nevertheless, a hydrolysis step to convert plant oils into free fatty acids is required to synthesize metal soap. A low-cost technology to overcome this problem is hydrolysis reaction using lipase enzyme. In this paper, a study was established to determine the effect of enzymatic pretreatment on biohydrocarbon product composition. Results show that enzymatic pretreatment increased selectivity towards diesel fuel fraction, although fractionation will enable utilization of shorter chain fractions.

1. Introduction

The impact of fossil fuel application on the environment has drawn the public's attention in the last decades. Aside from being the source of carbon dioxide which contributes to the greenhouse effect, fossil fuel is considered as unsustainable source of energy on account of its limited reserves. Consequently, the technology of renewable energy sources developed rapidly, such as biodiesel (based on fatty acid methyl ester, FAME) and bioethanol as a mixture for diesel fuel and gasoline. Nevertheless, these types of biofuel cannot be applied without blending with fossil fuel, because both FAME and ethanol are oxygenated compounds. To imitate the conventional diesel fuel and gasoline, hydrocarbon-based biofuel (or biohydrocarbons) has been synthesized in many kinds of research methods [1, 2]. The product, which is also called drop-in biofuel, is readily applied into automotive engine without the necessity of blending with fossil fuel.

Biohydrocarbons can be synthesized via decarboxylation of metal soap, which is saponification product from triglyceride [3]. Triglyceride is abundant resource, especially because it is the main component of plant oils, but synthesis of metal soap from triglyceride requires high pressure [4]. Severity of process increases operation cost, and in the case of biofuel products, economic feasibility is a strong consideration. Alternatively, synthesis of metal soap can also be established from free fatty acid in mild operating conditions [5]. Triglycerides, consequently, need to be hydrolyzed into free fatty acids prior to saponification. Hydrolysis of triglyceride can be established with noncatalytic thermal decomposition [6], which requires high temperature and high pressure (250 ^{0}C, 50 bar), nevertheless, mild process condition is preferable. Conversion of triglyceride into fatty acid can alternatively be done with enzymatic reaction, i.e., lipolysis. Using the lipase enzyme, the pretreatment step in biohydrocarbon synthesis is feasible to be established in atmospheric pressure and around room temperature.

Enzymatic synthesis biohydrocarbon from lipid has been reported before with *Escherichia coli* with yield of alkane under 0,5 g/L [7]. It requires active microorganism; thus oil in water emulsion is chosen for reaction medium. In this research, isolated enzyme was used instead of microorganisms, enabling the reaction to occur in oil-based medium and yields higher products.

This research aims to provide a comparison between biohydrocarbon product composition involving enzymatic pretreatment and without enzymatic pretreatment. Palm oil was selected as process feed, and crude plant lipase was selected as the lipolytic agent.

2. Material and Methods

2.1. Lipolysis of triglycerides
Lipolysis was conducted with crude plant latex lipase from frangipani (*Plumeria rubra*). Frangipani latex suspension was settled in 7°C and dried to obtain its dry, white powder precipitate. The latex lipase was mixed with palm oil and pH 8.25 buffer and stirred for 24 hours at room temperature. Lipolysis product, containing fatty acids and unreacted glycerides, was then converted into metal soap. In several experiments, sweet water was removed from lipolysis product prior to metal soap preparation. To separate sweet water, lipolysis product was heated until two layers were formed.

2.2. Preparation of metal soap
Metal hydroxide reagent was a mixture of calcium, magnesium, and zinc hydroxide, and was prepared using the procedure from reference [5]. In process without enzymatic pretreatment, palm fatty acid distillate (PFAD) was used as feed. In the process involving enzymatic pretreatment, lipolysis products were used as feed. Saponification products, which are in the form of dry, white powder, were then subjected to decarboxylation.

2.3. Decarboxylation of metal soap
Biohydrocarbon was produced via decarboxylation of metal soap in the previous step. Sample of 10 g metal soap was heated to 370°C and the temperature was held for 3 or 5 hours. The product obtained were clear yellow liquids, which was then analyzed using GC-FID.

2.4. Analysis of biohydrocarbon
The composition of biohydrocarbon was analyzed using gas chromatography Shimadzu 2010 (30 × 0.25 mm × 0.25 μm) equipped with an FID and a Restex-1 capillary column. Helium gas was used as a carrier. The temperature of injection was 340 °C. The column temperature was set from 40 to 300 °C with 5 °C/min rate, then raised to 315 °C at a rate of 1 °C/min and kept for 10 min. Alkane C8–C20 was used for calibration standards.

3. Results and Discussion
The objective of metal soap decarboxylation is the removal of carboxylate group from esters of fatty acids, leaving hydrocarbons as the product. Researches have been conducted to develop this technology, i.e., selection of metal oxide [8] and operating condition in saponification and decarboxylation [3].

Temperature rise in decarboxylation facilitates carbon chain cracking, which leads to various lengths of hydrocarbon products. The product is then fractionated so they can be applied based on their respective characteristics. Generally, a long carbon chain (C13-C17) the to diesel engines, medium carbon chain (C10-C14) is dedicated to aviation jet engine, and a shorter carbon chain (C7-C10) is dedicated for Otto engine.

IOP Conf. Series: Materials Science and Engineering **823** (2020) 012018 doi:10.1088/1757-899X/823/1/012018

This research aims to identify the effect of enzymatic pretreatment on product composition in biohydrocarbon synthesis via metal soap decarboxylation. Figure 1 shows the GC-FID chromatogram of biohydrocarbon synthesized. Figure 2 and Figure 3 shows experimental results with no enzymatic pretreatment. Biohydrocarbon was synthesized from palm fatty oil distillate (PFAD), which mostly consists of fatty acids. Figure 4 and Figure 5 shows the experimental result with enzymatic pretreatment. Plant oils, which mostly consists of triglyceride, were hydrolyzed using crude plant lipase to increase its fatty acids content prior to saponification.

In process without enzymatic pretreatment, product distribution was spread into C11-C15 biohydrocarbons. Isoalkane content was high, indicating a high reaction rate of isomerization during the process. The product was roughly suitable for aviation jet engine, though higher alkane content is preferable to comply with jet fuel specification. There was no significant difference between 3rd hour and 5th hour decarboxylation products. The results also confirm the feasibility of biohydrocarbon synthesis in atmospheric pressure.

PFAD, which is a by-product from the palm oil industry, mainly consists of free fatty acids (mostly palmitic and oleic acids). It is the top product of steam distillation in fatty acid removal, thus considered to be oleochemical industrial waste. Utilization of PFAD into biofuels is an effective way of increasing value of palm oil, nevertheless, biofuel is commonly produced in huge-capacity. It is desired to provide technology of biohydrocarbon synthesis from the more abundant resources, such as the plant oil itself.

Fatty acids, which is good substrate for metal soap synthesis in mild operation condition, has been commercially produced from plant oils through thermal hydrolysis of triglyceride (Colgate-Emery) process. This process is very effective and results in high yield; nevertheless, it requires high temperature and high pressure [6, 9]. The severity of process tends to increase product cost, and in the biofuel industry, it is expected to reduce operating costs as low as possible. Alternative technology of converting plant oil into fatty acids is enzymatic hydrolysis, which is feasible in room temperature and atmospheric pressure. Triglycerides in plant oils are converted into fatty acids with the utilization of lipase enzyme (thus the reaction is also known as lipolysis) prior to saponification and decarboxylation.

Biohydrocarbon produced using enzymatic pretreatment, as shown in Figure 4, contains a longer chain compared to biohydrocarbon from PFAD (Figure 2). It might be an effect of monoglyceride presence, which is produced from lipolysis. Monoglyceride is an intermediate substance in the hydrolysis of triglyceride. It might be easier to be saponified than triglyceride, thus metal soap from monoglyceride might be formed. When decarboxylation was performed into monoglyceride metal soap, long chains of hydrocarbons may have been produced more compared to decarboxylation of fatty acid metal soap.

Prolonged decarboxylation into 5 hours yields in more alkane (and alkene), suggesting that isoalkane was polymerized into a longer chain. The product composition was dominated by diesel fuel fraction (C14-C15 alkane). The shorter carbon chain fraction, which also contains isoalkane, is a potential source of jet biofuel and green gasoline. Compared to biohydrocarbon synthesis from PFAD, the enzymatic pretreatment process is more selective towards C15-C17 alkane formation, which is the theoretical product of metal soap decarboxylation derived from palm oil. Monoglyceride appeared to decarboxylate in different mechanism from fatty acids, which promotes diesel fraction more than pure metal soap from fatty acids.

This product was then improved by removing lipolysis lower product, which consists of glycerine and water (also known as sweet water). Glycerine and water were considered to be contaminant in saponification. The results are shown in Figure 6 and Figure 7. Separation of glycerine product, as identified in Figure 7, yields in a higher content of C14 and C15 alkane.

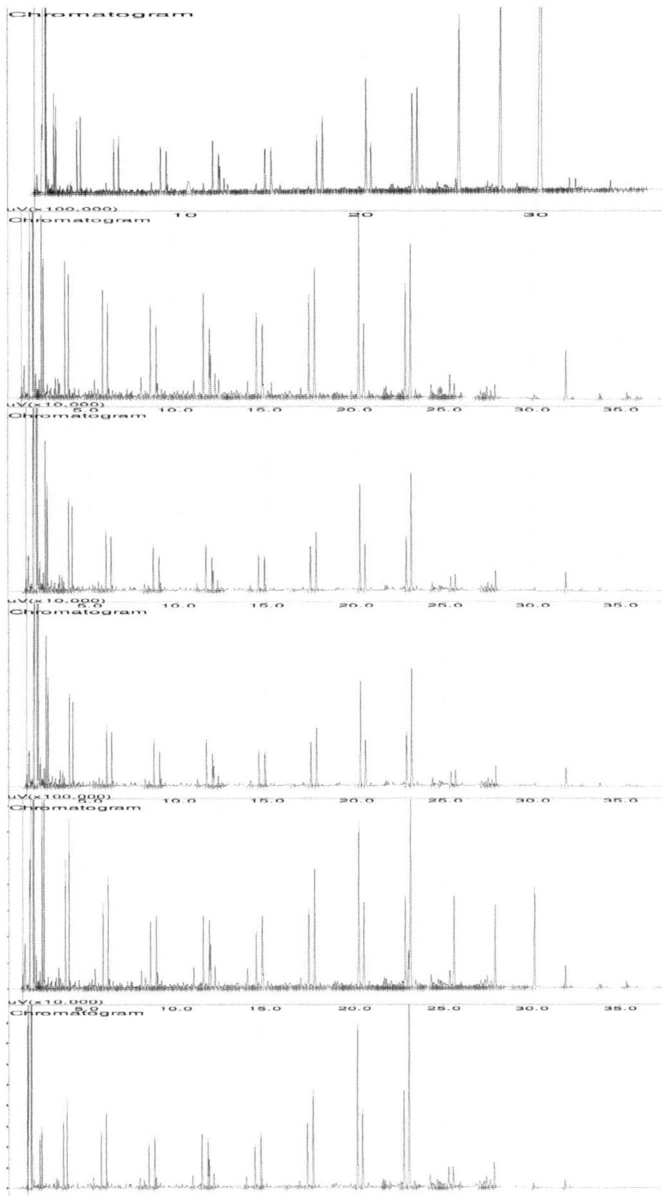

(a)

(b)

(c)

(d)

(e)

(f)

Figure 1. GC-FID chromatogram of of biohydrocarbon synthesized (a) from PFAD and 3 hours decarboxylation, (b) from PFAD and 5 hours decarboxylation, (c) from lipolyzed oil and 3 hours decarboxylation, (d) from lipolyzed oil and 5 hours decarboxylation, (e) from upper layer of lipolyzed oil and 3 hours decarboxylation, (f) from upper layer of lipolyzed oil and 5 hours decarboxylation

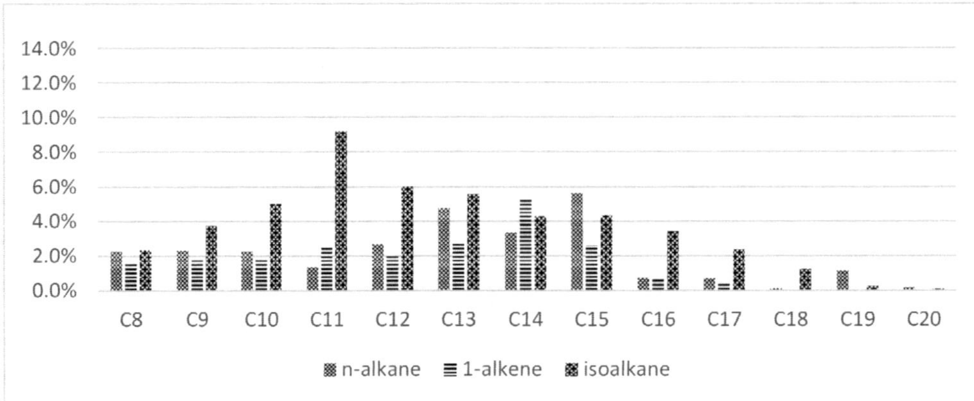

Figure 2. Product composition of biohydrocarbon synthesized from PFAD and 3 hours decarboxylation

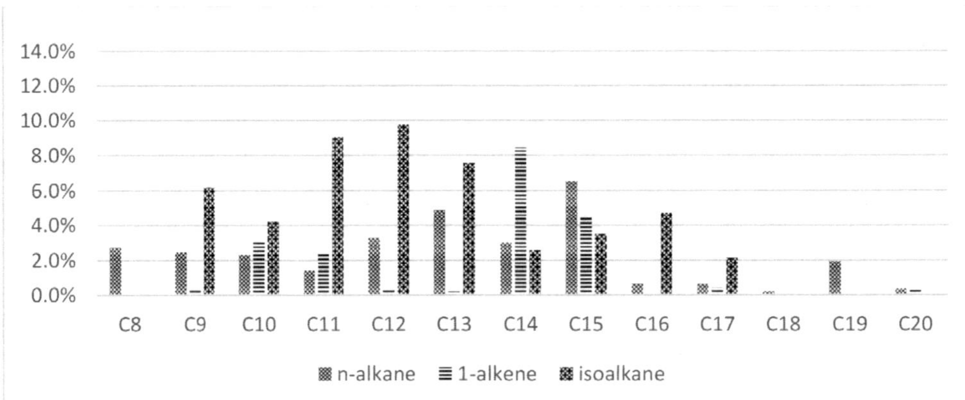

Figure 3. Product composition of biohydrocarbon synthesized from PFAD and 5 hours decarboxylation

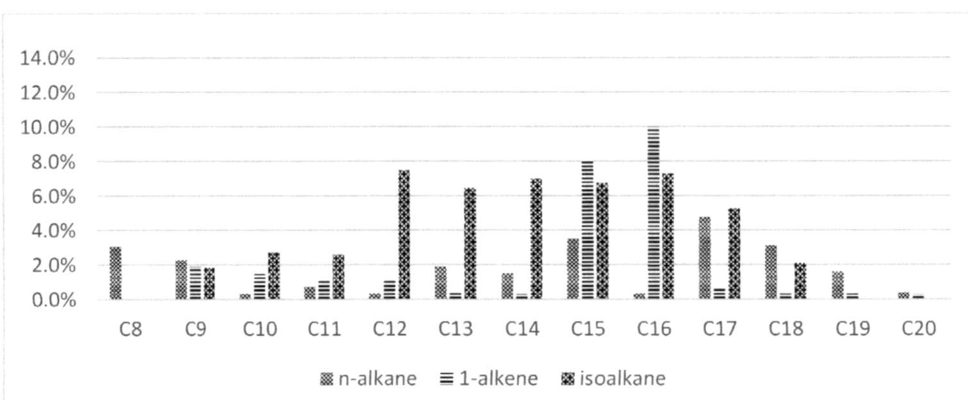

Figure 4. Product composition of biohydrocarbon synthesized from lipolysis product and 3 hours decarboxylation

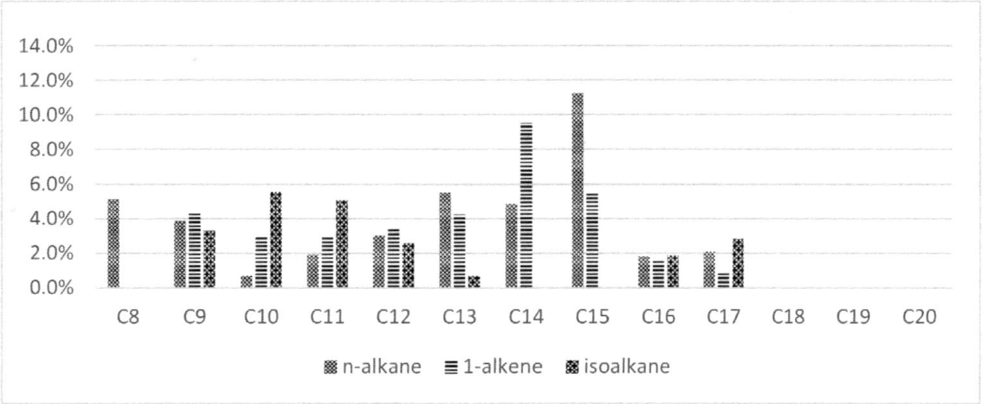

Figure 5. Product composition of biohydrocarbon synthesized from lipolysis product and 5 hours decarboxylation

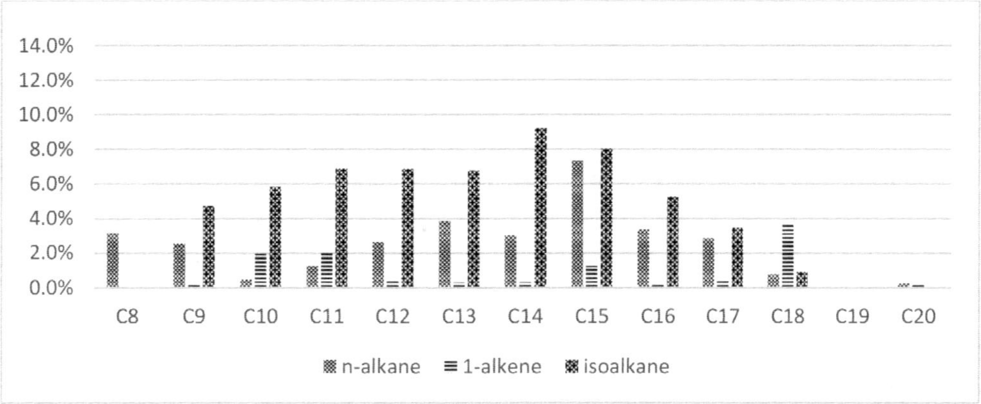

Figure 6. Product composition of biohydrocarbon upper layer of lipolysis product and 3 hours decarboxylation

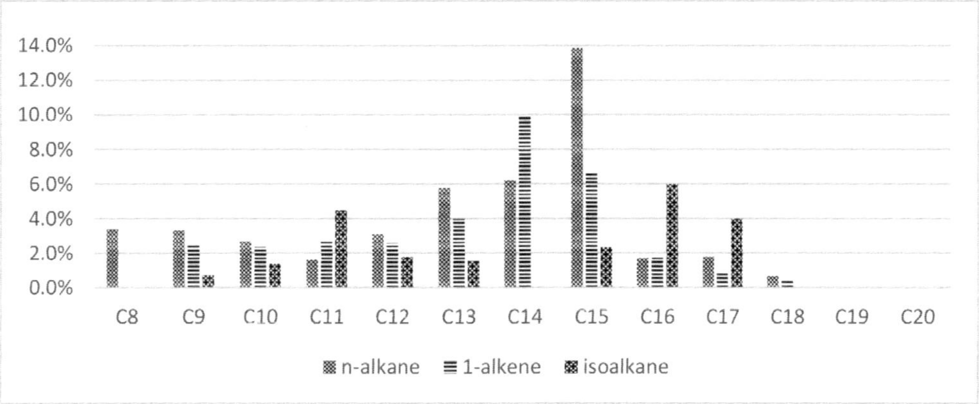

Figure 7. Product composition of biohydrocarbon upper layer of lipolysis product and 5 hours decarboxylation

4. Conclusion

In this research, an enzymatic approach was established to pretreat plant oils into an intermediate that enables metal soap synthesis in atmospheric pressure. The pretreatment appeared to increase the selectivity of biohydrocarbon product into a specific fraction, i.e., diesel fuel fraction. Monoglyceride presence was predicted to offer a different mechanism of saponification and decarboxylation. The decarboxylation period takes a role in deciding how far the reaction proceeds. Removal of sweet water in lipolysis product yields in higher selectivity towards diesel fuel fraction. Overall, biohydrocarbon production with enzymatic process in this research results in higher yield compared to previous references.

Reference

[1] Kubičková I, Snåre M, Eränen K, Mäki-Arvela P and Murzin D Y 2005 Hydrocarbons for diesel fuel via decarboxylation of vegetable oils *Catal. Today* **106** 197–200

[2] Wang W C, Thapaliya N, Campos A, Stikeleather L F and Roberts W L 2012 Hydrocarbon fuels from vegetable oils via hydrolysis and thermo-catalytic decarboxylation *Fuel* **95** 622–9

[3] Pratiwi M, Neonufa G F, Prakoso T and Soerawidjaja T H 2018 The synthesis of magnesium soaps as feed for biohydrocarbon production *MATEC Web Conf.* **156**

[4] Blachford J 1980 Manufacture of metallic soaps *US4316852A*

[5] Rogers J R H and Blew J W R 1956 Manufacture of metal soaps *US2890232A*

[6] Lascaray L 1952 Industrial fat splitting *J. Am. Oil Chem. Soc.* **29** 362–6

[7] Rahman Z, Nawab J, Sung B and Kim S 2018 A critical analysis of bio-hydrocarbon production in bacteria: current challenges and future directions *Energies* **11**

[8] Neonufa G F, Pratiwi M, Istyami A N, Elizabeth L, Dewi S S, Purwadi R, Prakoso T and Soerawidjaja T H 2018 An innovative method to produce drop-in fuel by alkaline earth-transition metals basic soap decarboxylation *MATEC Web of Conferences* **156**

[9] Barnebey H L and Brown A C 1948 Continuous fat splitting plants using the colgate-emery process *J. Am. Oil Chem. Soc.* **25** 95–9

Acknowledgment

The authors are thankful for support given by Institut Teknologi Bandung and BPDPKS (Indonesian Oil Palm Plantation Fund Management Agency) for the research fund awarded through P3MI 2019 and GRS 2018 program.

IOP Conf. Series: Materials Science and Engineering **823** (2020) 012019 doi:10.1088/1757-899X/823/1/012019

Phytoplankton (microalgae) as an alternative of renewable energy sources

Mery Rambu Djoru[1], Ricky Gimin[2], and Suwari[3]

[1]Faculty of Agricultural Technology, Universitas Kristen Artha Wacana Kupang, NTT 85000 Indonesia
[2]Faculty of Fisheries, Universitas Nusa Cendana Kupang, NTT 85000 Indonesia
[3]Faculty of Science and Engineering, Universitas Nusa Cendana Kupang, NTT 85000 Indonesia

E-mail: merry.hattu@gmail.com

Abstract. At present, primary energy resources for various industrial activities including transportation and electricity still rely on fossil sources, such as oil, gas, and coal. Fossil resources are increasingly depleted from time to time and cause the impact of erratic price increases, resources are increasingly limited and cause various environmental effects due to emissions of greenhouse gases produced due to burning these fossils. Various biological resources in Indonesia can be used as a source of raw materials for the production of renewable fuels as a substitute for fossils. Research on phytoplankton (microalgae) as an alternative to renewable energy sources has been carried out. This research was carried out in Teluk Kupang, which has been known as a habitat for Algae life and breeding. The purpose of this study is to obtain data on the existence and types of algae that can be utilized as an alternative renewable energy source. The research method used was observation and laboratory testing to determine the types of algae in Teluk Kupang. The results showed there were 34 types of microalgae, including species of *Nitzschia, sp* and *Thallassiosira, sp* which had cell compositions in the form of cellulose, glucan and oils (triglycerides) that had the potential to be extracted and utilized for the manufacture of renewable fuels and other valuable chemicals.

1. Introduction

The phenomenon of global warming and the depletion of energy sources of oil and gas at this time resulting in various studies to find potential alternative energy sources that can be developed from materials that can be renewed (renewable energy) which is also environmentally friendly. Rahmaniah et al. [1] states that to meet the level of consumption of oil and encourage the development and utilization of renewable alternative energy in the form of biofuel, such as Biodiesel. Chisty [2] states that biodiesel can be produced from various types of plants. Currently, the most commonly used sources of biodiesel are jatropha, palm oil, coconut, corn, as a diesel mixture. However, it is feared that there will be disruptions to world food stability because the market demand for biodiesel will compete with market demands for food crops so that one effort in several countries to increase energy independence is to prepare potential alternative fuels that are come from microalgae [3].

Research on microalgae, especially phytoplankton, as biodiesel based energy resources have been done. Microalgae is a unicellular species that can live in a colony or solitary and is a photosynthetic microorganism that has the ability to use sunlight and carbon dioxide to produce biomass and produces about 50% of oxygen in the atmosphere [4].

Microbial diversity on earth is very high, so with the high diversity, it is possible to get potential microalgae to produce biodiesel in large quantities. The content of fats and fatty acids in microalgae is a source of energy produced from the process of photosynthesis which is a hydrocarbon and is thought to produce energy that has not been extracted and fully utilized [5].

Table 1. Composition of oil content of several microalgae species in the stationary and exponential phases

Types of Fatty Acids	Fatty Acid Content (%)						
	Nitzchia,sp	*Thallassiosira,sp*	*Synechoccus, sp*	*Dictiosphaeruim pulchellum*	*Stichococcus,sp*	*Synechocystis,sp*	*Scenedesmus, sp*
Ac.miristat	3,15*	6,37*	26,09*	2,45*	1,54*	28,24*	1,03*
(C14:0)	2,67	4,59	25,96	2,38	2,12	13,34	1,12
Ac.palmitat	18,83*	20,67*	16,81*	11,41*	20,03*	5,70*	17,30*
(C16:0)	13,25	19,61	13,94	12,56	17,61	5,89	5,76
Ac.stearat	0,24*	0,27*	0,45*	0,61*	0,68*	0,36*	1,73*
(C18:0)	16,37	0,35	0,58	0,77	0,54	1,17	0,33

Source: [6], Note: * = stationary phase

According to Pratoomyot et al. [6], the diversity of microalgae species will make the fatty acid content in microalgae also varied. His further research shows that in general there are differences in the content of fatty acids in microalgae during the exponential phase and the phase of the stage, as shown in Table 1.

One reason for the development of biodiesel from microalgae by developed countries in Europe is because the oil content in microalgae is quite high, but it is also related to the environment. The composition of fatty acids in microalgae is very varied causing the characteristics of biodiesel produced also varies.

This study aims to obtain data on the existence of types of microalgae that have been studied containing oil so that they can be used as an alternative renewable energy source in Teluk Kupang, Kupang City, Nusa Tenggara Timur.

2. Material and Method

This research is an observational study and laboratory testing to obtain information about the types of phytoplankton in the waters of Teluk Kupang, so it can be identified several types of microalgae that have been studied in advance which are proven to have oil content so that it can be developed for further research.

2.1. Research Tools and Materials

The tools, materials, and uses used in this study are shown in Table 2. Determination of sampling locations using a purposive sampling method and sampling techniques using composite sampling and plankton net methods. Water and phytoplankton samples were then analyzed at the Biology Laboratory of the Faculty of Science and Engineering, Nusa Cendana University, Kupang.

Phytoplankton's identification to determine the types of microalgae refer to the book Identification of Plankton and references from Sachlan [7].

Table 2. Equipment, materials and use of research

Name of Equipment and Material	Use
Thermometer Hg	Measure temperature
Sample Bottle	Store water samples
pH meter	Measuring the pH of water
Label	Sample marker
Plastic bags	Storing sample bottles
Plankton net	Take and filter plankton samples
Balance	Weigh the sample
Filter paper	Filter sample
GPS (Global Positioning System)	Determine the coordinates of the sampling location
Ice Box	Save sample
Binocular Microscope	Identification Aids
Magnifier	View and identify samples
Identification Book	Identifying Plankton
Seawater	Observation object
Phytoplankton	Observation object
Lugol 4%	Preserve the sample
Acetone 90%, H_2SO_4, $HgCl_2$	Material analysis

2.2 Research procedures

This study used an observational method at two research stations with eight sampling points for microalgae collection in Teluk Kupang. The sampling points are shown in Table 3.

Table 3. Sampling points

No	Location	Station	Coordinate	
			S	E
1	Manikin	I A	10° 07'339"	123° 40'549"
2	Manikin	I B	10° 07'474"	123° 40'385"
3	Manikin	I C	10° 07'567"	123° 40'349"
4	Manikin	I D	10° 07'646"	123° 40'331"
5	Alak	II A	10° 10' 450"	123° 20'085"
6	Alak	II B	10° 10' 463"	123° 20'063"
7	Alak	II C	10° 10' 477"	123° 20'042"
8	Alak	II D	10° 10' 492"	123° 20'023"

Phytoplankton sampling using a 20mm plankton net [8] by filtering as much as 10 liters of water at each sampling point, netted phytoplankton samples are collected in a container and poured into a 250 ml sample bottle. Samples are then given 4 drops Lugol 4% and samples are taken to the laboratory for identification.

3. Results and Discussion

Land use along the Teluk Kupang coast is residential areas, ports or water transportation, fishing and aquaculture, hotels, restaurants, markets, shop houses, and agricultural land. These activities can affect the quality of coastal waters, where each coastal activity produces both organic and inorganic waste that can enter the bay waters which can be carried by rainwater or from irrigation and carried by river water or discharged directly by residents into the seas.

Table 4. Types of microalgae species at the research location

No	Station	Total Species	Number of Species	Species Type
1	IA	8	27	Triceratium americanum(7),Rhizosolenia alata(4),fragilaria, (4)Rhizosoloniaacicularis(1),Navicula(1),Liolomapacificum(1) Asteromphalus (1),Biddulphia(8)
2	IB	6	16	Chylindrotheaclosterium(1),Fragilaria(3),Coscinodiscus(1),Th allassiosira(1),Rhizosolenia alata(6), Guinardia(4)
3	IC	8	18	Stephanophyxis nipponica (2), Rhizosolenia setigera(3), Rhizosolenia sp(4) Rhizosolenia acicularis(2), Nitzschia sp (1), Lioloma pacificum(3),Flagilaria (2), Bacillaria(1)
4	ID	2	3	Pinnularia (1), Leptocylindrus danicus (2)
5	IIA	9	26	Triceratium (1), Streptooheca thamensis (1), Stephanophyxis palmeriana (3), Rhizosolonenia sp (12), Rhizosolonenia acicularis (1), Peridium (1), Nitzschia sp(1), Coscinodiscua (4), Navicula (2)
6	IIB	7	24	Sagita enflata (5), Rhizosolenia styliformis(6), Rhizosolenia antennata (1), Rhizosolenia alata(2), Rhizosolenia acicularis (1), Pinnularia (4), Nitzschia seriata (5)
7	IIC	6	15	Rhizosolenia alata (5), Navicula (1), Leptocylindrus danicus (1), Eucampia (2), Coscinodiscus (2), Leptocylindrus minimus (4)
8	IID	6	13	Rhizosolenia alata(5),Eucampia (2),Flagilaria(1), Navicula (1), Nitzschia seriata (2), Pinnularia (2)

Of all activities on the coast of Teluk Kupang, nitrate and phosphate are the most predominant predicted nutrients that can enter the waters where there is an agricultural activity directly adjacent to the coastal waters and applying chemical fertilizers namely Urea as a source of nitrate nutrients, SP36 as phosphate nutrients and KCl as an element of Potassium. The remnants of chemical fertilizers that are wasted from agricultural activities can enter the water and can increase water fertility, but excessive amounts can result in enrichment or eutrophication which triggers growth and population of phytoplankton (microalgae).

Microalgae are microscopic plants that belong to the class of algae and live in colonies and single-celled in all waters, both freshwater and seawater. Based on observations found 34 species of microalgae at eight observation points with a total number of species of 142 microalgae species including 4 classes are *Bacillariophyceae, Dinophyceae, Chlorophyceae,* and *Cyanophyceae.* Microalgae observations are shown in Table 4.

From the results of 34 species of microalgae found at the site, there are two types of microalgae namely *Nitzschia sp* and *Thallassiosira sp* that have been previously studied have cell walls containing cellulose, glucan, and oil that have the potential to be extracted and used for the manufacture of renewable fuels and valuable chemicals. *Nitzchia sp* and *Thallassiosira sp.* are microalgae that belong to the *Bacillariophyceae* class [9] and have an important role in aquatic ecosystems as primary producers. Environmental factors that influence the growth of microalgae include light, salinity, temperature, water pH, osmose pressure which may spur or inhibit growth [10]. According to Pratoomyot et al. [6], the diversity of microalgae species will make the content, fatty acids in microalgae also vary. Further research shows that in general, there are differences in the content of fatty acids in microalgae during the exponential and stationary phases (Table 1). In Botes [11], *Nitzschia sp* and *Thallassiosira sp* can be classified as follows (Table 5).

Table 5. Classification of *Nitzschia sp* and *Thallassiosira sp*

	Nitzschia sp	Thalassiosira sp
Division	*Bacillariophyta*	*Chrysophyta*
Class	*Bacillariophyceae*	*Bacillaruiphyceae*
Ordo	*Bacillariales*	*Centrales*
Family	*Bacillariaciae*	*Coscinodiscineae*
Genus	*Nitzschia*	*Thalassiosiraceae*
Species	*Nitzschia, sp*	*Thalassiosira sp*

Microalgae is a source of biomass containing high beneficial components such as protein, carbohydrates, fatty acids, including triglycerides, which can be converted into biodiesel. Several types of microalgae have potential as a source of oil. Microalgae oil content varies depending on the type of microalgae [12], it can be seen in Table 6 that the percentage of oil in *Nitzcia sp* is quite high at 45-47%, as well as the *Thalassiosira sp* percentage of 30%. Fat in microalgae consists of glycerol, saturated fatty acids or unsaturated fatty acids.

Table 6. Oil content in some microalgae

Microalgae	Oil content (%)
Botrycoccus braunii	25-75
Chorella sp	28-32
Crypthecodinium cohnii	20
Cylindrotheca sp	16-37
Dunaliella primolecta	23
Monallanthus salina	20
Nannocholoropsis sp	20-35
Neochlorpsis oleabundans	35-54
Nitzcia sp	45-47
Phaeodactylum tricornutum	20-30
Schizochytrium sp	50-77
Thalassiosira sp	30

(Source: [12])

The use of microalgae as raw material for biofuels has a higher profit compared with food crops. The advantages are phytoplankton growth that is faster than food crops, which is time to grow twice in 3.5 hours [2] microalgae can live in freshwater and seawater, and do not compete with food crops and also relatively more production costs low [13]. In addition, microalgae also produce at least 30 times more than land plants. Therefore, the cultivation of microalgae as a raw material for biofuels can save space and have high efficiency and effectiveness which is quite promising in the future.

Through several processes such as biophotolysis and fermentation, microalgae can produce hydrogen. Hydrogen is very easy to convert to heat, electricity, and fuel. The fat accumulation that occurs in the body of microalgae tends to increase when it is in a concentration and stressed environment [14]. So with the condition of Indonesia which is experiencing an energy crisis, microalgae can be an alternative development potential to produce promising biofuel fuels.

IOP Conf. Series: Materials Science and Engineering **823** (2020) 012019 doi:10.1088/1757-899X/823/1/012019

4. Conclusion

The climate in Indonesia is a tropical climate with sun exposure throughout the year, which is very suitable for the development and life of microalgae. The condition of the waters of the Kupang Bay, where there is a supply of nutrients from agricultural activities and domestic waste in the coastal areas, can be nutrients that increase water fertility. *Nitzschia sp* and *Thallassiosira sp* species that have the potential to become sources of energy or biodiesel raw materials have been found in Teluk Kupang, so they can be recommended for further research on microalgae culture to be extracted into biodiesel.

References

[1] Rahmaniah O, Setyarini R D and Maulida L 2010 Pemilihan metode ekstraksi minyak alga dari *Chorella sp* dan produksinya sebagai biodiesel *Seminar Teknik Kimia Soehadi Reksowardojo, Jurusan Teknik Kimia, Fakultas Teknologi Industri* (Surabaya: Intitut Teknologi Sepuluh November)
[2] Chisty Y 2007 *Biodiesel from Microalgae* (Palmerston North, New Zealand: Massey University)
[3] Halim R, Danquah M K and Webley P A 2011 Oil Extraction from microalgae for biodiesel production *Biores Technol* **102** 178-85
[4] Widjaja A 2009 Lipid production from microalgae as a promising candidate for biodiesel production *Makara Teknologi* **13**(1) 47-51
[5] Goswami and Kalita 2011 *Scenedesmus dimorphis* and *Scenedesmus quadricauda* two potent indegenous microalgae strains for biomass production and CO_2 Mitigation, A study on their growth behavior and lipid productivity under different concentration of urea as nitrogen source *Journal Algal Biomass*
[6] Pratoomyot J, Sriviles P and Noiraksar T 2005 Fatty acids composition of 10 microalgal species *Songklanakarin J. Sci Technology*
[7] Sachlan M 1982 *Planktonologi* (Semarang: Fakultas Peternakan dan Perikanan, Universitas Diponegoro)
[8] Romimohtarto K 2004 *Meroplankton Laut : Larva Hewan Laut yang Menjadi Plankton* (Jakarta: Djambatan)
[9] Tomas C R 1997 *Marine Plankton Identification* (London: Academic Press)
[10] Fogg G E and Thake B 1987 *Algal Cultures and Phytoplankton Ecology* 2nd edition (England: The University of Wisconsin Press) pp 12-36, 43
[11] Botes L 2001 Phytoplankton Identification Catalogue *Globallast Monograph Series No.7.* (Saldanha Bay, South Africa)
[12] Noer A H dan Dessy A 2015 Mikroalga sebagai sumber biomassa terbarukan *Teknik Kultivasi dan Pemanenan, Research Gate*
[13] Basmal 2008 Peluang dan tantangan pemanfaatan mikroalga sebagai biofuel *Squalen* **3**(1) 34-39
[14] Kawaroe M 2010 *Potensi Mikroalga dan Pemanfaatannya untuk Produksi Bio Bahan Bakar* (Bogor: IPB Press)

IOP Conf. Series: Materials Science and Engineering **823** (2020) 012020 doi:10.1088/1757-899X/823/1/012020

Biogas production using manure from KPBS Pangalengan's dairy farm and its role in reducing Citarum river pollution

Pramujo Widiatmoko[1], Jenny Rizkiana[2], Susilo Yuwono[3], Mohammad Taufiq[3], and Candra Purnama Hadi[3]

[1] Department of Chemical Engineering, Institut Teknologi Bandung, Bandung 40132 Indonesia
[2] Department of Bioenergy and Chemurgy Engineering, Institut Teknologi Bandung, Bandung 40132 Indonesia
[3] Cendikia Salim Mulia Foundation, Bandung, Indonesia

Email: pramujo@che.itb.ac.id

Abstract. Dairy farms have potential to be organic sources (i.e., cow dung) for producing biogas. This paper presents a study on biogas production from a dairy farm in Pangalengan, South Bandung Area, Indonesia. We installed the Tenari Model biogas reactor with 4 m^3 volume. The composition and amount of produced biogas were analyzed. Its effects on reducing water contaminants in Citarum River were studied by monitoring the reactor water effluent and mapping the river watershed. The produced biogas could supply daily cooking gas for the farm sufficiently.

1. Introduction

Pangalengan is a center of dairy farm in Southern Bandung, West Java, Indonesia. The dairy farmers established "The Koperasi Peternakan Bandung Selatan (KPBS)" in 1969 for collecting and further processing of their dairy products. The dairy products from The KPBS Pangalengan in 2017 reaches 85 tons/day from its 4500 members who are raising more than 15,000 cows [1]. However, there are concerns about the manure waste management system. Diawati et al. in 2014 reported that only 57% of the manure being processed i.e. 21% for compost and 14% for biogas [2]. The rest is discharged directly into water bodies. With 180 ton/day of manure released [3], it is expected to heavily contaminate rivers in Pangalengan and their downstream e.g. Cisurili, Cisangkuy, and Citarum Rivers.

The "Citarum Harum" in 2018 was aimed to restore the cleanness of The Citarum River in a long-term project. The project is a continuation of previous "Citarum Bergetar" (2001) and "Citarum Bestari" (2013) projects [4]. The effort to increase utilization of the dairy manure for biogas have been carried out in line with those government projects [5]. However, the utilization of manure for biogas still as low as 14% [2]. The main problems for biogas technology adoption are (1) high installation cost, (2) social culture that consider biogas is inappropriate for food processing, (3) improper design on type and layout of reactor which cause high maintenance cost (4) operator have limited time to maintain the biogas reactor, and (5) inadequate dissemination, training, and continuous monitoring program [2, 5-6]. Therefore, it is important to organize a continuous program that can increase the quantity of processed manure to biogas as well as improve the knowledge, skill, and culture of the farmers.

In this work, we initiate a project to utilize dairy manure for biogas production in Mekar Mulya Village, Margamulya, Pangalengan. The project is in collaboration with KPBS Pangalengan. The project is aimed to reduce direct discharging of dairy manure to rivers by increasing the usage of manure for biogas and to support the farmer economically by reducing the usage of daily cooking gas. To tackle the barrier of technology adoption, we designed the reactor with direct feeding, reducing operator involvement to maintain the reactor. Further, a concept of "Desa Binaan (fostered village)" has been pioneered from this project with the support from Student Association of Bioenergy and Chemurgy (HMTB) ITB. The program is aimed to increase operator knowledge as well as to ensure continuous monitoring of the program.

2. Methodology

A field survey was conducted to the dairy farming belong to the member of KPBS Pangalengan (Dairy farming of Mr. Adang Sholahudin). The water quality of Cisurili River close to the discharging point was analyzed to determine the effect of manure pollution on the water body. The upstream sampling point is about 15 m from the discharging point; meanwhile downstream ones are about 100 m after the point. A fiberglass biogas reactor of 4 m^3 in volume (Tenari model) was installed. Figure 1 shows schematic diagram of this project. Water quality after biogas reactor installation was analyzed. The biogas production was also analyzed to understand the trend and problem in the field. Further, the socialization of biogas project to KPBS farmers was conducted.

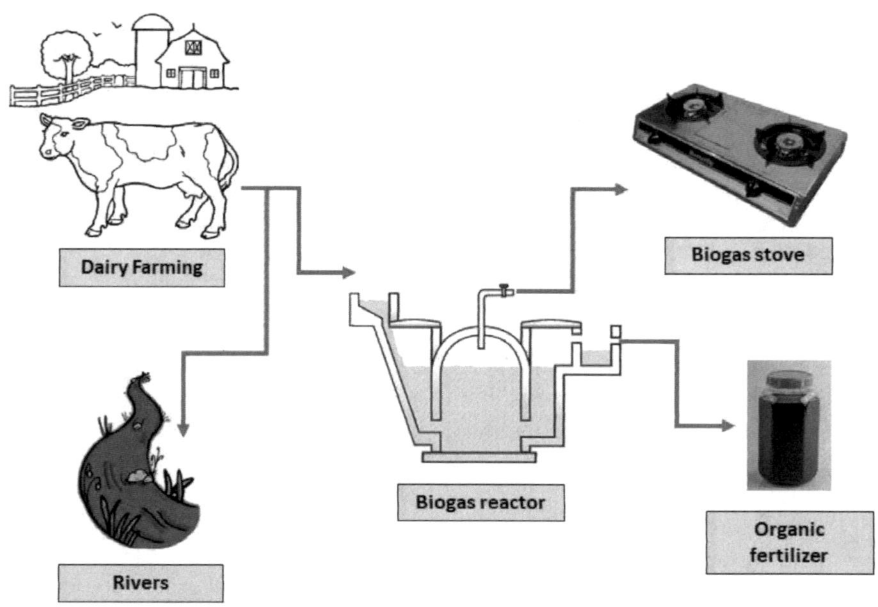

Figure 1. Schematic diagram of utilization of dairy manure

3. Result and Discussions

In this section, we discuss the safety of biogas reactor installation, water quality analysis, affected area of polluted water, produced biogas and organic fertilizer, and socialization of the project.

IOP Conf. Series: Materials Science and Engineering **823** (2020) 012020 doi:10.1088/1757-899X/823/1/012020

3.1. Safety aspects
The main components of biogas are methane (60%) and carbon dioxide (40%). In a worst-case scenario, the methane with a volume of 4 m^3 released from 2" hole and forms a flammable cloud. Based on analysis using Aloha®, the radius of the flammable area within 60% LEL (Flame pocket of 30000 ppm methane) is less than 10 m. There is a risk of explosion whenever open fire or other heat source presence within this area. However, Aloha® cannot determine the exact distance of this flame pocket due to near-field patchiness makes dispersion prediction less reliable. The area of 10% LEL has a radius of 22 m. Commonly, gas detectors were set to detect this concentration as an indicator of methane leaks.

In case vapor cloud explosion occurs due to 4 m^3 of methane is released, the radius of the affected area is estimated at 19 m with an overpressure of blast wave at 1 psig. This overpressure can cause shatter glass. Wall surrounds the reactor may act as protection for the blast wave. In real condition, the volume of methane is less than that worst-case scenario. Therefore, the explosion hazard in the biogas reactor location can be mitigated.

3.2. Water quality analysis and mapping of the affected area
The water quality of the Cisurili River was analyzed in the Water Quality Laboratory, ITB. Figure 2.c shows measured parameters of Total Dissolved Solids (TDS), Acidity (pH), and Biological Oxygen Demand (BOD) of both locations. The TDS and BOD increase, meanwhile the pH decrease as the manure enter the river. The TDS measures the dissolved content of organic and inorganic substances present in water. The BOD, which significantly increases from 26.2 to 62.5 mg/L, indicates high organic content added to the water body.

(a)

(b)

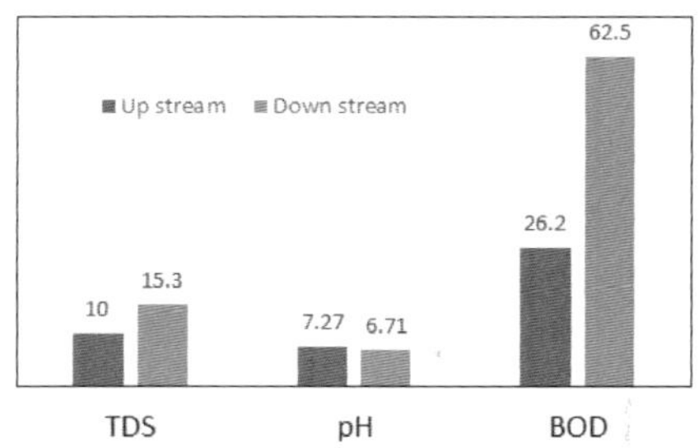

(c)

Figure 2. Sampling on Cisurili River (a) up-stream and (b) down-stream of discharging point (c) Instance parameters of the water quality. The units for Total Dissolved Solid (TDS) and Biological Oxygen Demand (BOD) are in mg/L.

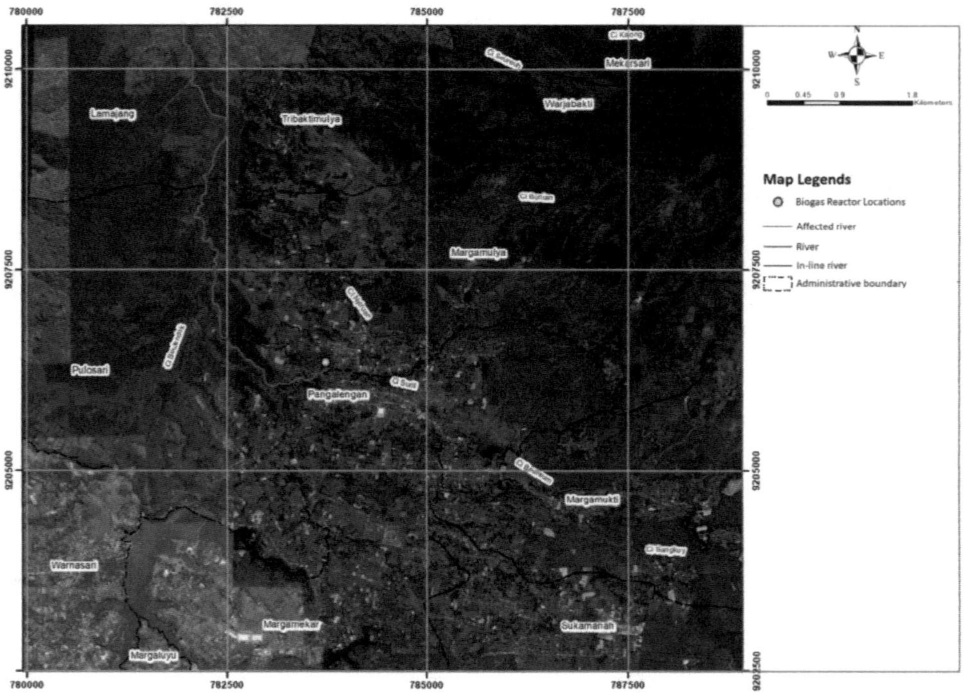

Figure 3. Maps of manure-polluted rivers from studied Dairy Farming

Figure 3 shows manure-polluted rivers in Pangalengan area. In our project, the farming is located about 10 m from Cisurili River. The river itself is already polluted by domestic wastes. Direct manure discharged into the river increases organic content in the water bodies. The polluted water then flows to Cisangkuy and Citarum Rivers. A dairy cow can produce 45.5 kg manure per day [13]. The studied dairy farming can produce 16.7 – 20 kg/day of compost. Therefore, half of the manure is still expected to be discharged into rivers even though composting process has been performed. As 57% of the manure have been processed, total manure pollutant from 15000 cows will reach 464 ton/day. This calculation result is much higher than the amount reported by Mulyana (2019) at 180 tons/day.

3.3. Biogas and organic fertilizer production

In this work, biogas is being produced from anaerobic decomposition of organic materials in the dairy manure by microbes. The biogas mainly composed of methane and carbon dioxide. Other traces gases also being produced such as H_2S and NH_3. Temperature, pH, and C/N ratio of biomass source affect produced biogas composition. Biogas production will be optimum at an operating temperature of 30 – 35, pH of 6.8 – 7.5, and C/N ratio of 20 – 30 [5]. Table 1 shows typical biogas composition from dairy manure. The early CO_2 concentration checking in the installed biogas reactor showed a higher value of 80.0%. It is expected that oxygen content in the early stage of reactor operation is high; thus reaction tends to form carbon dioxide instead of methane.

Table 1. Biogas composition from cow's manure

Biogas sources	CH_4 (%)	CO_2 (%)	H_2S (ppm)	*Ref.*
Dairy cow's manure	55-68	32-45	1500-8000	[7]
	50-75	25-45	<1000	[16]
	57	43	-	[10]

There are 20 dairy cows and 10 calves in the farm. Recebli et al. (2015) reported that bovine animal can produce 6.33 m^3 biogas in 20 days from a reactor with volume of 0.5 m^3 [9]. Sunaryo (2014) also reported that 0.8 – 1.6 m^3/day of biogas can be produced from 3 – 4 cows [14]. The available volume of our bioreactor (4 m^3) can only handle less than one-fourth of total manure, with expected biogas production of 2.5 m^3/day. Therefore, there is a very high potential of biogas production (6 – 12 m^3/day) if adequate number of biogas reactors are available (at least 6 reactors). Biogas volume of 1 m^3/day is equivalent to usage of 3 kg commercial LPG for 7 days cooking [5]. With the current condition (Figure 4), the reactor can supply biogas for up to three households with potential savings of IDR 69,000 – 295,000/month. The saving was calculated based on the price of 3 kg LPG at IDR 23,000. Higher saving potential could be expected as the price of the gas tends to increase year by year.

(a) (b) (c)

Figure 4. Photograph of (a) Tenari model of biogas reactor with 4 m^3 in volume (b) Biogas stove installation (c) Blue fire from the combustion of biogas.

Solid and liquid parts of biogas reactor effluent are separated by means of sedimentation. The solid part is being used for compost and worm growth substrate. Each dairy cow produces manure about 45.5 kg/day [13]. The studied dairy farm produces a total of 500 – 600 kg compost per day, which has been applied as crop fertilizer. Fermentation in the biogas reactor prior to application in the field increases the quality of the compost. Minde et al. (2013) reported that crop growth is getting better when using biogas reactor effluent than fermented manure due to lower pest attack and grass growth [14]. The production of biogas reduces C/N ratio in the compost from 10.7 to 7, thus increases its quality [15].

3.4. Challenges and prospects

The success of biogas technology adoption depends on how far we can handle both technical and socio-economic aspects. The technical aspects that should be considered are water availability and sustainability of reactor operation. Water to manure ratio is important parameter for biogas production. During the dry season (July to September), rainfall intensity reaches 76 - 120 mm (compared to 376 mm during the wet season). The studied farm experienced lack of water supply for a month. Therefore, the availability of water needs to be considered when installing and operating a biogas reactor.

Sustainability of operation depends on the operator's skill and time, the operability of the reactor, and dissemination. In this work, the reactor was designed for easy operation and maintenance. The previous-installed biogas reactor should be manually feed using a bucket, consuming much time and effort. The direct feeding of manure into biogas reactor in this work reduces time and effort for maintenance. Further, training and dissemination of the biogas technology were conducted to the field operator to increase the knowledge and skill to operate the reactor. To support sustainability, we introduce "Desa Binaan (fostered village)" with the support from the Student Association of Bioenergy and Chemurgy (HMTB) ITB. The students routinely visit the biogas location for training and maintain the reactor. In the future, they can improve the scale of project and technology to the most advanced ones.

Socio-economic aspects faced in the field are the mindset of people to use biogas for food processing and monetizing of biogas. People in the farm still use traditional wood stoves due to their advantage of warming the body during cold weather. For cooking, they already use gas stoves. Therefore, the application of biogas needs to consider this aspect. Further, the operating cost of the biogas reactor is the main reason to be considered. Increasing the economic value of having the reactor, such as by selling the biogas, can be used to cover the operating cost.

There is a bright prospect to increase the benefit of the biogas reactor installation. Further processing of biogas, installation of new biogas reactor, installation of dryer, filter, and pelletizing machines for processing biogas effluent are some projects that can be proposed to improve both technical and economic aspects of biogas reactor. Biogas bottling and selling will increase the bargaining position of biogas installation from the farmers' point of view. The compost can be used for worm growth substrate and use as fish feed which has a high economical value.

4. Conclusion and Future Works

We successfully install a biogas reactor in dairy farming. We found that direct discharging of manure to river significantly increases Total Dissolved Solid and Biological Oxygen Demand. Therefore, the installed biogas reactor is expected to reduce rivers pollution in Pangalengan area. With the prospect of biogas production up to 12 m^3/day and saving up to 295,000/month, the improvement on biogas production system is a promising project for dairy farming of KPBS Pangalengan.

Acknowledgments

This work was financially supported by the Community Service Program from the Ministry of Research, Technology, and Higher Education (Kemenristekdikti) of Indonesia 2019. Authors gratefully acknowledge the support from KPBS Pangalengan as a partner for this project. We thank Mr. Adang Shalahudin for providing the location for installation of biogas reactor in his dairy farm, and Bioenergy and Chemurgy Student Association (HMTB) for support.

References

[1] Mulijanti S L, Dianawati M and Rismayanti Y 2016 *Bul. Hasil Kajian* **6** 34
[2] Dianawati M, Mulijanti S L, Teddy S, Yulyatin A, Nurnayetti, Histifarina D and Riswita 2014 *Laporan Akhir m-P3MI pada Agroekosistem Lahan Kering Dataran Tinggi di Kabupaten Bandung* (Bandung: BPTP Jawa Barat)
[3] Mulyana 2019 *Limbah Kotoran Sapi Jadi Penyumbang Terbesar Pencemaran Sungai Citarum* (Jakarta: news.okezone.com, 5 September 2019)
[4] Presidential Regulation Republic of Indonesia No. 15 2018
[5] Dianawati M and Mulijanti S L 2015 *J. Lit bang Pert* **34** 125
[6] Hikmatian A M, Nurlina L and Alim S 2015 *Stud. e-J. Univ. Padjajaran* **4** 1
[7] Shelford T, Gooch C, Hines M, Choudhury A, Felton G and Lansing S 2018 *Farmer's Guide to Dairy-Derived Biogas Production, Treatment and Utilization* (South Burlington: Northeast SARE) *pp 6*
[8] Wahyudi J, Kurnani T B A and Clancy J 2015 *Int. J. Renew. Energy Dev.* **4** 219
[9] Recebli Z, Selimli S, Ozkaymak M and Gonc O 2015 *J. Eng. Sci. Technol* **10** 6
[10] El-Mashad H M and Zhang R 2010 *Bioresour. Technol.* **101** 4021
[11] Sunaryo 2014 *J. PPKM UNSIQ* **1** 21
[12] Haryati T 2006 *Wartazoa* **16** 160
[13] The Japan Institute of Energy 2008 *Buku Panduan Biomassa Asia* (Tokyo: The Japan Institute of Energy) pp 71
[14] Minde G P, Magdum S S and Kalyanraman V 2013 *J. Sust. Energy Environ* **4** 121
[15] Insam H, Gomez-Brandon M and Ascher J 2015 *Soil Biol. Biochem* **84** 1

IOP Conf. Series: Materials Science and Engineering **823** (2020) 012020 doi:10.1088/1757-899X/823/1/012020

[16] Al Seadi T, Rutz D, Prassl H, Kottner M, Finsterwalder T, Volk S and Janssen R 2008 *Biogas Handbook* (Esbjerg: University of Southern Denmark) p 41

IOP Conf. Series: Materials Science and Engineering **823** (2020) 012021 doi:10.1088/1757-899X/823/1/012021

H_2O_2 as a fungicide in the growth phase of green *Haematococcus pluvialis*

Judy Retti Witono, Valine Novianty, Y I P Arry Miryanti, Angela J Kumalaputri, Herry Santoso

Department of Chemical Engineering, Parahyangan Catholic University 40141, Indonesia

E-mail: judy@unpar.ac.id

Abstract. One of the potential sources for bioenergy and biochemical products is *Haematococcus pluvialis*. However, these microalgae are vulnerable to chytrid infection when the cells are transitioning from the green flagellates' stage to the resting stage. As a result, the productivity of their biomass is low. The objective of this research was to evaluate the effects of hydrogen peroxide as the fungicide, including those which were combined with fluazinam, on the growth of *H. pluvialis*. The variables used in this research were the dose of H_2O_2, the combination with fluazinam, and the frequency of H_2O_2 treatment. The initial cell number of *H. pluvialis* was 5 x 10^4 cells/mL culture medium. The 5 different treatments were done, i.e., (1) 0.028 mL H_2O_2/L liquid culture dosed every day; (2) 0.028 mL H_2O_2/L liquid culture dosed every day for only 3 days and then every two days combined with single dose 0.5 ppm of fluazinam on the first day; (3) same as (2) but without combination with fluazinam; (4) 0.035 mL H_2O_2/L liquid culture dosed every day for 3 days then every two days combined with single dose 0.5 ppm of fluazinam on the first day; (5) same as (4) but without combination with fluazinam. As a comparison, there was also a culture without any treatment as a control. The culture which was treated with hydrogen peroxide 0.028 mL H2O2/L liquid culture dosed every day for only 3 days and then every two days at the 7th day showed the highest cell number (45 x 10^4 cells mL^{-1} ~ 9 fold from the initial cell number), even compared to the culture without any treatment. In conclusion, the treatment with hydrogen peroxide is effective against chytrid infection. The combination with fluazinam does not affect more to the effectiveness of H_2O_2 as a fungicide. It looks that by adding fluazinam, the *H. pluvial* growth is depressed.

1. Introduction

The issue of limited petroleum resources in the world leads us to find a new material that can substitute it. One of the materials that can be sustainably produced in the future is microalga. Microalga contains macro- and micronutrients and bioactive components which are potential for human nutrition and promoting health. Furthermore, they are also a potential source for bioenergy and biochemical products.

The freshwater green microalga *Haematococcus pluvialis* has been receiving much attention nowadays because under stress condition it has the capacity to accumulate very high levels of the carotenoid astaxanthin [2, 3]. Therefore, it is a good source of ingredients for functional food. However, this microalga can be utilized more than just for that since it is rich of other bioactive components, as shown in Table 1.

The main problem on the green stage of *Haematococcus pluvialis* cultivation is parasites, which found growing epibiotically on algal cells and caused damage to the host cultures. The common microbial contaminants occurring in mass culture of *Haematococcus pluvialis* are fungal parasites, zooplankton predators (e.g., amoebas, ciliates, and rotifers), cyanobacteria, and several other eukaryotic microalgae. Of these contaminants, zoosporic fungi (chytrids) identified as the most dangerous one responsible for reduced astaxanthin productivity and frequent culture collapse [4-6].

Hydrogen peroxide is well known effective against all forms of microorganisms, including dormant forms with known high resistance such as bacterial spores and protozoal cysts [7]. However, the effective and safe use of hydrogen peroxide depending on the way it is used, in particular, the concentration [8, 9].

Therefore, the goal of this research is to determine the effectiveness of hydrogen peroxide concentration, which is combined with fluazinam as a micro biocide at the green stage.

Table 1. Typical common components of *Haematococcus pluvialis* [1]

Component	Minimum	Maximum	Mean
Protein (%)	17.30	27.16	23.62
Carbohydrates (%)	36.9	40.0	38.0
Fat (%)	7.14	21.22	13.80
Iron (%)	0.14	1.0	0.73
Moisture (%)	3.0	9.0	6.0
Magnesium (%)	0.85	1.4	1.14
Calcium (%)	0.93	3.3	1.58
Biotin (mg/lb)	0.108	0.665	0.337
L-carnitine (ug/g)	7.0	12	7.5
Folic acid (mg/100g)	0.936	1.48	1.30
Niacin (mg/lb)	20.2	35.2	29.8
Pantothenic acid (mg/lb)	2.80	10.57	6.14
Vitamin B1 (mg/lb)	<0.050	4.81	2.17
Vitamin B2 (mg/lb)	5.17	9.36	7.67
Vitamin B6 (mg/lb)	0.659	4.5	1.63
Vitamin B12 (mg/lb)	0.381	0.912	0.549
Vitamin C (mg/lb)	6.42	82.7	38.86
Vitamin E (IU/lb)	58.4	333	186.1
Ash (%)	11.07	24.47	17.71

2. Materials and Methods

2.1. Materials and culture conditions

Haematococcus pluvialis UTEX Collection Strain 2505 was purchased from Algae Analytics, US, and has been cultivated for almost one year in our laboratory. During that time, the inoculum was contaminated. Hydrogen peroxide 50% from Sigma Aldrich was standardized prior to use. Whereas fluazinam (Nando 500 SC) with active ingredient 500g/L was purchased from Nufarm, Indonesia. The stock cultures were grown photoautotrophically in Modified Bold's Basal 1N-3P medium with thiamine 0.1 mg/L to substitute vitamin B12 and biotin. Every run was prepared from 50% stock culture, which contains 5 x 10^4 cells/mL in 250 mL cylindrical glass bottles. The culture was illuminated with white-blue LED lamps at a light intensity of 3200 lux with dark:light cycle of 10:14 h. The cells were grown in batch cultures and aerated by bubbling air with the aeration rate of 2 L/min under the atmospheric temperature 25±1^{0}C.

2.2. Investigating the effect of hydrogen peroxide

Five different treatments and concentrations of hydrogen peroxide were given to the contaminated culture with the initial cell numbers of 5 x 10^4 cells/mL. The variations in each run can be seen in Table 2. The contaminated culture grown in the absence of hydrogen peroxide was used as a control. The cell growth based on cell number counting in every run was monitored for 11 days consecutively.

Table 2. Variation of runs

Run	$[H_2O_2]$ added	Fluazinam added	Frequency of H_2O_2 added
control	-	-	-
1	0.028 mL H_2O_2/L medium	-	daily
2	0.028 mL H_2O_2/L medium	0.5 ppm	daily for 3days then every two day
3	0.028 mL H_2O_2/L medium	-	daily for 3days then every two day
4	0.035 mL H_2O_2/L medium	0.5 ppm	daily for 3days then every two day
5	0.035 mL H_2O_2/L medium	-	daily for 3days then every two day

2.3. Growth measurement

Cell number was observed using an improved Neubauer hemocytometer (Marienfeld Superior, Lauda-Königshofen, Germany) under a binocular microscope (Olympus, Tokyo, Japan) at 400x magnification. The cells growth rate constant (μ) was estimated based on the following first-order equation:

$$\frac{dX}{dt} = \mu X \tag{1}$$

Where X and t represent the cell number (cells/mL) and time (days), respectively.

3. Results and Discussion

Figure 1 shows the growth of contaminated *Haematococcus pluvialis* cultures, either treated with or without hydrogen peroxide and fluazinam. In general, the cell number of microalgae in all variation decreased after 7 days. In the contaminated culture without micro biocides treatment (i.e., control run), the cell number of microalgae collapsed totally, mostly due to the fungus attack on the cell wall beside the lack of nutrients source for the biosynthesis of microalga. It can be observed from Figure 2(a), (b) and (c). Figure 2(a) shows the healthy cell of *Haematococcus pluvialis.* Figure 2(b) shows a green palmella cell that starts to be infected by chytrid sporangia and lysis. Figure 2(c) shows the empty cell. Spores grow rapidly and evolve into sporangia, which caused cells regeneration does not happen.

Induction of 0.028 mL hydrogen peroxide /L to the cultivation medium can reduce the average chytrid infection without negatively impacting the growth rate of microalga. The cell number can increase by 9 times from 5×10^4 to 45×10^4 on the 7th day (see Run 3). The same tendency occurs when fluazinam induced together with H_2O_2 into the medium culture (i.e., Run 2). But it seems that the use of H_2O_2 in combination with fluazinam suppresses not only the growth of fungi but also at the same time the growth of microalgae itself. The cell growth only reach 5 fold from 5×10^4 to 25×10^4 at the 7th day.

The frequency of H_2O_2 induction also affects the microalga growth. From the experiment Run 1 and Run 3, it was observed that the induction of H_2O_2 every day gave a lower cell number on the 7th day compared to the sample with the induction of H_2O_2 given only every two days after every day for 3 days. Hydrogen peroxide as an oxidizing agent, in the cultivation system, oxidizes biological macromolecules, namely carbohydrates, lipids, proteins, and nucleic acids. Cells become stressed which can cause DNA damage after such a long exposure to hydrogen peroxide. Furthermore, as the consequences, the growth and multiplication of the targeted microalga is inhibited [7].

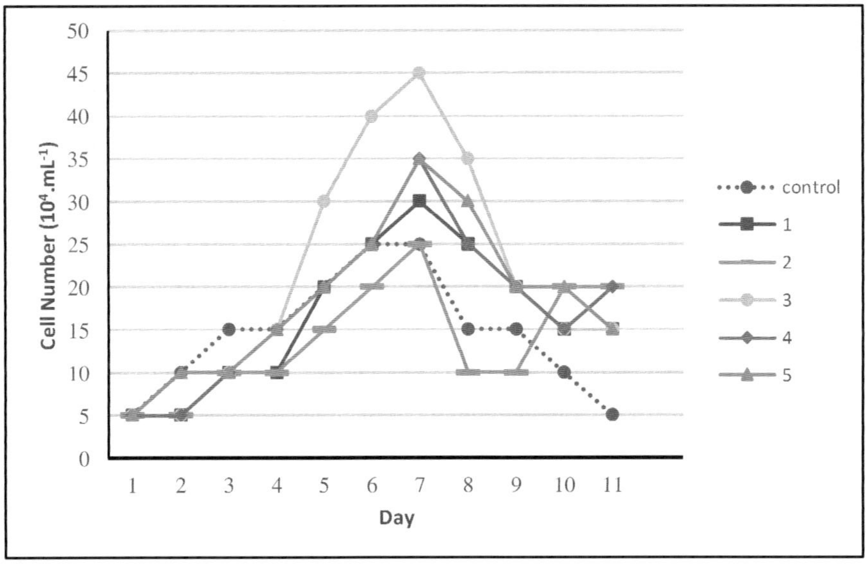

Figure 1. The growth of *Haematococcus pluvialis* in the green stage at the variation of fungicide H_2O_2 and fluazinam induction

In the matter of the amount of hydrogen peroxide concentration used (Run 3 and Run 5), the culture treated with 0.028 mL/L produces higher cell numbers than the culture treated with 0.035 mL/L H_2O_2 for the same treatment frequency. The high concentrations of hydrogen peroxide might be effective for reducing chytrid infections; nevertheless the cell lysis was also high.

Compare to our previous study [10], it is shown that the effect of H_2O_2 as a fungicide in the medium culture of *H. pluvialis* is more effective than fluazinam. In the use of fluazinam as a single fungicide, the *H. pluvialis* growth only reaches 5 times from the initial cell number. It is lower compared to the *H. pluvialis* growth in the use of H_2O_2 as a single fungicide.

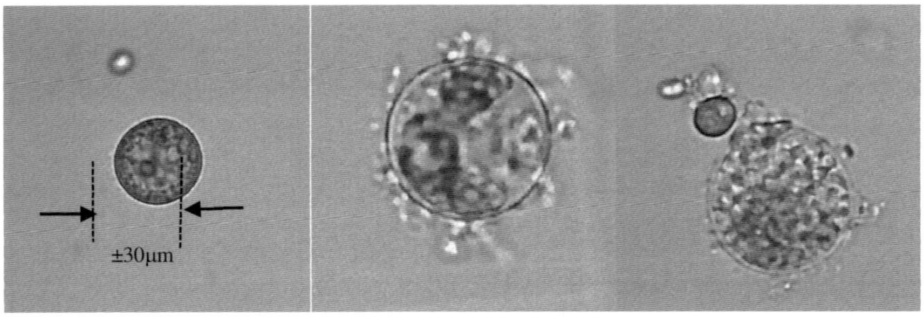

(a) (b) (c)

Figure 2. (a) The healthy motile macrozooid cell of *Haematococcus pluvialis*; (b) green pamella cell starts lysis (c) the empty cell

The cell growth rate constant μ (see Equation 1) for the first seven days was estimated using a nonlinear regression method that minimizes the sum of square error (SSE) of cell number X. The results are shown in Table 3.

Table 3. The cell growth rate

Run	μ (day^{-1})	SSE
control	0.296	97.562
1	0.307	24.527
2	0.270	6.602
3	0.384	113.050
4	0.327	9.653
5	0.327	15.328

The higher the value of the cell growth rate constant μ is the higher the cell growth rate of the culture in the system. The values of the cell growth rate constants μ estimated using the nonlinear regression method are consistent with the explanation given in the discussion above.

4. Conclusion

Hydrogen peroxide is a good fungicide for *Haematococcus pluvialis* cultivation. A combination of hydrogen peroxide and fluazinam as a fungicide is not necessary because it can suppress not only the growth of fungi but also the growth of microalga. Furthermore, the use of a higher concentration of H_2O_2, as well as the more frequent introduction of H_2O_2 can inhibit the microalga growth.

Acknowledgement

The authors would like to express their thanks to the Director General of Research and Development Reinforcement, Ministry of Research, Technology, and Higher Education (Direktorat Riset dan Pengabdian Masyarakat, Direktorat Jenderal Penguatan Riset dan Pengembangan RISTEKDIKTI), which finances this research and also Research Institutions and Community Development - Parahyangan Catholic University (LPPM – UNPAR) who assisted the authors in the preparation of this research. Special thanks are addressed to our research group who made this research can be carried out.

References

[1] Dr. Lorenz R T 1999 A technical review of Haematococcus algae *NatuRose^{TM} Technical Bulletin #060*

[2] Aflalo C, Meshulam Y, Zarka A and Boussiba S 2007 On the relative efficiency of two- vs. one-stage production of astaxanthin by the green alga Haematococcus pluvialis *Biotechnology and Bioengineering* **98**(1) 300-5

[3] Borowitzka M A 2018 *Biology of Microalgae* in *Microalgae in Health and Disease Prevention* I. A. Levine and J. Fleurence Eds. pp 23–72

[4] Carney L T and Lane T W 2014 Parasites in algae mass culture *Frontiers in Microbiology* **5** 278

[5] Gutman J, Zarka A and Boussiba S 2009 The host-range of *Paraphysoderma sedebokerensis,* a chytrid that infects *Haematococcus pluvialis European Journal of Phycology* **44**(4) 509–14

[6] Han D, Li Y and Hu Q 2013 *Biology and Commercial Aspects of Haematococcus pluvialis* in *Handbook of Microalgal Culture: Applied Phycology and Biotechnology* Richmond A and Hu Q Eds. (USA: John Wiley & Sons) pp 388-405

[7] McDonnell G 2014 The use of hydrogen peroxide for disinfection and sterilization applications *PATAI'S Chemistry of Functional Groups*

[8] Carney L and Sorensen K 2016 Methods for preventing lysis in a culture of *Haematococcus pluvalis* using hydrogen peroxide *US Patent 9,347,034 B1*

[9] Carney L T and Sorensen K 2015 Methods for treating a culture of *Haematococcus pluvialis* for contamination using hydrogen peroxide *US Patent 9,113,607 B1*

[10] Witono J R, Novianty V, Santoso H, Miryanti A and Kumalaputri A J 2019 Fluazinam potential as a fungicide in liquid culture system for the growth of Haematococcus pluvialis microalgae *IOP conference series: Materials Science and Engineering* **703** 1-7

IOP Conf. Series: Materials Science and Engineering **823** (2020) 012022 doi:10.1088/1757-899X/823/1/012022

Extraction of polyunsaturated fatty acids to reduce the iodine number of biodiesel products

Z Helwani[1], L Umar[2], G F. Neonufa[3], E Puspawiningtyas[4], T Prakoso[5], and G Lugito[5]

[1]Departement of Chemical Engineering, Universitas Riau, Pekanbaru 28293, Indonesia
[2]Department of Physic, Faculty of Mathematics and Natural Sciences, Universitas Riau, Pekanbaru 28293, Indonesia
[3]Department of Agricultural Technology, Faculty of Agricultural, Universitas Kristen Artha Wacana, Kupang, Nusa Tenggara Timur 85361, Indonesia
[4]Department of Chemical Engineering, Faculty of Engineering, Universitas Muhammadiah Purwokerto, Jawa Tengah 53182, Indonesia
[5]Departement of Chemical Engineering, Institute Technology of Bandung 40132, Indonesia

E-mail: zuchra.helwani@lecturer.unri.ac.id

Abstract. In the last ten years, Indonesia has succeeded in realizing the production and commercialization of oxygenated biofuel/bioethanol and bioethanol, even has applied mixing biodiesel into diesel to up to 20%-volume (or known as a B20 mixture). This success has at least helped reduce imports of fuel oil (BBM) such as diesel and gasoline, but no more than 20%. The rest (i.e., at least 80%) must still be imported and will cause severe pressure on the country's balance of payments because the current volume of fuel imports is no less than 50% of domestic needs. The quality of Indonesian biodiesel products generally still has a high cloud point and low oxidation stability. Diesel engine fuel consists of a mixture of methyl ester fatty acids (FAME: Fatty Acids Methyl Ester). Many constituent components are limited or not large (consist of 7-14 carbons). The essential constituents are the methyl esters of lauric, myristic, palmitic, stearic, oleic, linoleic, and linolenic acids. Methyl esters of polyunsaturated fatty acids (linoleic, linolenic, eleostearic) are inferior because they have a low cetane number and low oxidative stability. While methyl esters of saturated fatty acids have a superior demonic rate but have a high melting point (bad, higher than the requirements given by SNI). Related to the amount of unsaturated fat in biodiesel, it can reduce the quality of biodiesel, such as acid numbers, moisture content, and the emergence of sludge. Based on the current condition of biodiesel quality, to improve the quality of biodiesel produced can be done by reducing the number of unsaturated fats present in the biodiesel product. The initial step that has to be done is separation biodiesel products into saturated biodiesel with the characteristic iodine number <30-40 and unsaturated biodiesel with an iodine number > 70 through a fractionation process. Extraction was carried out using $AgNO_3$ with variations in the ratio of biodiesel feed to solvents. The best ratio of biodiesel:$AgNO_3$ to eliminate the levels of polyunsaturated free fatty acids in biodiesel products is 1:2 with an iodine value of 47.38 gr I_2/(100 gr oil)

1. Introduction

Biodiesel is an oxygenated biofuel for diesel engines consisting of a mixture of methyl esters from fatty acids. Until now, the use of biodiesel is still limited as a mixture of petroleum diesel. In trade terms, diesel oil mixed with biodiesel is stated in B-XX notation. For example, the B-10 shows that the fuel mixture contains 10% biodiesel volume and 90% diesel fuel volume. At present, Indonesia is the country that implements the mixing of biodiesel into diesel the highest in the world, which is 20% by volume (known as B20). The utilization of B20 has been approved by the Japanese Automobile Manufacturers Association (JAMA) has also made an open statement accepting and approving the use of B20 in tropical countries as long as the biodiesel is made from palm oil [1].

Although it is recognized as the best today, palm biodiesel has several weaknesses:

a. Cloud point is relatively high, which is around 18 °C. This makes palm biodiesel unusable at relatively high levels of blending in cold regions such as North Europe, North America, Japan, Korea, China, South Australia and New Zealand and (in Indonesia) the PT Freeport mining operations in Papua. This is a result of the fact that palm biodiesel contains many unbranched fatty acid groups (palmitate and stearic).
b. Compared to diesel fuel (B0), the oxidative stability of palm biodiesel is still much lower. The reason for this is that palm biodiesel still contains unsaturated fatty acid groups (especially polyunsaturated ones).

Soerawidjaja (2015;2017) states that the ideal biodiesel (or high-performance biodiesel) is biodiesel which has the following characteristics: (1) does not contain methyl esters from polyunsaturated fatty acids; (2) the chain of saturated fatty acids is not straight but branched methyl in the middle; and (3) preferably, containing several percent methyl esters of lauric oil (coconut, palm kernel) [2, 3].

Achievement of feature #3 can only be done by mixing a small amount of coconut biodiesel or palm kernel-core biodiesel into palm biodiesel, while feature #1 and feature #2 can only be achieved if appropriate chemical processing is done, which includes: (1) Elimination of groups polyunsaturated fatty acid groups through extraction and/or hydrogenation, and (2) Isomerization of the methyl branching of saturated fatty acid groups.

Polyunsaturated FAMEs, such as linoleic and linolenic, have low cetane rates and very low oxidative stability [4]. In addition, the presence of methyl esters of polyunsaturated fatty acids in biodiesel will have a catalytic effect significantly reducing the stability of biodiesel oxidation.

Methyl oleate has oxidation stability of about 14 hours while methyl linolenic oxidation has a stability of about 0.2 hours. If the mixing has a linear effect on the stability of oxidation, then the oxidation stability of the mixture of 88% methyl oleate - 12% methyl linolenic should be (0.88 x 14) + (0.12 x 0.2) = 12.34 hours. Figure 3 shows that the mixture has only about the same oxidation stability as methyl linolenic acid, thus showing the very poor catalytic effect of the presence of polyunsaturated FAMEs on the stability of biodiesel oxidation [5].

So to produce biodiesel with high oxidation stability, polyunsaturated FAMEs need to be removed from biodiesel. Furthermore, hydrogenation of these polyunsaturated FAMEs to at least monounsaturated FAMEs will allow it to be re-mixed into biodiesel components. The removal of polyunsaturated FAMEs can be done by extraction [6] or fractionation [7].

The purpose of this research is to determine the ratio between biodiesel feed and the amount of $AgNO_3$ used as a solvent so that polyunsaturated fatty acids can be removed entirely from feed biodiesel products and analyze the iodine number of raffinate products from the extraction process.

2. Methods

The study focused on testing the saturation FAME fractionation procedure (iodine number <30-40) and unsaturated (iodine number > 70) by extraction method using silver nitrate ($AgNO_3$) solvent [6]. The biodiesel used is obtained from domestic biodiesel producers, namely PT. Wilmar Bioenergy Indonesia. The raw material preparation stage consists of making an indicator of $MgSO_4$ desiccated, the FAME dehydration process (biodiesel), and the manufacture of $AgNO_3$ solvent reagents.

The FAME extraction step consists of the complex formation reaction phase between Ag^+ ions and the carbon double that is owned by unsaturated FAME. This reaction is estimated to reach equilibrium within 3 hours. After equilibrium is reached, the saturated fatty acid phase is separated from the aquatic phase (which contains the complex compounds produced) so that the saturated FAMEs can be separated from the unsaturated FAMEs. Then the washing and purification process is carried out for each saturated and unsaturated FAMEs that has been successfully separated.

The initial FAMEs (as received) and the saturated as well as unsaturated FAMEs that have been successfully separated were then analyzed with GC-MS to quantify their composition of the fatty acids. Their iodine numbers were also analyzed using the Wijs reagent follows the standard method of AOCS Cd 1-25. FAMEs composition analysis using GC-MS. Samples were diluted 20 times with hexane. The GC-MS used is the 2010 Shimadzu brand equipped with RTX-5 MS columns with dimensions of 30 m x 0.25 mm x 0.25 μm with AOC-20i auto-injector.

3. Results and Discussion

At the fractionation stage in connection with this series of studies, no direct oil and vegetable fats or fatty acid esters have been made thereof but using biodiesel obtained from PT Wilmar as a sample of raw material for the experiment. The GC-MS analytical results of fatty acid ester/acid compositions in feed biodiesel obtained from PT. Wilmar are presented in Table 1.

Table 1. Composition of fatty acids/esters in the original biodiesel (obtained from PT. Wilmar)

No	Fatty acid	%
1	Laurate acid (C12:0)	0.68
2	Miristate acid (C14:0)	2.15
3	Palmitate acid (C16:0)	39.17
4	Stearate acid (C18:0)	7.18
5	Oleate acid (C18:1)	42.41
6	Linoleate acid (18:2)	5.87
7	Others	2.54
	Total	100

The composition of saturated, monounsaturated, and polyunsaturated fatty acid ester metal (FAME) fractions based on GC-MS chromatogram data is presented in Table 2.

Table 2. Composition of saturated and unsaturated biodiesel in the original biodiesel

No	Biodiesel feed (FAME)	%
1	Saturated FAMEs	49.18
2	Monounsaturated FAMEs	42.41
3	Polyunsaturated FAMEs	5.87
3	Others	2.54
	Total	100

The biodiesel fractionation was carried out through a series of dehydration experiments of water molecules from the pristine biodiesel, which is carried out through the stripping of water molecules from the biodiesel sample using MgSO4 characterized in a desiccator. The process was then continued with the application of extraction treatment using silver nitrate ($AgNO_3$) as reported by Teramoto et al. (1994) [6]. Both series of experiments mentioned above, carried out on biodiesel initially which has an iodine number (AI) 57.22 gr I2/100 gr sample. The hope, with the extraction method Teramoto et al. (1994) will produce saturated biodiesel with an AI range of 30-40. From the beginning, the accent of this research was to produce high-performance/high-performance biodiesel, which is biodiesel with low cloud point (<< 18 °C) and high oxidation stability.

The results of the test for the composition of fatty acids/esters of fatty acids contained in saturated biodiesel (extraction results) have been carried out with GC-MS instruments. The results of the composition analysis are presented in Table 3.

Table 3. Composition of fatty acids/esters in the original biodiesel and saturated FAMEs

No	Fatty Acid	%-area			
		Biodiesel feed	Saturated FAMEs		
			ratio 2:1	ratio 1:2	ratio 1:2.5
1	Laurate acid (C12:0)	0.68	1.22	0.73	0.75
2	Miristate acid (C14:0)	2.15	3.47	2.34	2.23
3	Palmitate acid (C16:0)	39.17	42.73	42.48	41.37
4	Linoleate acid (C18:2)	5.87	4.27	0	0.03
5	Oleate acid (C18:1)	42.41	38.4	44.82	41.48
6	Stearate acid (C:18:0)	7.18	6.42	7.19	7.25
7	Others	2.54	3.49	2.44	6.89
	TOTAL	100	100	100	100

The data in Table 4 shows the progress of the separation of saturated and unsaturated biodiesel by the extraction method using silver nitrate ($AgNO_3$) solvent. There is a significant increase in C16 saturated fatty acids, along with a decrease in monounsaturated fatty acids (C18: 1) in three series of extraction experiments. This phenomenon implies that the extraction process using silver nitrate solvent takes place quite well, because at a ratio of 1: 2 all polyunsaturated fatty acids can already be removed from biodiesel feed products.

Table 4. Data of iodine number and oxidation stability of each extraction product in the variation of the feed ratio to the $AgNO_3$ solvent

No	Ratio of feed to solvent	Parameter	Biodiesel feed	Saturated FAMEs	Unsaturated FAMEs
1	2 : 1	Iodine number (gr I_2/100 gr sample)	57.22	52.62	155.71
2	1 : 2	Iodine number (gr I_2/100 gr sample)		53.21	159.04
3	1 : 2.5	Iodine number (gr I_2/100 gr sample)		52.21	157.42

3.1. Optimization of biodiesel extraction through the demetalization process

Biodiesel extraction is still carried out using silver nitrate solvent, $AgNO_3$, pro analysis with variation ratio of biodiesel volume: $AgNO_3$ solution (1: 2). As explained by the ratio of biodiesel volume: $AgNO_3$ of 1: 2, has been proven to be able to separate all FAMEs polyunsaturated (aquatic phase) from saturated and monounsaturated FAMEs (biodiesel phase). Optimization of the extraction process is carried out by continuously flowing inert gas (Nitrogen) into the reactor to prevent oxidation reactions, which will reduce the oxidation stability of the extracted biodiesel phase.

The data shown in Table 5 indicate that the presence of each water and Ag on biodiesel influences the iodine value of saturated FAME. This indicates that the high iodine value in the saturated biodiesel

IOP Conf. Series: Materials Science and Engineering **823** (2020) 012022 doi:10.1088/1757-899X/823/1/012022

volume ratio experiment to $AgNO_3$ solution (1: 2) occurs due to the presence of Ag^+ and water left in the extracted biodiesel phase. Thus, efforts should be made to eliminate Ag content (demetalization) and evaluation of the washing process in the extracted biodiesel phase. Therefore, the iodine number was tested for the extraction results. Data on iodine figures of the two experiments are presented in Table 5.

Table 5. Best experiment (biodiesel volume ratio: $AgNO_3$ = 1: 2)

No	Parameter	Biodiesel feed	Saturated FAMEs	
			Without demetalization	With demetalization
1	Iodine number (gr I_2/100 gr sample)	57.22	53.21	47.38

Table 6 shows that after the demetalization process, the iodine number figures decreased by 9.84 points from biodiesel feed while the reduction in Ion Figures in the extraction experiments without demetalization is only 4.01 points from the original biodiesel.

In the second phase of the experiment, the Cloud Point and Pour Point Analyses were carried out on the extracted saturated biodiesel in the ITB laboratory. Table 6 presents the results of the analysis.

Table 6. Cloud point and pour point data

Sample	*Cloud Point* (°C)		*Pour Point* (°C)	
Biosolar	4.2	4.1	-1.0	-1.0
Biodiesel Wilmar	12.8	12.7	9.0	9.0
Saturated Biodiesel (1:2) with demetalization	13.2		9.0	

4. Conclusion

Extraction of ethyl methyl fatty acids (FAME) with $AgNO_3$ solvent with biodiesel volume ratio: $AgNO_3$ which is 1: 2 can separate all polyunsaturated FAMEs from saturated and monounsaturated FAMEs. The iodine value of saturated biodiesel demetalization products is 47.38 gr I_2/gr sample, or 9.84 points reduced from feed biodiesel going to the expected Iod number range (30-40).

5. References

[1] Japan Automobile Manufacturers Association, Inc. *JAMA Position on the Use of Diesel Fuel Blended with FAME Concentrations in Excess of 5%* Available from: https://jama.or.jp/eco/wwfc/pdf/FAME_JAMA_Supplementary_Position_Statement_December2016

[2] Soerawidjaja T H 2015 *Toward High Performance Biodiesel* Workshop on higher blending of biodiesel (H-FAME) for automotive utilization in ASEAN (Thailand, September 17–18)

[3] Soerawidjaja T H 2017 *Menuju ke Biodiesel Berunjuk-Kerja (Berperforma) Tinggi* Forum Diskusi Pemanfaatan Biodiesel 30 % (B30) (Tangerang, 19-21 Desember)

[4] Moser B R 2009 Biodiesel production, properties and feedstocks *In Vitro Cellular and Development Biology – Plant* **45**(3) 229–66

[5] Goto S 2010 Benchmarking of biodiesel fuel standardization in East Asia *Biodiesel Fuel Trade Handbook* (ERIA Research Project Working Group)

[6] Teramoto M, Matsuyama H, Ohnishi N, Uwagawa S and Nakai K 1994 Extraction of ethyl and methyl esters of polyunsaturated fatty acids with aqueous silver nitrate solutions *Ind. Eng. Chem. Res.* **33**(2) 341-5

[7] Aktiebolaget 1952 Separator method of fractionating fatty oils *Paten Inggris Raya* No. GB682797

IOP Conf. Series: Materials Science and Engineering **823** (2020) 012023 doi:10.1088/1757-899X/823/1/012023

The effect of demineralization towards gasification performance of low-rank coal

Jenny Rizkiana[1], Reyhan Fitri Ananda[1], Nasywa Kamilah[1], Ghiffary Azka Nur Aulia[1], Sandy Fajar Maulana[1], Aghietyas Choirun Az Zahra[1], Winny Wulandari[1], and Dwiwahju Sasongko[1]

[1] Chemical Engineering Department, Institut Teknologi Bandung, Jalan Ganesha 10 Bandung 40132

E-mail: jr@che.itb.ac.id

Abstract. Gasification process is an alternative way to utilize low-rank coal. However, gas production is not very high due to the high content of volatile matters and high mineral content. One interesting way to increase gas production is by reducing the mineral content in the coal by acid leaching process. In the present study, acid leaching is occupied to pre-treat the coal by using HF solution so that the mineral content in the coal can be leached out. The results show that acid leaching process successfully decreased the ash content in the coal. However, in terms of gas production, it appears that acid leaching slightly decreases the total gas production. It could happen since acid processing of coal might increase its crystallinity and thus it becomes more difficult to be gasified.

1. Introduction

The increase in the world's population becomes a great factor in the rise of energy demand [1]. The world's energy consumption was at 13,105 million tons in 2015 and reached 13,276.3 million tons in 2016, while Indonesia's energy consumption increases at 164.8 million tons and 175 million tons respectively [2]. One alternative in generating energy to meet the energy demand is through oxidation of hydrogen gas. Currently, 98% of hydrogen is produced from coal and 50% of it comes from the natural gas reforming process [3]. Coal places second after crude oil as an energy source until 2016 [2]. Hydrogen gas from coal can be utilized to generate a turbine to produce energy [4] and potentially used as fuel cell feedstock that is environmentally friendly in the future [5].

Since 2016, the use of coal as a fuel to generate energy has been declining to comply with the regulation to reduce CO_2 emission in order to suppress environmental damage [6]. Indonesia Government in the United Nations Framework Convention on Climate Change (UNFCC). In Indonesia National Energy General Plan (RUEN), coal is viewed as a primary energy source that is capable of supplying energy for more than 30% of total energy in 2025. Based on its coal rank, lignite and sub-bituminous coal dominate in Indonesia at 24,641 million tons with bituminous and anthracite as minorities at 3,814 million tons [7]. Thermal decomposition of lignite and sub-bituminous coal discharges coal ash that contains high silica and sulfur content. Both components give off inhibition effect toward the coal utilization process.

Gasification is a conversion method of low ranked coal that allows maximum energy to be extracted with a minimum amount of residue and high carbon conversion [8-9]. Gasification emits less CO_2 compared to other coal utilization methods [10]. Gasification has high efficiency and flexibility in producing gas to generate electricity and synthesize liquid fuel [11-12]. Steam as gasification agent to gasification process intensifies hydrogen gas production. Potassium carbonate gives a favorable effect toward gasification process with steam as the gasification agent [13]. The increase of reaction rate results in raise of gas yield. However, SiO_2 content in coal ash is not preferable in gasification process as it inhibits gasification process.

Coal ash removal can be accomplished by using acid, base, alcohol, or combination of acid, base, and alcohol [14-18]. Literature concludes that acid solution gives the most effective result in coal mineral removal contained in coal ash. The coal ash removal process is called demineralization. Demineralization process improves the gasification process in 2 methods, silica removal, and pore formation to increase surface area. Several factors need to be considered in the demineralization process, they are type of acid solution, concentration, temperature, demineralization time, and coal particle size. The effect of factors mentioned previously can be seen in the reduction of coal ash produced after demineralization. Furthermore, the silica content in demineralized coal will diminish compared to the one in raw coal. This paper covers the effect of coal demineralization using HF solution toward coal gasification in removing silica and sulfur content.

2. Methodology

The experiment methodology includes material preparation, coal demineralization, coal gasification, and analysis and data interpretation.

2.1. Material preparation

Coal is crushed into sizes less than 232 μm (> 48 mesh) and then dried in an oven at 110°C for 14 hours. The determination of ash content in coal is achieved through combustion in a furnace at 750°C or 2 hours. Element content in ash is analyzed using EDS, while the composition of coal is determined through proximate and ultimate analysis. The two analyses of raw lignite (RL) and demineralized lignite (DL) are shown in Table 1.

Table 1. Proximate and ultimate analysis of RC and DC

Parameter	RC	DC
Proximate	(%-weight) dry basis	(%-weight) dry basis
Ash	9.54	1.63
Volatile matter	47.38	51.53
Fixed carbon	43.08	46.84
Ultimate		
Sulfur	1.07	1.07
Carbon	64.38	64.38
Hydrogen	3.74	3.74
Nitrogen	1.14	1.14
Oxygen	19.71	19.71

2.2. Coal demineralization

10g of dried coal is dissolved in a 250 mL HF solution in a Teflon glass. The solution low-rank for 4 hours at 70°C. Then, the solution is neutralized using aqua DM before being filtered. The solid residue from the filtration is later dried in an oven for 14 hours at 110°C before being stored in a desiccator.

2.3. Coal gasification

The gasification process is carried out with a fixed bed downdraft reactor for 80 minutes using steam as a gasification agent with steam production temperature at 90-92°C. 3g of each raw lignite and demineralized lignite is used for the gasification process. Nitrogen, as gas carrier, is flown with a rate of 85 cm^3/minute. The furnace used is heated beforehand with a heating rate of 6°C/minute until the operating temperature in the reactor reaches 700°C. The outlet gas is later condensed with cooling water at 14-16°C. Condensed gas becomes tar while incondensable gas is stored in a gas bag every 10 minutes. The amount of incondensable gas produced is compared with the amount of fixed carbon in the feed sample per gram.

2.4. Analysis and data interpretation

Data obtained in material preparation and coal demineralization are processed, analyzed, and presented in graphs. Raw lignite and demineralized lignite contents are determined in proximate and ultimate analysis, while ash content determination from feed is done through ASTM D.3174. Synthesis gas compositions are analyzed using gas chromatography (GC) with thermal conductivity detector (TCD) that is connected with 2 types of columns, namely Porapak Q and Molesieve 5A. The GC result is further analyzed for graph fabrication. Surface area and char sizes obtained from gasification process are analyzed through nitrogen isothermal adsorption method.

3. Results and Discussion

3.1. Ash element content of Raw Lignite (RL) and Demineralized Lignite (DL)

Coal demineralization process is a pre-treatment one to reduce coal ash elements which are silica and alumina. Ash mineral content distribution in LC and DL is determined through EDS mapping analysis shown in Table 2.

Table 2. Coal ash mineral content in RL and DL

Element	Compositions (%-weight) dry basis	
	RL	DL
Carbon (C)	3.13	0.66
Oxygen (O)	4.34	0.67
Sodium (Na)	0.26	0.04
Magnesium (Mg)	0.20	0.04
Aluminum (Al)	0.41	0.05
Silica (Si)	0.37	0.03
Sulfur (S)	0.07	0.02
Potassium (K)	0.02	0.002
Calcium (Ca)	0.02	0.004
Phosphor (P)	0.05	0.01
Iron (Fe)	0.67	0.11
Total	9.54%	1.63%

Coal that has been demineralized with HF has lower ash content compared to the raw lignite. HF not only dissolves silica but also almost all minerals in raw lignite. The result shows 91.7% silica content reduction, 90.3% aluminum content reduction, and also 88.9% potassium content. According to Steel et al. (2001) [19], HF solution is very reactive in dissolving all mineral components in coal but pyrite. At 3 M, HF is able to make aluminosilicate to be soluble in water.

3.2. Single fuel gasification: Raw Lignite (RL) and Demineralized Lignite (DL)
Feed characteristics determine the distribution of a carbon fuel gasification product [20]. Lignite coal contains complex organic compounds, namely vitrinite and liptinite, oxidized organic compounds, and water [21].

3.2.1. Gasification of RL and DL product conversion. The main products of coal gasification process generally composed of synthesis gas (H_2, CO, CO_2, and CH_4), liquid product (volatile matter and condensable gas, and char [21].

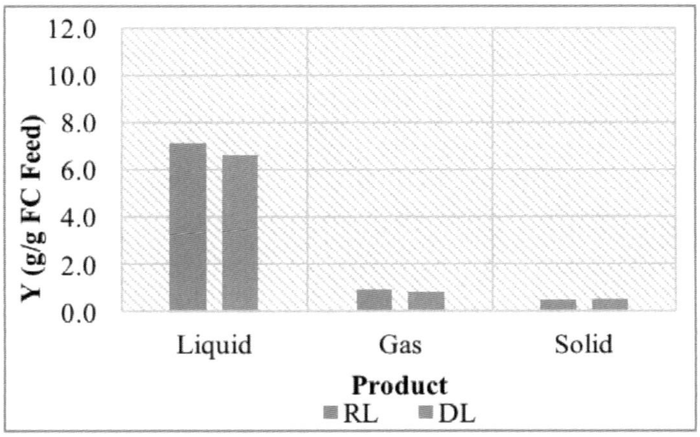

Figure 1. Gas, liquid, and solid products of gasification process of RL and DL

Figure 1 shows the distribution of RL and DL gasification products. The main components of liquid product include large aromatic compound (mixture of oxygenation, phenolic esters, phenolic alkyls, heterocyclic ethers, polyaromatic hydrocarbons, etc.) and alkali metal compounds [21]. As both components are soluble in HF solution, DL has lower liquid product conversion compared to RL. As for the gas product, DL synthesis gas conversion is less than the RL one, which means that demineralization treatment does not enhance synthesis gas conversion of RL gasification. Synthesis gas conversion depends on several factors, one of them is the char structure [22].

Carbon conversion of the DL gasification is higher than the RL one due to the demineralization process that improves lignite carbon conversion to char compared to RL. In gasification with H_2O medium, lignite ash with high silica content helps molecular breakdown of large hydrocarbons to smaller compounds such as methane and ethane [21]. Demineralization using HF and HCl upgrades the crystallinity of the char structure; therefore it is hard to break compared to coal without any pre-treatments [23]. This results in aromatic polycyclic hydrocarbons to have high densities to form bonds that are very strong and resistant to heat, making it difficult to decompose into liquid and gas products.

3.2.2. Gasification of RL and DL synthesis gas profile and trend. Based on Basu [22], solid-gas reactions that are likely to happen in char gasification process include Boudouard, methanation, and primary water-gas. After the synthesis gas is formed, a gas-gas reaction takes place, which forms a new gas compound. Common reactions include steam reforming and water gas shift. In determining the tendency of the gasification reaction, gas compositions of every 10 minutes are analyzed using Shimadzu Gas Chromatograph 2014 that has been calibrated with industrial standard synthesis gas. This experiment covers gasification process at low temperature and is kept constant at 700 ± 40°C. Figure 2 shows synthesis gas profile and H_2/CO ratio of RL gasification for 80 minutes.

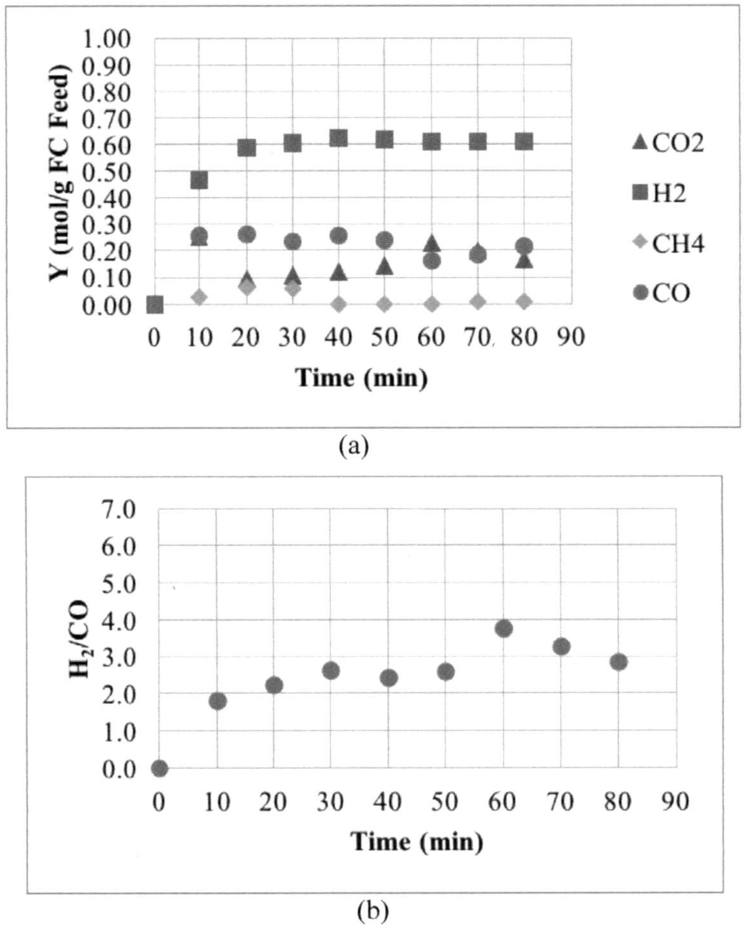

Figure 2. (a) Synthesis gas profile and (b) H_2/CO ratio every 10 minutes of RL gasification

Generally, H_2 and CO_2 contents increase while CO gas yield decreases. This causes H_2/CO ratio to relatively rise until the 80th minute. At the 10th minute, H_2 gas yield is at its lowest while CO_2 and CO gas yield are quite high. This means that in the beginning, Boudouard reaction and water gas reaction dominate. The high CO_2 content is derived from residual pyrolysis result when a new gasification agent, steam, is released in the 0th minute. There is an increase in methane yield from the 0th minute to the 30th minute. This shows that a methanation reaction takes place from RL carbon and H_2. From the 0th minute to the 20th minute, a decrease in CO gas identifies Boudouard reaction rate drops and reaction sifts toward the primary water gas reaction, which is indicated by a rapid rise in H_2 concentration. Char conversion reaction is dominated by the primary water gas reaction marked by the increase in H_2 concentration, a drop in CO concentration, and constant CO_2 content.

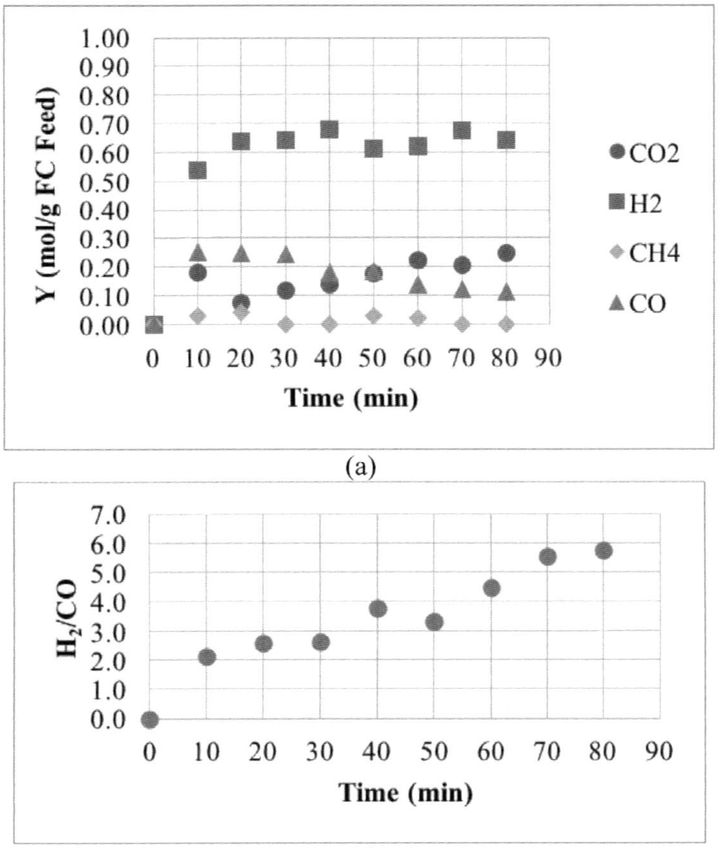

(b)

Figure 3. (a) Synthesis gas profile and (b) H_2/CO ratio every 10 minutes of DL gasification

Figure 3 shows the synthesis gas profile and H_2/CO ratio from gasification of DL. The results of synthesis gas profile from gasification of RL and DL have a similar pattern. Both results show a constant H_2 concentration until the 80th minute. The synthesis gas profile of DL gasification fluctuates more compared to the RL one due to higher alkali and alkaline earth metal content in RL that gives a more stable gasification rate [24]. Methane gas that appears at the beginning of DL gasification is a residual gas produced from lignite pyrolysis, such as RL gasification [25-26]. However, DL gasification produces more methane than RL as methane cracking process in DL gasification is lower than the one in RL gasification [27]. According to Sun et al. (2007) [28] and Wei et al. (2011) [29], coal demineralization lessens catalytic effect toward the methane cracking process by Fe and Mg in coal ash. At the 40th minute of DL gasification and the 60th minute of RL gasification, a drip of steam condensate occurs that makes the primary water gas reaction producing more H_2 and CO_2.

3.2.3. Gasification of RL and DL synthesis gas yield. Synthesis gas compositions depend on characteristics of carbon, type of gasification medium, and operating condition [20]. Figure 4 presents the yield of each component in synthesis gas from gasification of RL and DL.

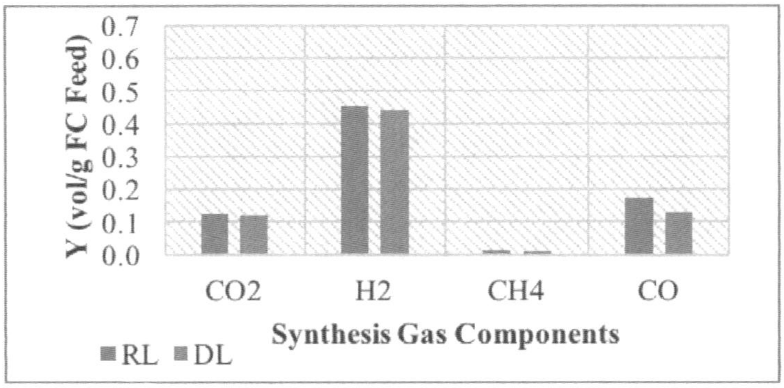

Figure 4. Gas component distribution of gasification of RL and DL

RL gives better H_2, CO, CO_2, and CH_4 yield compared to DL. This shows that coal demineralization by HF leaching method slightly lessens the synthesis gas yield. The result is supported by research done by Murata et al. (1996) [30] that states HF leaching degrades conversion rate of char in coal gasification. It is possible that when the coal is pre-treated by a strong acid, the coal tends to be more crystalline; as such, it becomes more difficult to be gasified. Even though more pores are formed, the crystallinity of the carbon makes it harder to be gasified compared with the raw coal.

4. Conclusions

Acid leaching pre-treatment have been done for low rank coal by using HF solution. It appears to be the good method to demineralize coal. However, pre-treating the coal reduces gasification performance in terms of a decrease in synthesis gas and H_2 yield. An increase in char yield is caused by a drop in coal reactivity significantly that affects the catalytic effect on the overall gasification process. It is possible that pre-treatment using acid increases coal crystallinity as such it becomes more difficult to be gasified.

5. References

[1] Pinto F, Franco C, André R N, Tavares C, Dias M, Gulyurtlu I and Cabrita I 2003 Effect of experimental conditions on co-gasification of coal, biomass and plastics wastes with air/steam mixtures in a fluidized bed system *Fuel* **82** 1967–76

[2] British Petroleum 2017 *BP Statistical Review of World Energy 2017* pp 1–52 (British Petroleum)

[3] Dunn S 2003 Hydrogen futures: toward a sustainable energy system *Int J Hydrogen Energy* **27** 235-64

[4] Weiland N T, Means N C and Morreale B D 2012 Product distributions from isothermal co-pyrolysis of coal and biomass *Fuel* **94** 563–70

[5] Midilli A, Ay M, Dincer I and Rosen M A 2005 On hydrogen and hydrogen energy strategies I : Current status and needs *Renewable and Sustainable Energy Reviews* **9** 255–71

[6] Ellis N, Masnadi M S, Roberts D G, Kochanek M A and Ilyushechkin A Y 2015 Mineral matter interactions during co-pyrolysis of coal and biomass and their impact on intrinsic char co-gasification reactivity *Chemical Engineering Journal* **279** 402–8

[7] Direktorat Jenderal Mineral dan Batubara 2017 *Laporan Kinerja Direktorat Jenderal Mineral dan Batubara* (Jakarta: Kementerian Energi dan Sumber Daya Mineral)

[8] Ahmed I I, Nipattummakul N and Gupta A K 2011 Characteristics of syngas from co-gasification of polyethylene and woodchips *Applied Energy* **88** 165–74

[9] Cormos C.-C 2012 Hydrogen and power co-generation based on coal and biomass/solid wastes co-gasification with carbon capture and storage *International Journal of Hydrogen Energy* **37** 5637–48

[10] Krerkkaiwan S, Fushimi C, Tsutsumi A and Kuchonthara P 2013 Synergetic effect during co-pyrolysis/gasification of biomass and sub-bituminous coal *Fuel Processing Technology* **115** 11–18

[11] Zhang Z, Pang S and Levi T 2017 Influence of AAEM species in coal and biomass on steam co-gasification of chars of blended coal and biomass *Renewable Energy* **101** 356–63

[12] Pinto F, André R N, Carolino C and Miranda M 2014 Hot treatment and upgrading of syngas obtained by co-gasification of coal and wastes *Fuel Processing Technology* **126** 19–29

[13] Shahbaz M, Yusup S, Inayat A, Patrick D O and Ammar M 2017 The influence of catalysts in biomass steam gasification and catalytic potential of coal bottom ash in biomass steam gasification: A review *Renewable and Sustainable Energy Reviews* **73** 468-76

[14] Steel K M, Besida J, O'Donnel T A and Wood D G 2001 Production of Ultra Clean Coal Part I—Dissolution behaviour of mineral matter in black coal toward hydrochloric and hydrofluoric acids *Fuel Processing Technology* **70** 171-92

[15] Mukherjee S and Borthakur P C 2001 Chemical demineralization/desulphurization of high sulphur coal using sodium hydroxide and acid solutions *Fuel* **80** 2037–40

[16] Behera S K, Chakraborty S and Meikap B C 2017 Chemical demineralization of high ash Indian coal by using alkali and acid solutions *Fuel* **196** 102-9

[17] Ratanakandilok S, Ngamprasertsith S and Prasassarakich P 2001 Coal desulfrization with methanol/water and methanol/KOH *Fuel* **80** 1937-42

[18] Wu Z and Steel K M 2007 Demineralization of a UK bituminous coal using HF and ferric ions *Fuel* **86** 2194-200

[19] Steel K M, Besida J, O'Donnell T A and Wood D G 2001 Production of Ultra Clean Coal - Part II - Ionic equilibria in solution when mineral matter from black coal is treated with aqueous hydrofluoric acid *Fuel Processing Technology* **70** 193–219

[20] Mendiburu A Z, Carvalho J A and Coronado C J R 2014 Thermochemical equilibrium modeling of biomass downdraft gasifier: Stoichiometric models *Energy* **66** 189–201

[21] Mallick D, Mahanta P and Moholkar V S 2017 Co-gasification of coal and biomass blends: Chemistry and engineering *Fuel* **204** 106–28

[22] Basu P 2013 *Gasification Theory, Biomass Gasification, Pyrolysis and Torrefaction* 2nd edition (Elsevier Inc.)

[23] Zhu W, Song W and Lin W 2008 Catalytic gasification of char from co-pyrolysis of coal and biomass *Fuel Processing Technology* **89** 890–96

[24] Bell A T and Veraa M J 1977 Effect of alkali metal catalysts on gasif ication of coal char *Fuel* **57** 194–200

[25] Fermoso J, Arias B, Plaza M G, Pevida C, Rubiera F, Pis J J, García-Peña F and Casero P 2009 High-pressure co-gasification of coal with biomass and petroleum coke *Fuel Processing Technology* **90** 926–32

[26] Lee W J, Kim S D and Song B H 2002 Steam gasification of an australian bituminous coal in a fluidized bed *Korean Journal of Chemical Engineering* **19** 1091–6

[27] Ramarao M, Kumar S M, Vivekanandan S, Suresh D V, Kumar S P and Jayaprakash D 2017 Effect of Biomass Blending on Gasification : A Review *Advanced in Natural and Applied Sciences* **11** 300–10

[28] Sun Z Q, Wu J H, Haghighi M, Bromly J, Ng E, Wee H L, Wang Y and Zhang D K 2007 Methane cracking over a bituminous coal char *Energy and Fuels* **21** 1601–5

[29] Wei L, Tan Y S, Han Y Z, Zhao J T, Wu J and Zhang D 2011 Hydrogen production by methane cracking over different coal chars *Fuel* **90** 3473–79

[30] Murata S, Murakami A and Nomura M 1996 Steam gasification of coal: The effects of acid- and alkali leaching of coal on its gasification rate *The 211th ACS National Meeting* **41**(1) 237-40 (USA, March 24-28)

IOP Conf. Series: Materials Science and Engineering **823** (2020) 012024 doi:10.1088/1757-899X/823/1/012024

Production of biofuel by low temperature Fischer-Tropsch using Co-K/γ-Al_2O_3

I.G.B.N Makertihatha, Fadhli, Zaki Al Fathoni, Subagjo

Department of Chemical Engineering, Institut Teknologi Bandung, Bandung 40132 Indonesia

E-mail: makertia@che.itb.ac.id, fadhli@che.itb.ac.id

Abstract. Fischer Tropsch is a heterogeneous catalytic chemical reaction which converts a mixture of hydrogen and carbon monoxide (syngas) into a hydrocarbon product with varying chain length by polymerization reaction on the surface of the catalyst. The hydrocarbon produced from Fischer Tropsch reaction using bio-syngas is biofuel (diesel, kerosene, gasoline) that can replace petroleum-based fuels. Cobalt catalyst with potassium promoter and γ-Al_2O_3 support has been successfully synthesized in Catalysis and Reaction Engineering (CaRE) laboratory, Institut Teknologi Bandung (ITB). Co-K/γ-Al_2O_3 catalyst was prepared by dry impregnation method on the γ-Al_2O_3 support under alkaline conditions. Catalysts were characterized using X-Ray Diffraction (XRD), temperature program reduction (TPR), and N_2 physisorption measurements such as Barrett-Joyner-Halenda (BJH) and Brunauer-Emmett-Teller (BET) methods. Co-K/γ-Al_2O_3 catalyst activity was evaluated using fixed bed reactor with various flow rates and temperatures which is still classified as Low-Temperature Fischer Tropsch (LTFT) process. The best results were obtained at minimum syngas flow rate and highest reaction temperature with a total pressure of 20 bar. The results show CO and H_2 conversions were 96.6% and 82.31% respectively. Selectivity value of the hydrocarbon product was calculated using the Anderson Shultz Flurry (ASF) equation. The greatest selectivity value was obtained for C_{5+} product with selectivity value was 86.07 % wt.

1. Introduction

Dependence on fuel derived from fossil becomes a serious problem in the next few decades. It is estimated that the total world energy consumption is 13,864.9 Mtoe (Million tons of oil equivalent) in 2018, the percentage of fossil fuel use reaches 84.7% [1]. This led the researchers to be able to find synthetic liquid fuels that could be used generally and commercially. Several fuel synthesis technologies based on conversion technology XTL (X to Liquid) began to be developed, one of which is the Fischer Tropsch process.

The Fischer Tropsch process is a syngas conversion process ($CO + H_2$) derived from natural gas, coal, or biomass into long chain hydrocarbons that can be upgraded to diesel or kerosene fractions [2]. Processing of Fischer Tropsch synthesis products can be done by hydrocracking reactions to form fuel products according to what is needed, such as jetfuel, diesel or gasoline [3]. Fischer Tropch process has become an important key in XTL technology producing fuel, chemical compounds, and other hydrocarbon products [4].

Fischer Trospch reaction is a catalytic reaction by metals such as Fe, Co, Ru, Rh and Ni. The choice of catalyst in this process will determine the condition of the process and the type of hydrocarbon of the resulting product. Currently, only Fe and Co catalysts are already used in industrial-scale [5]. The resulting product is a wide variety of organic compounds, especially n-paraffins and 1-olefins depending on the type of catalyst and operating conditions. Other products that may be formed are branched hydrocarbons, 2-olefins, and various oxygenate compounds (alcohols, aldehydes, ketones, etc.). The resulting product may comprise one carbon atom up to hundreds of carbon atoms in long-chain hydrocarbons [6,7]. Fischer Tropsch synthesis is a polymerization reaction on the surface of the catalyst with a CHx monomer. The CHx monomer is formed from the hydrogenation process of CO adsorbed on the surface of the catalyst to produce hydrocarbons with a wide chain range and wide function [8]. In this polymerization, the main carbon chain will increase with the addition of monomers one by one carbon atom. Every single addition step, the termination process can occur and will determine the product produced and is desorbed from the surface of the catalyst [9,10]. Reactions that may occur during the Fischer Tropsch process are presented in Table 1 [11].

Table 1. Fischer Tropsch Reaction

	Main Reaction
Paraffin Formation	$nCO + (2n+1)H_2 \rightarrow C_nH_{2n+2} + nH_2O$
Olefin Formation	$nCO + 2nH_2 \rightarrow C_nH_{2n} + nH_2O$
WGSR	$CO + H_2O \rightarrow CO_2 + H_2$
	Side Reaction
Alcohol Formation	$nCO + 2nH_2 \rightarrow C_nH_{2n+1}O + (n+1)H_2O$
Boudard Reaction	$2CO \rightarrow C + CO_2$

Cobalt catalyst in the Fischer Tropsch reaction is known to provide good performance, high activity, high degree of polymerization, and the formation of byproducts such as small oxygenate compounds (<5%) when compared with iron catalysts. The cobalt-based catalyst is also not selective to WGSR reactions as well as iron catalysts. The suitable Cobalt catalyst used for the production of the target product is diesel. Diesel is obtained from hydrocracking/isomerization of a long chain of paraffin compounds [12]

Fischer Tropsch process is very sensitive to operating conditions. Temperature has an effect on desorption effect and speed of hydrogenation process. This may affect the chain polymerization process and the degree of hydrogenation of the product [13]. In the industry are known 3 operating conditions of Fischer Tropsch reaction based on reaction temperature. High-Temperature Fischer Tropsch (HTFT) is operated at temperatures above 320^0C, usually using an iron-based (Fe) catalyst producing a straight chain of olefin with a low molecular mass. Medium Temperature Fischer Tropsch (MTFT) at temperature 270^0C and Low Temperature Fischer Tropsch (LTFT) at temperature less than 250^0C [14]. In LTFT process used Cobalt (Co) catalyst producing a hydrocarbon with big molecular mass (wax). The resulting wax will be cracking with hydrogenation process to produce fuel [15]. Therefore the aim of this study was to synthesize the Co-K/γ-Al_2O_3 catalyst for the Fischer Tropsch reaction and study the influence of condition operation on the conversion and selectivity of the Fischer Tropsch reaction.

IOP Conf. Series: Materials Science and Engineering **823** (2020) 012024 doi:10.1088/1757-899X/823/1/012024

2. Experimental Section

2.1. Synthesis of Co-K/γ-Al_2O_3 catalyst

Catalyst Co-K/γ-Al_2O_3 synthesized by alkaline impregnation method using an ammonia solution. The support used is γ-Al_2O_3. $Co(NO_3)_2.6H_2O$ was used as a precursor of Co and KNO_3 was used as a metal precursor of K. The impregnant solution was prepared by dissolving the precursor compound into the ammonia solution. The impregnant solution was impregnated on a γ-Al_2O_3 support with stirring for ± 15 minutes. Impregnation process was carried out in stages for a total of 15% metal loading. Impregnated support was dried at 120^{0}C for 3 hours and continued with calcination at 400^{0}C for 6 hours. Co-K/γ-Al_2O_3 catalysts were characterized using XRD, BET and TPR methods.The title is set 17 point Times Bold, flush left, unjustified. The first letter of the title should be capitalized with the rest in lower case. It should not be indented. Leave 28 mm of space above the title and 10 mm after the title.

2.2. Activity test of Co-K/γ-Al_2O_3 catalyst

The catalyst activity test was performed on the fix bed reactor by using syngas as feed. Testing of catalyst activity was carried out under various operating conditions which are still included in LTFT category. An Analysis of gas composition in inlet and outlet of the reactor was carried out by using gas chromatography with Thermal Conductivity Detector (TCD) detector. The product formed was analyzed with GC equipment with FID detector. Gas carrier was used He and H_2. The activity testing tool scheme is described in Figure 1.

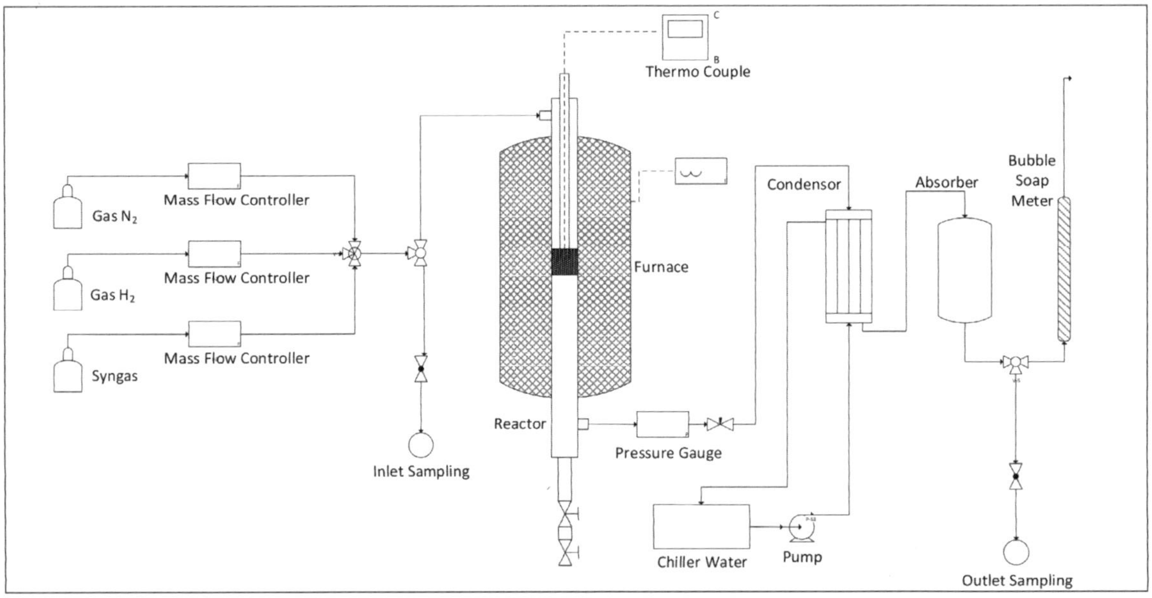

Figure 1. Fixed bed reactor scheme

The catalyst activity testing process begins with an activation process of catalyst using H_2 gas with a flow rate of 90 mL/ min at a 5 bar reactor pressure. Activation of the catalyst was carried out at 450^{0}C for 8 hours. The catalyst activity test was carried out at 20 bar syngas pressure and varying syngas flow rate. Temperature reactions used were 230^{0}C, 240^{0}C, and 250^{0}C to determine the effect of temperature on the Fischer Tropsch process. Catalyst activity test carried out until it reaches a steady-state condition. Gas sampling was taken every hour and analyzed by GC-TCD. The liquid product was taken at the end of the testing process and analyzed by GC FID. CO and H_2 conversions were calculated based on the area ration of CO and H_2 peaks in gas chromatograph.

3. Results and Discussion

3.1 Catalyst characterization

The XRD diffractogram pattern of the Co-K/γ-Al_2O_3 catalyst is shown in Figure 2. Referring to the crystallographic open data, the specific peak of Co appears on the 2θ = 36,559 with the highest intensity. This peak represents the crystal phase [311] of Co_3O_4 crystals [16]. Conaro et al, 2012, explained that K_2O crystals have a specific peak at 2θ = 39.59 with the highest intensity [17]. At the Co-K/γ-Al_2O_3 catalyst diffractogram the K_2O peak was not seen due to the small concentration of K promoter and low sample crystallinity. The catalyst particle size Co was calculated using the debye scherer equation at the highest peak that appears on the diffractogram. Co particle size obtained is 8.45 nm. According to Shimura et al. (2014), Co particle size that has good performance for the Fischer Tropsch reaction is 8-12 nm [18].

The physisorption test of nitrogen on the catalyst was carried out to determine the surface area, volume and pore diameter of the catalyst. The volume and pore diameter of the catalyst were calculated using Brunauer-Emmet Teller (BET) and Barret-Joyner-Halenda (BJH) isotherm methods. Measurement results are presented in table 2. The measurement results show a decrease in values on surface area, pore volume and pore diameter. This is due to the impregnation process which causes the filling of the support pore by metal Co.

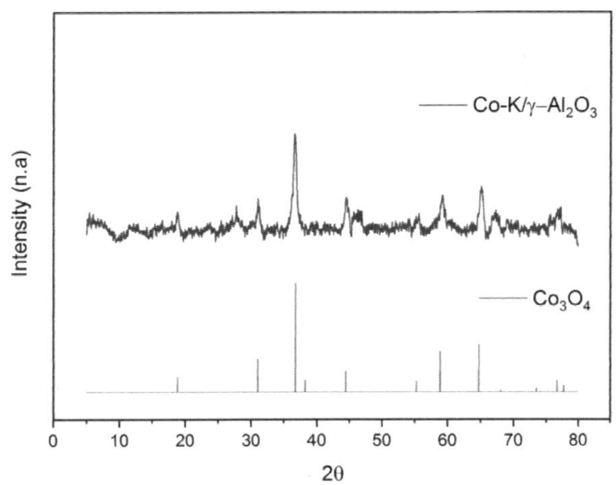

Figure 2. X-Ray diffraction patterns of the catalyst

Table 2. Pore volume, pore diameter, and surface area of catalyst

	Conc. K	BET suface area (m^2/g)	Pore Volume (cm^3/g)	Pore Diameter (nm)
γ-Al_2O_3	-	227.3	0.46	8.40
Co-K/γ-Al_2O_3	1	176.5	0.36	8.20

TPR analysis was used to determine the characteristics of the reduction of Co_3O_4 to Co^0 which is the active phase in the Fischer Tropsch reaction. Cordoba, et al 2017, explained that the reduction of Co_3O_4 compounds through 2 stages in different temperatures. Co_3O_4 was reduced to CoO at temperatures of 250-300^0 C and CoO reduced to Co^0 at temperatures of 400-500^{0}C [19]. TPR Analysis of catalyst Co-K/γ-Al_2O_3 showed a double peak at temperatures of 250-480^{0}C and 500-800^{0}C. The first peak is interpreted as a reduction process from Co_3O_4 to Co^0 in 2 stages. In contrast to the study of Cordoba et al 2017, the peak that appears is only one but wide enough. Wide peak models were interpreted that the reduction peak of Co_3O_4 to CoO shifted and merged with the second peak which described the reduction of CoO to Co^0. TPR peaks at temperatures of 500-800^{0}C represent the reduction of Cobalt aluminate compounds formed during the calcination process. Aluminate compounds have a stronger bond with metals and require a much higher reduction temperature.

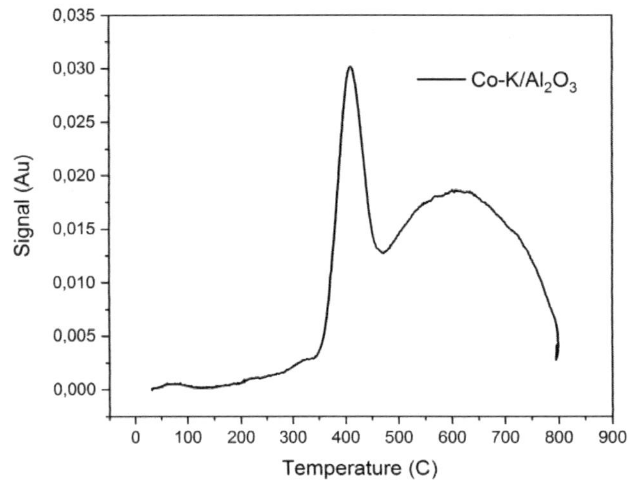

Figure 3. H_2-TPR profile of the catalyst

3.2 Activity test of Co-K/γ-Al_2O_3

Testing of the catalytic activity of the Co-K/γ-Al_2O_3 catalyst was carried out in a variety of operating conditions to determine its effect on catalyst performance. All experiments were carried out at a pressure of 20 bar and a CO: H_2 ratio of 1:2. Conversion of CO and H_2 is based on the ratio of the area of CO and H_2 to the area of the N_2 inert gas. The selectivity value of the hydrocarbon product formed is calculated using the Anderson Shultz Flurry equation to obtain the probability value of the extension of the hydrocarbon chain (α). Test results of catalyst activity and calculation of alpha values are shown in Table 3.

Table 3. A slightly more complex table with a narrow caption.

T (^{0}C)	Q (mL/min)	XCO (%)	XH_2 (%)	α^a	SC_1 (%wt)	SC_2 (%wt)	SC_3 (%wt)	SC_4 (%wt)	SC_{5+} (%wt)
230	25	75.33	69.99	0.85	2.12	3.63	4.65	5.30	84.30
	30	60.19	58.87	0.84	2.53	4.25	5.36	6.01	81.84
	35	45.70	43.12	0.81	3.57	5.80	7.05	7.62	75.96
	40	32.61	31.88	0.77	4.92	7.66	8.94	9.27	69.21
240	25	87.71	77.52	0.83	2.76	4.60	5.76	6.40	80.48
	30	67.01	65.15	0.80	3.78	6.08	7.35	7.90	74.89
	35	57.93	57.3	0.76	5.49	8.40	9.65	9.85	66.60
	40	44.41	42.64	0.74	6.60	9.80	10.93	10.83	61.85
250	25	95.19	81.83	0.82	2.95	4.88	6.07	6.70	79.40
	26.7	91.36	78.72	0.80	3.77	6.07	7.34	7.89	74.93
	30	81.48	74.76	0.80	3.78	6.08	7.35	7.90	74.89
	35	100	100	-					
	40	100	100	-					

[a] probability value of the extension of the hydrocarbon chain.

Catalytic cracking process of waste cooking oils using iron (Fe) catalysts to produce biofuel

Martha Aznury[1], Ahmad Zikri[1], Adi Syakdani[1], and Erlinawati[1]

[1]Department of Chemical Engineering, Politeknik Negeri Sriwijaya, Palembang 30139 Indonesia

E-mail: ahmad.zikri@polsri.ac.id

Abstract. Generally, there are 2 (two) methods of converting waste cooking oils into biofuel, namely by esterification of triglycerides into methyl esters (biodiesel), and utilizing thermal cracking of catalysts and hydrogenation to produce green diesel. The presence of oxygen content in waste cooking oils-free fatty acids will cause the formation of CO or CO_2 gas during the cracking process, so it is necessary to inject hydrogen to prevent it. In this study, an analysis of operating conditions, the characteristics of the biofuel produced, and the effect of the use of iron (Fe) catalysts were studied to convert used-cooking oil to biofuel. The hydrogenation of waste cooking oil is carried out at pressures of 30 and 40 psia, with heating temperature conditions of 250-350°C and operational pressures of 800-1000 psia. The maximum yield obtained was 31.7% (v) at 350°C and the use of 0.1% wt Fe catalyst. The resulting biofuel had a density of 0.8314 gr/ml and calorific value of 10152.66 cal/gr. By analysis contain of biofuel had C_{10}-C_{15} 5.5%(v), C_{15}-C_{18} 40.27%(v), C_{18}-C_{20} 38.78%(v), and > C_{20} 15.45%(v), respectively.

1. Introduction

A global environment that demands a better level of environmental quality encourages various energy experts to develop the energy that is more environmentally friendly and supports the security of sustainable supply. One solution to overcome this problem is by replacing petroleum with green diesel obtained from waste cooking oil with hydroprocessing processes which can be alternative energy.

Cooking oil is one of the basic human needs as a processing tool for frying. The waste of cooking oil in Indonesia reaches more than 6.43 million tons per year [1] so that the right technology is needed to handle this waste by converting waste cooking oil to green diesel.

There are generally two ways to produce diesel from waste cooking oil, namely utilizing esterification which converts triglycerides to methyl-esters with the help of methanol and using of thermal treatment which includes thermal cracking, catalysts, and hydrotreating. The hydrotreating is considered as the most efficient method in waste cooking oil cracking. This is because the hydrogenation process consumes less heat energy and minimizes coke formation due to the presence of hydrogen [2] It involves the hydrogenation of the double bonds of the chain and removes Oxygen with metal as the catalyst. The hydrotreating of waste cooking oil leads to C_{15}-C_{18} hydrocarbons production as green diesel [3].

IOP Conf. Series: Materials Science and Engineering **823** (2020) 012025 doi:10.1088/1757-899X/823/1/012025

2. Experimental

2.1. Material

Waste cooking oil was obtained from some small restaurants and used without any purification. Waste cooking oil was used 2 L for 1 run. Iron Fe was purchased from Sigma-Aldrich Chemical Company (St. Louis, MO, USA) as a catalyst.

2.2. Functional Design approach

The Catalytic Hydrogenation method several components that have their respective roles, namely:

1. Green Diesel Reactor; Serves to convert used-cooking oil and hydrogen through a hydraulic process into green diesel. This reactor is made of stainless steel with a height of 40 cm, an outer diameter of 2.5 inches and a thickness of 10 mm.
2. Heater; Serves as a source of heat for the reaction needs of the reactor. This heater type belt heater with a height of 30 cm and a diameter of 2.5 inches. This heater has a capacity of 1000 Watt-hour 220 Volt.
3. Cooler; Serves to change the phase of green diesel products from gas to liquid. This cooler is divided into two parts: shell and tube. The cooling fluid in the shell is water while the cooled product flows in the tube

2.3. Characterization

The elemental composition of gree diesel was determined using an analyzer GC-MS (Agilent Technologies 7890A gas chromatograph 5975C mass selective detector). The density of green diesel was determined by ASTM D4052 and for heating value used by ASTM D240-02

3. Results and Discussion

The long chain of hydrocarbons crack to short-chain have three methods. The thermal cracking with high temperature for the lighter product as the first method. The second method uses catalytic cracking. The last method is catalytic hydrocracking and occurs in the presence of a bifunctional catalyst and a high hydrogen pressure. The catalytic hydrocracking process consumes less thermal energy and the presence of hydrogen minimizes coke formation for liquid product [4].

In this study, the authors have conducted various operating conditions to find the optimum conditions from the hydrogenation of waste cooking oil to green diesel. One of the parameters reviewed by the author is the volume of green diesel. In this research, the product consisted of two phases, gas and liquid. The gas phase is believed to be propane, water, carbon dioxide, and hydrogen that does not react. The influence of operating conditions on green diesel production can be seen in Figures 1.

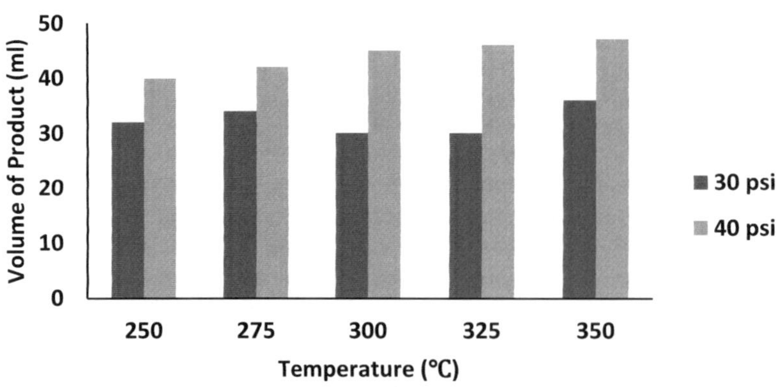

Figure 1. Temperature and H_2 pressure on product of green diesel

Based on Figure 1 to get the highest volume of green diesel is influenced by the higher hydrogen pressure operational and reaction temperature. High temperatures favored making waste cooking oil to liquid hydrocarbons consisting mainly of gasoline. That initial hydrogen pressure in a reactor made for the catalytic cracking step, and hydrogenation and hydrocracking were possible for the cracking and rearrangement to yield light hydrocarbon molecule [5].

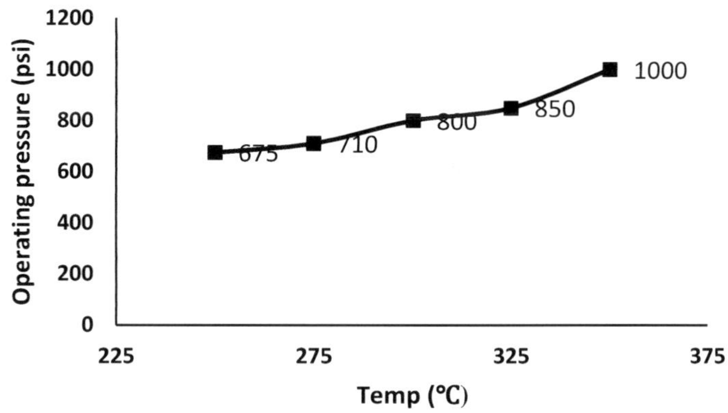

Figure 2. Temperature and catalyst relation to operating pressure

Figure 2 also proves the effective function of the catalyst to accelerate the reaction due to a large amount of gas in the reactor.

In addition to getting the relationship between product temperature and volume, the authors also get the effect of temperature on the density of the product. To determine the effect of operating conditions on product density in Figure 3.

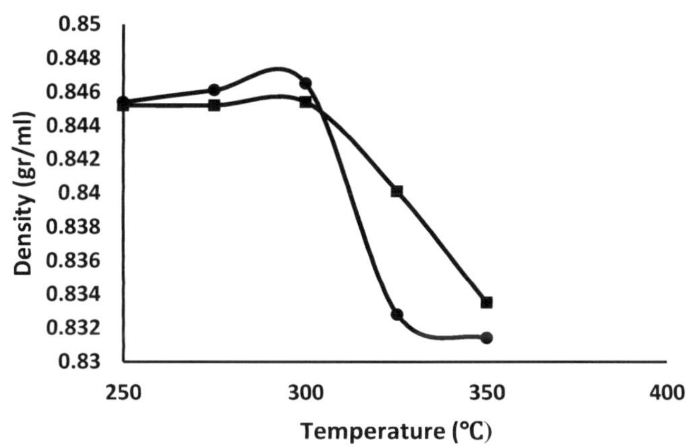

Figure 3. The density of green diesel

The results are known from Figure 3 that the lowest product density is in the operating conditions where hydrogen supply and high reaction temperatures occur. This is because in these operating conditions hydrogenation works best. The hydrogenation process is best because the low density proves that the chemical chain is more straight than the others, which means that hydrogenation has structure C_{15}-C_{18}.

That hydrogenation works better at higher temperatures; the results of GC-MS analysis have also proven it. In GC-MS analysis, the samples taken are samples in operating conditions where the hydrogen pressure enters 30 psi and the reaction temperature is 300°C and 350°C. It can be seen in Table 1.

Table 1. Effect of temperature on green diesel composition

Component (%)	Temperature (°C)	
	300	350
C_{10}-C_{15}	10,47	5,5
C_{15}-C_{18}	5,64	40,27
C_{18}-C_{20}	38,5	38,78
>C_{20}	45,42	15,45

Products at low temperature operating conditions have a hydrocarbon composition with atomic number C more than 20. Green diesel products at high temperature operating conditions have C_{15}-C_{18}. Results with the Fe catalyst process indicated that chains of C_{15}-C_{18}were 40.27% that established with thermal cracking in 350°C. C_{15}-C_{18}chains produced by the hydro-deoxygenation (HDO) reaction. The Fe/NZ catalyst process in producing of C_{16} or C_{18} chains produced by the HDO reaction [6]. The addition of metals promises to be useful catalysts, as they can act as Lewis acid sites on yields products. To indicate the presence of new Lewis acid sites on the Fe sites of Fe/Beta catalysts [7]. Fe direct to HDO direct C–O bond cleavage [8].

The following reaction mechanism that occurs in the hydrogenation process of waste cooking oil into green diesel can be seen in Figure 4.

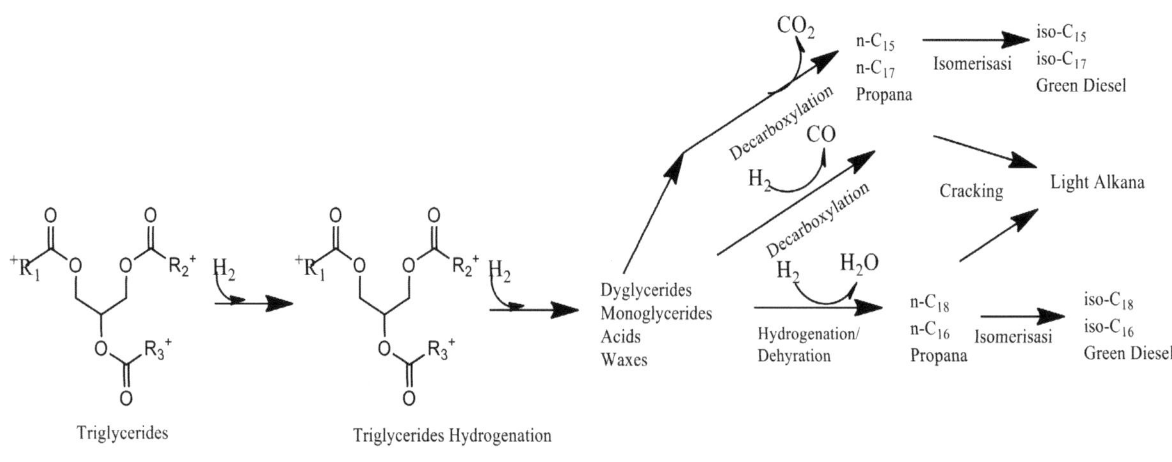

Figure 4. Reactions in the hydrotreatment of generic vegetable oils and fatty acids [9]

One of the chemical properties that determine the quality of fuel is the amount of energy contained in the fuel. This value is usually termed the heating value. So the authors analyze the heating value of green diesel products. The heating value was used by the Bomb Calorimeter (ASTM D240-02). The test results can be seen in Table 2 also compare with previous research.

Table 2. Summary on studies of different hydrodeoxygenation reaction systems

Sample	Reactor Type	React Cond H_2	Catalyst	Main Product	Perform.	Density (g/ml)	Heating value (cal/gr)	Ref
Waste Cooking Oil A	Batch	T = 350°C P = 40 psi H_2 = 1 h	Fe = 0.1 gr	C_{15}-C_{18}=40.27 C_{18}-C_{20}=38.78 >C_{20}= 15.45	Yield 31.70 %v	0.8314	10152.67	This work
Waste Cooking Oil B	Batch	T = 300°C P = 30 psi H_2 = 1 h	Fe = 0.1 gr	C_{15}-C_{18}=40.27 C_{18}-C_{20}=38.78 >C_{20}= 45.42	Yield 20.57%vl	0.8454	10121.76	This work
Vegetable oil	Ecofining	1.5-3.8			88-89%v	0.78	10516.3	[10]
Waste Cooking Oil	Fixed batch	T = 350°C P = 2 MPa H_2/oil ratio = 400Nm^3/m^3	Ru = Al13-monmorillonite	C_{15}-C_{18} n-parraffins	Yield:82.1-84%	0.78	-	[2]
Palm Oil	Batch	T = 375°C P = 12Bar H_2 = 2 h	Nat. Zeolit = 3% Fe/Nat. Zeolit =3%	C_{15} =33.04% C_{15} =59.19%	Yield:58 % Yield:89 %	-	-	[6]

Thermal cracking with hydroprocessing used Fe metal catalysts from waste cooking oil to diesel range hydrocarbon C_{15}-C_{18}. However, Green diesel is produced yield 31.70% and still goes further to get efficiency to compare the other research. Waste cooking oil with a ruthenium catalyst supported on Al-polyoxocation-pillared montmorillonite used bio-hydrogenated diesel by hydrotreatment. The process was simultaneously deoxygenated on hydrocarbon products (98.9 wt %) were n-$C_{18}H_{38}$, n-$C^{17}H_{36}$, n-$C_{16}H_{34}$, and n-$C_{15}H_{32}$.[2].

The obtained materials were tested for the conversion of palm oil into diesel-range hydrocarbons (C_{15}-C_{18}) under conditions of 375°C and 12 bar H_2 for 2 h. NZ and Fe/NZ produced a liquid hydrocarbon with straight-chain (C_{15}-C_{18}) alkanes as the most abundant products. The liquid yield product exhibited a higher conversion into diesel-like hydrocarbons, reaching more than 58% and 89% when the natural zeolite (NZ) and Fe/NZ metal loading (3 wt.%) catalysts, respectively [6].

4. Conclusions
Based on the results of research on the production of green diesel from waste cooking oil with catalytic hydrogenation method. The product volume obtained it is concluded that the higher the operating temperature (350°C), the volume of green diesel products coming out is increasing. H_2 pressure is also very influential because of the higher the pressure, the higher the operating temperature produced.

5. Acknowledgments
The authors would like to acknowledge the financial support from the directorate of research and community service, ministry of research, technology and higher education for providing funding research project grants.

6. References
[1] Kuncahyo P, Zuhdi A, Fathallah M and Semin 2013 Analisa prediksi potensi, bahan baku biodiesel sebagai suplemen bahan bakar motor diesel di Indonesia *Jurnal Teknik Pomits* **2**(1) ISSN: 2337-3539 (2301-9271 Print)
[2] Liu Y, Sotelo-Boyás R, Murata K, Minowa T and Sakanishi K 2012 Production of biohydrogenated diesel by hydrotreatment of high-acid-value waste cooking oil over

ruthenium catalyst supported on al polyoxocation-pillared montmorillonite *Catalysts* **2**(1) 171-90

[3] Choudhary T V and Phillips C B 2011 Renewable fuels via catalytic hydrodeoxygenation *Applied Catalysis A: General* **397**(1-2) 1-12

[4] Satterfield C N 1991 *Heterogeneous Catalysis in Industrial Practice* (New York: McGraw-Hill)

[5] Sotelo-Boyás, Rogelio, Trejo-Zárraga F and Hernández-Loyo F 2012 *Hydroconversion of Triglycerides into Green Liquid Fuels* doi: 10.5772/48710

[6] Putra R, Lestari W, Wibowo F R and Susanto B H 2018 Fe/Indonesian natural zeolite as hydrodeoxygenation catalyst in green diesel production from palm oil *Bulletin of Chemical Reaction Engineering & Catalysis* **13**(2) 245-55

[7] Deng J, Liu J, Song W and Zhao Z 2017 Selective catalytic reduction of NO with NH_3 over Mo-Fe/Beta catalysts: The effect of Mo loading amounts *RSC Adv.* **7** 7130-39

[8] Hong Y and Wang Y 2017 Elucidation of re-action mechanism for m-cresol hydrodeoxy genation over Fe based catalysts: A kinetic study *Catal. Commun.* **100** 43-47

[9] Huber G W, O'Connor P and Corma A 2007 Processing biomass in conventional oil refineries: Production of high quality diesel by hydrotreating vegetable oils in heavy vacuum oil mixtures *Applied Catalysis A-General* **292** doi: 10.1016/j.apcata.2007.07.002

[10] Kalnes T N, Koers K P, Marker T and Shonnard D R 2009 A technoeconomic and environmental life cycle comparison of green diesel to biodiesel and syndiesel *Environmental Progress & Sustainable Energy* **28**(1)

IOP Conf. Series: Materials Science and Engineering **823** (2020) 012026 doi:10.1088/1757-899X/823/1/012026

Green diesel production from Crude Palm Oil (CPO) using catalytic hydrogenation method

Ahmad Zikri[1] and Martha Aznury[1]

[1] Department of Chemical Engineering, Politeknik Negeri Sriwijaya, Palembang 30139 Indonesia

E-mail: martha_aznury@polsri.ac.id

Abstract. Green diesel is an alternative solution for solving problems using biomass energy as a fuel source. The advantages of this second-generation green diesel or biodiesel (Gen-2nd) are capable of reaching cetane numbers 70-90, far higher than the Gen-1st biodiesel performance of cetane numbers 50-65, respectively. The production process is through hydrogenation reactions with hydrogen injection of 4-9 MPa, the use of heterogeneous $NiMo/Al_2O_3$ catalysts, and takes place in temperatures of 280 - 380°C. The need to design this catalytic hydrogenation reactor to convert crude palm oil (CPO) into green diesel fuel is good and safe when operating at high pressures and temperatures. The optimum operation was obtained by varying the amount of CPO oil raw material and $NiMo/Al_2O_3$ catalyst used in producing the best percent yield and green diesel characterization. At a temperature of 315°C, the highest yield was 68.2%, where the number of products began to decline above these temperature conditions. The green diesel specifications obtained have met diesel oil standards (Directorate General of Oil and Gas, 2016) by testing density, kinematic viscosity, water content, flash point, calorific value, and cetane numbers.

1. Introduction

Technological products are needed in a continuous search for improved sources of energy, and green diesel is an alternative solution for solving problems using biomass energy as a fuel source. This renewable green diesel fuel has efficiently, the same energy densities as petroleum, derived fuels due to the removal of oxygen contents and equal carbon-carbon percentage [1]. The advantages of this second-generation biodiesel or green diesel (Gen-2nd) are capable of reaching cetane numbers 70-90, far higher than the Gen-1st biodiesel performance of cetane numbers 50-65, respectively.

Green diesel is produced with hydrodeoxygenation vegetable oil (triglycerides) or animal fats through catalytic processing with hydrogen, produced a mixture of straight-chain and branched saturated hydrocarbons which typically contain 15 to 18 carbon atoms per molecule (C15 to C18) [2]. Hydrodeoxygenation can be a very expensive process due to hydrogen injection of 4-9 MPa, the use of heterogeneous $NiMo/Al_2O_3$ catalysts, and takes place in temperatures of 280–380°C. In hydrogenation process, the C-O bond will be cut using H_2 gas, the carboxylic functional group is disconnected into the form of CO_2 compound by decarboxylation process, and CO and H_2O compounds were released through decarbonylation process [3].

Compared to the transesterification of triglyceride reactions to the Gen-1st biodiesel, methyl or ethyl esters compounds, catalytic cracking and hydrogenation have several advantages such as low

IOP Conf. Series: Materials Science and Engineering **823** (2020) 012026 doi:10.1088/1757-899X/823/1/012026

processing costs, and energy consumption for the process separation. In addition to producing diesel fuel types, this reaction (fig.1) also produces light hydrocarbon such as propane, gasoline, and kerosene. However, this cracking process usually requires high operating conditions (temperature and pressure), and it becomes a major consideration on its process safety [4].

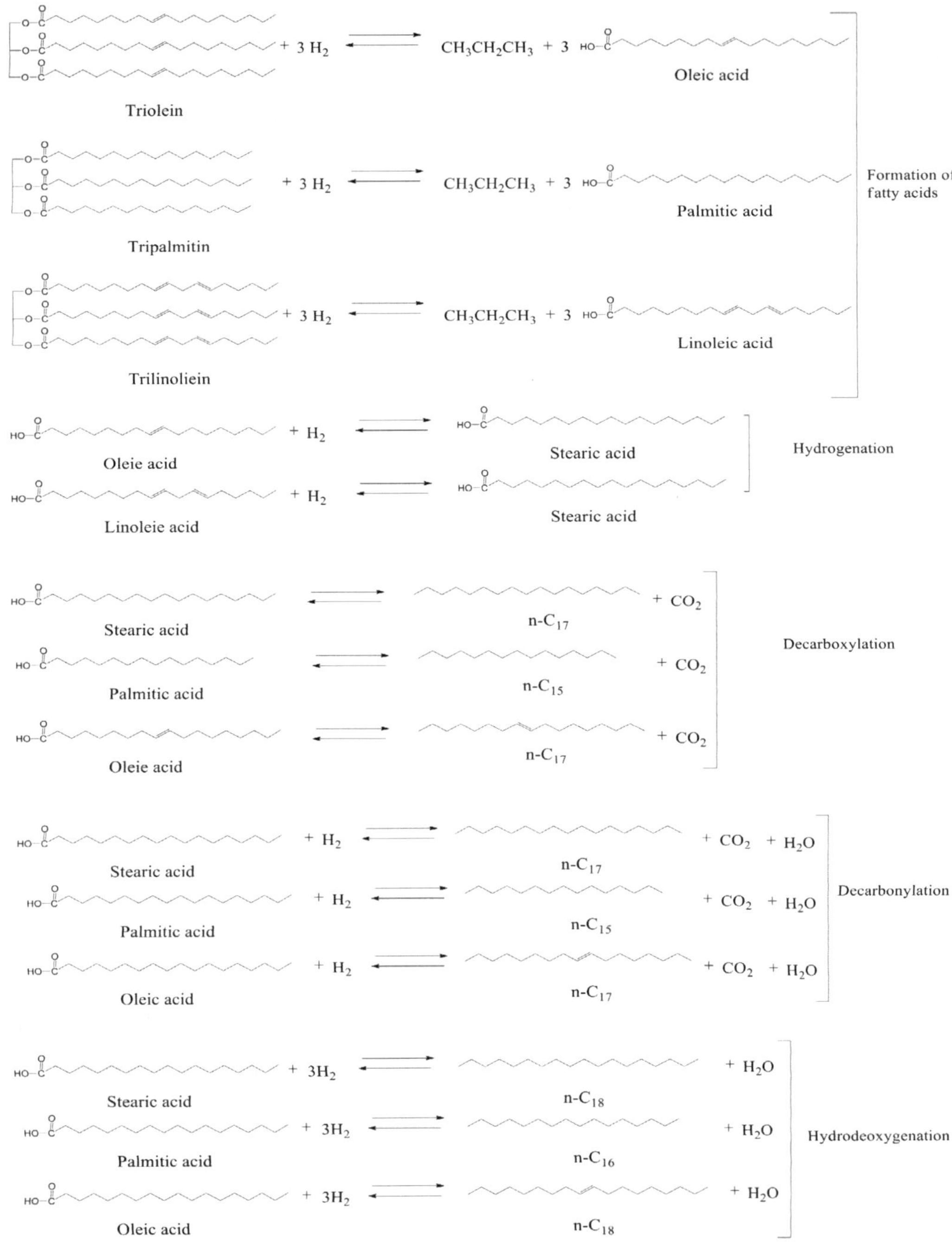

Figure 1. The synthesis of triglycerides into green diesel [5]

Future biorefineries are expected to be projected to enhance sustainable and integrated bioenergy products. Therefore, research continues to be conducted, and catalysts play an important role in increasing the conversion of vegetable oil in producing the desired product. The selection of the catalysts for the hydrodeoxygenation process depend upon the activity, stability, and importantly the selectivity of metal and support catalyst as well as its high surface area and extensive range of high porosity. The new catalysts to process vegetable oils may be synthesized taking into account the following considerations: a) high activity toward deoxygenation, b) minimization of coke formation, c) water resistance, d)capability to regenerate in single processes, e) high tolerance to chemical poisons, f)scalability in any commercial process [6].

The activity of monometallic Ni/Al_2O_3, Mo/Al_2O_3, and $NiMo/Al_2O_3$ sulphided catalysts were studied for comparative hydroprocessing activity of sunflower oil. Harnos et al. [7] compared the activity of sulphided catalysts with the non-sulphided catalysts. They tested both sulphided and non-sulphided Pd/activated C, Pd/Al_2O_3, Ni/Al_2O_3, and bimetallic solid acid catalysts $NiMo/Al_2O_3$. The reaction conditions were set at 613.15 K under the pressure of 21 bars of H_2 gas. The major products distribution were found to be C17, C18 and small alkanes with different catalyst.

The content of $NiMo/\gamma\text{-}Al_2O_3$ in each catalyst was about waste cooking oil 5 wt%. A maximum of 77.97% green diesel yield was achieved at nearly complete conversion of WCO using $NiMo/\gamma\text{-}Al_2O_3$ at a temperature of 400°C, a pressure 60 bar, and 4 hours of reaction time. The oxygen content was decreased from 14.25 wt% to 13.35 wt%, at a temperature of 400°C, a pressure of 30 bar and 1 hour of reaction time [8].

2. Methodology

2.1. Material

The materials used in this research are Crude Palm Oil (CPO), hydrogen gas, Ammonium Heptamolibdat Tetrahidrat, Nickel (II) Nitrate Heksahidrat, Aluminium Oxide. CPO produced from PT Perkebunan Mitra Ogan which is circulating in the area of Palembang, South Sumatera.

2.2. Experimental procedure

This study began with the process of making catalysts, nickel molybdenum with support of alumina ($NiMo/Al_2O_3$). The raw materials used are Nickel (II) Nitrate Hexahydrate ($Ni\ (NO_3)_2.6\ H_2O$) as for Ni, Ammonium Heptamolybdate Tetrahydrate ($(NH_4)\ 6Mo_7O_{24}.4\ H_2O$) as for Mo, and alumina commercial Al_2O_3. The process of making this catalyst consists of impregnation, drying, and calcining.

The Aluminium Oxide (Al_2O_3) commercial is heated at 120°C for 2 hours and then cooled in the desiccator. Ammonium heptamolybdate tetrahydrate ($NH_4)\ 6Mo_7O_{24}.4\ H_2O$) 0.87 M as much as 10 ml and Nickel (II) Nitrate Hexahydrate ($Ni\ (NO_3)_2.6\ H_2O$) 2M as much as 10 ml is impregnation with 12 gr of Al_2O_3 which has been activated. The mixture is stirring and heated at a temperature of 80°C to dry, then more drained in the oven with a temperature of 120°C for ± 4 hours. The mixture is calcined in a furnace with a temperature of 600°C for 5 hours. After that, the mixture can be used in hydrodeoxygenation reaction as a $NiMo/Al_2O_3$ catalyst.

Hydrodeoxygenation reaction of 300 ml CPO uses 0.15 gr $NiMo/Al_2O_3$, and 25 psia hydrogen, the process is performed during 1 hour of operation in the temperature in the range 270 – 330°C. Analysis condensate of green diesel products through the analysis of physical (qualitative and quantitative) and chemical properties.

3. Results and Discussion

The cracking product consists of organic liquid products, water, gas, and coke. The organic liquid product contains a large number of components of liquid hydrocarbons [9]. This hydrodeoxygenation reaction occurs at constant pressure in 55 bars, Figure 2 shows the percent yields of green diesel products that continue to rise with the higher operating temperature, the temperature of 315 °C is the optimum condition of the result at 68.2%. Yield is the ratio of the weight of condensate products obtained by the amount of crude palm oil used as raw material.

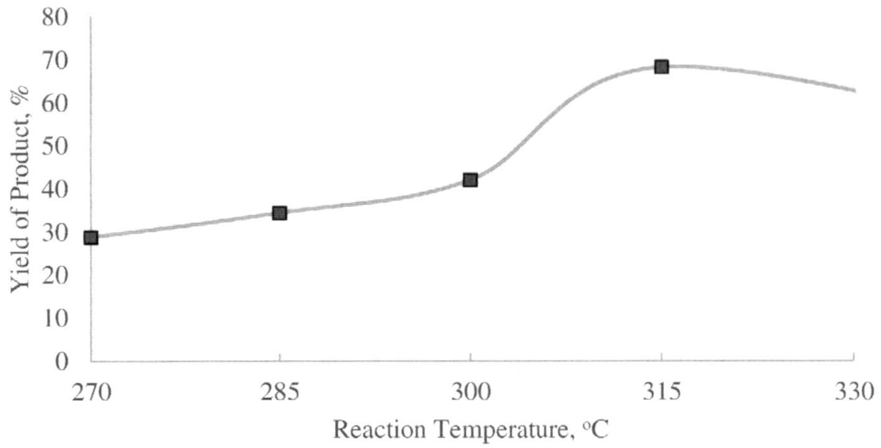

Figure 2. Effect of reaction temperature on the percent yield of green diesel

Under optimum temperature, CPO has not been fully converted into green diesel products, so that while the condensation product is excreted many of the vapor is detached. However, at a temperature of 330 °C there is a decrease in the quantity of products produced, this is because the temperature is too high resulting in the occurrence of hydrocarbons that have formed carbon deposits (coke) and aromatic compounds, this is shown from the more blackish colors and odors of products. Coke formation was observed for all catalytic cracking of used palm oil [10] and will affect the density (Fig.3-a) and viscosity (Fig.3-b) of the product.

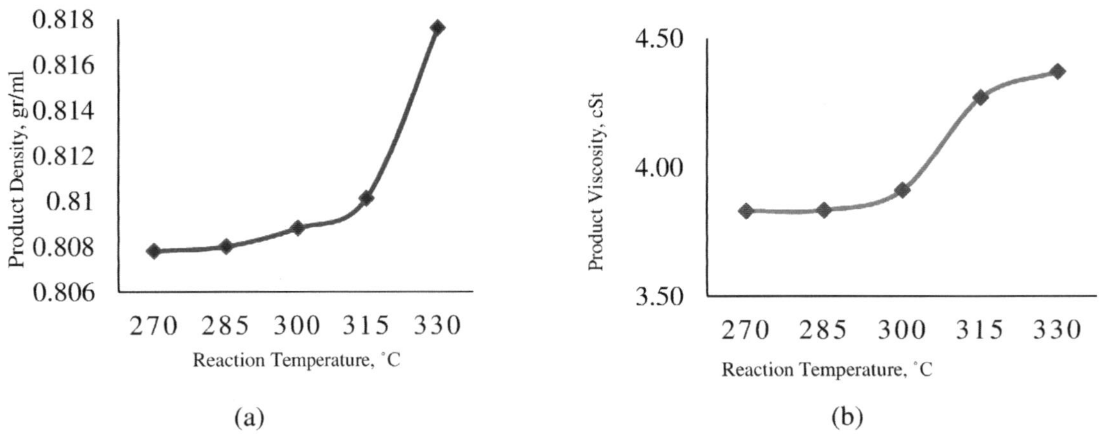

Figure 3. Effect of reaction temperature on the product density (a) and viscosity (b) of green diesel

IOP Conf. Series: Materials Science and Engineering **823** (2020) 012026 doi:10.1088/1757-899X/823/1/012026

Table 1. Green diesel product specifications

Characteristics	Unit	Standard*	Research	Test Method
Density @15 °C	gr/ml	0.8150 – 0.8600	0.8101	ASTM D 1298
Kinematic Viscosity @ 40 °C	cSt	2.0 – 4.5	4.27	ASTM D 445
Flash Point	°C	52	58	ASTM D 93
Cetane Number		48	75	ASTM D 613

* The standard and quality of diesel fuel from the Directorate General of Oil and Gas RI 2016 [11]

The density and viscosity of the product are strongly influenced by the number of paraffin hydrocarbons that are formed, the shorter the chain then the more the value. Figures 3 are shown, at the optimum temperature of the density is 0.8101 gr/ml and the viscosity is 4.27 cSt, and this is still in the specification of the diesel fuel (Table 1). The amount of value is increased with the high operating temperature, coke has started to form. Increased density of thermal cracking product fluid followed by increased viscosity of the fluid. The test of the flashpoint product (ASTM D-93) is still above standard, which is 58°C, but still in the limit of 52–96°C from diesel fuel. The ignition delay is caused by still not pure diesel product that is analyzed and the content of oxygen. The presence of oxygen, in water molecules and acid, reduces the stability and quality of the fuel. Deoxygenation process does not last perfectly, can be caused by the amount of hydrogen injected is still limited. At the test of the heating value, obtained the value of 10693 cal/gr. The heating value product is close to the standard of green diesel fuel, 44 MJ/kg [8] or equivalent to 10516 cal/gr, the resulting fuel products has approached its standard value.

In chemical composition testing with GC-MS analysis (Fig.4), it was obtained that the resulting product was dominated by the paraffin chain C15-C18 39.63%, range of diesel fuel. While other compositions are gasoline, kerosene, and naphtha (17.55%), lubricating oils (6.28%), paraffin (3.82%), and other compounds (33.44%).

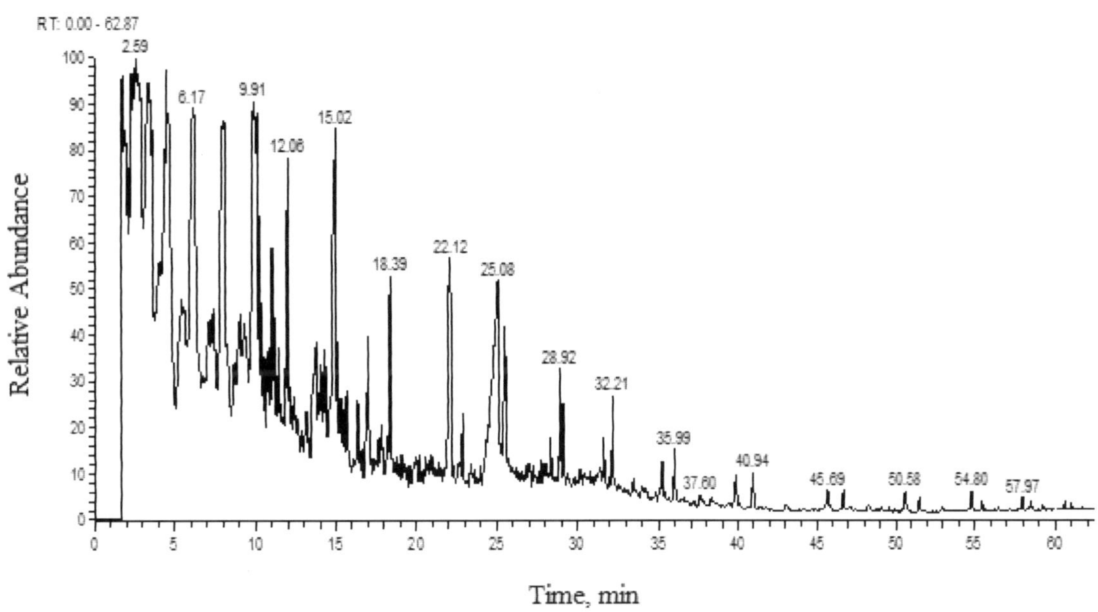

Figure 4. GC-MS analysis of green diesel product

Compounds from the C5-C14 chain contained in the product due to the continuous thermal cracking reaction to convert the compound into a lighter fraction, and also formed a myristic acid which has a chemical formula $C_{14}H_{28}O_2$. Meanwhile, the hydrocarbon > C18 is due to the large molecular fatty acids in the feed such as oleic acid (fig.1) not yet fully crack by the process. Da Mota, et al. (2014) in their research on the scale pilot plant produces green diesel by using CPO, through GC-MC analysis product obtained including 91.38% (w/w) of hydrocarbons such as normal paraffin, naphthenic, Olefin and 8.62% (w/w) oxygenation compounds [12].

4. Conclusion

The optimum temperature of the hydrogenation of crude palm oil using $NiMo/Al_2O_3$ catalyst to produce the green diesel fuel is 315 °C, with a product yield of 68.2%. The physical and chemical properties have fulfilled the specification of diesel fuel, which is density 0.8101 gr/ml, kinematic viscosity 4.27 cSt, flash point 58 °C, cetane number 75, and calorific value is10693 cal/gr.

Acknowledgment

This research was supported by Applied Research 2019, Ministry of Research, Technology and Higher Education. The authors were thankful to Energy Laboratory, Department of Chemical Engineering Polytechnic of Sriwijaya for the provision of equipment.

References

[1] Ameen M, Azizana M T, Yusup S, Ramli A and Yasira M 2017 Catalytic hydrodeoxygenation of triglycerides: An approach to clean diesel fuel production *Renewable and Sustainable Energy Reviews* **80** 1072–88 doi: 10.1016/j.rser.2017.05.268

[2] Douvartzides S L, Charisiou N D, Papageridis K N and Goula M A 2019 Green diesel: biomass feedstocks, production technologies, catalytic research, fuel properties and performance in compression ignition internal combustion engines *Energies* **12** 809 doi: 10.3390/en12050809

[3] Susanto B H, Nasikin M, Sukirno and Wiyo A 2014 Synthesis of renewable diesel through hydrodeoxygenation using Pd/zeolite catalysts *Procedia Chemistry* **9** 139-50 doi: 10.1016/j.proche.2014.05.017

[4] Mouljin J A, Makkee M and Van Diepen A E 2013 *Chemical Process Technology 2nd Edition Production of Light Alkenes* (John Wiley & Sons Ltd.)

[5] Sotelo-Boyás R, Trejo-Zárraga F and Hernández-Loyo F D 2012 Hydroconversion of triglycerides into green liquid fuels *IntechOpen* **8** 199 doi : dx.doi.org/10.5772/48710

[6] Choudhary T V and Phillips C B 2011 Renewable fuels via catalytic hydrodeoxygenation *Applied Catalysis A: General* **397**(1-2) 1-12 doi: 10.1016/j.apcata.2011.02.025

[7] Harnos S, Onyestyák G and Kalló D 2012 Hydrocarbons from sunflower oil over partly reduced catalysts *React Kinet Mech Catal* **106** 99–111

[8] Heriyanto H, Murti S, Heriyanti S, Sholehah I and Rahmawati A 2018 Synthesis of green diesel from waste cooking oil through hydrodeoxygenation technology with nimo/ɣ-al2o3 catalysts *MATEC Web of Conferences* doi: 10.1051/matecconf/201815603032

[9] Dewanto M A R, Januartrika A A, Dewajani H and Budiman A 2017 Catalytic and thermal cracking processes of waste cooking oil for bio-gasoline synthesis *International Conference on Chemistry, Chemical Process and Engineering (IC3PE) AIP Conf. Proc.* **1823** 020099-1-8 doi: 10.1063/1.4978172

[10] Chang W H and Tye C T 2013 Catalytic Cracking of Used Palm Oil Using Composite Zeolite *The Malaysian Journal of Analytical Sciences* **17**(1) 176–84

[11] Directorate General of Oil and Gas 2016 *Standard and Quality (Specification) of Diesel Fuel Type Marketed Domestically* (Ministry of Energy and Mineral Resources RI)

[12] Da Motaa S A P, Mancioa A A, Lhamasa D E L, de Abreua D H, da Silvaa M S, dos Santosa W G, de Castroa D A R, de Oliveirab R M, Araújoa M E, Borgesc L E P and Machadoa N T 2014 Production of green diesel by thermal catalytic cracking of crude palm oil (elaeis

guineensis jacq) in a pilot plant *Journal of Analytical and Applied Pyrolysis* **10** 1 – 11 doi: 10.1016/j.jaap.2014.06.011

IOP Conf. Series: Materials Science and Engineering **823** (2020) 012027 doi:10.1088/1757-899X/823/1/012027

Effect of metal type on basic soap pyrolysis produce bio-gasoline

E Puspawiningtiyas[1,2], M Pratiwi[1], G F Neonufa[3], R Purwadi[1], A N Isyami[1], L Elizabeth[1], T H Soerawidjaja[1], Subagjo[1], and T Prakoso[1]

[1]Department of Chemical Engineering, Institut Teknologi Bandung, Ganesha Street No 10, Bandung, West Java, Indonesia
[2]Department of Chemical Engineering, University Muhammadiyah of Purwokerto, Purwokerto 53182, Center Java, Indonesia.
[3]Department of Agriculture Product Technology, Universitas Kristen Artha Wacana, Kupang 85000, Indonesia

E-mail: endartiyas@yahoo.com

Abstract. The basic metal soap from oleic acid submitted to pyrolysis to produce gasoline-like hydrocarbon fuels (biogasoline). The pyrolysis was carried out in a semi-batch reactor at 450°C and a feeding rate of 5 g/ 15 min. The influences of various metal (Ca, Mg, Zn) hydroxides, which are used to made basic soap, have been examined. The bio hydrocarbon of basic metal soap was characterized by GC-FID and FT-IR, showing the formation of paraffin, iso-paraffin, olefin, ketone, and aromatics. The liquid fraction derived from the pyrolysis of basic metal soap was dominated by gasoline (C7-C11) fraction hydrocarbon. The maximum yield of light C7-C11 hydrocarbon was 82.3 %-mol fraction of basic Ca soap and the highest yield of biohydrocarbon reached 34.77 wt.% from pyrolysis of basic Mg soap.

1. Introduction
The conversion of biomass, vegetable or animal oils, and fat to obtain renewable fuels has been studied. There are many potential biohydrocarbon production methods [1] which products having compatible characteristics to be used as renewable fuels. Pyrolysis is one of the most extensive methods to convert vegetable oil or fatty acids to hydrocarbon. Pyrolysis of various oils to produce hydrocarbon has been investigated, including bio-oil [2], macauba fruit oil [3], waste cooking oil [4], cooking oils [5], waste oil [6], palm mesocarp fiber oil [7]. Asomaning et al. [8], Billaud et al. [9], and Wan and Chang [10] have pyrolyzed fatty acids into hydrocarbon. However, these processes have disadvantages, which are the yield of biohydrocarbon still low relatively and contains many oxygenated compounds [11], high acid value [10], undergo polymerization, much form sludge and high viscosity [12]. Furthermore, it was not stable and formed abundant wax [13]. In their study, they compared biohydrocarbon from triglycerides or fatty acid pyrolysis and its metal soap pyrolysis. The results of these investigations showed that metal soap pyrolysis could overcome the weaknesses of biohydrocarbon from triglycerides or fatty acids pyrolysis.

IOP Conf. Series: Materials Science and Engineering **823** (2020) 012027 doi:10.1088/1757-899X/823/1/012027

According to Kaisha (1923) [14], a good metal soap for use as raw material for making bio hydrocarbons via pyrolysis is basic soap. The report showed that stoichiometric soap produces not only short-chain hydrocarbon but also ketone and aldehyde, while on basic soap, the liquid product of pyrolysis contains just hydrocarbon. This statement was supported by Hites and Biemann (1972) [15] who found there was always ketone and carbonyl group presented in the liquid product of stoichiometric soap pyrolysis and it disappeared at a temperature more than 500 °C.

Types of triglyceride or fatty acids affect the hydrocarbon chain in liquid product pyrolysis. Mahera and Bressler in [16] pyrolyzed stearic acid into hydrocarbon that the biohydrocarbons are comprised mainly of a homologous series of hydrocarbons from carbon numbers C1 to C4, while least of alkane-alkene ≥ C8 (5 %). However, pyrolysis of oleic acid and linoleic acid produce biohydrocarbon contains much more hydrocarbon from carbon number C6 – C10. These showed that the pyrolysis of unsaturated fatty acids produces much more gasoline fraction hydrocarbon than saturated fatty acids pyrolysis [17].

For this study, we proposed to generate gasoline-like hydrocarbon fuels via pyrolysis of basic metal soap from oleic acid and to further investigate the effect of metal types (Ca, Mg, Zn) in the liquid product of oleic metal basic soap pyrolysis.

2. Method

The oleic acids, 78 %, were obtained from PT Energi Sejahtera Mas. Calcium chloride dihydrate [$CaCl_2.2H_2O$], zinc chloride tetrahydrate [$ZnCl_2.4H_2O$], and magnesium chloride hexahydrate [$MgCl_2.6H_2O$] were supplied by Merck, used to prepare metal hydroxide by co-precipitation with sodium hydroxide [18]. The Ca, Mg, and Zn soap was obtained by the saponification reaction of oleic acids with metal hydroxide that refers to [19], [20] and [21]. Pyrolysis of the basic soap was performed in a semi-batch reactor at 450 °C, atmospheric pressure and feeding rate 5 g/ 15 minutes. The pyrolysis liquid products were analyzed by GC-FID that following [22].

2.1. *Preparation of metal hydroxide*

Metals (Ca, Mg, Zn) hydroxide were prepared by the co-precipitation method. The first burette filled with 50 ml of 0.2 mol sodium hydroxide [NaOH] and the second burette with 50 ml of 0,1 mol solution of $CaCl_2.2H_2O$, $MgCl_2.6H_2O$ and $ZnCl_2.4H_2O$ severally. The process was controlled by the droplet rate of A burrete to keep the pH of the solution in 9.6. When the co-precipitation was finished, the suspension was kept stirred for 30 minutes. The metal hydroxide which formed was separated by filtering; then, it was washed with water. Thereafter, the metal hydroxide was dried at 105 °C for 1 hour.

2.2. *Preparation of basic soap*

The metals hydroxide 0.05 mol and 0.1 mol oleic acids were mix in saponification reactor. The mixture was heated to 40 – 45 °C, added 0.18 of formic acid 98- 100 % and 4 ml of water. The mixture was mixed at least 30 minutes, basic soap that formed was dried at 60 °C for 12 hours.

2.3. *Pyrolysis experiments*

The Pyrolysis of basic soap was conducted in a stainless steel reactor unit equipped with a vessel (ID: 8 cm, L: 36 cm), two thermocouples for the liquid and vapor phases, feeder, an air condenser, and liquid collector. The reactor was heated to 450 °C, then N_2 gas was passed into the reactor for about 10 minutes to drive out O_2 from the reactor. Basic soap samples (50 mg) were fed to the reactor at a feeding rate of 5 g/ 15 min. The temperature was kept constant during the feeding process. Afterward, reactor heating continued for 2 h. The pyrolytic vapor was condensed and collected as pyrolytic oil in liquid collector for GC-FID and FTIR analysis.

2.4. Variable analysis

The variables observed were yield of biohydrocarbon, identify hydrocarbon fraction distribution and various functional groups and compounds contained in pyrolysis liquid product. The yield hydrocarbon was calculated by comparing the weight of the actual liquid product to the theoretical product. Analysis of biohydrocarbon composition used a Shimadzu GC-FID 2010 instrument. The GC-FID is equipped with a capillary column (RTX-1) with a dimension of 30 m x 0.25 x 0.25 μm with a flame ionization detector. The sample (one microliter) was injected into GC with a split ratio of 1:50 with a carrier gas flow rate (Helium) of 42.9 ml/minute. The injector and detector temperature of 430°C. The GC-FID temperature program used for analysis was initially of 40°C, 5°C/minute to 300°C, then 1 °C/ minute until 340°C. The Fourier Transform Infra-Red (FT-IR) analysis was carried out with the Bruker FT-IR instrument, ALPHA-Platinum FT-IR spectrometer with Platinum Diamond sampling.

3. Result and Discussion

The effect of metal type on yield and composition hydrocarbon by pyrolysis of oleic basic soap has been investigated. The pyrolysis temperature 450 °C with tolerance ± 5 °C. In the previous studies, pyrolysis of palm stearin soap was carried out at the temperature 400 – 500 °C [23] while the pyrolysis of stearic, oleic and linoleic sodium soap was carried out at the temperature 450 – 750 °C. The composition of oleic acids used as renewable material on the saponification reaction, as shown in table 1. The Table shows that oleic acids as material content, total unsaturated fatty acid 92.9 %-wt (oleic acid 80.6 %-wt).

Table 1. The composition of oleic acids

Fatty acids	%-wt
≤ C18	6.4
Oleic acids (C18:1)	80.6
Linoleic acids (C18:2)	12.3
Others	0.7

To ensure that Ca, Mg, Zn-Oleic soap were basic soap, Ft-IR analyses have to be done. The peak intensity of the hydroxyl group (-OH) usually appears in the range of wavenumber regions between 3200-3570 cm^{-1}. Figure 1 shows that at all of metal oleic detected a hydroxyl (-OH) group. The detection of the –OH group implies that the soap produced by the three metals is basic soap.

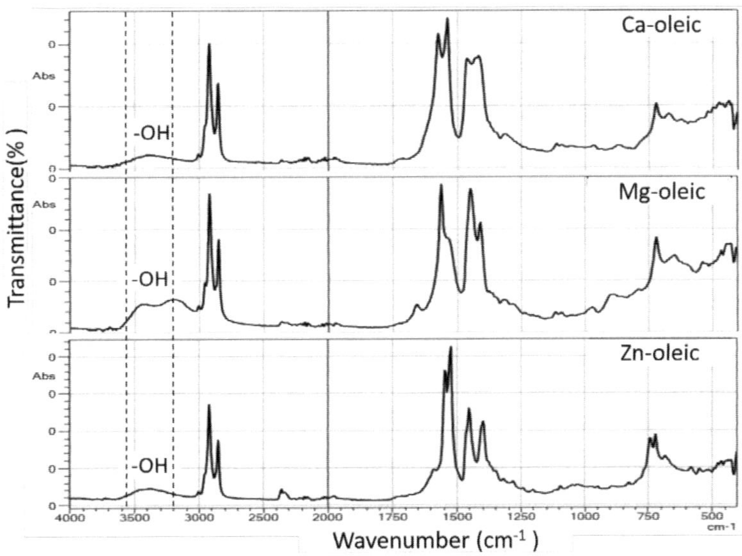

Figure 1. FT-IR spectra of the basic soaps on Ca, Mg, and Zn-oleic

3.1. Yield of biohydrocarbon

Mass percentage yields were calculated based on bio hydrocarbon theoretical yields, as depicted in Figure 2. The mass percent yield of bio hydrocarbon from Mg-oleic soap is 34.78 %, higher than Ca-oleic soap (23.23 %). The observed could be explained decomposition temperature of $Mg(OH)_2$ (285 - 380 °C) is lower than $Ca(OH)_2$ (325 – 415 °C), which results in earlier on formed bio hydrocarbons. The mass percent yield of bio hydrocarbon from Zn-oleic soap is 8.4 %, lower than Mg-oleic and Ca-oleic soap, although decomposition temperature of $Zn(OH)_2$ is the lowest among the others. This fact also was not supported by the previous researcher that reported Zn affects increasing hydrocarbon yield on decarboxylation of stearin MgZnCuFe soap [24]. The possible explanation for the observed difference could be in temperature reaction. The temperature in this study was higher (450 °C) than Neonufa, in [24]

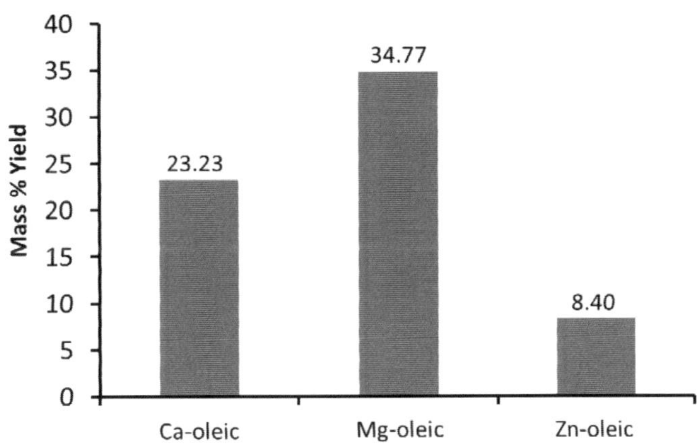

Figure 2. Mass percentage yields of different basic soap

3.2. Characteristic of biohydrocarbon distribution

The pyrolysis of oleic basic soap resulted in a series of products observed as periodic sets of peak when analyzed by GC-FID. Figures 3 and 4 show the results of GC-FID analysis of biohydrocarbon in the liquid fraction product of oleic basic soap pyrolysis and composition as a function of carbon number at various metal. The results indicated the presence of hydrocarbon molecules ranging from C7 (heptane) to C19 (nonadecane) in the pyrolysis liquid product. All of the metal soap variation, the most abundant hydrocarbon product of in the pyrolysis liquid fraction, were C7 to C11 alkane and alkene, suggesting that the cracking process has been going well. Figure 4 shows that the liquid fraction product of Mg-oleic soap and Ca-oleic soap were dominated C10 (decane), while Zn-oleic soap was C11 (undecane).

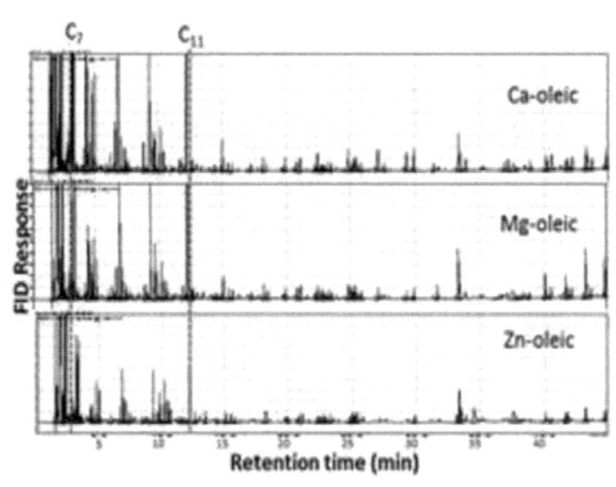

Figure 3. GC-FID chromatogram of hydrocarbon in the liquid fraction product of oleic basic soap pyrolysis at 450 °C

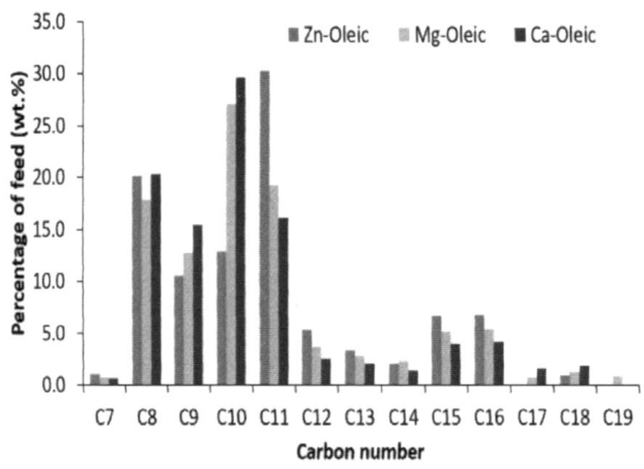

Figure 4. Composition as a function of carbon number at various metal.

The mechanism of cracking reactions in metal oleic basic soap pyrolysis start with decarboxylation basic soap :

$$C_8H_{17}\text{-}CH{=}CH\text{-}C_7H_{14}\text{-}CO\text{-}O\text{-}M\text{-}OH \rightarrow C_8H_{17}\text{-}CH{=}CH\text{-}C_7H_{15} + MCO_3$$

Metal–oleic Basic soap Heptadecene Metal carbonate

Then, the following types of reactions are possible :
Cracking of heptadecene

$$C_7H_{15} \wr CH_2\text{-}CH{=}CH\text{-}CH_2\text{-}C_6H_{13} \rightarrow C_7H_{14} + CH_3\text{-}CH{=}CH\text{-}C_7H_{15}$$

Heptadecene Heptene Decene

$$C_7H_{15}\text{-}CH_2\text{-}CH{=}CH\text{-}CH_2 \wr C_6H_{13} \rightarrow C_7H_{15}\text{-}CH_2\text{-}CH{=}CH\text{-}CH_3 + C_6H_{12}$$

Heptadecene Undecene Hexene

Then, hidrogenation of alkene

$$C_7H_{14} + H_2 \rightarrow C_7H_{16}$$

Heptene Heptane

$$CH_3\text{-}CH{=}CH\text{-}C_7H_{15} + H_2 \rightarrow CH_3\text{-}CH_2\text{-}CH_2\text{-}C_7H_{15}$$

Decene Undecane

$$C_7H_{15}\text{-}CH_2\text{-}CH{=}CH\text{-}CH_3 + H_2 \rightarrow C_7H_{15}\text{-}CH_2\text{-}CH_2\text{-}CH_2\text{-}CH_3$$

Undecene Undecane

$$C_6H_{12} + H_2 \rightarrow C_6H_{14}$$

Hexene Hexane

3.3. Quantitative analysis

The identification and quantification of chemical compounds contained in bio hydrocarbon reported in Figure 5. The classification of bio hydrocarbon in Fig.5 based on hydrocarbon fraction type and chemical species depicted in Fig.6. Figure 5 showed the most abundant hydrocarbons fraction were light fraction (C7-C11) in all of the oleic metal soap. The light fraction Ca-oleic (82.3 wt. %) is higher than Mg-oleic (77.7 wt.%) and Zn-oleic (74.9 wt.%).

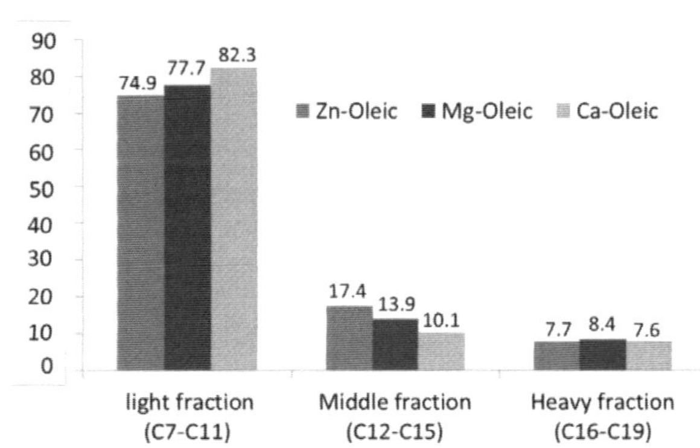

Figure 5. Fraction composition of hydrocarbon in the liquid fraction of oleic soap pyrolysis at 450 °C for various

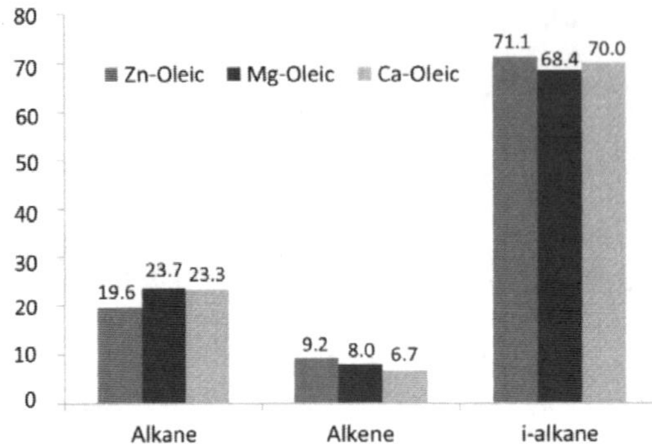

Figure 6. Alkane, alkene, and i-alkane composition as a function of carbon number at various metal.

According to [10], generally, decomposition of oleic acid into short alkane begins decarboxylation to reduce CO_2 and formed a long alkene chain, afterward cracking to formed two shorter alkenes and finally hydrogenation to obtained short alkane. Lappi and Alén in [25] reported that radicals formed in these reactions then undergo various other reactions such as disproportionation, β-scission, isomerization, hydrogen abstraction, and aromatization. Figure 6 illustrates product distribution and selectivities of alkane, alkene, and iso-alkane (i-alkane) in bio hydrocarbon. All of metal oleic soaps produce i-alkane were higher than alkane and alkene. This could be explained isomerization rate was higher than the hydrogenation rate. Zn-oleic soap obtained the highest i-alkane (71.1%) among others due to that Zn is a transition metal that could be isomerized hydrocarbon [26].

3.4. Oxygenate compound analysis

The presence of oxygenate compounds in hydrocarbons is not expected because these cause an auto-oxidation/ thermal oxidation reaction that forms a polymer that could be damage the engine. Analysis to detect the presence of oxygenate compounds in bio hydrocarbon carried out used FT-IR. The results of FT-IR analysis bio hydrocarbons that obtained pyrolysis of oleic basic soap is shown in Fig.7

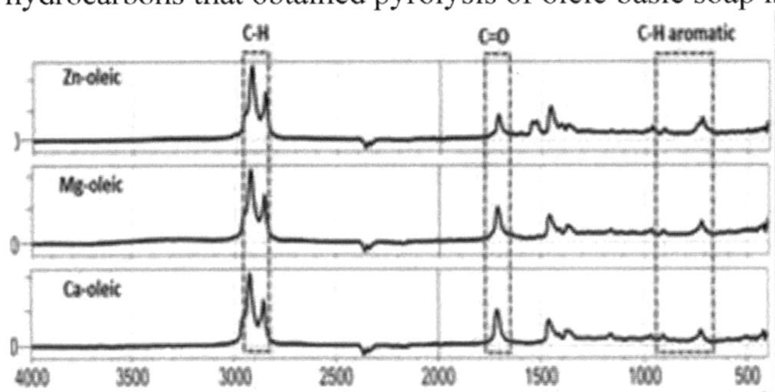

Figure 7. FT-IR of biohydrocarbon of oleic basic soap pyrolysis

Figure 7 shows the functional group that detected in bio hydrocarbon were alkane group (C-H) appear in the range of wavenumber regions between 2845 – 2935 cm^{-1}, ketone group (R-CO-R) in 1705 – 1725 cm^{-1} and aromatic ring group (aryl) in 670 - 900 cm^{-1} [27]. The alkanes group that has been detected supported the result of GC-FID analysis that presences of alkane and i-alkane, while the ketone group indicated still detected oxygenate in bio hydrocarbon. The presence of ketones was caused by the persistence of fatty acids in the oleic basic soap or the soap that was made still stoichiometry. The aromatic ring group in biohydrocarbon showed aromatization reaction occurred during pyrolysis.

4. Conclusions

The oleic basic soap can be converted to liquid biohydrocarbon, especially gasoline (C7-C11) fraction hydrocarbon. The maximum of light C7-C11 hydrocarbon fraction 82.3 %-mol from Ca basic soap and the highest yield of biohydrocarbon reached 34.77 wt.% from pyrolysis of Mg basic soap.

5. Acknowledgments

This work was supported by the Indonesian Oil Palm Estate Fund Agency (BPDPKS) under GRS-16 Grant.

References

[1] Demirbaş A 2002 Diesel fuel from vegetable oil via transesterification and soap pyrolysis *Energy Sources* **24**(9) 835-42

[2] Adjaye J D and Bakhshi N N 1995 Production of hydrocarbons by catalytic upgrading of a fast pyrolysis bio-oil, Part I: Conversion over various catalysts *Fuel Processing Technology* **45**(3) 161-83

[3] Fortes I C P and Baugh P J 2004 Pyrolysis-GC/MS studies of vegetable oils from Macauba fruit *Journal of Analytical and Applied Pyrolysis* **72**(1) 103-11

[4] Wang J, Zhong Z, Zhang B, Ding K, Xue Z, Deng A and Ruan R 2016 Upgraded bio-oil production via catalytic fast co-pyrolysis of waste cooking oil and tea residual *Waste Management*

[5] Chiaramonti D, Buffi M, Rizzo A M, Prussi M and Martelli F 2015 Bio-hydrocarbons through catalytic pyrolysis of used cooking oils: Towards sustainable jet and road fuels *70th Conference of the Italian Thermal Machines Engineering Association ATI 2015* (Elsevier Ltd)

[6] Lam S S, Liew R K, Jusoh A, Chong C T, Ani F N and Chase H A 2016 Progress in waste oil to sustainable energy with emphasis on pyrolysis techniques *Renewable and Sustainable Energy Reviews* **53** 741-53

[7] Khanday W A, Kabir G and Hameed B H 2016 Catalytic pyrolysis of oil palm mesocarp fibre on a zeolite derived from low-cost oil palm ash *Energy Conversion and Management* **127** 265-72

[8] Asomaning J, Mussone P and Bressler D C 2014 (a) Pyrolysis of polyunsaturated fatty acids *Fuel Processing Technology* **120** 89-95

[9] Billaud F, Guitard Y, Tran Minh A K, Zahraa O, Lozano P and Pioch D 2003 Kinetic studies of catalytic cracking of octanoic acid *Journal of Molecular Catalysis A: Chemical* **192**(1-2) 281-88

[10] Wan S W and Chang C C 1947 China's Motor Fuels from Tung oil *Industrial and Engineering Chemistry* **39** 1543-8

[11] Lappi H and Alén R 2009 Production of vegetable oil-based biofuels-thermochemical behavior of fatty acid sodium salts during pyrolysis *Journal of Analytical and Applied Pyrolysis* **86**(2) 274-80

[12] Hiebert D R 1985 Decarboxylation and hydrogenation of safflower and rapeseed oils and soaps to produce diesel fuels *Master Thesis* (Montana: Montana State University)

[13] Joonwichien S A D A 2006 The study of preparation of biodiesel from pyrolysis of palm stearin and soap of palm stearin over catalyst *International Conf. on Green and Sustainable Innovation*

[14] Kaisha N G K K 1923 A Method of Manucafturing Hydrocarbon oils from oils, fats or fatty acids *US Patent* No GB175974

[15] Hites R A and Biemann K 1972 On the mechanism of ketonic decarboxylation Pyrolysis of calcium decanoate *Journal of the American Chemical Society* **94**(16) 5772-7

[16] Maher K D and Bressler D C 2007 Pyrolysis of triglyceride materials for the production of renewable fuels and chemicals *Bioresource Technology* **98**(12) 2351-68

[17] Asomaning J, Mussone P and Bressler D C 2014 (b) Thermal deoxygenation and pyrolysis of oleic acid *Journal of Analytical and Applied Pyrolysis* **105** 1-7

[18] Zhao R., Yin C, Zhao H and Liu C 2003 Synthesis, characterization, and application of hydotalcites in hydrodesulfurization of FCC gasoline *Fuel Processing Technology* **81**(3) 201-9

[19] Hirsch A and Fleischer E1990 Process for The Production of Basic Soaps of Divalent Metals in Powder Form *US Patent* No 4.927.548

[20] McAskie W 1989 Ruminant Feedstuffs, Their Production and Apparatus for Use There *US Patent* No 4.826.694

[21] Rogers R H and Blew W R 1956 Manufacture of Metal Soaps *US Patent* No 2.890.232 31 Serial No 588.265

[22] Neonufa G F, Pratiwi M, Soerawidjaja T H and Prakoso T 2017 High Selectivity of Alkanes Production by Calcium Basic Soap Thermal Decarboxylation *24th Regional Symposium on Chemical Engineering (RSCE)* (Semarang Indonesia: C-62)

[23] Joonwichien S 2007 Bio oils From Palm Stearin by Pyrolysis Over Catalysts *Thesis*

[24] Neonufa G F 2018 Drop-in Fuel Production Technology of Diesel and Bioavtur Type via Catalytic Thermal Decarbpxylation Basic Soap Base on Magnesium and Transition Metal Combination *Ph.D Thesis* (Institut Teknologi Bandung)

[25] Lappi H and R Alén 2011 Pyrolysis of vegetable oil soaps - Palm, olive, rapeseed and castor oils *Journal of Analytical and Applied Pyrolysis* **91**(1) 154-8

[26] Fontaine M F, Fishkill M, Riordan D, Beacon N Y, Ryer J and New Brunswick N J 1961 Treatment of Reformed Hydrocarbons with A Zinc Oxide-Zinc Chromite Catalyst *US Patent* No 2.967.143

[27] Coates J 2000 *Interpretation of Infrared Spectra A Practical Approach* (New Jersey: John Wiley & Sons Ltd) pp 10815-37

IOP Conf. Series: Materials Science and Engineering **823** (2020) 012028 doi:10.1088/1757-899X/823/1/012028

Effects of coal and biomass types towards the quality of hybrid coal produced via co-torrefaction

J Rizkiana[1], A C A Zahra[1], W Wulandari[1], W H Saputra[1], R Andrayukti[1], A M Sianipar[1], D Sasongko[1]

Department of Chemical Engineering, Institut Teknologi Bandung, Bandung, Indonesia

E-mail: jr@che.itb.ac.id

Abstract. Coal is an abundant source of energy in Indonesia, but the reserves are dominated by low-rank coal, which is less favorable to be utilized. Indonesia also has various biomass that has potential to be used as an alternative energy source. Combining low-rank coal and biomass in a co-torrefaction process resulted in hybrid coal: an upgraded solid fuel that has higher calorific value than its original coal. The aims of this research are to analyze the effect of different biomass and coal blends on hybrid coal yield, calorific value, energy yield, and non-neutral CO2 emission reduction. This study utilized two types of coal that are X and Y, and three different types of biomass: (a) sugarcane bagasse, (b) rubberwood, and (c) empty palm fruit bunch that are being most widely found in Indonesia. The blends consist of 30%wt biomass were co-torrefied in a vertical tubular furnace reactor for 60 minutes with temperature 300°C in an inert environment and ambient pressure. Solid yields of hybrid coal founded in the range of 57.0 to 63.8% which different types of biomass gave significant effect. Calorific value was increased 37.6-44.1% to 5681-6288 kcal/kg from its original coal. The energy yield ranges from 77.0 to 89.0%. The product reduced non-neutral CO_2 emission within the range of 18.1-22.2%.

1. Introduction

Indonesia is one of the largest coal-producing countries in the world. Based on the Performance Report of the Directorate General of Mineral and Coal, the Indonesian Ministry of Energy and Mineral Resources, coal reserves and resources until September 2018 are estimated to be 37 and 166 billion tons, respectively [1]. However, Indonesia's coal reserves and resources are dominated by low-rank coal by around 70%. The common problems of low-rank coal which less favorable for direct combustion purposes are its high moisture content resulting in low calorific values, low efficiency, and high transportation costs [2]. The low-rank coal also not environmentally friendly because it emits a huge amount of CO_2 gas during the process, which considered the main contributor to global warming.

Indonesia also has an abundant resource of biomass waste but has not been utilized properly. Biomass comes from various sources of waste, such as from forestry, agriculture, livestock, industry, and settlements and the aquatic environment [3]. Biomass is a potential and new renewable energy source because it is more environmentally friendly, considering that CO_2 emissions from biomass are neutral and low in sulfur content [4]. However, biomass has a disadvantage at its low heating values due to its high water and oxygen content.

A method that can be used to improve the quality of low-rank coal is by co-torrefaction with biomass, producing new solid fuel called hybrid coal. The co-torrefaction process is expected to provide several benefits such as (i) an increase in coal calorific value, (ii) decreased coal-water content, (iii) increased biomass hydrophobicity which prevents decay, (iv) decreased non-neutral CO_2 emission through the utilization of biomass waste for energy generation, (v) increased utilization of low-rank coal and biomass waste [4][5][6]. Several previous studies have been carried out to obtain hybrid coal by mixing and processing coal and various types of biomass, such as *Larix leptolepis* wood sawdust [4], yellow poplar wood [6], sugarcane juice, and molasses [5]. The research on using various types of coal and native Indonesian biomass waste has not been done much.

This study aims to determine the effect of coal and biomass types towards hybrid coal quality based on solid yield, proximate components, constituent elements, calorific value, energy yield, and non-neutral CO_2 emissions. In this study, the two types of lignite coal used, namely X and Y which are obtained from the Mineral and Coal Technology Research and Development Center (Puslitbang tekMIRA). Biomass waste used are empty palm fruit bunch obtained from Mulawarman University, rubberwood sawdust obtained from one of the furniture industries in Pati and sugarcane bagasse obtained from the Trangkil Sugar Factory, Pati. Torrefaction is done in a vertical tubular furnace reactor with a 1.6 L/minute nitrogen rate, at ambient pressure and temperature of 300°C for 60 minutes. The composition of biomass in the mixture of coal and biomass is 30%-w/w. The results of these experiments can be used in determining which type of coal and biomass used as raw materials that produce the best quality of hybrid coal. It also serves as a reference for a further scaled-up co-torrefaction reactor.

2. Methodology

The two types of coal used named X and Y. Coal was reduced to 28-100 mesh in diameter, while the three types of biomass used, which were sugarcane bagasse (BG), rubberwood sawdust (K), and empty palm fruit bunches (TKS), were ground to 48-100 mesh. Binder made from a starch solution was used to make coal-biomass granules sized 1-1.5 cm in diameter. Coal and biomass type variation used were shown in Table 1.

Table 1. Experiment variations

Variation	Coal Type	Biomass Type	Code
1		Empty palm fruit bunch	X-TKS
2	X	Rubberwood sawdust	X-K
3		Sugarcane bagasse	X-BG
4		Empty palm fruit bunch	Y-TKS
5	Y	Rubberwood sawdust	Y-K
6		Sugarcane bagasse	Y-BG

The experiment was carried out using a vertical tubular fixed bed reactor with a diameter of 5 cm and a length of 50 cm which was equipped with a temperature controller to maintain the co-torrefaction temperature. The reactor was connected to a nitrogen tank with a flow rate regulator to measure the nitrogen gas flow rate. The reactor was heated and maintained under conditions of low-temperature co-torrefaction (300°C) and ambient pressure for one hour. Raw samples, which have been weighed and characterized early, were placed in stainless basketball sieves and inserted into the reactor. Updraft nitrogen gas flowed to the furnace so that co-torrefaction done in an inert environment and nitrogen also acts as a carrier gas that prevents volatile matters build up during the process. After co-torrefaction for 1 hour at a temperature of 300°C, the furnace was cooled down into room temperature. Hybrid coal product was removed from the furnace, weighed, and characterized. The characterization of raw materials included proximate analysis, ultimate analysis, analysis of calorific value, and analysis of lignocellulose content and characterization of hybrid coal products included proximate analysis, ultimate analysis, and analysis of calorific value. Proximate analysis including quantifying moisture content, volatile matter, carbon, and ash using ASTM D.3173, D.3175, D.3172 and D.3174 respectively.

Meanwhile, characterization using ultimate analysis measures carbon, hydrogen, nitrogen using ASTM D.5573 as a basis and also oxygen and sulfur using ASTM D.3176 and D.4239. ASTM D.5865 used to analyze the calorific value of each material. All characterizations were carried out at the Mineral and Coal Technology Research and Development Center (tekMIRA), Bandung except for the analysis of lignocellulose content, which is carried out at the Pulp and Paper Center (BBPK), Bandung. Other than that, the actual yield of hybrid coal products obtained from the weight ratio of solid products vs. the blends of coal and biomass inserted into the reactor.

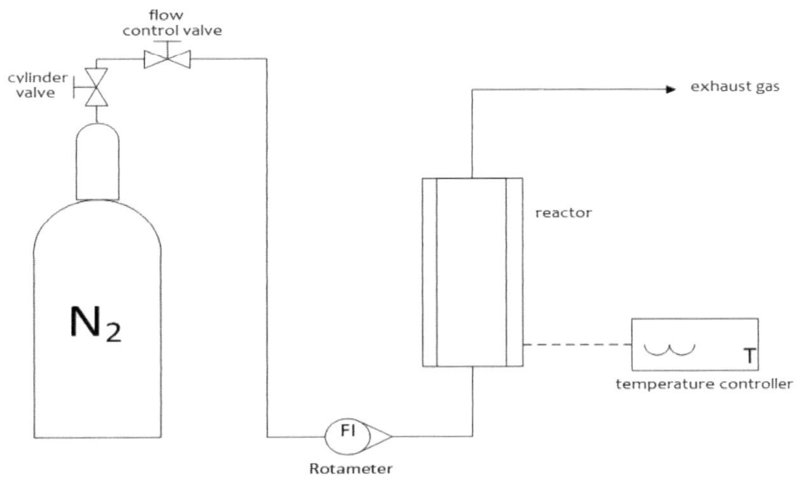

Figure 1. The scheme of hybrid coal co-torrefaction experimental equipment.

3. Results and Discussions

3.1. Effects of coal and biomass types to solid yields

Solid yield is defined as the mass ratio between the co-torrefied solid product and the raw material. Hybrid coal solid yield for various coal and biomass samples are shown in Figure 2. Based on Figure 2, the hybrid coal solid yield obtained was in the range of 57.04-63.8% with errors between the experimental and theoretical results in the range of -9.85-12.76%. The theoretical solid yield was calculated based on the solid yield data obtained through torrefaction of each raw material, which shown in Figure 3.

Based on Figure 2, the type of coal affects the synergistic effect on hybrid coal solid yield. A positive synergistic effect occurs on all hybrid coal with Y coal raw material, while a negative synergistic effect occurs on all hybrid coal with coal raw material X. The negative synergistic effect on solid yield is likely to occur because of the H and OH radicals released by biomass increase the coal's aromatic ring breakdown, considering the high H content of biomass [7]. Another possibility is the presence of a catalytic effect of inorganic components on biomass ash [8]. The positive synergistic effect on solid yield is likely to occur because the residual thermal decomposition of biomass accumulates on the surface of coal molecules then undergoes polymerization and condensation reactions to produce tar which clogs the pores of coal molecules [9]. This event causes blockage in coal volatile matters path so that the thermal decomposition of coal is inhibited [10]. The results of the analysis of variance (ANOVA) on solid yield data indicate that there is a significant effect on the types of biomass and there is no significant effect on the types of coal. The insignificant interaction between coal and biomass on the hybrid coal solid yield is also identified based on the interaction plot.

IOP Conf. Series: Materials Science and Engineering **823** (2020) 012028 doi:10.1088/1757-899X/823/1/012028

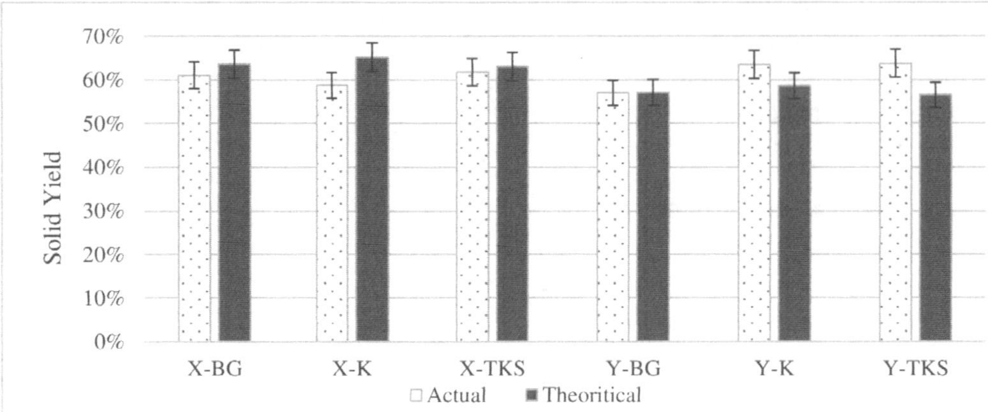

Figure 2. Hybrid coal solid yield

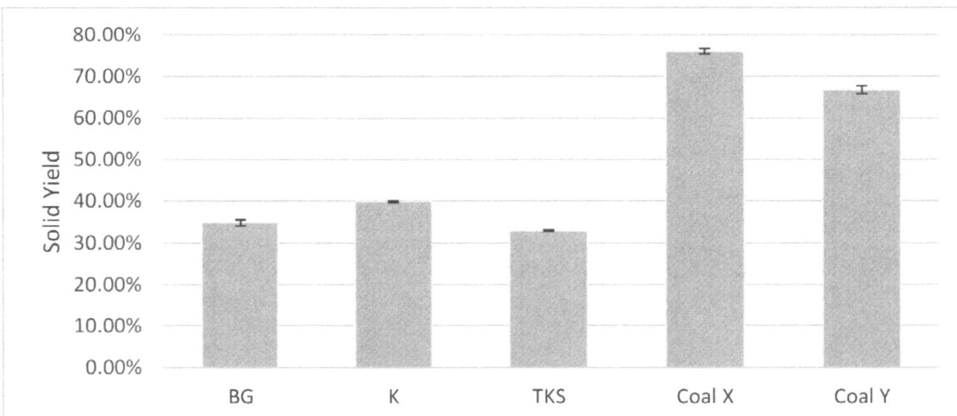

Figure 3. Raw materials solid yield

Hybrid coal solid yield can be influenced by the content of lignin, hemicellulose, and cellulose in biomass. At a temperature of 300°C, the largest decomposition occurs in the hemicellulose component, while cellulose and lignin components decompose in small amounts [11]. This indicates that the hemicellulose component becomes an important parameter in biomass torrefaction at a temperature of 300°C. The results of the analysis of the biomass lignocellulose component are shown in Table 2. The solid yield from coal and biomass raw material results are shown in Figure 3. Based on Table 2, the empty palm fruit bunch should provide the least solid yield because it has the highest hemicellulose content. However, the experimental results in Figure 2 show that the least solid yield is achieved by the coal X-rubberwood mixture, which is different from the estimation based on Table 2. The experimental results in Figure 3 show that the least solid yield is found on hybrid coal made from empty palm fruit bunch, which is the same as estimation made based on Table 2. This happens because lignocellulose constituents in biomass consist of several complex compounds.

Table 2. Biomass lignocellulose composition data

No	Parameter	Unit	Sample		
			K	TKS	BG
1	Lignin	%-w/w	27.82	20.78	21.52
2	Hemicellulose		23.64	31.65	31.26
3	Cellulose		48.54	47.57	47.21

In the torrefaction mechanism, the reactions that occur in biomass vary and can produce different substances. Hardwood and softwood biomass with similar hemicellulose compositions under identical torrefaction conditions may give different solid yield [11]. Biomass components, especially hemicellulose, contain xylan in varying amounts. Xylan (4-O-methylglucuronoxylan) is the most reactive component in the torrefaction process, and its decomposition rate is faster than other biomass components so that mass loss in biomass is more influenced by xylan content in biomass than overall hemicellulose content [11]. To validate the loss of mass obtained, an analysis of xylan content needs to be conducted.

3.2. Effects of coal and biomass types towards proximate components

Proximate components describe general characteristics of the distribution of coal, biomass, and hybrid coal components[12]. Proximate components consist of moisture, volatile matter, fixed carbon, and ash. The proximate analysis results of raw materials and hybrid coal products are shown in Figure 4. The moisture contents of hybrid coal were low and in a narrow range, which is 0.6-2.1%-w/w. When biomass undergoes torrefaction, some hydroxyl groups are broken through dehydration [13]. It prevented the formation of hydrogen bonds so that hybrid coal becomes hydrophobic. Moreover, the heat absorbed by the feed made the moisture contents to vaporized at temperature 108°C and continued by the vaporization of volatile matter at temperature more than 200°C [14].

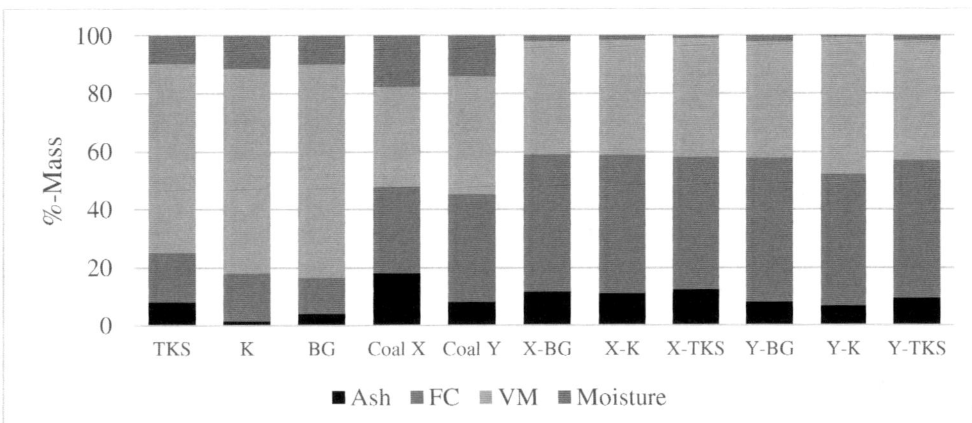

Figure 4. Proximate analysis results

The volatile matter of hybrid coal varies and is in the range of 39.2-47.1%-w/w, while the fixed carbon ranges from 45.3-47.7%-w/w. The biomass component, mainly hemicellulose, decomposes, and is released as volatile matter during torrefaction [11]. It causes a decrease in volatile matter accompanied by an increase in fixed carbon of hybrid coal. In raw X coal, sugarcane bagasse biomass causes the largest decrease in volatile matter content of hybrid coal, followed by rubberwood and empty palm fruit bunch. In raw Y coal, sugarcane bagasse biomass causes the largest decrease in volatile matter content of hybrid coal, followed by empty palm fruit bunch and rubberwood. This difference is caused by synergistic effect differences between each type of coal and biomass, causing different thermal decomposition processes [8].

Hybrid coals containing raw Y coal have higher volatile matter content than hybrid coals containing raw X coal, which are similar to volatile matter content trends in raw coal. Hybrid coals containing raw Y coal have higher carbon content than hybrid coals containing raw X coal, which are similar to fixed carbon content trends in raw coal. The lignocellulose component has decomposed naturally in coal so that only a small portion of the volatile matter of coal is released during torrefaction. It is proven by

IOP Conf. Series: Materials Science and Engineering **823** (2020) 012028 doi:10.1088/1757-899X/823/1/012028

solid yields of raw coal torrefaction are higher than solid yields of raw biomass torrefaction, based on Figure 3.

Hybrid coals containing empty palm fruit bunch have the highest ash content, followed by sugarcane bagasse and rubberwood. Hybrid coals containing raw X coal have higher ash content than hybrid coals containing raw Y coal. They are similar to ash content trends of the raw materials. This happens because the mass of ash in the sample does not change during torrefaction [10].

3.3. Effects of coal and biomass types towards constituent elements

The constituent elements of coal, biomass, and hybrid coal are obtained from the ultimate analysis results shown in Figure 5. An increase in C content followed by a decrease in O and H content of hybrid coal products compared to the raw materials occurs. Decreases in the moisture content and volatile matter content followed by an increase in the fixed carbon content during the torrefaction change the elemental composition of hybrid coal because water and volatile matter contain many O and H, while fixed carbon contains many C [13].

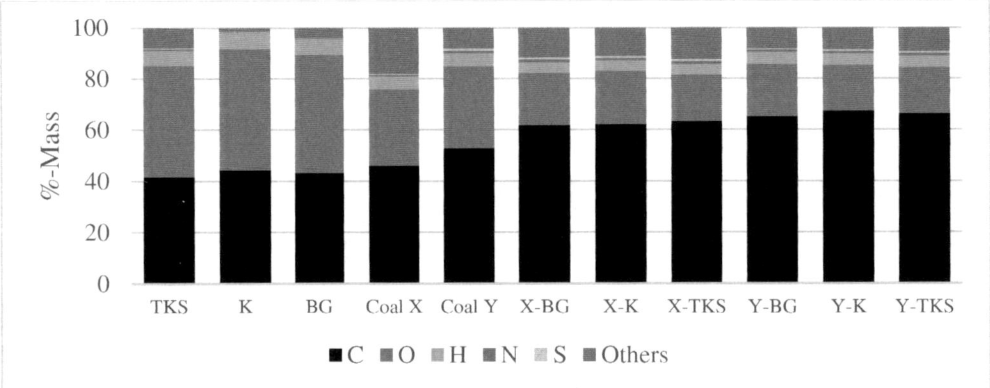

Figure 5. Ultimate analysis results

Van Krevelen diagram of raw materials and hybrid coal products in Figure 6 shows a decrease in the ratio of atomic O/C and H/C in hybrid coal products when compared to its constituent coal. It indicates an increase in coal rank after being processed into hybrid coal. Hybrid coals containing empty palm fruit bunch have the smallest O/C and H/C values. It indicates that the types of empty palm fruit bunch's volatile matter are the most easily decomposed, as proven by the data of solid yield of raw empty palm fruit bunch torrefaction that have the lowest value. Raw X coal has a smaller O/C and H/C atomic ratio than raw Y coal so that it can be said that X coal rank is better than Y coal. However, the rank of hybrid coal produced does not follow the trend of its constituent raw coal. Hybrid coals containing raw X coal have a smaller H/C atom ratio than hybrid coals containing raw Y coal, while the reverse result is obtained for O/C atomic ratio value. It shows that X coal is more difficult to be volatilized than Y coal, as proven by the data of solid yield of raw X coal torrefaction X is higher than the solid yield of raw Y coal torrefaction.

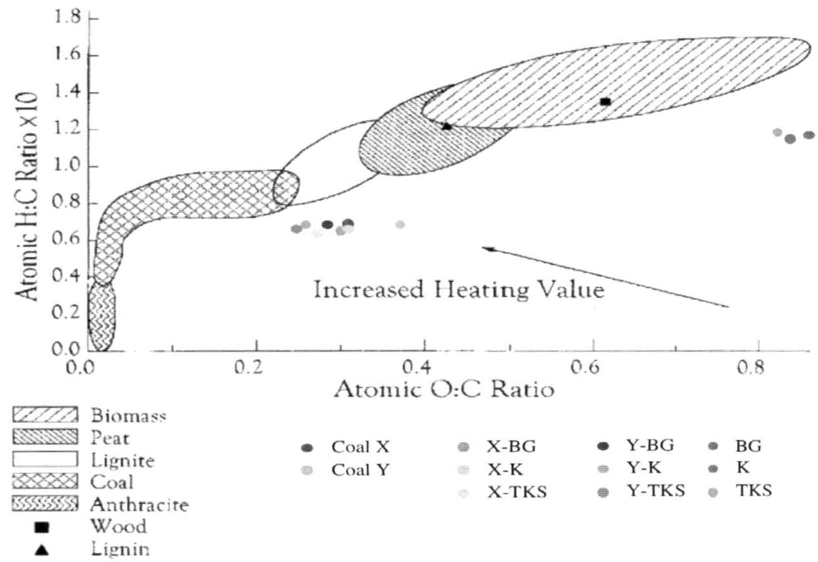

Figure 6. Van Krevelen Diagram

3.4. Effects of coal and biomass types towards calorific value

The calorific value of raw materials, hybrid coal products, and several coal standards in the form of gross calorific value (GCV) are shown in Figure 7. It shows increases in the calorific values of hybrid coals in the range of 37.6-44.1% compared to their constituent raw materials. Decreases in the moisture and volatile matter contents followed by an increase in the fixed carbon content during torrefaction cause an increase in calorific value because the carbon content is the main source of heat energy released when burning a fuel [13]. Hybrid coal products have calorific values in the range of 5681-6288 kcal/kg. Hybrid coals containing raw Y coal have higher calorific value than hybrid coals containing raw X coal, which are similar to calorific value trends in raw coal. It indicates that the calorific value of hybrid coal is mainly affected by the calorific value contributed by raw coal. Meanwhile, biomass types do not show a trend towards the calorific values of hybrid coal because the calorific values of the raw biomass used are in a narrow range, which is 3925-3979 kcal/kg.

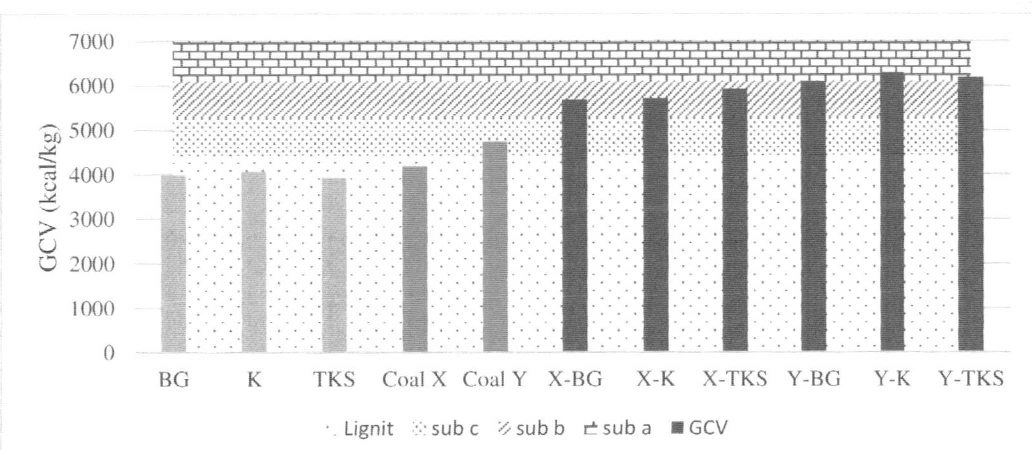

Figure 7. Calorific values of raw materials, products, and coal standards (based on ASTM D-388)

3.5. Effects of coal and biomass types towards energy yield

Energy yield is a ratio of energy amount of torrefied solid product to the energy amount of its raw materials. Hybrid coal energy yield in Figure 8 varies and ranges from 77.0-89.0%-w/w. The sequence of hybrid coal energy yield is similar to the sequence of hybrid coal solid yield. It indicates that solid yield is a more dominant factor than calorific value, so that the effects of biomass and coal on energy yield are similar to the effects of biomass and coal on solid yield [13].

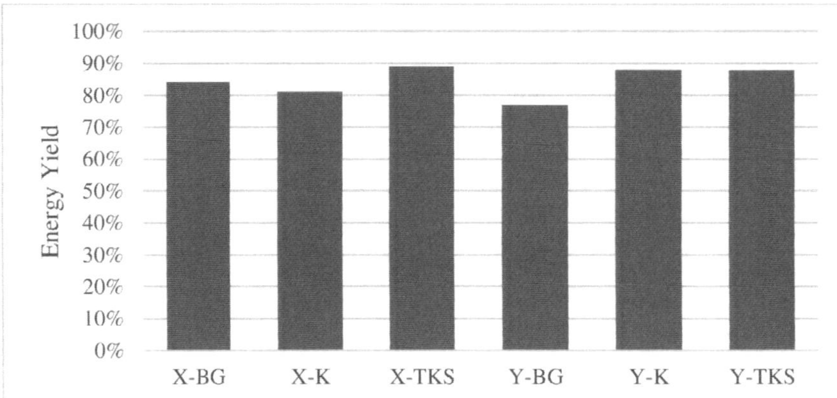

Figure 8. Hybrid coal energy yield

3.6. Effects of coal and biomass types towards CO_2 emissions reduction

Non-neutral CO_2 emissions reductions of hybrid coal when they are used as fuel are shown in Figure 9. The reduction values vary and are in a narrow range, which is 18.1-22.2%. These results indicate that hybrid coal is more environmentally friendly than raw coal when used as fuel. The highest reduction value is achieved by hybrid coal from rubberwood mixtures. Hybrid coals containing raw Y coal have higher reduction values than hybrid coals containing raw X coal do. CO_2 emissions reduction of hybrid coal is proportional to the calorific value of hybrid coal and inversely proportional to the composition of non-neutral C in the product. A high calorific value reduces the mass of hybrid coal needed. A low composition of non-neutral C lowers the non-neutral CO_2 emitted.

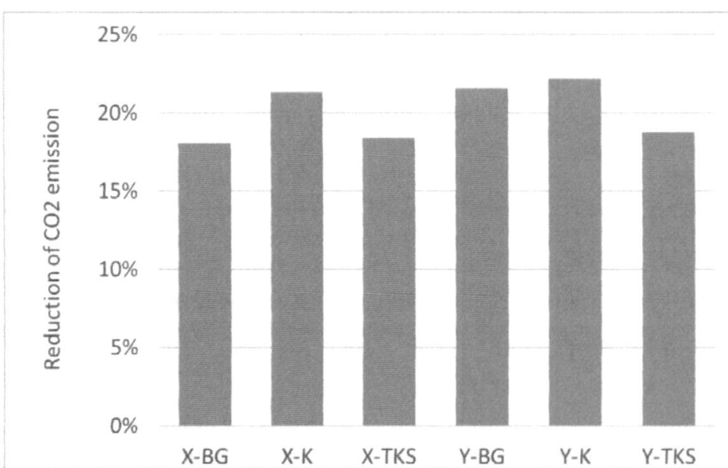

Figure 9. Hybrid coal non-neutral CO_2 emissions reduction

4. Conclusion

From this study, it can be concluded that hybrid coal solid yield is in the range of 57.0-63.8%. Coal types do not have a significant effect on solid yield while biomass types have a significant effect on solid yield. In addition, it is found that there are no significant interactions of coal and biomass on solid yield. From the result of proximate and ultimate analysis, there is an increase in fixed carbon, a decrease in moisture content, a decrease in the volatile matter and a decrease in the atomic ratio O/C and H/C in hybrid coal. It causes an increase in hybrid coal calorific value by 37.6-44.1% to be 5681-6288 kcal/kg. Y hybrid coals have higher calorific value than X hybrid coals do. Meanwhile, the energy yield of hybrid coal was in the range of 77.0-89.0%. Coal and biomass types give varied effects on hybrid coal energy yield with the same sequence as hybrid coal solid yield. All of coal and biomass combinations reduces non-neutral CO_2 emissions in the range of 18.1-22.2%. Y coal and rubberwood sawdust produce the highest CO_2 emissions reduction in hybrid coal.

5. References

[1] Ministry of Energy and Mineral Resources, Republic of Indonesia (ESDM RI) 2018 *Rekonsiliasi data, sumber daya batubara indonesia kini 166 miliar ton, cadangan 37 miliar ton* Available from: https://www.esdm.go.id/en/media-center/arsip-berita/rekonsiliasi-data-sumber-daya-batubara-indonesia-kini-166-miliar-ton-cadangan-37-miliar-ton (2019 February)
[2] Song Y, Tahmasebi A and Yu J 2014 *Bioresour. Technol.* **174** 204-11
[3] Tripathi M, Sahu J N and Ganesan P 2016 *Renew. Sust. Energ. Rev.* **55** 467-81
[4] Park D K, Kin S D, Lee S H and Lee J G 2010 *Bioresour. Technol.* **101** 6151-6
[5] Lee D W, Bae J S, Lee Y J, Park S J, Hong J C, Lee B H, Jeon C H and Choi Y C 2013 *Environ Sci Technol* **37** 1704-10
[6] Jeong H M, Seo M W, Jeong S M, Na B K, Yoon S J and Lee J G 2014 *Bioresour. Technol.* **155** 442-5
[7] Sonobe T, Worasuwannarak N and Pipatmanomai S 2008 *Fuel Process. Technol.* **89** 1371-8
[8] Quan C and Gao N 2016 *BioMed Res* **2016** 1-11
[9] Chen C, Ma X and He Y 2012 *Bioresour. Technol.* **117** 264-73
[10] Darmstadt H, Garcia-Perez M, Chaala A, Cao N Z and Roy C 2001 *Carbon* **39** 815-25
[11] Basu P 2013 *Biomass Gasification, Pyrolysis and Torrefaction: Practical Design and Theory 2nd Ed* (San Diego: Elsevier Inc) p 88
[12] Speight J G 1994 *The Chemistry and Technology of Coal* (Boca Raton: CRC Press) p 225
[13] Chen W H, Peng J and Bi X T 2015 *Renew. Sust. Energ. Rev.* **44** 847–66
[14] Wulandari W, Sasongko D, Rizkiana J, Anggayana K, Widayat A H, Zahra A C A and Azhari M 2019 *AIP Conf. Proc* **2085** 020022

Acknowledgments

The authors would like to thank Program Penelitian, Pengabdian kepada Masyarakat dan Inovasi, Institut Teknologi Bandung year 2019 for providing financial support for this research.

IOP Conf. Series: Materials Science and Engineering **823** (2020) 012029 doi:10.1088/1757-899X/823/1/012029

Biohydrocarbon production for jet fuel from palm oil derivative products

G F Neonufa[1], M Pratiwi[2], E Puspawiningtiyas[2], R Purwadi[2], and L Elizabeth[2]

[1] Department of Agricultural Product Technology, Universitas Kristen Artha Wacana, Kupang, East Nusa Tenggara, Indonesia
[2] Department of Chemical Engineering, Institut Teknologi Bandung, West Java, Indonesia

E-mail: godliefneonufa@gmail.com

Abstract. Jet biofuel can be produced from vegetable oils and fats that are rich in lauric acid and myristic acid. Palm kernel oil is known to contain 47.8% lauric acid and 16.3% myristic acid. Jet biofuel production by the thermochemical method through decarboxylation of basic soap made from palm kernel oil has been extensively investigated. The objective of this study was to produce jet biofuel from basic soap made from palm kernel oil. Several metal compounds from the acetate group were selected for use in making basic soap. Decarboxylation of soap was carried out at 350°C and atmospheric pressure for 5 hours in the semi-batch reactor. Zinc metal enhances the yield of jet biofuel after decarboxylation of soap. Approximately 40 and 62 weight% of the jets biofuel had been obtained in this experiment.

1. Introduction

Today, most of the energy we use comes from fossil fuel. According to the definition of chemistry, fossil fuels consist of hydrocarbon molecules. Petroleum is the fossil fuel that is most in danger of running out. According to Demirbas [1] peak in global oil (petroleum), production may occur between 2015 and 2030. Figure 1 shows global oil production scenarios based on today's production.

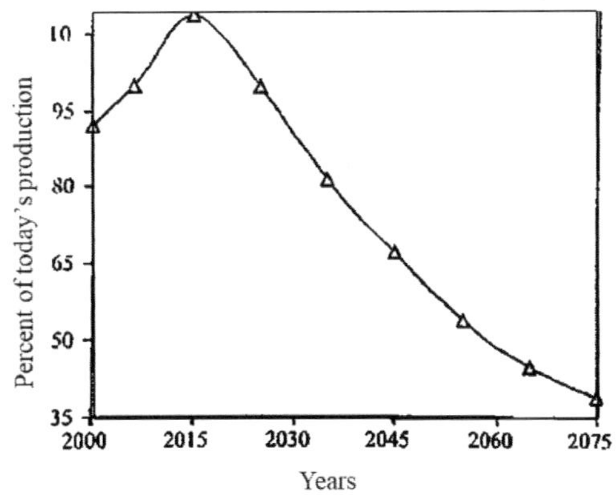

Figure 1. Global oil production scenarios based on current production [1]

The increase in the rate of consumption of fossil fuels was triggered by the rapid growth of the manufacturing and transportation industries. To ensure the availability of liquid fuels for various industries of the future, of course, the production of fossil replacement fuels needs to be done. There are today many reasons for search and find the replacement of fossil fuel, that is: 1) The limited oil reserves that are estimated to remain for the needs of 80-100 years [1]; 2) uncertain of fossil fuel price; and 3) environmental protection [2]. Renewable fuels are also often called alternative fuels that have a big potential for meeting the future needs of liquid fuel [3]. Renewable fuel is a promising alternative solution because it is clean and environmentally safe. It also has a source that is more evenly distributed than fossil resources [4]. Biomass of vegetable oils and fats has been shown may produce renewable fuel by converting it into bio-hydrocarbon that compatible with fossil fuel [5]. However, Fossil fuels can also be termed biomass since they are the fossilized remains of plants that grew millions of years ago [6]. The use of vegetable oils as fuel for internal combustion engines is not new. Rudolf Diesel, the inventor of the diesel engine, used peanut oil to fuel one of his engines at the Paris Exhibition of 1900 [7]. It is possible because the vegetable oils and fats are components of triglycerides that contain fatty acids that have a molecular structure similar to hydrocarbons (fossil fuels); however, contaminated with oxygen. Such similarity is shown through the molecular structure of lauric acid and dodecane hydrocarbons below [8]:

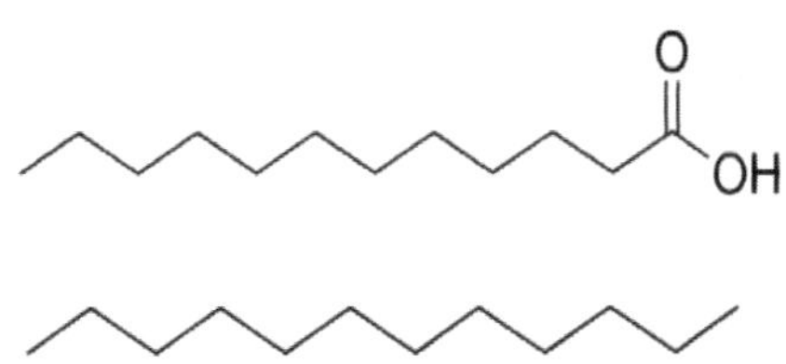

(a)

(b)

Figure 2. Molecular structure of: (a) lauric acid and (b) dodecane

According to Mancini et al. [9], palm kernel oils contain many fatty acid components, especially lauric acid (47.8%) and myristic acid (16.3%). The two types of fatty acids mentioned above can be sources of biohydrocarbon production, especially for jet fuels [10]. However, to obtain biohydrocarbons from fatty acids, oxygen in the chain of molecules of fatty acids must be removed. Two reaction pathways that have commercial applications are decarboxylation [11] and hydrodeoxygenation [12] already been proved to be effective in eliminating oxygen from fatty acid molecules. Nevertheless, the more expected technology is to have a minimum to moderate production capacity, relatively easy to operate, and have a broad implementation scope to various regions producing raw materials. The Hydrodeoxygenation reaction pathway consumes a lot of hydrogen gas to suppress or force the oxygen to form water molecules and leaving fatty acid molecules in the form of biohydrocarbon. Its reaction can only take place effectively under high-pressure conditions in the presence of Pd and Pt catalysts [13]. Meanwhile, the decarboxylation reaction pathway for converting basic soap (derivatives of fatty acids) can be effective without Pd and Pt catalysts, not consuming hydrogen and operating well under atmospheric conditions [14]. With the reaction pathway, the catalytic process of basic soap is effectively carried out by various alkaline earth and transition metals used for soap making. The yield and characteristics of biohydrocarbon products obtained through decarboxylation of basic soap are largely determined by the type and combination of metals and model compounds (the type of vegetable oil) that used for produce of the basic soap [10]. The objective of this study was to investigate the biohydrocarbon production for jet fuel from palm oil derivative products.

2. Material and Method

2.1. Materials and reagents
Magnesium acetate tetrahydrate [$Mg(OOC_2H_5)_2.4H_2O$] and zinc acetate dihydrate [$Zn(OOC_2H_5)_2.2H_2O$] supplied by Merck (analytical grade) were used to prepare the metal soaps. The $C_{12/14}$ and $C_{12/16}$ methyl esters derived from palm kernel oils were purchased from PT. Ecogreen Batam-Indonesia. The ultrahigh-purity (>99.999%) N_2, and He used in the analysis of the gas and liquid products were purchased from PT. Brataco, Indonesia. Ethanol (purity ≥ 96%) and sodium hydroxide (anhydrous, pro analysis) were purchased from PT. Brataco, Indonesia.

2.2. Experimental set-up
Two types of methyl esters derived from palm kernel oil have been used in this study are $C_{12/14}$ and $C_{12/16}$. The composition of $C_{12/14}$ and $C_{12/16}$ methyl esters were identified by fragmentation patterns from a Shimadzu 2010 Plus Mass Spectrometric (MS) detector and by matching gas chromatography retention times with known standards. Product separation was achieved using a restex-5 capillary column with a length of 30 m and an inner diameter of 0.25 mm. The injector and detector port temperature was 300 °C. The column temperatures were programmed from 150 to 200 °C at a rate of 5 °C/min and then raised to 290 °C at a rate 10 °C/min and were kept at that temperature until no appearance of peak. The biohydrocarbons production was performed by decarboxylation process of the basic soaps which has been resulted from saponification process. Basically, the order of each step included in biohydrocarbons production can be explained as follows: the ME was firstly mixed with hot ethanol in a glass batch reactor with a stirrer to get a uniform mixture. Subsequently, an aqueous solution of sodium hydroxide at 20% by weight was added to the mixture while continue stirring. As soon as the mixture reaches a good level of consistency, a solution consists of magnesium – zinc acetate combination at a ratio of 9:1 in distilled water was gradually added to the mixture while stirring. The Mg-Zn basic soaps produced by ion exchange were insoluble in water and easily separated from their soluble sodium counterparts.

The produced basic soaps from this process were filtered, washed with hot water, and then dried in an oven at 70 °C for about 48 hr. The dried basic soaps were subsequently used as the material for decarboxylation process. The Mg-Zn basic soap decarboxylation process was performed under a condition of destructive distillation at 350 °C and atmospheric pressure in a glass batch reactor for 5 hours without any catalysts. The catalytic activities of Mg-Zn combination in basic soaps were evaluated by determining the yield of liquid jets biofuel (or bio-hydrocarbons) and their fraction distribution. The fraction distribution was analyzed by gas chromatography Shimadzu 2010 equipped with a flame ion detector (FID) and a restex-1 capillary column, using Helium as a carrier gas. The injection port temperature and the temperature program were the same as those used with the GC-MS. Standard alkane of n-C_{12}, n-C_{16}, and n-C_{18} was used for identification of retention times and calibration of response. The analysis method of GC-FID was used to quantify the biohydrocarbon concentration of liquid product (sample). The determination of the freezing point has been tested with ASTM D-2500 procedures.

IOP Conf. Series: Materials Science and Engineering **823** (2020) 012029 doi:10.1088/1757-899X/823/1/012029

3. Results and Discussion

3.1. Main content of fatty acids in the model compounds

The analysis results of fatty acids content in $C_{12/14}$ and $C_{12/16}$ methyl esters are presented in Table 1. The results show that the contents of dodecanoate ($C_{12:0}$) as main fatty acid in $C_{12/14}$ methyl esters are 74.3 wt%, while those in $C_{12/16}$ methyl esters are 67.1 wt%. Therefore, the content of dodecanoate in $C_{12/14}$ methyl esters 7.2 wt% higher than $C_{12/16}$ methyl esters. Otherwise, $C_{12/14}$ methyl esters have the content of hexadecanoic ($C_{16:0}$) fatty acid lower than $C_{12/16}$ methyl esters. Meanwhile, both $C_{12/14}$ and $C_{12/16}$ methyl esters have relatively the same tetradecanoate content. The data in Table 1 show that the two types of methyl esters are renewable feedstock which suitable to be converted into hydrocarbons that are compatible with jet fuel (consist of C10-C14 hydrocarbons).

Tabel 1. Main fatty acids of model compounds

Model compounds	Fatty acids, (wt %)		
	$C_{12:0}$	$C_{14:0}$	$C_{16:0}$
ME $C_{12/14}$	74.3	25.1	0.4
ME $C_{12/16}$	67.1	25.5	7.4

3.2. GC-FID chromatograms of liquid product sample

GC-FID chromatograms of the liquid bio-hydrocarbons product sample are shown in Figure 3. According to chromatograms the liquid bio-hydrocarbons that were resulted by basic soaps had a relatively wide range of carbon chain length, i.e., between C_8 - C_{18}. It shows that Mg-Zn metals combination which was expected to act as a catalyst in basic soap decarboxylation do not selective to only the main product i.e., n-undecane (n-C_{11}).

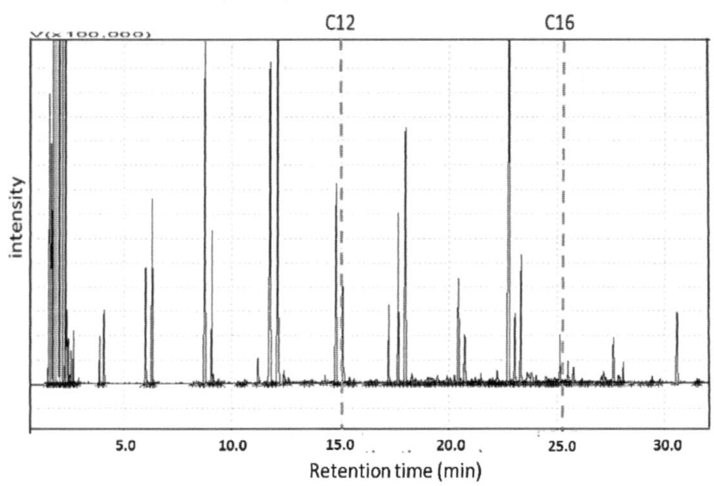

Figure 3. GC-FID chromatograms of biohydrocarbon derived from $C_{12/16}$ ME basic soaps

Variation of the carbon chain length in the liquid bio-hydrocarbon suggested that the conversion process of basic soaps Mg-Zn(OH)(ME $C_{12/16}$) were predicted to not only involve the decarboxylation process, but also others processes such as cracking (generating the small and light products), dehydrogenation (generating unsaturated bio-hydrocarbons) and polymerization (generating long-chain bio-hydrocarbons).

3.3. Effect of metal types in the basic soap on the yield of biohydrocarbon (jet biofuel)

The influence the metal types in the basic soap on the yield of jet biofuel products were studied in these experiments. The analysis results of jet biofuel yield of basic soap decarboxylation which produced from some types of metal are shown in Figure 4. The result shows that the effect of zinc metal in basic soap is relatively significant, while those in other metals are insignificant. The data in Figure 4 show that the catalytic effect of zinc metal enhances the yield of liquid biohydrocarbons of basic soap decarboxylation. Therefore, from the catalysis point of view, the metal which has high catalytic activity for basic soap decarboxylation was Zn metal

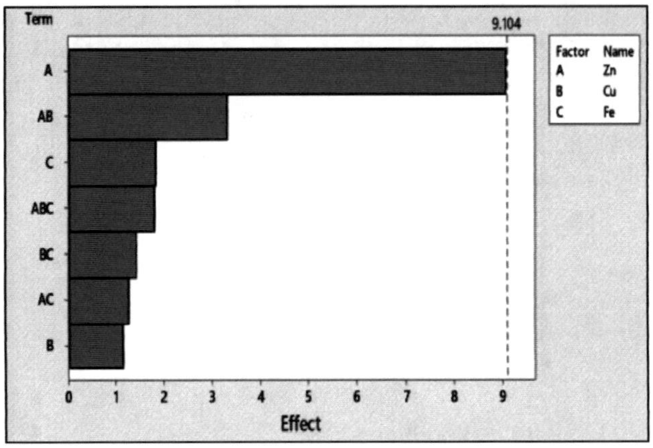

Figure 4. Pareto chart of the metals effect (respons is Yield of jet biofuel, α=0.05)

3.4. Yield of jet biofuel from decarboxylation of methyl esters soaps

Both ME compounds with profiles as presented in Table 1, had been converted to basic soaps and were decomposed through a series of decarboxylation process to result in jets biofuel. Experiment result showed that there was a significant difference on the yield of liquid biohydrocarbons derived from decarboxylation of basic soap $C_{12/14}$ and $C_{12/16}$ (Figure 5). Percent yield is the percent ratio of actual yield to the theoretical yield. It is calculated to be the experimental yield divided by theoretical yield multiplied by 100%, as indicated by the equation below:

$$Y = \frac{y(actual)}{y\,(theoretical)} \times 100\% \quad (1)$$

The highest yield of jets biofuel was generated by basic soap decarboxylation of the Mg-Zn(OH)(ME $C_{12/16}$) i.e. 62.17 wt% than the Mg-Zn(OH)(ME $C_{12/14}$) i.e. 40.06 wt%. These difference may has correlated with the lauric acid ($C_{12:0}$) and Palmitic acid ($C_{16:0}$) content in methyl esters. Apparently, the yield of jets biofuel increases when methyl esters contain significant palmitic acids. While, the yield of jets biofuel are relatively low with increasing lauric acids content.

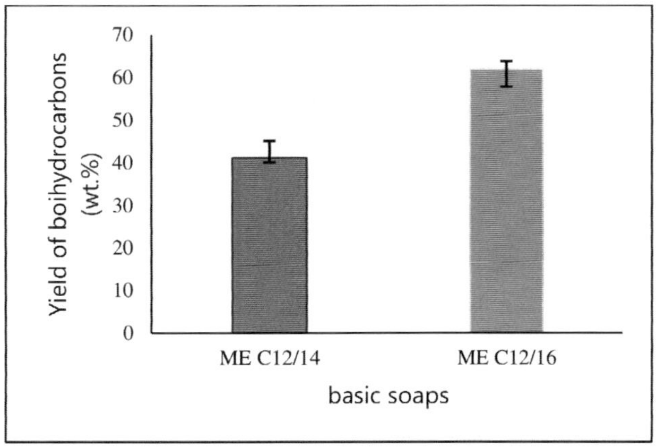

Figure 5. Biohydrocarbon (jets biofuel) of basic soap decarboxylation

3.5. Distribution of carbon chain length from the n-alkane fractions

The n-alkane content according to the carbon chain length is shown in Figure 6. The results show that the n-alkane content in the liquid biohydrocarbons of both (Mg-Zn(OH)(ME $C_{12/14}$)) and Mg-Zn(OH)(ME $C_{12/16}$) basic soaps tend to be similar. However, for the Mg-Zn (OH)(ME $C_{12/14}$) basic soap, the yield of n-C11 was found the highest followed by n-C15. Whereas for the Mg-Zn(OH)(ME $C_{12/16}$) basic soap, the yield of n-C15 was found highest and followed by n-C11. It seemed that the short-chain n-paraffin compounds, e.g., n-C_8, n-C_9, and n-C_{10}, which was found the highest was generated by decarboxylation of the Mg-Zn(OH)(ME $C_{12/14}$) basic soap. The behavior of increasing the yield of n-C_8 to n-C_{11} and n-C_{15} alkanes chain length shown in Figure 6 can be explained as follows: lauric acid ($C_{12:0}$) is the main component of ME $C_{12/14}$ and ME $C_{12/16}$ (see Table 1). Ideally, it can be decarboxylated to produce n-C_{11} hydrocarbons. However, if cracking accompanies the decarboxylation reaction, then short n-alkane hydrocarbons such as n-C_{10}, n-C_9 and n-C_8, even n-C_7, n-C_6, and n-C_5 can be produced.

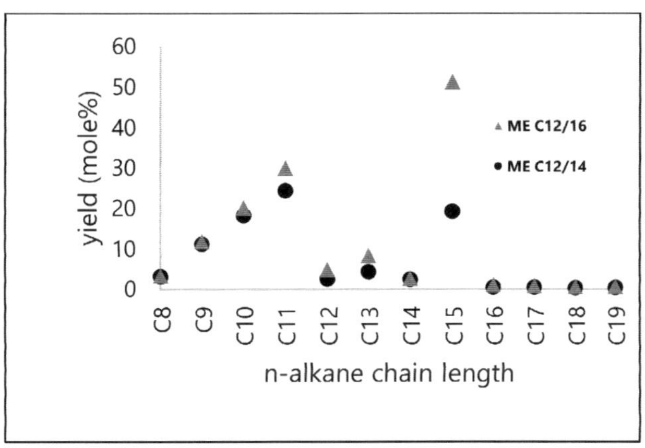

Figure 6. Relationship between yield and chain length of the n-alkane

The impact of cracking reactions resulted in light alkane hydrocarbons such as n-C_1 to n-C_4 as well [13]. Therefore, n-C_8 to n-C_{10} hydrocarbons are more suitable to be called products from cracking than decarboxylation reactions. It seems that the cracking process is effective, so the products produced are relatively high compared to n-C_{11}. Meanwhile, n-C_{15} hydrocarbons are estimated as a result of polymerization of 1-C_{13} hydrocarbons, i.e., decarboxylation products from tetradecanoate ($C_{14:0}$) with 1-C_2 produced both dodecanoate ($C_{12:0}$) and n-C_{11} hydrocarbons cracking. In addition, the n-C15 hydrocarbon is the result of hexadecanoate ($C_{16:0}$) decarboxylation, especially from ME $C_{12/16}$ as well. These analysis results indicated that the secondary reaction such as cracking had accompanied the decarboxylation process of Mg-Zn(OH)(ME $C_{12/14}$) basic soap. In that case, the cracking reaction producing more short n-alkane (< n-C_8) was most likely to occur slowly; therefore, n-C_{10} was still found more than other short n-alkanes. On the other hand, for Mg-Zn (OH)(ME $C_{12/16}$) basic soap, it was very likely that other reactions such as cracking and polymerization had occurred rapidly as soon as the decarboxylation process began. As a result, n-C_{15} products were the most resulted (by polymerization) as well as the small and light n-alkanes products under n-C_8 (by cracking); thus it reduced the yield of the n-C_{11} and n-C_{10}.

3.6. Grouping of liquid biohydrocarbons fraction

The grouping of liquid biohydrocarbons according to the alkane and alkenes fractions is shown in Figure 7.

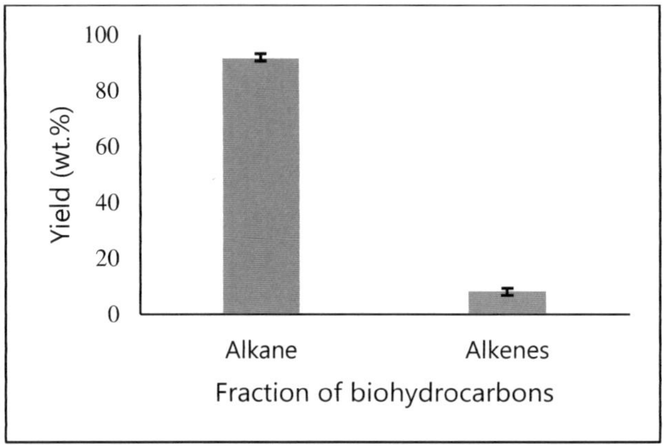

Figure 7. Faction of alkane and alkenes in biohydrocarbons

The data in Figure 7 show that the alkane fraction in liquid biohydrocarbons of basic soap decarboxylation much higher than alkenes. The results mentioned above, corresponding to the product expected from the basic soap decarboxylation, as shown by the reaction equation scheme follow:

$$(RCOO)\mathbf{M}(OH) \rightarrow RH + \mathbf{M}CO_3 \qquad (2)$$

Where **M** are metal or metal combination.

4. Conclusion

Thermal decarboxylation of methyl esters basic soaps derived from palm kernel oils under 350°C and atmospheric pressure yielding liquid bio-hydrocarbons corresponding to the jets biofuel. Decarboxylation reactions have been effective in this study, evidenced by the formation of n-alkanes in large numbers. The basic soaps decarboxylation reaction was always accompanied by cracking and polymerization reactions. Mg-Zn(OH)(ME $C_{12/14}$) basic soap was a better choice as decarboxylation feed because it resulted in liquid bio-hydrocarbons that contain more n-undecane as the main components of jets biofuel. This study has proven that the production of jet biofuel can take place without consuming hydrogen and operating at atmospheric pressure.

Acknowledgments

This work was supported by Indonesian Oil Palm Estate Fund Agency (BPDPKS) under GRS-16 Grant.

References

[1] Demirbas A 2008 *Energy Conv. Manag.* **49** 2106-16
[2] Jimenez E S and Crocker M 2012 *J. Chem. Technol. Biotechnol.* **87** 1041-50
[3] Choundhary T V and Phillips C B 2011 *Appl. Catal. A: Gen.* **397** 1-12
[4] Wang W C, Thapaliya N, Campos A, Stikeleather L F and Roberts W L 2012 *Fuel* **95** 622-29
[5] Neonufa G F, Soerawidjaja T H and Prakoso T 2017 *J. Eng. Techol. Sci.* **49** 575-86
[6] Zailong Z, Yanhua S, Yongxin L and Wuzheng L 1999 *Chin. Sci. Bull.* **44** 1523-27
[7] Demirbas A 2003 *Energy Sourc.* **25** 457-66
[8] Anneken D A, Both S, Christoph R, Fieg G, Steinberner U and Westfechtel A 2012 *Fatty acids*, Encyclopedia of Industrial Chemistry (Weinheim: Wiley-VHC Verlag GmbH & Co. KGaA)
[9] Mancini A, Imperlini E, Nigro E, Montagnese C, Daniele A, Orrù S and Buono P 2015 *Molecules* **20** 17339-61
[10] Neonufa G F 2017a Drop-in fuel production technology of green-diesel and bioavtur type via catalytic thermal decarboxylation of basic soap base on magnesium and transition metal combination *Doctoral Dissertation* (Bandung, Indonesia: Institut Teknologi Bandung)
[11] Sari E, Dimaggio C, Kim M, Salley S O and Simon Ng K Y 2013 *Ind. Eng. Chem. Res.* **52** 11527-36
[12] Simacek P, Kubicka D, Kubickova I, Homola F, Pospisil M and Chudoba J 2011 *Fuel* **90** 2473-79
[13] Fu J, Lu X and Savage P E 2011 *Chem. Sus. Chem.* **4** 481-86
[14] Neonufa G F, Pratiwi M, Purwadi R, Prakoso T, Soerawidjaja T H and Elizabeth L 2019 *IOP Conf. Ser.: Mater. Sci. Eng.* **543** 012069

Tauc Plot Software: Calculating energy gap values of organic materials based on Ultraviolet-Visible absorbance spectrum

Albert Zicko Johannes, Redi Kristian Pingak and Minsyahril Bukit

Department of Physics, Faculty of Science and Engineering, University of Nusa Cendana, Jl. Adi Sucipto-Penfui, Kota Kupang, Nusa Tenggara Timur, Indonesia

Email: zickojohannes@staf.undana.ac.id

Abstract. Studies on bio-organic materials for optoelectronic devices have increased recently, mainly due to the low cost of the materials. Electrical properties of the materials are important to review whether a material is suitable for optoelectronic applications. The energy gap is an important parameter to investigate the electrical properties of materials, and therefore, the band gap value has to be determined accurately. The energy gap of a semiconductor material can be calculated from its absorption spectrum using the Tauc plot method. In this article, we designed software based on the Tauc plot method (Taucplot4dotz) and ran a trial test using data from our previous studies on Sukun (*Artocarpus Altilis*) leaves extract and Gamal (*Gliricidia Sepium*) leaves extract. Results obtained were more accurate than our previous calculations since here we used both horizontal dan vertical linear extrapolation lines compared to only one in our previous calculations. We also used re-fitted data of ZnO [26]. Our accurate and fast calculation using the software showed that energy gap values of Sukun leaves and Gamal leaves extract were 3.033 eV and 1.833 eV, respectively. For ZnO, the energy gap was found to be 3.6205 eV, slightly different from the original data (3.625 eV) [26]. From the results, it can be concluded that the software provides an easier and faster calculation of energy gap values of materials, from which one can further investigate the potential of the materials suitable for optoelectronic devices based on bio-organic materials. **Keywords**: tauc-plot method, organic materials, energy gap, Absorbance, UV-Vis Spectrum.

1. Introduction

Research on optoelectronic devices has increased recently. One of many important factors to investigate whether a material is suitable for optoelectronic applications is the electrical property of materials. In the field of material physics, one way to determine the electrical properties of a material is energy gap values. An energy gap is an energy region where the density of states of electrons is zero. As an energy region where electron cannot exist, the energy gap separates two allowed electronic energy states. These electronic energy states in materials instead of having discrete energies as in the case of a single atom, they form bands, namely the valence band and the conduction band. The valence band is the region where electrons occupy the outer states in atomic structure and is fully filled with electrons. Meanwhile, the conduction band is the region where electrons can move freely to generate electrical current and has higher energy than the valence band. Usually, the conduction band is empty,

IOP Conf. Series: Materials Science and Engineering **823** (2020) 012030 doi:10.1088/1757-899X/823/1/012030

but in the case of metals, it is partially filled. The energy region between these two bands forms a forbidden band known as the energy gap. The energy gap value is defined as the difference between the top of the valence band and the bottom of the conduction band and it can be used to determine whether a material is categorized as a metal, a semiconductor or an insulator. Metals do not have energy gap, meaning that the conduction band and the valence band overlap with each other so the electron can move freely from the valence band to the conduction band. On the other hand, semiconductors and insulators have energy gap, with values being less than 4 eV for the former and greater than 4 eV for the latter [1].

Calculations of material energy gaps have been carried out in many studies using various materials, including bulk materials, thin layers, bio-organic materials and nanomaterials [2-17]. There are various methods of calculating them, including the photoluminescence (PL) method [2-5], the Cody plot method [6,7,8], the Tauc plot method [9-20], and so on. The Tauc plot method is one of the most popular methods in calculating energy gap values. This is because the calculations are simple and accurate for amorphous materials [19, 20]. Calculation of energy gap using this method requires only UV-Vis Absorbance spectrum data.

In this study, software to calculate energy gap using the Tauc plot method is designed. The purpose of this design is to make the calculations faster and easier.

Tauc Plot Method was first developed by Tauc in 1968 [9]. This method gained popularity when research on thin layers began to develop rapidly [14-17]. At that time, scientists were competing to conduct experiments in searching for new materials.

Calculation of energy gap value using the Tauc Plot method is started by calculating the absorption coefficients (α), namely the equation:

$$\alpha = \frac{A \ln(10)}{l} \quad (1)$$

where A = Absorbance value from UV-Vis spectrometer and l = sample width (cm).

In addition, the conversion of wavelength λ (nm) into energy units hv (eV) is also calculated using equation:

$$1\,\text{Joule} = 1\,hv = \frac{hc}{\lambda} \times 6.241509 \times 10^{18}\ \text{eV} \quad (2)$$

According to Tauc, the energy gap values can be obtained from the following equation:

$$(\alpha hv)^{1/m} = B\left(hv - E_g\right) \quad (3)$$

where h = Planck's constant (4.135 × 10-15 eV.s), v = frequency (s^{-1}), B = a comparative constant, E_g = energy gap (eV).

m indicates the type of electronic transition, with values for different transitions being:

- ½ = Direct
- 3/2 = Direct forbidden
- 2 = Indirect
- 3 = Indirect forbidden

These m values are used in the calculations and the one that yields the best fitting results is chosen as the 'correct' m value [18], which can be used as a preliminary indication about the type of electronic transition in materials. However, to determine the type of transition accurately we need to use other experimental methods.

To evaluate equation (3), values of $(\alpha hv)^{1/m}$ and hv are calculated first. Then, the next step is to plot the results. The plot yields an irregular curve but there are some parts that are in the form of a straight or linear vertical line. In accordance with equation (3) where the value of E_g must be directly proportional to the value $(\alpha hv)^{1/m}$ then the equation applies to this linear part of the curve.

Next, in the linear part, extrapolation is carried out so that it cuts the *hv* axis. The intersection of the extrapolation line with the *hv* axis is the energy gap value (E_g) of the material of interest. In the modified Tauc plot method [19,20], a second extrapolation line is drawn from the horizontal direction, then the intersection between the two lines determines the E_g value. This leads to more accurate E_g values. Therefore, we apply this method in developing our software.

Extrapolation is a technique of estimating data from previously owned data. This extrapolation technique has been used in various fields of science [21-24]. The simplest extrapolation technique is a linear extrapolation technique where the value we want to find $f(x)$ is approached using the equation:

$$f(x) = f(x_n) + \frac{f(x_{n+1}) - f(x_n)}{(x_{n+1} - x_n)} x \tag{4}$$

with x_n = the value of x to n, and x_{n+1} = the value of x to n + 1.

In applying the modified Tauc plot method, two extrapolation lines are needed so that calculations with two equations are carried out. Then the intersection is searched, when the two lines intersect, we will get a combination of two equations, and the value of x as E_g can be determined :

$$f(x_1) + \frac{f(x_2) - f(x_1)}{(x_2 - x_1)} x = f(x_3) + \frac{f(x_4) - f(x_3)}{(x_4 - x_3)} x \tag{5}$$

$$f(x_1) + Mx = f(x_3) + Nx \tag{6}$$

$$E_g \approx x = \frac{f(x_3) - f(x_1)}{M - N} \tag{7}$$

where 1 and 2 are data points for extrapolating from the horizontal direction, whereas 3 and 4 are data points for extrapolation from the vertical direction.

2. Research Method

The development of the TaucPlot4dotz software [25] is divided into two stages, namely: Design and Trial.

2.1. Design

At this stage, a flow chart is created to apply the Tauc plot method. This flow chart can be seen in Figure 1. Calculations begin by entering input data in the form of wavelength (λ) and absorbance (A) data files which can be obtained from the results of UV-Vis spectrophotometer characterization of the sample. The data is then displayed and plotted on the screen.

The next step is to calculate the absorption coefficient values (α) using equation 1 and also do the calculation of the wavelength conversion value from nanometer (nm) to the energy unit (eV) using equation 2. In addition, at this stage, the software calculates the $(\alpha h v)^{1/m}$ values in equation 3. The results of this calculation are displayed and then plotted vs energy (E).

The last step is to determine the linear area of the curve plotted and to choose four points (X1, X2, X3, and X4) which will be used to make two straight lines extrapolating from the horizontal and vertical directions (equation 4). The intersection of the two lines is calculated by equation 8 to obtain the energy gap value.

Graphic User Interface (GUI) of the software design can be seen in Figure 2.

2.2. Trial

At this stage software testing is carried out using sample data: ZnO (re-fitted data from [26]) and from previous studies on leaves extract [10,11], using the following steps:

- First enter the data by pressing the INPUT button and selecting the data format * .csv and obtain the first plot as shown in figure 3.A.
- Then calculate the absorption coefficients, the conversion of wavelength and (ahv) 1 / m, by selecting the value of m and determining the sample length value first, then pressing the CALCULATE button. The results obtained are shown in figure 3.B, where the results of calculations and the second plot are shown.
- After the results are obtained, it is followed by determining the linear area of the curve from the second plot and calculating two extrapolation lines by specifying four points (specify the point from the table and press the X1H button to select, repeat for the next X2H, X1V and X2V). Choose these four points from the "linear area" of the curve. X1H and X2H are chosen from the horizontal part while X1V and X2V are chosen from the vertical part.
- After the selection, press the Energy Gap button to obtain the energy gap values. Examples of some results can be seen in Figure 3.C.

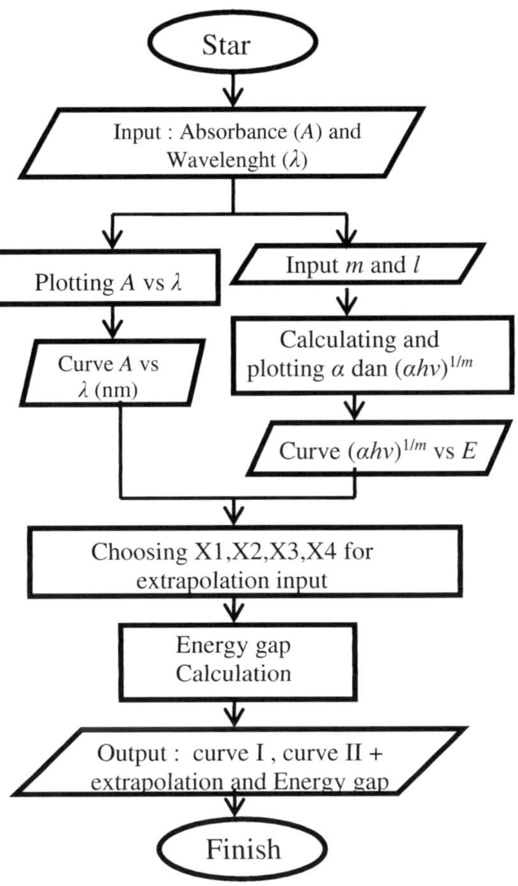

Figure 1. Flowchart of designing the TaucPlot4dotz software [24].

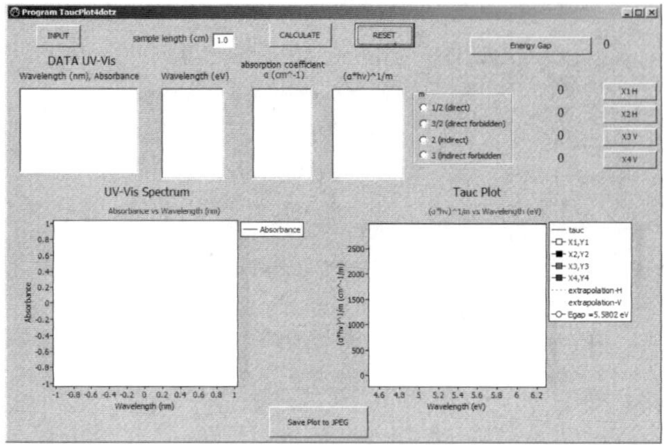

Figure 2. GUI of the TaucPlot4dotz software [24].

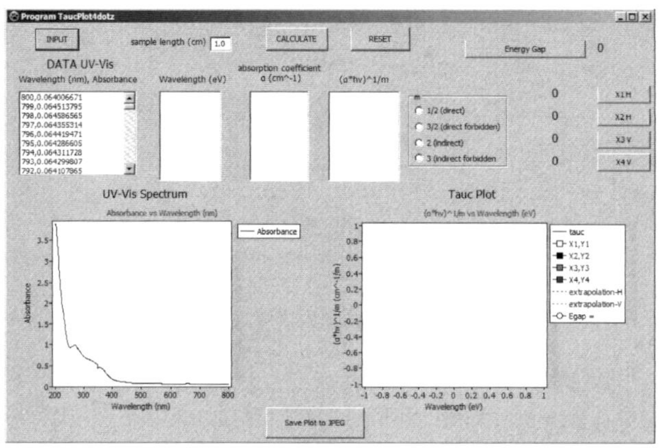

(A)

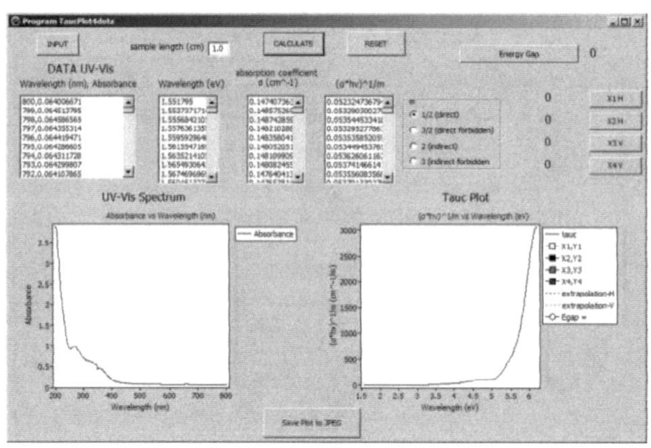

(B)

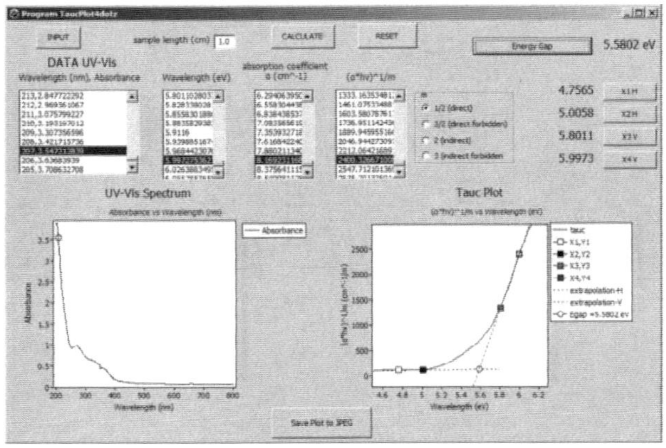

(C)

Figure 3. (A) Input of input data. (B) Example of calculation results of absorption coefficient, conversion of wavelength and (ahv) 1 / m., (C) Example of the results of Energy Gap Calculation.

3. Results and Discussion

Calculation results for ZnO (re-fitted data from [26]) can be seen in Figure 4. ZnO absorbance spectrum was chosen because it has well-known energy gap value of 3.1 eV - 3.5 eV and the type of transition is direct transition [18]. Energy gap value of ZnO obtained from our calculation was 3.6206 eV, which is only slightly different from the original data result of 3.625 eV [26]. This result clearly showed that the software has good accuracies.

Other results of the data testing can be seen in Figure 5 and Table 1. The table shows six experimental data, namely three experimental data on the absorbance of Sukun (*Artocarpus Altilis*) leaves extract from Kassa, et al. [10] and three experimental data on the absorbance of Gamal (*Gliricidia Sepium*) leaves extract from Tedju, et al. [11]. From the data, the energy gap values were determined using the TaucPlot4dotz software.

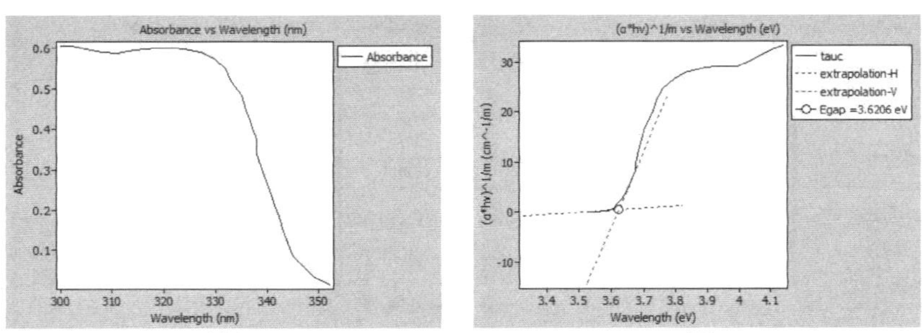

(A) (B)

Figure 4. (A) ZnO Absorbance re-fitted data from [26], (B) ZnO Energy Gap Calculation Result using our software

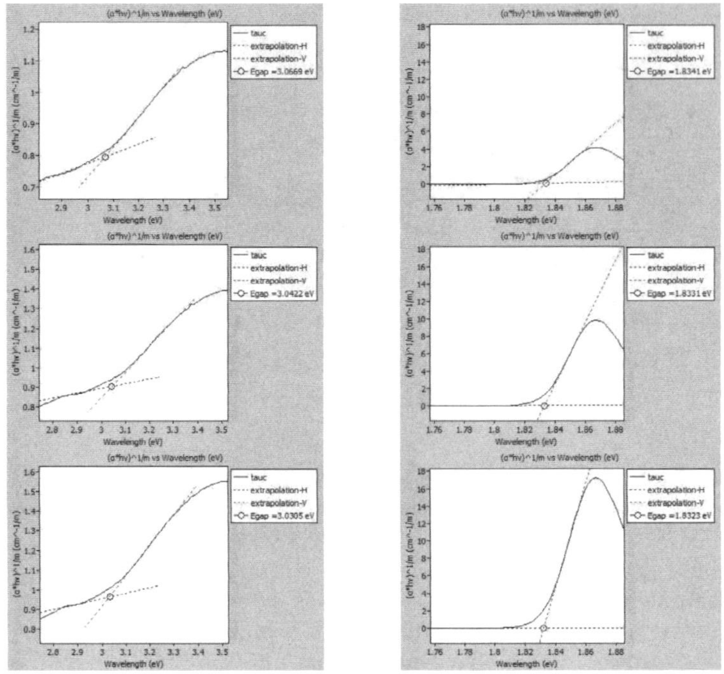

(A) (B)

Figure 5. (A) Sukun (*Artocarpus Altilis*) leaves extract, (B) Gamal (*Gliricidia Sepium*) leaves extract.

Table 1. Energy gap values comparison between calculation from Kassa et al. [10] and Tedju et al. [11] with Taucplot4dotz software.

Types of extract	ppm in ethanol	Previous Calculation (eV) [10,11]	taucplot4dotz (eV)
Artocarpus Altilis	100	3,03	3,066
	200	3,03	3,042
	300	3,03	3,030
Gliricida Sepium	200	1,832	1,834
	300	1,832	1,833
	400	1,831	1,832

Energy gap values compared with the energy gap values from our previous calculations which only use one extrapolation line in [10,11] show that the values are quite similar. However, one of the main results here that differ from our previous results is that our results presented in this article are more accurate. According to Ehsan, et al [19] and Russel, et al. [20], the use of two extrapolation lines also provide more accuracy compared to one extrapolation line calculations.

To the best of our knowledge, there is no "Tauc plot specific" software, but Tauc plot method can be implemented with various plotter-calculation software. One of the advantages of the software we develop here over other software is that there is no need to apply the equation and plot the curve. Our software is faster and easier to use.

4. Conclusions

Results show that energy gap values can be calculated faster and easier using our software. The results also provide us with information of the possible types of electronic transitions. Finally, the results obtained here are also more accurate because of the use of the modified Tauc plot method [19,20].

Acknowledgments

We would like to thank Renaldo A. Kasa and Julius B. Tedju for letting us use previous experimental data so we could run the software testing in this research.

References

[1] Babu V S 2010 *Solid State Devices and Technology* 3rd edition (New Delhi: Pearson Education India)
[2] Monemar B 1974 *Phys Rev B.* **10**(2) 676
[3] Ketterer B, Heiss M, Livrozet M J, Rudolph A, Reiger E and Morral A F *Phys. Rev. B* **83**(12) 125307
[4] Ghetmiri S A, Du W, Margetis J, Mosleh A, Cousar L, Conley B R, Domulevicz L, Nazzal A, Sun G, Soref R A and Tolle J 2014 *App. Phys. Lett.* **105**(15) 151109
[5] Jiang S, Fang Y, Li R, Xiao H, Crowley J, Wang C, White T J, Goddard III W A, Wang Z, Baikie T and Fang J *Angewandte Chemie Int. Ed.* **55**(22) 6540-4
[6] Cody G D, Brooks B G and Abeles B 1982 *Solar Energy Mat.* **8**(1-3) 231-40
[7] Mok T M and O'Leary S K 2007 *J. App. Phys.* **102**(11) 113525
[8] Sánchez-Vergara M E, Alonso-Huitron J C, Rodriguez-Gómez A and Reider-Burstin J N 2012 *Molecules.* **17**(9) 10000-13
[9] Tauc J. 1968 *Mat. Res. Bull.* **3**(1) 37-46
[10] Kasa R A, Bukit M and Johannes A Z 2017 *J. Fis. Fisika Sains dan Aplikasinya* **2**(1) 10-6
[11] Tedju J B, Bukit M and Johannes A Z 2018 *J. Fis. Fisika Sains dan Aplikasinya* **3**(3) 142-6
[12] Dandara M, Pingak R K and Johannes A Z 2019 *J. Fis. Fisika Sains dan Aplikasinya* **4**(1) 48-51
[13] Missa M M, Pingak R K and Sutaji H I 2018 *J. Fis. Fisika Sains dan Aplikasinya* **3**(2) 86-90
[14] Daniyati R, Zharvan V, Ichsan N, Pramono Y H and Yudoyono G 2015 *Pros. Semin. Sains Dan Teknol. vol 1* (Surabaya, Indonesia)
[15] Dongol M, El-Nahass M M, El-Denglawey A, Elhady A F and Abuelwafa A A 2012 *Current App. Phys.* **12**(4) 1178-84
[16] Tumuluri A, Naidu K L and Raju K J 2014 *Int. J. Chem. Tech. Res.* **6**(6) 3353-6
[17] Scafetta M D, Cordi A M, Rondinelli J M and May S J 2014 *J. Phys. Cond. Matter* **26**(50) 505502
[18] Vizbicke, Brian D, Patel, Shane, Davis, Benjamin E and Birnie D P 2015 *Phisica Status Solidi B* **252**(8) 1700-10
[19] Ehsan M F, Ashiq M N, Bi F, Bi Y, Palanisamy S and He T 2014 *RSC Advances* **4**(89) 48411-18
[20] Russell H B, Andriotis A N, Menon M, Jasinski J B, Martinez-Garcia A and Sunkara M K 2016 *Scientific reports* **6** 20822
[21] Colwell R K and Coddington J A 1994 *Philosophical Transactions of the Royal Society of London Series B: Biological Sciences* **345**(1311) 101-18
[22] Abutaqiya M I, Sisco C J and Vargas F M 2019 *Fluid Phase Equilibria* **483** 52-69
[23] Tomić A Š, Antanasijević D, Ristić M, Perić-Grujić A, Pocajt V 2018 *Sci. of the Total Env.* **610** 1038-46
[24] Liu B, Johnson S G, Joannopoulos J D, Lu L 2018 *J. Optics* **20**(4) 044005
[25] Johannes A Z 2019 *Taucplot4dotz* online: https://tinyurl.com/taucplot4dotz
[26] Abraham N and Dekany I 2010 *Colloids and Surfaces a-Physicochemical and Engineering Aspects* **364**(1-3) 26-33

IOP Conf. Series: Materials Science and Engineering **823** (2020) 012031 doi:10.1088/1757-899X/823/1/012031

Coating of mercapto modified silica on iron sand magnetic material for Au(III) adsorption in aqueous solution

Martasiana Karbeka[1*], Nuryono[2], Suyanta[2]

[1]Department of Chemisty, Universitas Tribuana Kalabahi, Alor, 85812 Indonesia
[2]Department of Chemistry, Universitas Gadjah Mada, Yogyakarta, 55284 Indonesia

E-mail: martasianakarbeka@untribkalabahi.ac.id

Abstract. The functionalized silica coating mercapto on iron sand magnetic material (MM/SiO_2/MBI)has been done through sol-gel process. The successfully synthesized material is used as adsorbent (MM/SiO2/MBI) in the Adsorpi Au (III) process. Material characterization is carried out with an adsorbent stability test against acids. Adsorption is performed in a batch system and the unadsorped Au (III) ion is analyzed with the Atomic Absorption Spectrophotometer (AAS). Characterization results indicate the synthesis of PB/SiO_2-MBI adsorbent has been successfully performed. The coating of iron sand magnetic material enhances stability to acid. ADsorbent (MM/SiO_2/MBI) is capable of adjudisorate ion Au (III) with the highest adsorption occurring at pH 1. Kinetics review indicates that the adsorption of the ion (III) is following the second-order pseudo kinetics with a value of k 3.57 × 10-3 g/mg. Minute and adsorption isotherm follows the isotherm pattern of Langmuir with an adsorption capacity of 125 mg/g.

1. Introduction

The surface of the silica gel containing silanol functional groups (≡Si-OH) and siloxane (≡Si- O-Si≡) that has been widely used as adsorbent heavy metal ions [1]. However, silica gel has drawbacks when applied as an adsorbent for soft metal. The weak interaction is caused by the low capacity of the oxygen atom in the silanol and siloxane in donating electrons to the soft metal ions [2,3,4].

Gold is one type of metal ion growing software utilization in various fields, especially the use of gold in electronic devices. This should be a concern for electronic waste produced continues to rise and the electronic waste is precious metals that they can be utilized such as Au, Ag, Pd and Pt as well as basic metals constituent electronic devices such as Cu, Al, Ni, Sn, Zn and Fe [5,6,7]. Gold in electronic waste is mixed with other metals that we need a method to process the collected (recovery) of gold from e-waste that is by adsorption methods. Selection of adsorption method is used because the equipment is simple, economical, easy to operate and high efficiency of metals with low concentrations. Silica is used as an adsorbent to adsorb ions of gold metal by modifying the surface using specific functional groups. Modification of the silica surface adsorption determines the selectivity of the gold metal ions. Generally, modifier compounds having functional groups such as -OH, -NH, -SH and -COOH (Takeuchi et al., 2007). Based on the theory of hard-soft acid-base (HSAB), gold is classified as soft acids and tends to form strong bonds with functional groups such as -NH soft alkaline, -CN and

- SH. Based on the theory HSAB the silica surface modified with compounds containing basic groups to become more malleable gold is selectively adsorbed soft acid. Modified silica functional groups proved to be selective for the adsorption soft gold [4,5,8,9]. Modifications of adsorbents have been carried out to produce adsorbents that are selective and specific in absorbing gold. Some researchers, use the -NH group as a modifier in the adsorbent to absorb gold. More and more - NH groups are expected to get more gold metal ions absorbed. Some modifiers used include the use of chitosan which has many -NH groups, the use of aminobenzimidazol with two -NH groups and tolylbiguanide with three -NH groups [3,4,15]. Modification of the adsorbent with mercaptobenzimidazole has one -SH group. This research was conducted to determine the effectiveness of the -SH group in adsorption gold.

However, the adsorption method has drawbacks in the separation process for the separation of the adsorbent from the adsorbate performed with conventional filtering techniques, decantation or can also utilize the instrument via centrifugation. It will be very difficult if applied in the neighborhood so attractive to develop the silica material that has magnetic properties to facilitate the separation process only with the help of an external magnet. Easily oxidized magnetite but with their coating with silica and magnetite functional groups are protected from the acidic or alkaline [10,11,12]. The commonly used magnetic source is a synthesis $FeCl_2{\cdot}6H_2O$ and $FeCl_3{\cdot}4H_2O$ to produce magnetite (Fe_3O_4). The use of magnetic materials can also be used from natural iron sand. The sand iron content of magnetite (Fe_3O_4) in nature has a fairly high [13,14]. The advantage of this research is to use magnetite from natural sand to produce adsorbent materials that have magnetic properties. Thereby reducing the use of chemicals for magnetite synthesis. Also, the adsorbent produced is an adsorbent that has been modified with the thiol (-SH) functional group which is expected to have the ability of adsorption with gold metal ions according to the principle of HSAB.

2. Objects and Methods

2.1. Adsorption of metal ions Au (III)

2.1.1 Variations in pH. The solution of Au (III) 1000 ppm was diluted to 100 ppm in a 100 mL flask. The solution of Au (III) of 100 ppm was diluted to 50 ppm in a 10 ml flask. 1-6 pH adjustment is done by adding 1 M NaOH and 1 M HCl and then diluted with a buffer of pH 1-6. Solution with a variety of pH placed in a plastic bottle and put 10 mg of adsorbent MM/SiO_2/MBI after it was shaken for 60 minutes using a 200 VRN shaker (130 rpm), The next stage is separated from the solution adsorbent using an external magnet (Neodymium, N35 grade, size 40 mm x 30 mm x 10 mm) then the solution was analyzed using FAAS, Analytic Jena contrAA 300 to determine the optimum pH solution of Au (III) adsorbed on the adsorbent MM/SiO_2/MBI.

2.1.2 Variation of contact time. Adsorption was done by varying the contact time of 15, 30, 45, 60, 90, 120 and 150 minutes at optimum pH. Adsorbent MM/SiO_2/MBI has added as much as 10 mg and shaken on every variation of a predetermined time using a 200 VRN shaker (130 rpm), Adsorbent MM/SiO_2/MBI separated from the solution of Au (III) using external magnetic assistance. The solution that has been separately analyzed using FAAS, Analytic Jena contrAA 300.

2.1.3 Variations in the concentration of metal ions Au (III). Solution of Au (III) to be varied concentration of 10 ppm, 20 ppm, 50 ppm, 100 ppm, 150 ppm, 200 ppm and 300 ppm at optimum pH and dissolved in 10 mL flask using optimum pH buffer (pH meters), Further included 10 mg of adsorbent MM/SiO_2/MBI on every variation of the concentration and shaken for 90 minute using a 200 VRN shaker (130 rpm)then adsorbent MM/SiO_2/MBI separated from the solution of Au (III) using an external magnetic assistance. The solution that has been separately analyzed using FAAS, Analytic Jena contrAA 300.

3. Results and Discussion

3.1. Effect of pH

pH plays an important role in determining the ability of the adsorbent to adsorb metal ions due to changes in pH affects the surface of the adsorbent and is followed by changes in metal ion species. Therefore, to determine the effect of pH on the adsorption of Au (III) by MM/SiO_2/MBI must perform variations of pH 1-6. The effect of pH changes on the number of ions Au (III) adsorbed can be seen in Figure 1.

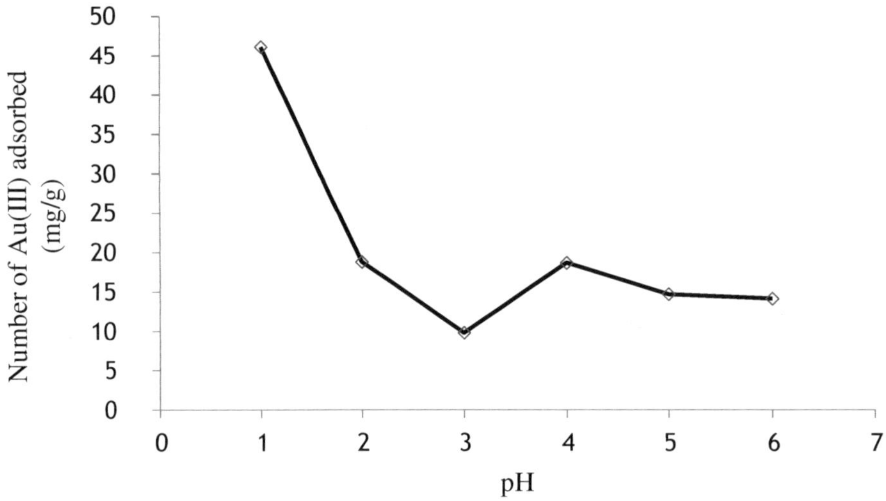

Figure 1. Graph effect of pH on the adsorption of Au (III)

Based on Figure 1 is known that the higher the pH, the number of ions Au (III) adsorbed decline. MBI group has a pKa value of 2.38 and a pH <pKa associated with the formation of the pair RNH_3^+/RNH_2. The decrease in pKa increases the number of protonated amines. Protonated at pH 1-2 occurs quickly because the value of pH <pKa, making it possible to pH 1 occurs perfect protonated amine groups on MBI. Thus despite the anion Cl competition with $[AuCl_4]^-$ to be bound by the -SH group but the presence of the anion Cl no effect because $[AuCl_4]^-$ in addition to interacting with the -SH group may also interact electrostatically with the amine group via the Cl atom in the $[AuCl_4]^-$. The more the amine groups are protonated at low pH led to the increasing number of electrostatic forces formed by the anion $[AuCl_4]^-$ so that at pH 1 provides the highest adsorption amount. Therefore, at low pH (pH 1), the amine group (-NH) and mercapto group (-SH) play a role in the process of adsorption of Au (III) so that the number of metal ions Au (III) adsorbed to be very high.

At pH 2-3 decrease the number of ions Au (III) is adsorbed. This indicates that not only the S atom that plays a role in the process of adsorption of Au (III), but it is possible that a reduced number of protonated N atom caused a decrease in adsorption capacity. Ion absorption Au (III) at pH 4 slightly increased and then remained constant at pH> 4. In the area of pH 4-6 speciation occurred metal hydroxide ions Au (III) to $[AuCl_3(OH)]^-$, $[AuCl_2(OH)_2]^-$ and $[AuCl(OH)_3]^-$. The existence of OH- ligands with O atoms which are rich in pairs of free electrons and a large electronegativity and also the presence of Cl-rich electrons lead to increased repulsion between the metal ion complexes Au (III) with the MBI group. Speciation of metal ions to hydroxide causes an increased number of hydroxide ions resulting in a decrease in the adsorption of Au (III) by the adsorbent MM/SiO_2/MBI. N atoms in the pH range 4-6 MBI have not experienced so protonated N atoms still has a free electron pair causing electrostatic repulsion. The repulsion leads to a decrease in the number of ions Au (III) adsorbed. The existence of OH- ligands with O atoms which are rich in pairs of free electrons and a large electronegativity and also the presence of Cl-rich electrons lead to increased repulsion between the metal ion complexes Au (III) with the MBI group. Speciation of metal ions to hydroxide causes an increased number of hydroxide ions resulting in a decrease in the adsorption of Au (III) by the adsorbent MM/SiO_2/MBI.

3.2. Metal ion adsorption kinetics Au (III)

The adsorption kinetics explain the rate of absorption of solute by the contact time of the reaction is an essential characteristic absorption efficiency. The longer the contact time Au (III) will also be increased to at a certain time to reach equilibrium. Figure 2 shows the influence of variations in the time on the efficiency of the adsorption of Au (III) by MM/SiO_2/MBI.

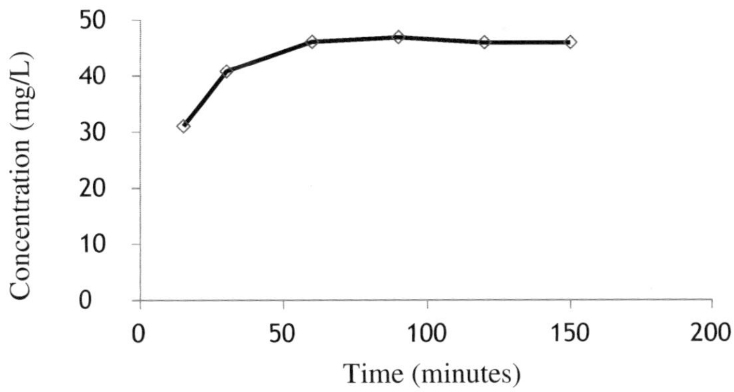

Figure 2. Graph effect of contact time on the adsorption of Au (III)

Figure 2, the kinetic curve shows that the first 60 minutes happen speed of absorption Au (III) which is fast and the number of ions Au (III) adsorbed increasing. Adsorption speed slows down until it reaches equilibrium at 90 minutes so the addition of interaction time does not give effect to an increase in the number of metal ions Au (III) are adsorbed. Increasing the amount of Au (III) happens quickly because the surface of the MM/SiO_2/MBI is -SH group with a high affinity for Au (III) so that when there is contact between the Au (III) with an adsorbent then will quickly -SH group donate electron pairs. Also, there is the amine group (-NH) protonated thus accelerating the process of adsorption of metal ions Au (III), which is in the form of $[AuCl_4]^-$. This interaction occurs electrostatically. When the contact time of the longer visible reaction was slow. In this process the -SH group is estimated to have been used to interact with the ion Au (III) so that the slow phase is due to the interaction of Au (III) with silanol and siloxane contained in the material MM/SiO_2/MBI.

This reaction takes place slowly because the ions Au (III) is acidic soft and O atoms in silanol - OH group is alkaline but -SH group more alkaline so that Au (III) more effectively interact with -SH and when interacting with -OH become less effective. Silanol role in the slow reaction because it is located on the relatively inside so that when the interaction of Au (III) with silanol takes longer and silanol is also less effective in donating an electron pair to interact with Au (III). In this process the -SH group is estimated to have been used to interact with the ion Au (III) so that the slow phase is due to the interaction of Au (III) with silanol and siloxane contained in the material MM/SiO_2/MBI.

Adsorption kinetics MM/SiO_2/MBI on Au (III) was studied with the first-order reaction kinetics model, second-order, pseudo-1 order and pseudo-2 order. Kinetic parameters are shown in Table 1.

IOP Conf. Series: Materials Science and Engineering **823** (2020) 012031 doi:10.1088/1757-899X/823/1/012031

Table 1. Parameters of adsorption kinetics of Au (III) in MM/SiO_2/MBI

Material	Reaction kinetics model	The parameters of reaction kinetics		
		K	q_e	R^2
MM/SiO2/MBI	Order 1			0.52
	Order 2			0.5
	Pseudo-1 order	0,026 min^{-1}	0,929 $mg.g^{-1}$	0.708
	Pseudo-2 order	0.003 g / mg.menit	50 $mg.g^{-1}$	0.998

Selection of kinetic model based on the degree of linearity for any kinetic model used. Based on Table 1 it can be seen that the greatest value of R^2 lies in the pseudo second-order kinetic model. Based on the price of R^2 can be interpreted that the appropriate kinetic model to describe the adsorption of Au (III) by MM/SiO_2/MBI is a Pseudo-2 order kinetic model. Order kinetic model pseudo-2 order based on the assumption that the rate-limiting step is the chemical adsorption involving power valence electron exchange between the metal ions with the adsorbent. The adsorbent is not in excessive quantities to adsorb metal ions. The number of active sites on the adsorbent decreases linearly during the adsorption process takes place. The kinetic data of Au (III) adsorption with MM / SiO2 / MBI were compared with the kinetic data of pseudo-2 order adsorbent of chitosan silica magnetic material (MMSC) of 51.020 mg/g, aminobenzimidazole (MM@SiO_2/AB) kinetic data of 9.50 mg/g and tolylbigunide (MM@SiO_2/TB) magnetic material of 6.83 mg/g [3,4,15]. Based on the kinetic parameters it can be seen that the modified adsorbent of the thiol group (-SH) has a higher gold adsorption ability than the -NH group. Judging from the number of groups in the modifier, it can be seen that the -NH groups in chitosan, aminobenzimidazol, and tolylbiguanide are more than the -NH groups in mercaptobenzimidazole. The results of this study indicate that the -SH group is more effective at adsorption gold metal ions.

3.3 Metal ion adsorption isotherms Au (III)
Total Au (III) adsorbed on MM/SiO_2/MBI as a function of concentration is shown in Figure 3. The increased concentration of Au (III) the amount of adsorbed also increased.

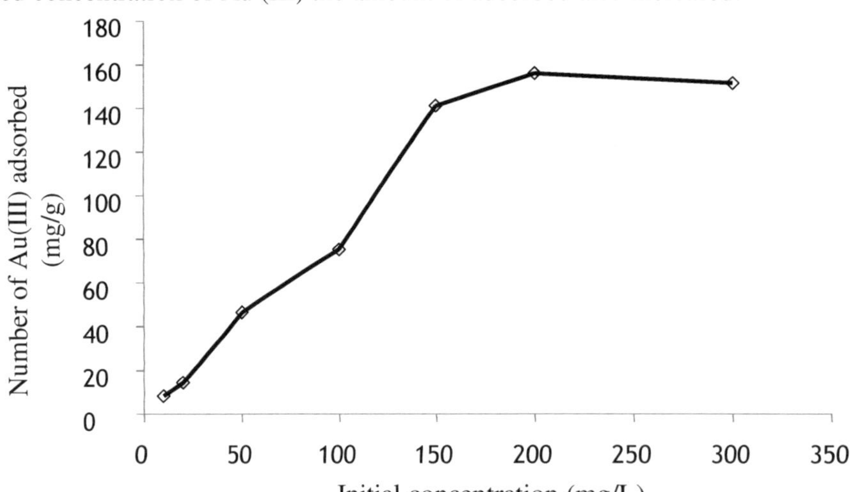

Figure 3. Graph ion concentration influences Au (III)

Based on the results of the adsorption of Au (III) in MM/SiO_2/MBI with various concentrations obtained by the equilibrium concentration at 200 mg/L. Then contacted the equilibrium concentration of silica without modification. Comparative results are presented in Table 2 adsorption.

Table 2. Adsorption comparison of Au (III) in silica unmodified (MM/SiO_2) and silica modified ($MM/SiO_2/MBI$)

Material	The initial concentration Au (III) (mg/L)	Total Au (III) adsorbed (mg/g)
MM/SiO_2	200	45.833
$MM/SiO_2/MBI$	200	156.111

The results proved that the adsorption of Au (III) with modified silica ($MM/SiO_2/MBI$) 3.5 times greater than the silica without modification (MM/SiO_2). According to the principle, HSAB - SH group and the amine group can interact better with Au (III) because of the -SH group and amine alkaline than the soft -OH groups on MM/SiO_2 alkaline hard. Low ability O atoms in silanol and siloxane as an electron donor is a result of the bond is between the O atoms with atomic Si in the structure of silica (possibility occurs bond partially between O and Si through bonding πd-p) so that the electron density of the O atoms is reduced and effect on the low capacity of O atoms as electron donors. Based on the principle HSAB-OH group is less effective as a donor atom. Therefore, it can be stated that the modification of the silica with the functional groups ($MM/SiO_2/MBI$) was able to increase the adsorption capacity to Au (III). Predictions made possible interaction between Au (III) with $MM/SiO_2/MBI$.

The interaction of Au (III) with $MM/SiO_2/MBI$ adsorbents can be hypothesized as shown in Figure 4. The adsorption process is carried out in an acidic atmosphere (pH 2). At pH 2 it is possible to protonate the -NH group in the form of $-NH^+$ which is present in the mercaptobenzimidazol modifier and at pH 2 also the gold metal ions are in the metal ion complexes Au (III). Thus, in an acid solution, the surface of iron sand is coated with silica through and the mercaptobenzimidazol modifier interacts with silica through the CPTMS 'bridge' through the formation of siloxane bonds. The silane group in CPTMS is needed in bond formation because the mercaptobenzimidazole modifier does not have a silane group to be able to interact directly with silica. This hypothesis is supported by research on Coatings of L-Arginine Modified Silica by using GPTMS as a 'bridge' connecting [16]. In contrast to the mercaptopropyltrimethoxysilan modifier (MPTMS) which already has a silane group, it does not need a 'connecting bridge' [17,18]. When there is an interaction of the gold metals with $MM/SiO_2/MBI$ adsorbents in an acidic atmosphere, the gold metals in the form of Au (III) interacts with the -SH group by giving electron pairs to form covalent bonds. Another possible interaction is the electrostatic interaction between the group -NH which undergoes protonation to become $-NH^+$ with anion $[AuCl_4]^-$ in an acidic solution.

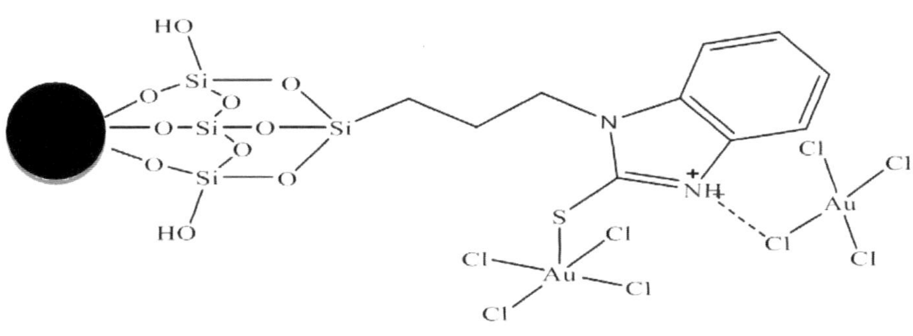

Figure 4. Predicted interactions Au (III) with $MM/SiO_2/MBI$

The adsorption process with ion concentration variations Au (III) in $MM/SiO_2/MBI$ then studied the adsorption isotherm. The equilibrium isotherms are used to explain the nature of the adsorbent surface and affinity $MM/SiO_2/MBI$ on Au (III). Two types of isotherm are used to describe the data in Table 3 is the Langmuir isotherm and Freundlich isotherm. The most important part in the adsorption process is a site owned by the adsorbent located on the surface, but the number of these sites will be reduced if a closed surface is increasing. Freundlich isotherm models suggest that the process of adsorption on the surface is heterogeneous. Freundlich isotherm models show the adsorbate layer formed on the surface of the adsorbent is a multilayer. The relationship between the parameters of Langmuir and Freundlich shown in Table 3.

Table 3. Langmuir and Freundlich isotherm parameters on the adsorption of Au (III)

parameter Langmuir			parameter Freundlich		
R^2	q_m (mg / g)	K_L (L/mg)	R^2	K_F mg / g	N
0.66	333.33	0,022	0.97	2.62	1.67

Data Table 3 shows the adsorption isotherms patterns Au (III) tends to follow the Freundlich adsorption isotherm. This is an indication that the adsorption of Au (III) in the adsorbent MM/SiO_2/MBI is a multilayer. Which follows the pattern adsorption isotherm Freundlich, when the surface is saturated or nearly saturated with adsorbate, there can be two things: the adsorption layer is formed over the second and so on adsorbate that has been bound to the surface, this phenomenon is called multilayer adsorption.

4. Conclusion

Based on the results of the research showed that the maximum adsorption of ions Au(III) adsorbent occurs at pH 1 with an optimum contact time of 60 minutes. Mercapto modified silica has an ability to adsorb ions Au (III) 3.5 times higher than unmodified silica. Adsorption capacity obtained was 333.33 mg/g (based on the Langmuir isotherm models). Kinetics study follows the model Pseudo-2 order with constant adsorption $3x10^{-3}$ g/mg.menit. Further studies can be done by ion desorption and adsorption Au(III) in multi-metal system.

5. References

[1] Jal P K, Patel S and Mishra B K 2004 Chemical modification of silica surface by immobilization of functional groups for extractive concentration of metal ions *Talanta* **62** 1005–28
[2] Buhani, Narsito, Nuryono and Kunarti E S 2009 Amino and mercapto-silica hybrid for Cd(II) adsorption in aqueous solution *Indones. J. Chem* **9** 170-6
[3] Muflikhah M, Rusdiarso B, Putra E G R and Nuryono N 2017 Modification of silica coated on iron sand magnetic material with chitosan for adsorption of Au (III) *Indones. J. Chem* **17** 264–73
[4] Rettob A L, Nuryono, Pasaribu Y N, Buyang Y and Waremwa R S 2019 Adsorption of $[AuCl_4]^-$ on iron sand magnetic material coated with aminobezimidazol modified silica **1** 277-80
[5] Kim E-Y, Kim M-S, Lee J-C and Pandey B D 2011 Selective recovery of gold from waste mobile phone PCBs by hydrometallurgical process *J. Hazard. Mater* **198** 206–15
[6] Vats M C and Singh K 2015 Assessment of gold and silver in assorted mobile phone circuit boards (PCBs): Original article *Waste. Manage.* **45** 280-88
[7] Yi Q, Fan R, Xie F, Min H, Zhang Q and Luo Z 2016 Selective recovery of Au(III) and Pd(II) from waste PCBs using ethylenediamine modified persimmon tannin adsorbent **31** 185-94
[8] Wahyuni E T, Roto, Nissah F A and Aprilita N H 2018 Modified silica adsorbent from volcanic ash for Cr(VI) anionic removal *Indones. J. Chem* **18** 428-33
[9] Roto, Yusran and Kuncaka A 2016 Magnetic adsorbent of Fe_3O_4@SiO_2 core-shell nanoparticles modified with thiol group for chloroauric ion adsorption *J. Apsusc* **377** 30-36
[10] Nuryono, Rosiati N M, Rusdiarso B, Candra S, Sakti W and Tanaka S 2014 Coating of magnetite with mercapto modified rice hull ash silica in a one-pot process *SpringerPlus*

[11] Melnyk I V, Pogorilyi R P, Zub Y L, Vaclavikov M, Gdula K, Dąbrowsk A, Seisenbaeva G A and Kessler V G 2018 Protection of thiol groups on the surface of magnetic adsorbents and their application for wastewater treatment *scientificreports Open* 20-31
[12] Kuzminska M, Carlier N, Bancov R and Gaigneaux 2015 Magnetic nanoparticles: improving chemical stability via silica coating and organic grafting with silanes for acidic media catalytic reactions J. *Apcata*
[13] Prasdiantika R and Susanto 2017 Preparasi dan penentuan jenis oksida besi pada material magnetik pasir besi lansilowo *Teknosains* **6** 7
[14] Rettob A L and Karbeka M 2019 Pengaruh konsentrasi larutan HF pada proses preparasi terhadap kadar unsur bahan magnetik pasir besi *Walisongo Journal of Chemistry* **3** 1
[15] Nuryono, Rosiati N M, Rettob A L, Suyanta and Arryanto Y 2018 Coating of 2-aminobenzimidiazole and 1-(o-tolyl)biguanide functionalized silicas on iron sand magnetic material for sorption of $[AuCl_4]^-$ *Indones. J. Chem* **19**(2) 395-404
[16] Amaria, Suyanta and Nuryono 2017 Coating of L-arginine modified silica on magnetite through two different sol-gel routes *Indones. J. Chem* 256-63
[17] Nuryono, Muliaty E, Rusdiarso B, Sakti S C W and Tanaka S 2014 Adsorption of Au(III), Cu(II) and Ni(II) on magnetite coated with mercapto groups modified rice hull ash silica *J. Ion Exch* 114–21
[18] Nuryono, Rosiati N M, Rusdiarso B, Sakti S C W and Tanaka S 2014 Coating of magnetite with mercapto modified rice hull ash silica in a onepot process *SpringerPlus* **3** 515

IOP Conf. Series: Materials Science and Engineering **823** (2020) 012032 doi:10.1088/1757-899X/823/1/012032

Lithium and Calcium recovery by activated carbon from coconut shell char

C W Purnomo[1,2] and A Widyadhana[1]

[1]Department of Chemical Engineering, Universitas Gadjah Mada, Indonesia
[2]Agrotechnology Innovation Center, PIAT UGM, Indonesia

E-mail: chandra.purnomo@ugm.ac.id

Abstract. Spent rechargeable batteries and electronic waste become a future challenge. The precious metal inside should be recovered for pursuing a circular economy society. Here, lithium phosphate battery has been dismantled and leached the cathodes by acidic solution to produce metal solutions. The metal content will be separated by a simple adsorption mechanism using activated charcoal. Coconut shell char has been activated by KOH in several temperature settings and then tested to adsorb lithium and calcium ions in the leaching solution. The capacity of adsorption and selectivity between the ions were investigated. The adsorption capacity is strongly related to the activation temperature and KOH/char ratio. Higher activation temperature and agent ratio tend to provide better-activated carbon for metal ion adsorption. It is found that Li ions have lower affinity toward the surface of carbon especially in low concentration than Ca ions.

1. Introduction

Lithium-ion batteries (LIB) are popular batteries used as a power source because of their energy density. At present, the LIB is estimated to constitute 37% of the market for rechargeable resources. Batteries constitute 39% of overall lithium usages. The prospect of an electric vehicle in the future that is increasingly apparent automatically increases the need for LIB in a larger size. Light vehicles that use electric hybrid systems, hybrids and electric hybrid plug-ins will need a lightweight battery.

Meanwhile, many other industries such as ceramics, rocket propellant, nuclear, pharmaceutical, lubricants, vitamin A production, and the synthesis of organic compounds also require lithium compounds. Lithium mining from salt or ore in the future cannot meet market demand so that lithium mining from wastewater sources and seawater starts to be pursued [1].

Bioleaching is another method used in lithium recovery efforts. Spodumene, a source of lithium, is a mineral that is difficult to be processed and requires a high-temperature treatment. This is clearly not profitable from an industrial perspective. The role of microorganisms is interesting to use because it requires lower energy, but a longer time required. The microorganisms such *as Penicillium purpurogenum, Aspergillus niger,* and *Rhodoterula rubra* has been observed for this application.

The absorption of lithium through electro-sorption has also been investigated. Cathodic polarization increases sorption, while, conductivity and pH variations are easily optimized to recover lithium electrochemically [2]. In his research, the simple adsorption via activated coconut shell char with KOH activator will be observed. It is expected that activated carbon with KOH at a certain temperature and chemical ratio will be able to provide high lithium adsorption. Since the activating method provides important surface properties and structure of carbon to be used for selected purposes [3].

2. Methods

2.1. Activated carbon preparation
Coconut shell charcoal is crushed and sieved using a standard sieve to get particle size between 10 and 14 mesh. Then, a mixture of KOH and charcoal powder with weight ratio of KOH/char at 0.5: 1, 1: 1, and 1.5: 1 is prepared. All three samples were activated at 600 °C to study the effect of the activating agent ratio. For the effect of activation temperature study, the equal ratio of activator and charcoal was used. Activation temperature variations are 500 °C, 600 °C, and 700 °C using a horizontal tube furnace under nitrogen flow and heating rate of 20 °C/min. All six samples were washed using distilled water until the pH was near neutral, then filtered and dried. Samples were characterized through analysis using Fourier Transform Infrared Spectroscopy (FTIR) and also the Surface Area Analyzer. Activated carbon sample identification is summarized in Table 1.

Table 1. Activated carbon ID

ID	Activation	
	KOH/Char ratio	Heat Treatment Temp., °C
C500/1	1:1	500
C600/1	1:1	600
C700/1	1:1	700
C600/0.5	0.5:1	600
C600/1.5	1.5:1	600

2.2. Activated carbon adsorption
LiFePO4 battery leachate solution was prepared by soaking cathode layer with sulfuric acids. After leaching, the Fe ion was separated from the solution by precipitation using $CaOH_2$. Thus, inside the solution remain Li and Ca as the major ions. In taking the batch adsorption data, the weight of charcoal added to 50 mL solution was varied from 1 to 15 gram inside Erlenmeyer flasks. Then the mixture is then shaken using a shaker bath for 1 day at room temperature. The concentration of the solution after adsorption was analyzed using ICP-EAS (Perkin Elmer).

3. Results and Discussion
FTIR analysis was carried out to identify functional groups on the carbon surface. In Figure 1, we can see the absorption bands owned by coconut shell charcoal and activated char at wavenumbers around 3448.72 cm-1 (OH), 2924 cm-1 (CH_n), and 1600 cm-1 (C = O). At an increase in temperature, it can be seen as a shift in absorption wavenumbers. KOH activator also seems to cause a shift in absorption wavenumbers. It can be said that there is no new functional group that appears by the activation method. All the peaks come from the origin of char but in the lower intensity for the activated ones.

For the intensity of the peaks, the temperature of activation plays an important role in reducing the intensity. Higher temperature will create carbon with less functional groups. The highest activation temperature sample (C700/1) shows almost flat spectra. Meanwhile, for KOH ration the effect is minor. The increase of the ratio from 0.5 to 1.5 seems does not give any impact on the chemical properties of the samples since the three spectra are similar (C600/0.5 to C600/1.5).

For porosity characterization, the results are provided in Table 2. It is clearly seen that the raw char has very small porosity. Then, the increase of treatment temperature to 600 °C does not significantly increase the surface area, while, for the highest temperature the surface area starts to improve rapidly. It can be assigned that the carbon is very hard and solid that is not easy to be activated. Meanwhile, for

KOH ratio, using low ratio of 0.5 did not alter so much the char while for the highest ratio the porosity is increase significantly. It can be concluded that only two samples (C700/1 and C600/1.5) can be considered as activated carbon. In other words, the activation can be achieved by increasing the temperature higher than 600 °C or using more KOH at lower temperature.

Table 2. Porosity profile of the chars

ID	Porosity		
	BET Surface area, m^2/g	Pore volume x100, cc/g	BJH Pore average diameter, A
Raw char	0.7	0.7	26.5
C500/1	12	1.2	26.6
C600/1	71.5	7	26.8
C700/1	502	30.5	26.9
C600/0.5	21.5	2.2	26.9
C600/1.5	226.8	15	27.1

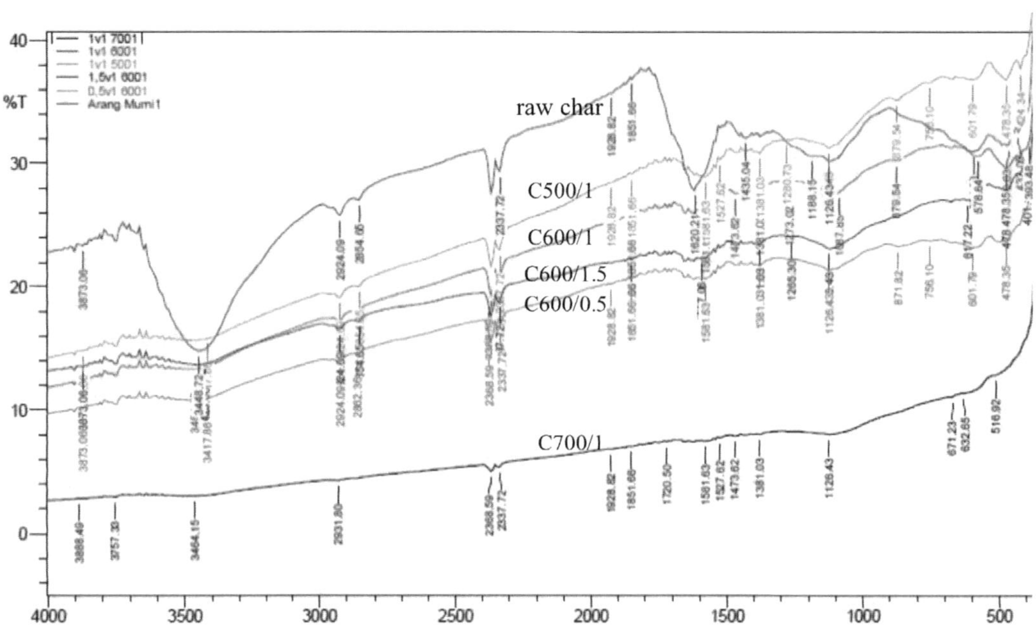

Figure 1. FTIR analysis results from coconut shell charcoal

In Figure 2, it can be seen that the lithium adsorption isotherm of carbon with different activation temperatures (500 °C, 600°C, and 700°C) and activated agent ratio (0.5, 1.0 and 1.5). It appears that the adsorption capacity is relatively low in lower concentration. It indicates that Li-ion has low affinity toward the carbon surface. The increase of the capacity in high concentration has a positive correlation with the porosity. The larger surface area of carbon tends to adsorb Li-ion more. In IUPAC classification of isotherm, this isotherm is considered as type III isotherm which is typical for non-porous or macroporous adsorbent. Thus, the lithium ions seem does not attach inside the micropores but rather than outer and large pores surfaces. The low affinity of carbon toward Li-ions also has been reported elsewhere [4,5]. Low interaction between carbon atoms in the surface of char and Li ions in the solution could be due to the low atomic mass of the ions. The interatomic potential interaction suggests that the force has a positive correlation with the atomic mass of the interacting atoms as well as the distance among the atoms.

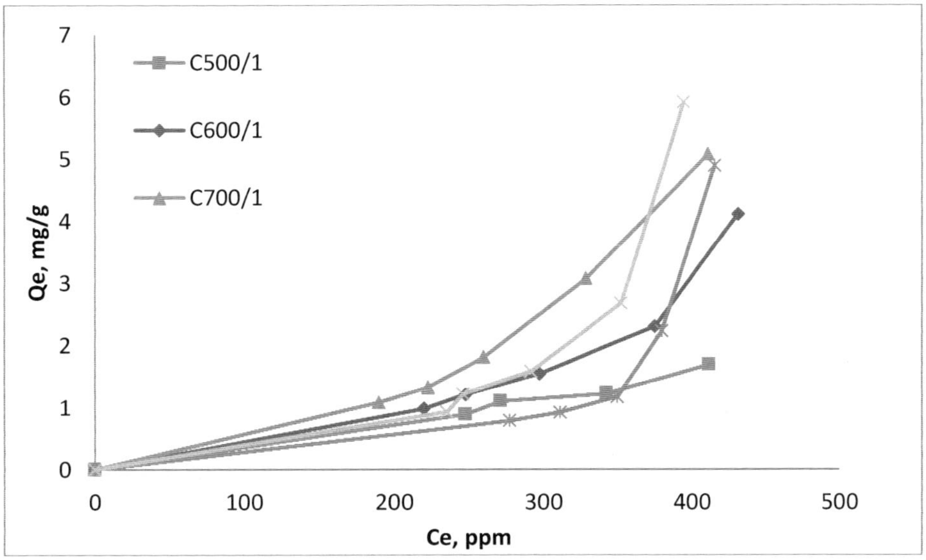

Figure 2. Adsorption isotherm of Lithium ions

Figure 3 shows the adsorption isotherm for Ca ions. In general, the isotherms are different from Li-ion, the isotherm represent type I or II of the IUPAC classification. The small shoulder in lower solution concentration can be a hint for monolayer adsorption saturation of micropore filling of Ca ions. It indicates a relatively strong interaction between pore surface and Ca ions. The low porosity of carbon (C600/05) still represents the type III isotherm since there is no significant micropore that can be filled by Ca ions instead of the outer surface and macropores.

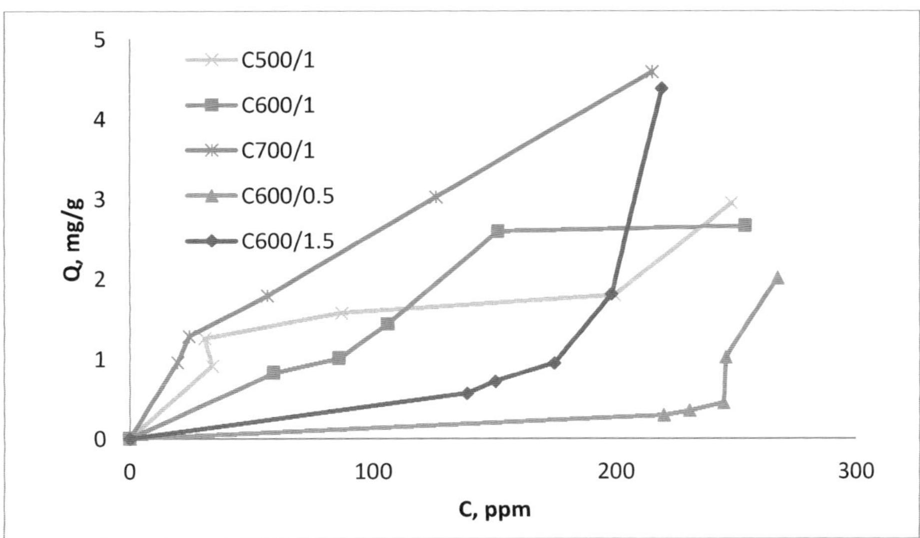

Figure 3. Adsorption isotherm of Calcium ions

4. Conclusions

The method of activation strongly related to the porosity development of the char that will lead to adsorption characteristics. The prepared activated carbon has different behaviour in adsorbing Li and Ca ions. Li-ions tend to be less attracted to the pores of carbon. In low concentration, Ca ion will be absorbed larger than Li-ions while in the concentrated solution, the capacity is relatively similar.

5. References

[1] Chung W J, Torrejos R E C, Park M J, Vivas E L, Limjuco L A, Lawagon C P and Nisola G M 2017 *Chem. Eng. J.* **309** 49
[2] Alfarra A, Frackowiak E and Béguin F 2002 *Electrochimica Acta* **47**(10) 1545
[3] Jeong J M, Rhee K Y and Park S J 2015 *J. Ind. Eng. Chem.* **27** 329
[4] Purnomo C W, Kesuma E P, Perdana I and Aziz M 2018 *Waste Manage.* **79** 454
[5] Purnomo C W, Kesuma E, Wirawan S K and Hinode H 2017 *ASEAN J. Chem. Eng.* **1** 91

Acknowledgments

Financial supports from RTA UGM and PUIPT 2019 Grant are highly appreciated.

Kapok fibre as potential oil-absorbing material: Modification mechanism and performance evaluation

Graecia Lugito[1], Arip Kustiana[1], Reyhant Martuani[1], and I G Wenten[1]

[1]Department of Chemical Engineering, Institut Teknologi Bandung, Bandung 40132 Indonesia

E-mail: graecia@che.itb.ac.id

Abstract. Kapok fibre has gained considerable attention as a potential oil-absorbing material due to its lipophilicity, hydrophobicity, and configuration of its central lumen which is the key component of its high oil absorption capacity. However, the weak structural integrity of the fibre greatly retards its application. This study aims to develop a reinforcement method for kapok (*Ceiba pentandra*) fibre to enhance its reusability and oil recovery. Oil-sorption capacities of several modified kapok fibres have been evaluated to resolve the oil-polluted water issues. Oxidation using sodium chlorite ($NaClO_2$) followed by sol-gel coating with tetraethylorthosilicate (TEOS) and sponge formation via freeze-drying method successfully enhance the structural integrity of the fibre without significantly reduce its hydrophobicity. Further coating the sponge with dodecyltrimethoxysilane (DTMS) using the chemical vapor disposition method results in superhydrophobic fibre. The best performance achieves an oil-sorption capacity of 48.6 g-oil/g-fibre and can maintain 62.6% of its oil-sorption capacity even after ten reuse cycles.

1. Introduction

Kapok fibre is a natural cellulosic fibre that is light (low density), hydrophobic, and oleophilic. Unlike other fibre, kapok has a lumen (air cavity) in the middle of its waxy fibre wall is the key to its high oil absorption capacity and good buoyancy [1]. Kapok fibres are characterized by having a high level of acetyl groups (13.0%), which make them superhydrophobic and do not get wet with water [2].

Pristine kapok has shown relatively stable performance in reclaiming oil from oil-in-water emulsion as it shows little change in morphology after being used as an oil absorbent. It also selectively absorbs oil from water or seawater with an oil intake of 40 g/g fibre [2]. Kapok fibre has shown high oil-to-water sorption (O/W) ratios, up to more than 200 times [3]. In kapok fibre, there is two-steps oil sorption mechanism: (i) oil is absorbed by utilizing the similarity of hydrophobic properties and van der Waals forces between oil and waxy layer on the wall surface of kapok while water is rejected; (ii) the absorbed oil then passes through the fibre cellulosic wall and enter the lumen by utilizing the capillarity effect. The oil sorption capacity and oil retention of kapok fibre are highly dependent on the oleophilic fibre wall and the physical properties of the oil (e.g., density, viscosity, surface tension, as well as contact angle with kapok surface) [4]. Furthermore, Abdullah et al. have found out that the kapok fibre can maintain its ability to absorb oil up to 15 cycles with a capacity reduction of only 30% compared to the new kapok fibre [5].

Although detected as a good oil absorbent, the utilization of kapok fibre in the field is still underdeveloped due to the challenges in modulation and reusability. The short length and waxy smooth surface of kapok fibre cause poor inter-fibre cohesion; the wax content on the fibre surface allows the oil to diffuse in the lumen but hardly holds the oil not to diffuse out of the lumen (low retention) [6]. Therefore, surface modification on kapok fibres has to be done to enhance their inter-fibre cohesion as well as oil retention. The aim of this study is to evaluate several surface modification methods for removing the wax on the kapok fibre surface reinforcing its structural integrity as well as oil sorption capacity and recovery without losing its affinity to oil and water repellency.

2. Materials and Methods

Kapok fibre was purchased from a farmer in Kadipaten Village, Tasikmalaya Regency, Indonesia. Sodium chlorite ($NaClO_2$), Acetic acid (CH_3COOH), ammonium hydroxide solution ($NH_3.H_2O$), n-hexane, methanol, ethanol, and cooking oil were purchased from local suppliers. Tetraethylorthosilicate (TEOS) was purchased from Merck & Co., while dodecyltrimethoxysilane (DTMS) and sodium dodecylbenzene sulfonate (SDBS) were purchased from Shandong Bovi Energy Technology Co., China.

Super-hydrophobic kapok fibre was prepared by a sol-gel silanation technique adapted from Wang et al. [7]. The fibre was separated from seeds and cleaned from dust by aeration before pretreated using $NaClO_2$ in a glass reactor (equipped with overhead stirrer and condenser) under acid condition (pH 4.5, temperature 80°C, stirring rate 400 rpm) to eliminate the waxy layer from the kapok surface exposing the hydroxyl group on the cellulosic material. The concentration of $NaClO_2$ was varied from 0.25%, 0.5%, 3%, 4%, and 5% to give the optimum oil sorption capacity vs. weight loss [8]. The effect of concentration variations from 0.5 to 3 %-w have been reported previously and are not reconsidered in this study.

The pretreated fibres were mixed with TEOS (4%-w) and surfactant SDBS (1.2 mmol/L) at room temperature for 20 minutes with slow addition of $NH_3.H_2O$ (1.8%-w) to perform a homogeneous sol-gel reaction on the kapok surface. The resulting fibre was then washed with methanol and dried in a vacuum oven. The hydrophilized TEOS coated fibre was freeze-dried prior to the hydrolyzation with dodecyltrimethoxysilane (DTMS) via chemical vapor deposition in oven preset at 120°C for an hour to form a superhydrophobic layer. Variations of pretreatments and coating methods were conducted to gain optimum results.

The chemisorption of the TEOS and DTMS on the kapok fibre was characterized using attenuated total reflectance Fourier transform infrared (FTIR ATR, Bruker) and compared to the spectra of pristine kapok fibre summarized in Table 1. The hydrophobicity of the pristine and modified kapok fibre was determined by water contact angle measurements. Oil was recovered from the fibre without severe disruption of the appearance of the fibre. The oil sorption capacities, intakes and recovery of pristine and modified kapok fibre were analyzed and presented in gram oil per gram dried kapok. The sorption/desorption cycle was repeated ten times to evaluate the recyclability of the fibre.

Table 1. FTIR spectra of kapok fibre [8,9]

Wavenumbers (cm^{-1})	Assignment
3419	Non-free O-H stretching
2929 and 2850	Asymmetric and symmetric C-H stretching in CH_2 and CH_3
1741	C=O stretching of ketones, carboxylic and ester groups in lignin; acetyl ester in xylan
1504 and 1602	Aromatic ring C-C skeletal stretching
1375	C-H bending
1245	C-O bending
1065	C-C stretching

3. Results and Discussion

Hydroxyl groups on cellulosic fibre are important for the purpose of modification [10]. Therefore, the purpose of pretreatment is to distort the waxy layer, exposing the hydroxyl groups on the wall surface of kapok fibre to be chemically bonded with silica precursors at TEOS coating stage without damaging the wall structure that may result in the reduction the oil sorption capacity. Figure 1 shows the visible appearance (A) as well as O-H stretching FTIR spectra (B) of pristine and pretreated kapok fibre of different concentrations of $NaClO_2$. The increasing concentration of $NaClO_2$ enhances the whiteness and compactness of kapok fibre, as well as the hydroxyl group concentrations on the fibre surface. However, severe weight loss of the treated fibre may indicate wall structure damage. Therefore, further assessment of oil sorption capacity has to be done to determine the optimum $NaClO_2$ concentration for pretreatment process.

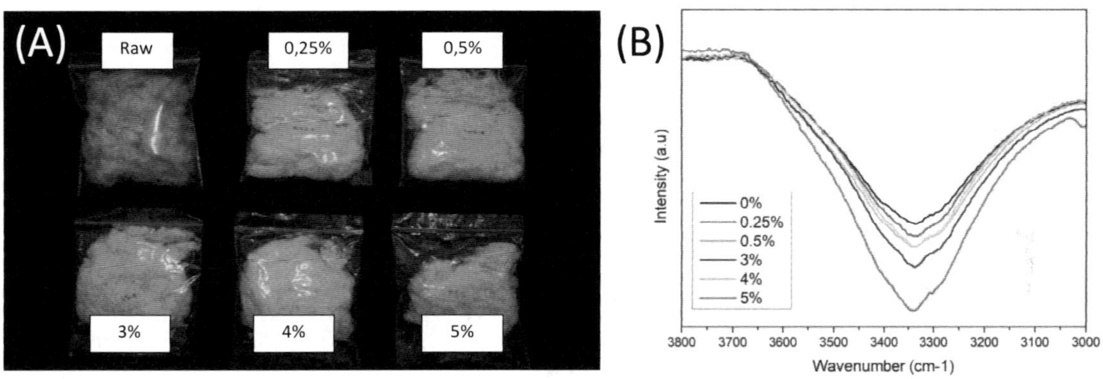

Figure 1. Kapok fibre before and after pretreatment using different concentration of $NaClO_2$

Oil sorption capacity of the $NaClO_2$-oxidized fibres have been measured and results are shown in Figure 2. Compared to the pristine kapok, the oxidized kapok loses its oil sorption capability indicating the reduction of the surface wax content. The oil sorption capacity of the fibre was first increased with the increment concentration of $NaClO_2$ to 0.5%-w but gets decreased afterward. Thus, kapok fibre pretreated with $NaClO_2$ concentration of 0.5%-w is chosen to be further studied.

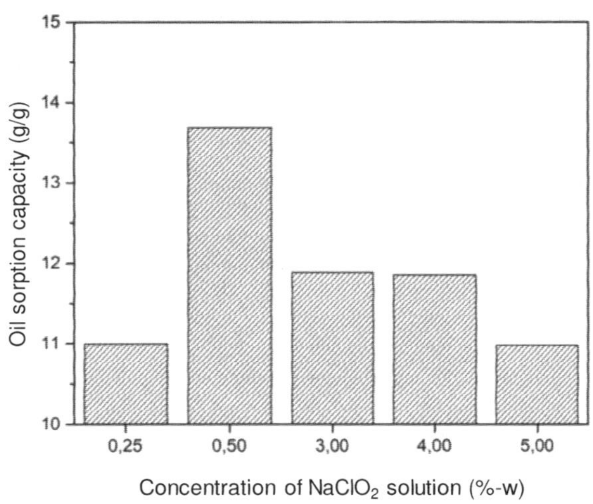

Figure 2. Oil absorption capacity of Kapok fibre after pretreatment with different concentration of $NaClO_2$

Table 2 summarizes the variations of the modification process conducted in this study. With average fibre length of 1.8 cm [6], kapok fibre is poorly modulated; therefore, coating is an effective way to reinforce the cellulosic fibres and enhance their inter-fibre cohesion. The kapok fibre is coated with organosilane DTMS with or without TEOS as the precursor. To retain their structural integrity, the coated kapok fibres have been soaked and stirred in water with different weight ratios (0.2 and 0.6 %-w) then freeze-dried. Strand scission was done prior to the freeze-drying process to promote a homogeneous fibre dispersion in water.

Table 2. Variations of modification processes on kapok fibre samples

Code	Variations
Var.0	Raw kapok fibre
Var.1	Kapok fibre sponge after $NaClO_2$-pretreatment, TEOS coating, strand scission, freeze-dried with fibre to water concentration 0.6 %-w, followed by DTMS coating
Var.2	Kapok fibre sponge after $NaClO_2$-pretreatment, TEOS coating, strand scission, and freeze-dried with fibre to water concentration 0.2 %-w, followed by DTMS coating
Var.3	Kapok fibre sponge after $NaClO_2$-pretreatment, TEOS coating, freeze-dried with fibre to water concentration 0.6 %-w (without strand scission), followed by DTMS coating
Var.4	Kapok fibre sponge after $NaClO_2$-pretreatment, freeze-dried with fibre to water concentration 0.6 %-w (without strand scission), followed by DTMS coating

The FTIR spectra of the pristine and modified kapok fibre have been recorded to assess the success of each modification process. Figure 3 shows the FTIR spectra of pristine and modified kapok fibre as indicated on the graph. TEOS coating reduces the intensity of the hydroxyl group (at 3342 nm^{-1}) but brings up the Si-O-Si spectra (at 1051 nm^{-1}); while DTMS coating increases the intensity of spectra at 2923 and 2856 that belong to $-CH_2$ and $-CH_3$, respectively [8].

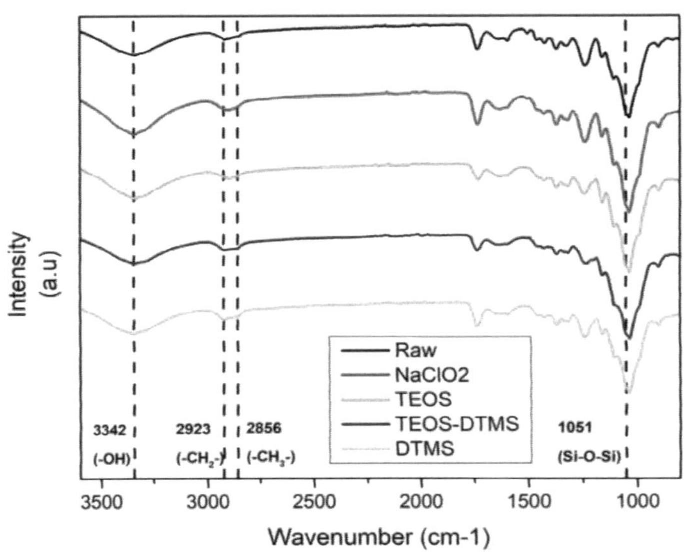

Figure 3. FTIR Spectra of pristine and modified kapok fibres

Figure 4 shows the water contact angle (WCA) on pristine and modified kapok fibres. Chemical modification using TEOS and DTMS does not significantly affect the WCA. Freeze-dried modified kapok fibres, however, show higher WCA compared to those without structural integrity enhancement. Superhydrophobic kapok sponges (WCA = 152.3 ± 3.5) could be achieved after $NaClO_2$-pretreatment,

TEOS coating, freeze-drying with fibre to water concentration 0.6 %-w (without strand scission), then followed by DTMS coating.

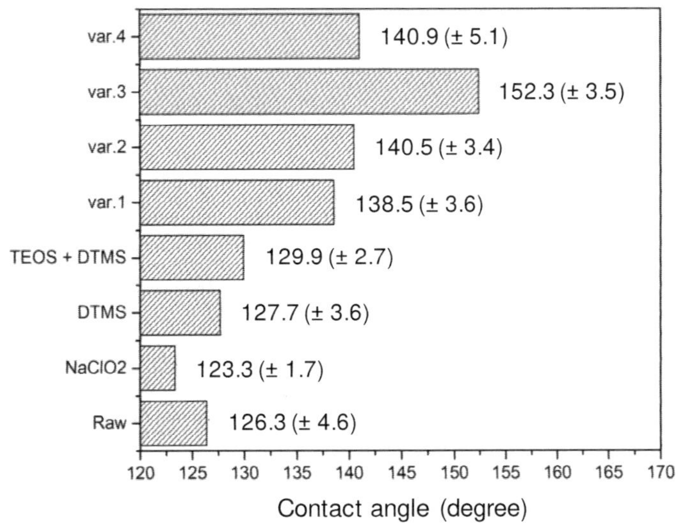

Figure 4. Water contact angle of raw kapok fibre, $NaClO_2$-pretreated, DTMS coated, and TEOS-DTMS coated samples, respectively

Reusability analyses have been carried out to test the structural integrity enhancement of the sponge during ten times recycle. Figure 5 shows the oil sorption capacity of pristine kapok fibres vs. modified sponges. The pristine kapok loses 55% oil sorption capacity at the 4th cycle and could only retain 29.8% of its oil sorption capacity at the end cycle. The pristine kapok fibres tend to crumble and hard to be recovered as a whole. Most modified sponges have better reusability than the pristine kapok.

Var. 2 (strand scission, 0.2 %-w fibre in water) shows the highest oil sorption capacity at the first cycle due to the highest pore volume; although the sponge is not crumbled, it directly loses its integrity in the second cycle. Comparing Var. 1 and Var. 3, strand scissions enhance the air cavities inside the sponge that can be substituted with oil. In terms of their structural integrities throughout the cycles, the sponge without strand scission shows slightly better reusability compared with that with strand scission.

The effect of TEOS as a precursor for DTMS coating can be analyzed from the comparison of Var.3 and Var.4. Overall, the addition of TEOS results in a better reusable sponge. TEOS itself enhances the hydrophilicity of the kapok fibre, making the fibre suspension more homogeneous for the freeze-drying process. A homogeneous fibre suspension causes the fibres to intertwine and results in stronger structural integrity of the sponge. Var.3, therefore, is the best variation with oil-sorption capacity of 48.6 g-oil/g-fibre and oil-sorption retention of 62.6% even after ten reuse cycles.

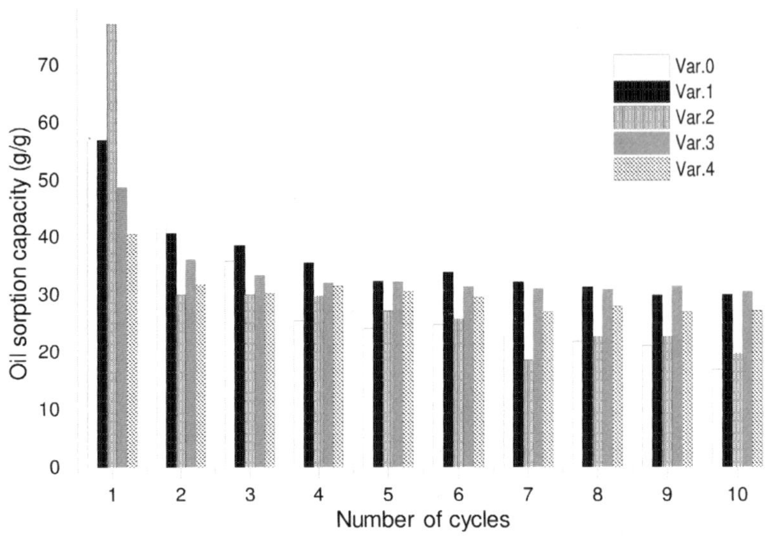

Figure 5. Oil sorption capacity of pristine kapok fibre and modified sponges after ten reuse cycles

4. Conclusion

Kapok fibre sponge shows the optimum oil sorption capacity after oxidation using 0.5 %-w sodium chlorite ($NaClO_2$) solution followed by sol-gel coating with tetraethylorthosilicate (TEOS) and sponge formation via freeze-drying method successfully enhance the structural integrity of the fibre without significantly reducing its hydrophobicity. Further coating the sponge with organosilane such as dodecyltrimethoxysilane (DTMS) using the chemical vapor disposition method recovers the superhydrophobicity of the fibre. The best performance achieves an oil-sorption capacity of 48.6 g-oil/g-fibre and can maintain 62.6% of its oil-sorption capacity even after ten reuse cycles.

References

[1] Rengasamy R S, Das D and Karan C P 2011 Study of oil sorption behavior of filled and structured fibre assemblies made from polypropylene, kapok and milkweed fibres *J. Hazard. Mater.* **186**(1) 526-32

[2] Hori K, Flavier M E, Kuga S, Lam T B T and Iiyama K 2000 Excellent oil absorbent kapok [Ceiba pentandra (L.) Gaertn.] fibre: Fibre structure, chemical characteristics and application *J. Wood Sci.* **46** 401-4

[3] Ali N, El-Harbawi M, Jabal A A and Yin C Y 2012 Characteristics and oil sorption effectiveness of kapok fibre, sugarcane bagasse and rice husks: oil removal suitability matrix *Environ. Technol.* **33**(4–6) 481-86

[4] Lim T T and Huang X F 2007 Evaluation of hydrophobicity/oleophilicity of kapok and its performance in oily water filtration: comparison of raw and solvent-treated fibres *Ind. Crop. Prod.* **26**(2) 125–34

[5] Abdullah M A, Rahmah A U and Man Z 2010 Physicochemical and sorption characteristics of Malaysian Ceiba pentandra (L.) Gaertn. as a natural oil sorbent *J. Hazard. Mater.* **177**(1–3) 683-91

[6] Dhanabalan V and Laga S K 2015 Kapok Fibre: A Perspective Fibre *Semantic Scholar Proceeding*

[7] Wang J T, Zheng Y A and Wang A Q 2012 Superhidrofobik kapok fibre oil-absorbent: Preparation and high oil absorbency *Chemical Engineering Journal* **213** 1-7

[8] Wang J T, Zheng Y A, Kang Y R and Wang A Q 2012 Effect of kapok fibre treated with various

solvents on oil absorbency *Ind. Crops Prod.* **40** 178-84
[9] Wang J T, Zheng Y A, Kang Y R and Wang A Q 2013 Investigation of oil sorption capability of PBMA/SiO_2 coated kapok fibre *Chem. Eng. J.* **223** 632-7
[10] Li S H, Zhang S B and Wang X H 2008 Fabrication of superhydrophobic cellulose-based materials through a solution-immersion process *Langmuir* **24** 5585-90

Acknowledgments
This research has been financially supported by Research Programs, Community Services, and Innovation, Institut Teknologi Bandung (P3MI ITB).

IOP Conf. Series: Materials Science and Engineering **823** (2020) 012034 doi:10.1088/1757-899X/823/1/012034

The influence of Ag concentration on the antibacterial properties of plastic made from silica immobilized with EDTA-Ag combined with chitosan

Hermania Em Wogo[1,2], Marini T. Tanuab[1], Luther Kadang[1], Antonius R.B.Ola[1]

[1]Chemistry Department, Faculty of Science and Engineering, Nusa Cendana University 85118 Indonesia
[2]Laboratory of Chemistry, Nusa Cendana University 85118 Indonesia

E-mail: hermania_wogo@yahoo.com

Abstract. Antibacterial plastic is made by using composites silica gel immobilized EDTA-Ag combined with chitosan. Silica gel was formulated with three variations of Ag concentrations (i.e., 10^{-3}, 10^{-4}, and 10^{-5} M) and then tested for its antibacterial activity against *Escherichia coli* and *Staphylococcus aureus*. Result of the antibacterial test showed that the immobilized EDTA-Ag silica gel product could inhibit the growth of *Staphylococcus aureus* and *Escherichia coli* with the diameter of zone inhibition around 15.9 and 15.6 mm respectively at the lowest Ag concentration of 10^{-5}M. This result showed that higher concentration of Ag led to the decrease of antibacterial activity. The obtained products were then synthesized into plastic with variations of chitosan weight of 0.3 and 0.7 grams. The antibacterial test confirmed that the plastic had strong antibacterial properties with the diameter of zone inhibition for *Staphylococcus aureus* and *Escherichia coli* was 16.7 and 15.9 mm, respectively. The percentage of degradation was observed at 6.02 to 63.17% while water absorption test for plastic with 0.3 gram chitosan was 86.51% and for plastic with 0.7 gram chitosan was 16%. This indicated that more chitosan additions were able to expand antibacterial properties and reduce the absorption of plastic against water.

1. Introduction

Bacteria are microorganisms that can live and affect the life of cells in the human body. A lactic acid bacterium is an example of probiotic bacteria that play an essential role in human intestine [1]. In contrast, *Escherichia Coli* and *Staphylococcus aureus* bacteria are examples of pathogenic bacteria that can cause diseases in the human body. One of the efforts to overcome bacterial growth is by using antibacterial materials that can be synthesized from organic or inorganic materials. Inorganic metal ions such as Ag, Zn, Cu and Hg which are combined with porous materials like silica and clay are often used to produce antibacterial products such as antibacterial glass and medical tools [2].

Silica gel produced from rice husk has several advantages compared to other porous materials such as having finer grains, affordable at low cost, and high abundance of raw materials. Silica can be used as an antibacterial material through modification stages using other bactericidal substances.

Adoe *et al.* [3] used silica gel immobilized EDTA-Ag as an antibacterial material against *E-coli*. The results showed that silica gel immobilized EDTA-Ag was bactericidal against *E-coli* up to 99% at the concentration of 10 mg/mL. Wogo *et al.* [4] using silica gel immobilized EDTA-Ag and chitosan to produce products in the form of antibacterial plastic, which can inhibit the growth of *S. aureus* and *E. coli* up to 99.99%.

Biodegradable plastic is a type of plastic that can be decomposed in the environment and has good waterproof properties. A research conducted by Firdaus and Anwar [5] utilizes solid-liquid waste tapioca flour as a raw material for making biodegradable plastic films obtained that their plastic products can be degraded perfectly on the second week. Therefore the presence of biodegradable plastic will reduce industrial waste. Moreover, biodegradable plastic that has antibacterial properties will add the higher economic value of plastic especially in food and health packagings.

2. Experimental Procedures

2.1. General tools and materials

The tools used in this research are glassware, petri dish, 60 mesh sieve, magnetic stirrer, Laminar water flow, and Colony counter (SCAN® 500, version 6.1.2.0). While the materials used in this research are the rice husk, HCl, NaOH, CH_3COOH, $AgNO_3$, EDTA, Chitosan, the bacteria *Escherichia coli,* and *Staphylococcus aureus* strains.

2.2. Procedures

a. Rice husk ash preparation

Clean and dry rice husks were roasted to black and ignited for 4 hours at 700 °C. The ash obtained was crushed and then sifted with a 60 mesh sieve. 20 g of rice husk ash was washed with 120 mL of HCl 6 M continued with stirring for 1 hour, then rinsed with distilled water until neutral and then dried in an oven at 110 °C for ± 2 hours.

b. Preparation of sodium silicate

As much as 20 g of rice husk prepared were dissolved with 330 mL of 2M NaOH and then boiled up to thicken and were heated for 30 minutes at 500 °C.After that 200 mL of distilled water was added cooled overnight and filtered.

c. Synthesis of EDTA-Ag immobilized silica gel

A total of 20 mL sodium silicate solution was put into a plastic container, 5 mL of EDTA 0.005 M solution and 5 mL of $AgNO_3$ 10^{-3} solution, 10^{-4} M, 10^{-5} M was added. The solution mixture was homogenized and HCl solution 3 M was added dropwise until gel was formed. The gel was left overnight, filtered, washed with distilled water until neutral and dried in an oven for 1 hour at 110 °C. The obtained product was crushed, sifted with a 60 mesh sieve and tested for antibacterial activity.

d. Antibacterial assay on *Escherichia coli* and *Staphylococcus Aureus*

A total of 25 g NA was dissolved in distilled water and heated to thicken and placed in an Erlenmeyer. Furthermore, the tools and materials were sterilized at 121 °C for 30 minutes and stored in Laminar airflow. Pure cultures of *Escherichia coli* and *Staphylococcus aureus* bacteria were dissolved in a test tube containing 2 mL NA media and grown in an incubator for 18-24 hours at 37 °C. Bacteria produced by culture were taken 1 eye dose and put in 9 mL NaCl 0.9% then diluted. The results of the dilution were transferred 1 mL into the petri dish and then poured the media so that it was 1 cm thick and left.

e. Preparation of antibacterial plastics

A total of 0.3 g and 0.7 g chitosan each dissolved in 50 mL CH_3COOH 1% stirring for 24 hours then added silica gel solids result of the procedure c. The mixture solution was filtered and poured into a petri dish (15 cm in diameter) and dried at 65 °C for 7 hours. The formed plastic was released from the petri dish using 3 M NaOH and washed with distilled water until the pH is neutral then dried at room temperature. Plastics were tested for antibacterial activity using the paper disc method.

f. Qualitative test of plastic products

1. Biodegradation test

   Carried out invitro in the ground for a certain period of time. The samples were then removed, cleaned with distilled water, dried, weighed by the weight of the remaining samples and calculated as percent degradation using the formula [6] :

$$\%\,Degradation = \frac{w_1 - w_2}{w_1} x\ 100\ \% \tag{1}$$

   With w_1 is the weight of the plastic before the degradation process, and w_2 is the weight of the plastic after the degradation process.

2. Test of water absorption on plastic products

   The samples were weighed and then dripped with water 5 times along the surface of the sample and left to stand for 2 minutes, then drained with tissue and weighed the final mass.

$$\%\,Excessweight = \frac{w_1 - w_2}{w_1} x\ 100\ \% \tag{2}$$

   With w_1 is the initial sample weight and w_2 is the final sample weight.

3. Results and Discussion

a. Rice husk and ash preparation

The rice husk was collected from Noelbaki Village in Kupang eastern Indonesia. Changing the black color from the results of casting (Figure 1) indicates that these organic compounds have not been oxidized completely.

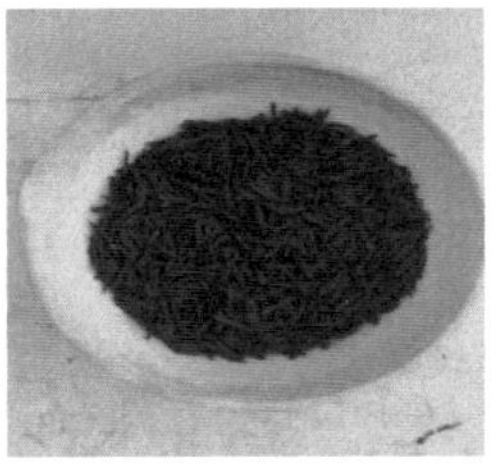

Figure 1. Results of rice husk processing

According to Kurniawati [7] ignition at temperatures above 700 °C will produce silica in the form of stable crystals. While combustion at temperatures below 700 °C will produce very low content of silica. Kalapathy and Proctor [8] stated that silica reactivity is influenced by the temperature of ignition with its optimum reactivity at a temperature of 550 - 700 °C. Therefore ignition was carried out at 700 °C resulting in unstable amorphous silica. The result of white ignition as shown in Figure 2 according to Bolle's statement [9] that the white color of the husk ash produced indicates high silica content.

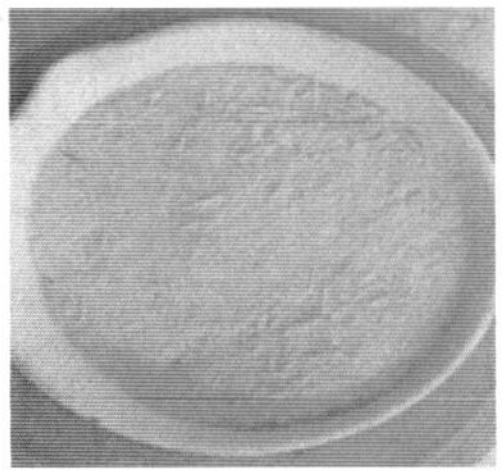

Figure 2. Ignition result of rice husk

Rice husk ash was activated using HCl 6 M to dissolve metal and non-metal oxides which are impurities in husk ash such as Na_2O, MgO, K_2O, CaO, P_2O_5 and Fe_2O_3 [10]. Then wash it to neutral pH with distilled water to release ash from chloride ions.

b. Preparation of sodium silicate solution
The rice husk ash was reacted with 2 M NaOH solution. According to Ngatidjo*et al.,* [11] the initial indicator of sodium silicate solution was that it was slippery when the solution hits the surface of the skin. The mechanism for the reaction of sodium silicate formation is illustrated in Figure 3.

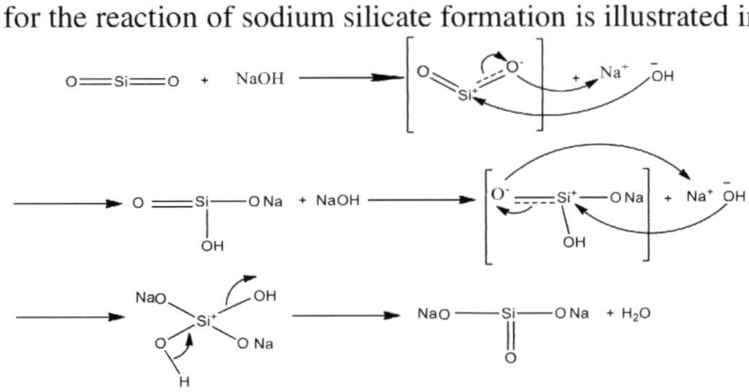

Figure 3. Formation reaction mechanism

c. Manufacture of EDTA-Ag immobilized silica
Immobilized silica gel EDTA-Ag was prepared through a sol-gel process because it has several advantages including high levels of thermal and mechanical stability, good solvent durability and easy surface modification. HCl was added dropwise until the gel was formed. Mechanisms of reaction for the formation of Silica Gel is depicted in Figure 4.

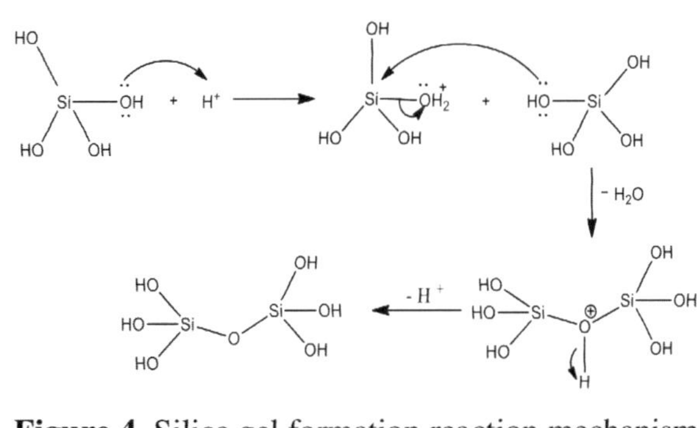

Figure 4. Silica gel formation reaction mechanism

In the silica gel, there is a coordination bond between the metal Ag and the electronegative oxygen atom from the OH-group on the surface of silica and EDTA. EDTA also acts as a ligand carrying Ag^+ ions into the bacterial environment so that the exchange of EDTA ligand will occur with bacterial biological ligands [12]. Oxygen atom acts as a Lewis base that will donate a pair of electrons to the metal Ag which acts as a Lewis acid by providing empty orbitals and forming Ag-O complex bonds. The estimated mechanism of binding of Ag metal to silica and EDTA is illustrated in Figure 5.

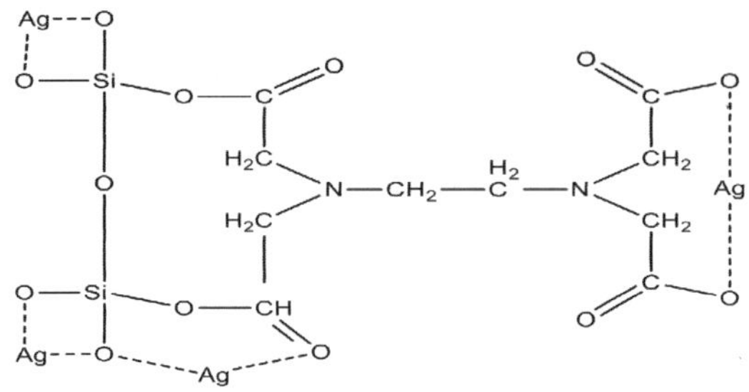

Figure 5. Estimated bound mechanism metal Ag on silica and EDTA

d. Antibacterial test against *Staphylococcus aureus* and *Escherichia coli*
Tests were carried out using the paper disc method with the aim of knowing the Large Area Diameter Obstacle (DDH). The results of the antibacterial activity test can be seen in Table 1.

Table 1. Regional diameter inhibits bacterial growth in EDTA-Ag immobilized silica gel products.

Bacterial Type	Inhibited Area Diameter (mm)				
	Sample type			Positive control (Tetracycline)	Negative Control (Aquades)
	A	B	C		
Staphylococcus aureus	13,7	14,8	15,9	17,8	0
Escherichia coli	12,2	14,1	15,6	17,4	0

Information: A, B, and C respectively are immobilized EDTA-Ag products with variations in concentration Ag 10^{-3} M, 10^{-4} M, and 10^{-5} M.

Based on the available data it can be seen that the inhibitory power in *S. aureus* bacteria is greater than *E. coli*. This is due to differences in the structure of cell walls that determine the amount of activity of antibacterial compounds. Gram-positive bacteria have a simpler cell wall structure than gram-negative bacteria, making it easier for antibacterial compounds to enter the cell. Gram-negative bacteria contain outer membranes that can block the entry of large molecules [13].

The results obtained have the potential as a strong antibacterial in accordance with the classification of responses to bacterial growth inhibition areas proposed by Pratama [14] that if the inhibition area diameter <5 (weak), 5-10 (medium), 10-20 (strong),> 20 (very strong).

e. Antibacterial plastic products and antibacterial tests on *Staphylococcus aureus* and *Escherichia coli.*

1. Antibacterial plastic products

The results of antibacterial testing with the greatest inhibitory power were then used to inhibit or kill the growth of *Staphylococcus aureus* and *Escherichia coli* bacteria. Each chitosan amount was dissolved in 50 mL of 1% acetic acid with stirring for 24 hours with the aim that chitosan could dissolve completely in acetic acid. The use of acetic acid as a solvent was based on the character of chitosan which is alkaline and mostly dissolves in the acidic type organic or dilute minerals through protonation of free amino groups at pH> 6.5.

The next step is to add each mixture of acetic acid - chitosan solution with silica solids immobilized EDTA-Ag with the optimum concentration of Ag in antibacterial testing is 10^{-5} M. Then filtering is carried out in order to separate the solution from the sample's sediment that may not dissolve. Then printed into a petri dish and dried in an oven at 65 °C for 7 hours so that the solution can form plastic, then released using 3 M NaOH. The resulting plastic is then washed using distilled water periodically in order to neutralize the plastic sample previously removed using NaOH. The plastic is then dried at room temperature so that it is free of excess water content caused by the previous washing process, in addition to not damaging the plastic structure produced. The plastic obtained can be seen in Figure 6.

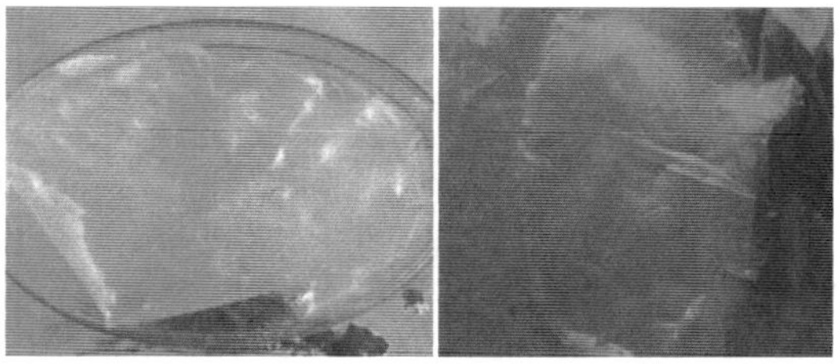

(a) (b)

Figure 6. Plastic (a) 0.3 gram chitosan and (b) 0.7 gram chitosan

Estimates of the mechanism that occurs between immobilized EDTA-Ag and chitosan products is illustrated in Figure 7.

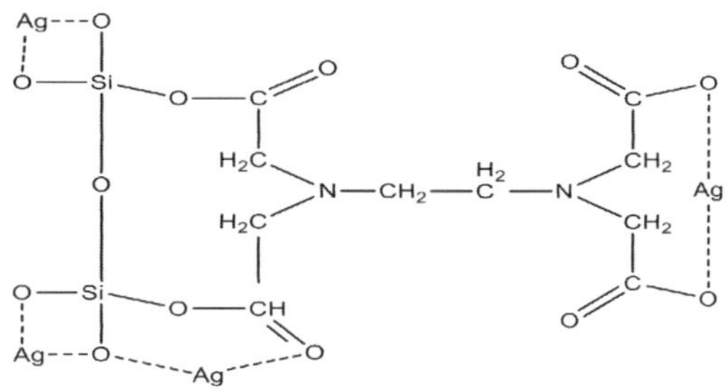

Figure 7. Estimated reaction mechanism between immobilized EDTA-Ag and chitosan

2. Antibacterial test against *Staphylococcus aureus* and *Escherichia coli*

The synthesized plastic then tested for antibacterial activity against *Staphylococcus aureus* and *Escherichia coli* bacteria. The complex formation reaction between chitosan and metal Ag occurs through covalent coordination by N atoms from NH_2 groups and –OH groups from chitosan. The Ag metal ion acts as a Lewis acid (electron acceptor), while nitrogen from chitosan acts as a Lewis base that will donate a pair of electrons to coordinate with the metal. After complex formation, the positive charge density on chitosan will increase and cause the antibacterial activity of plastic to increase. The results of the antibacterial activity test of plastic can be seen in Table 2.

Table 2. Area diameter inhibits bacterial growth in plastic products

Bacterial Type	Inhibited Area Diameter (mm)			
	Sample Type A (0,3 gr)	Sample Type B (0,7 gr)	Positive Control (Tetracycline)	Negative Control (Aquades)
Staphylococcus aureus	13	16,7	17,8	0
Escherichia coli	14,1	15,9	17,4	0

Based on the data in Table 2 shows that chitosan is one of the ingredients that have the potential as an antibacterial agent. This can be seen from the addition of chitosan which enlarges the inhibitory area of the bacteria. The factors that cause large inhibition in both bacteria are the cell wall structure factors of bacteria. Gram-negative bacteria have a double membrane containing little peptidoglycan which makes the cell wall thinner but has thick lipopolysaccharide resulting in difficult interactions between bacterial cell walls and antibacterial compounds while gram-positive bacteria are composed of 90% peptidoglycan and have a single membrane that makes the cell wall easily attacked by antibacterial compounds.

f. Plastic sample testing
In this study, two tests were carried out, namely biodegradation testing which aims to determine the time of degraded plastics using the method of landfilling in the soil and testing the absorption of water on plastic products which was carried out to determine the amount of absorption of plastic against water. Data from the biodegradation test and plastic product resistance test on water are shown in Tables 3 and 4.

Table 3. Biodegradation test

Plastic type	Planting Variations								
	Initial Plastic Weight (w1), (gram)			Plastic Weight After Degradation Process (w2), (gram)			Degradation (%)		
	1	2	3	1	2	3	1	2	3
A	0,0399	0,0822	0,1413	0,0375	0,0665	0,0981	6,02	19,10	30,57
B	0,1129	0,0723	0,2354	0,0964	0,0456	0,0867	14,61	36,93	63,17

Information: A and B are plastic with chitosan variations of 0.3 grams and 0.7 grams. While the numbers 1,2 and 3 are plastic that is placed on the ground surface, piled up at a depth of 5 cm and 10 cm.

Table 4. Water absorption test on plastic products

Plastic type	Plastic Size (cm)	Initial Weight (w1), (gram)	Final Weight (w2), (gram)	Excess Weight(%)
A	2 x 2	0,0341	0,0636	86,5103
B	2 x 2	0,0488	0,0567	16,1885

Information: A and B are plastic with chitosan variations of 0.3 grams and 0.7 grams.

Based on the data in Table 3, it shows that the more variations in plastic are planted, the more plastic mass will decrease. This is because the depth of the soil is one of the factors that influence the degradation process. In Table 4, testing is carried out with the intention of obtaining plastic products that have low water absorption. Based on existing data, plastic with 0.7 gram chitosan content has lower water absorption than plastic with 0.3 gram chitosan content. In plastics with chitosan content, there is less chance that air is trapped in plastic, making plastic more absorb water.

4. Conclusion

The optimum concentration of Ag in antibacterial plastic products was 10^{-5}M, with a large diameter inhibition zone of 15.9 mm for *Staphylococcus aureus* and 15.6 mm for *Escherichia coli*. Higher amount of chitosan led to the greater diameter of zone inhibition of *Staphylococcus aureus* from 13 mm to 16.7 mm and *Escherichia coli* from 14.1 mm to 15.9 mm. The absorption of water by plastic was 86.51% for plastic with 0.3 gram chitosan and 16.18% for 0.7 gram chitosan. Plastic made from immobilized silica gel composites EDTA-Ag and chitosan was able to be degraded in the environment within 14 days with the percentage of degradation around 6.02 - 63.17%.

5. References

[1] Anastiawan 2014 Isolasi dan karakterisasi bakteri probiotik yang berasal dari usus itik pedaging Anas domesticus *Skripsi* (Makassar: Universitas Hasanuddin)
[2] Li B, Yu S, Jim Y H and Shi S 200 Antibacterial vermiculite nano-material *Journal of Minerals and Materials Characterization and Engineering* **1** 61-8
[3] Adoe K A, Wogo H E and Siburian R A F 2016 Sintesis Silika Gel Terimobilisasi EDTA-Ag sebagai Bahan Antibakteri Escherichia coli *Prosiding Seminar Nasional 1 Laboratoium Riset Terpadu Universitas Nusa Cendana* (Kupang) pp 187-96
[4] Wogo H E, Kedjo G Y, Selan, O Th E and Ola A R B 2017 Sintesis Plastik Antibakteri *Prosiding Seminar Nasional 2 Laboratorium Riset Terpadu Universitas Nusa Cendana* (Kupang) pp 15-23
[5] Firdaus F and C Anwar 2004 Potensi limbah padat-cair industri tepung tapioka sebagai bahan baku film plastik biodegradabel *Logika* **1**(2) 38-44
[6] Rohaeti E 2009 Karakterisasi Biodegradasi Polimer *Prosiding Seminar Nasional Penelitian, Pendidikan dan Penerapan MIPA* (Yogyakarta) pp 248-57
[7] Kurniawati W 2003 Sintesis dan karakterisasi silika gel kering dari abu sekam padi menggunakan natrium hidroksida dan asam sitrat *Skripsi* (Yogyakarta: FMIPA UGM)
[8] Kalapathy U and Proctor A 2000 A simple method for production of silica from rice hull ash *Bioresource Technology* **73** 257-62
[9] Bolle T C M 2010 Sintesis silika gel terimobilisasi dithizon dari abu sekam padi *Skripsi* (Kupang: Universitas Nusa Cendana)
[10] Nenohaifeto A N, Wogo H E and Tawa B D 2016 Pemanfaatan Silika Gel Terimobilisasi EDTA-Ag sebagai Bahan Antibakteri Staphylococcuss aureus *Prosiding Seminar Nasional 1 Laboratorium Riset Terpadu Universitas Nusa Cendana* (Kupang) pp 99-106
[11] Ngatidjo, Faried F and Lestari I 2011 Pemanfaatan Abu Sekam Padi (ASP) Payo dari Kerinci sebagai sumber silika dan aplikasinya dalam ekstraksi fasa padat ion tembaga (II) *Jurnal Penelitian Universitas Jambi Seri Sains* **13**(2) 47-52
[12] Rizzotto M 2012 Metal complexes as antimicrobial agents *A Search for Antibacterial Agent* (Argentina: National University of Rosario) pp 82-5
[13] Kalemba A and Kunicka A 2003 Antibacterial and antifungal properties of essential oil *Current Medical Chemistry* **10** 813-29
[14] Pratama M R 2005 Pengaruh ekstrak serbuk kayu siwak (Salvadora persica) terhadap pertumbuhan bakteri Streptococcus mutans dan Staphylococcus aureus dengan metode difusi agar *Laporan Hasil Penelitian Program Studi Biologi Fakultas MIPA Teknologi Sepuluh Nopember* (Surabaya: Institut Teknologi Sepuluh Nopember)

IOP Conf. Series: Materials Science and Engineering **823** (2020) 012035 doi:10.1088/1757-899X/823/1/012035

Oil palm empty fruit bunch ash valorization through potassium extraction

Tjokorde W. Samadhi[1], Winny Wulandari[1] and Kezia R. Tirtabudi[1]

[1]Chemical Engineering Department, Faculty of Industrial Technology, Bandung Institute of Technology, Jl. Ganesha 10, Bandung, Indonesia

E-mail: winny@che.itb.ac.id

Abstract. Oil palm empty fruit bunch (OPEFB) is a by-product of palm oil processing that is considered solid waste. It is produced in a large amount, counting up to 23% of the fresh fruit bunch (FFB). The ash content in OPEFB comprises a potassium compound. Potassium can be potentially recovered as potassium carbonate by extracting it with water. This study aims to investigate a method to recover potassium from OPEFB ash as potassium carbonate for further processing into potassium-based fertilizer. This research consists of pyrolysis process of OPEFB into char, then the char is analyzed by TGA in order to determine the ashing temperature. The temperature of 250°C, 400°C and 550°C are chosen for the ashing (combustion) process. The resulted ash is then extracted by water with the Feed to Solvent ratio of 1:4, 1:5, and 1:6, respectively. From the AAS analysis result, the highest potassium concentration is obtained at temperature of 400°C with F:S ratio = 1:6, with the potassium concentration of 49,94 %-w/w. This value is suitable for liquid fertilizer.

1. Introduction

Indonesia is listed as the country with the largest of palm oil production in the world [1] and it is estimated that the amount of production will continue to increase. The area of oil palm plantations in Indonesia is estimated at 12.3 million Ha in 2017 [2], while the increase of production (crude palm oil) reaches 35.4 million tons [2].

The processing of fresh fruit bunch (FFB) into crude palm oil (CPO) produces several types of waste, including oil palm empty fruit bunch (OPEFB), mesocarp fiber (MF), palm kernel shell (PKS), palm kernel meal (PKM), and palm oil mills effluent (POME) [3]. There has been some usage of PKS and MF to be the boiler fuel used in the palm oil processing plant, and it produces biomass waste as oil palm ash. Other oil palm wastes have established downstream use. Fronds and trunks are used for animal feedstock, while kernel shells are used as fuel for the steam boiler in palm oil mills [4].

It has been reported that oil palm ash has been used as crude fertilizer, novel adsorbent, concrete pozzolanic cement material [5] and blend cement paste [6]. However, OPEFB alone has not been used as fuel to the boiler in oil plant processing plant due to its alkali content that promotes slagging and fouling [7,8]. Alkali-induced slagging in biomass-fired furnaces is the unwanted combustion side effect caused by alkali metal aerosols released from biomass [7]. Typically, OPEFB is burned and partially spread over empty land as a mulch.

Among the several solid wastes from oil palm plantation, OPEFB waste has the highest number proportion with a total of 23% for each FFB [9]. OPEFB is also known to contain potassium, and it has potential to serve as the sources of potassium, including a potential source for fertilizer. From the several works of literature, it is reported that OPEFB contains 30-46 % of K_2O [10], [11], [12] mainly in the form of carbonate.

Potassium extraction from wood ash has been studied, for example, is a study regarding potash production from bamboo ash [13]. From this study, is it known that extraction from bamboo ash produced 0.7 %-mass of K_2O in solution and it is suggested that bamboo ash extract has a potential to be used as the intermediate for fertilizer product [13]. Therefore, OPEFB's ash is also expected to be used as a source for potassium fertilizer which can be used for oil palm plantations themselves. This study aims to investigate a method to recover potassium from OPEFB ash as potassium carbonate for further processing into potassium-based fertilizer.

2. Materials and Method

The materials used in this study were oil palm empty fruit bunch (OPEFB) demineralized water. OPEFB was obtained from East Kalimantan.

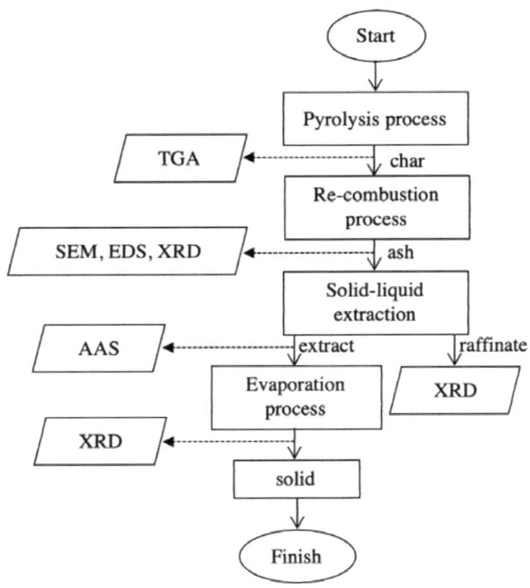

Figure 1. Block diagram of OPEFB extraction process.

Figure 1 shows the block diagram of OPEFB extraction process. This study consisted of several stages, which are as follows:

(1) Pyrolysis process of OPEFB. Pyrolysis of OPEFB was carried out in a 40 cm diameter pyrolysis reactor. The char was then characterized using TGA (Thermogravimetry Analysis/Linseis) in order to determine the melting point and temperature range of the OPEFB ash.

(2) Ashing/re-combustion process. The ashing is carried out in order to convert char into ash. The temperature chosen for the ashing process was determined from the TGA result. The ashing process was carried out in a muffle furnace (Nabertherm). This further combustion process aims to remove substances as the volatile matter and fixed carbon contained in the OPEFB char. Then the ash is crushed into 100 mesh. The ash was then analyzed using SEM/EDS (SU3500).

(3) Solid-liquid extraction process of ash. Ash was extracted with demineralized water. The extraction process is carried out in a water-bath with a temperature of 80 °C. The extraction is a single-stage solid-liquid extraction. Firstly, 50 g of each ash sample was weighed and

IOP Conf. Series: Materials Science and Engineering **823** (2020) 012035 doi:10.1088/1757-899X/823/1/012035

thoroughly mixed with required volume (200 ml, 250 ml, and 300 ml) of demineralized water. Then extraction process is carried out for 4 hours at temperature of 80 °C. Subsequently, filtration is carried out by using vacuum filtration to separate the extract solution from raffinate. The extract solution obtained is then analyzed for its potassium content using Atomic Absorption Spectrophotometry (AAS, Perkin-Elmer).

(4) Evaporation of extract solution. Subsequently, the extract solution is evaporated by heating the solution above a hotplate until it formed solid phase.

(5) Characterization of OPEFB ash and potassium salt. The phase characterization of the ash and the resulted salt was carried out by XRD analysis (Bruker).

The parameters conducted in this study include ashing temperature in muffle furnace and the ratio of feed (ash) to solvent (demineralized water) (F:S). The parameter that is investigated the concentration of potassium in the solution. The variable parameters are as follows:

- ashing temperature : 250 °C, 400 °C, and 550 °C
- Ratio of ash to demineralized water (F:S) : 1:4, 1:5 and 1:6

3. Result and Discussions

3.1. OPEFB pyrolysis

Pyrolysis of OPEFB is carried out by pyrolysis using the furnace in the open area. This process aims to convert the biomass of OPEFB into char. The average yield of char to the mass OPEFB biomass (db) is 11.6 %. The char of OPEFB is then ground using ball mill to reduce its size and refine its texture to become powder. Subsequently, the char of OPEFB was analyzed using TGA. The TGA and DSC curves of OPEFB can be seen in Figure 2.

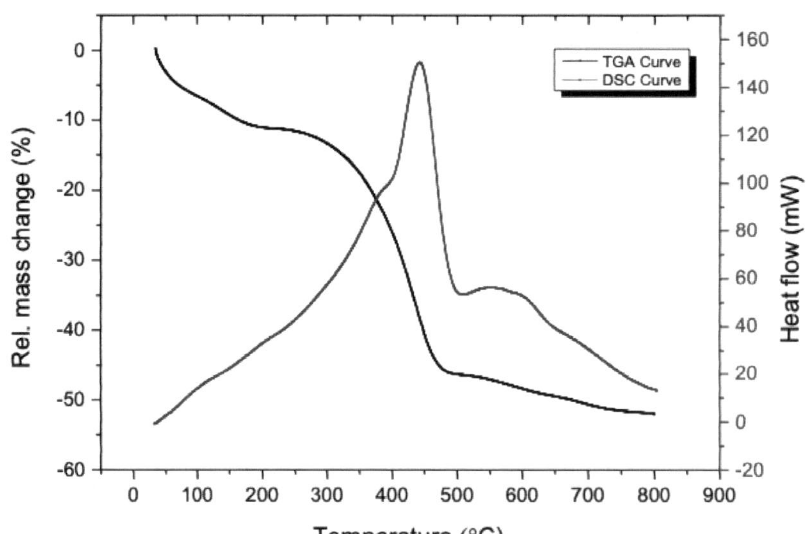

Figure 2. TGA and DTG analysis result of OPEFB's char.

The TGA curve in Figure 2 it is stated in black while the DSC stated in blue. The first peak in DSC curve shows the mass loss of 11 % seen in TGA curve. This loss of mass shows the loss of moisture content in OPEFB's char. Then the second peak is seen at a temperature of 435 °C and is the maximum rate of around 0.054 %-mass /°C. The loss of mass at this peak (230 °C – 500 °C) shows the loss of volatile matters and fixed carbon contained in the OPEFB's char. Then the pyrolysis process suddenly decreased at temperature of 500 °C to 800 °C, the process continued at a very slow rate.

OPEFB char was subsequently processed into char by ashing. Figure 3 shows the photograph of OPEFB ash in different ashing temperature: 250 °C, 400 °C, and 550 °C. The ash resulted from 250 °C is darker compared to 400 and 550 C due to its carbon or tar content.

The ash then analyzed by using XRD to determine the potassium mineral phase that found in the ash of OPEFB. Figure 4 shows a graph of XRD analysis result from the three ashing temperature. This is similar to ash from boiler (Oil Palm Ash /OPA), which contains KCl, sylvite [14]. For OPA, however, it also contains a metal-based substance of potassium molybdenum selenium ($K_2Mo_3Se_{18}$).

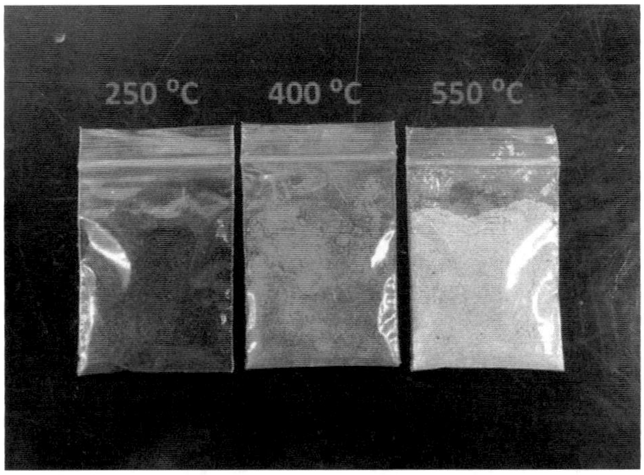

Figure 3. The OPEFB ash in different ashing temperature.

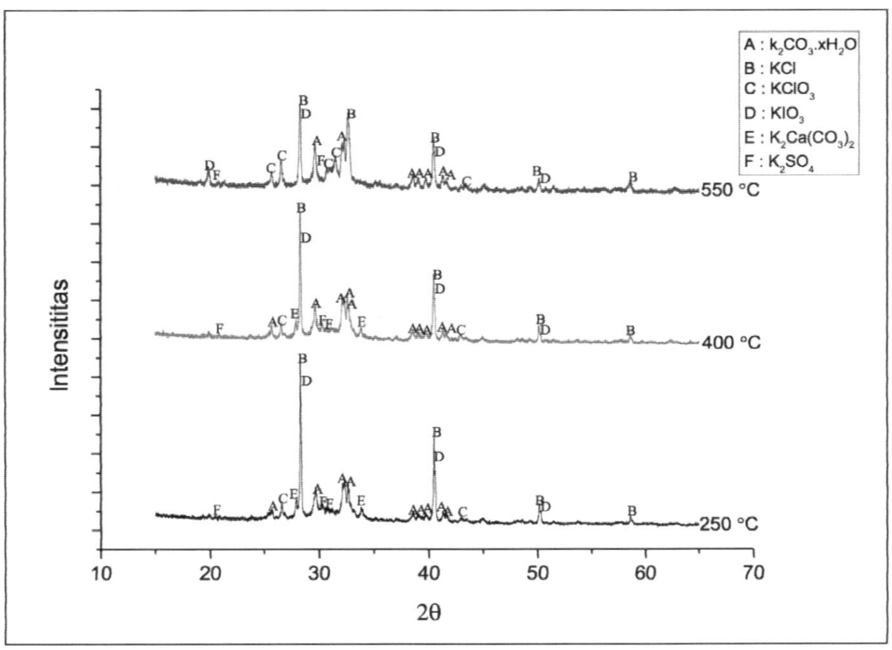

Figure 4. XRD analysis of OPEFB ash at different ashing temperature

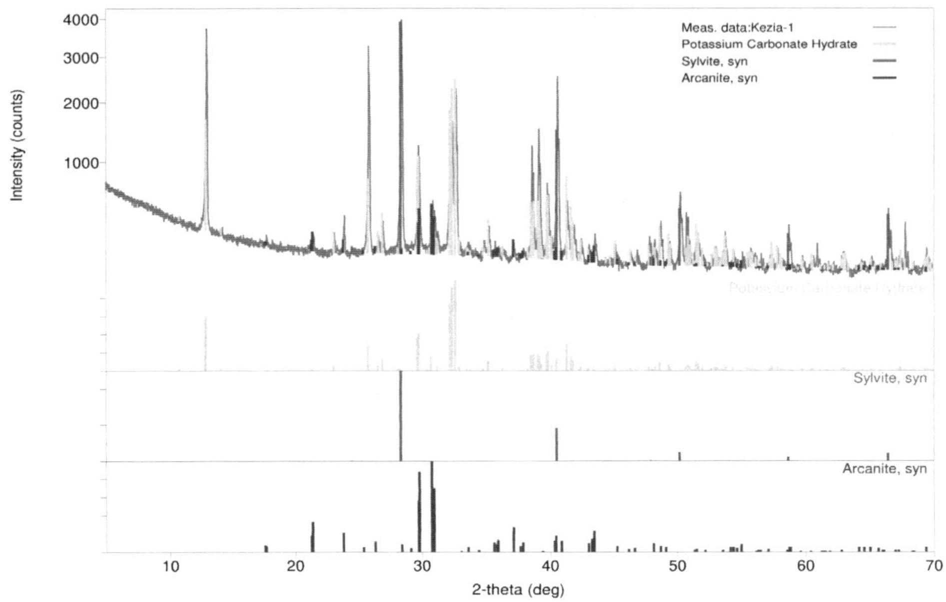

Figure 5. The XRD pattern of OPEFB's ash at ashing temperature of 400 °C and the solid to liquid ratio of 1:4.

Table 1. The XRD analysis of OPEFB's ash at ashing temperature 400 °C and the solid to

Phase name	Formula	Figure of merit	Phase reg. detail	DB card number
Potassium Carbonate Hydrate	K_2CO_3 $(H_2O)1.5$	0.655	ICDD (PDF2.DAT)	01-073-0470
Sylvite, syn	KCl	0.790	ICDD (PDF2.DAT)	00-041-1476
Arcanite, syn	K_2SO_4	1.601	ICDD (PDF2.DAT)	00-044-1414

Figure 5 gives additional information on XRD results, where potassium carbonate, sylvite (KCl) dan arcanite (K_2SO_4) are the major phase in the OPEFB ash. Based on Figure 4, its seen that the three types of ash with different ashing temperatures show a similar configuration. Based on the XRD result, the mineral phases found in OPEFB's ash are $K_2CO_3.xH_2O$, KCl, $KClO_3$, KIO_3, and $K_2Ca(CO_3)_2$.

Then the three ashes with each ashing temperature were analyzed by SEM/EDS to analyze its alkali content semi-quantitatively. Potassium (K) and K_2O compounds from the result of EDS analysis for each temperature are shown in Figure 6.

The overall result of EDS analysis of OPEFB's ash showed that the ash contained high potassium. At the ashing temperature of 250°C, the potassium content in OPEFB ash is approximately 58 % as K_2O. It increased to 67% as K_2O at ashing temperature of 400°C; and lowered to 57% as K_2O at higher ashing temperature, i.e., 550°C. This shows that extraction of OPEFB's ash has the potential to produce extracts with high potassium content.

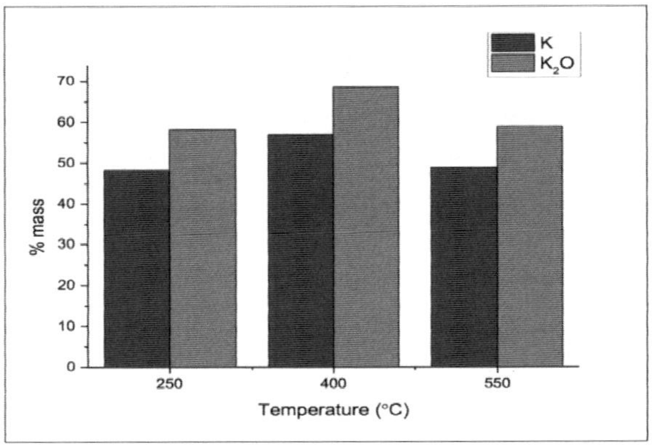

Figure 6. The EDS analysis of Potash (K) and K_2O compound in OPEFB's ash.

3.2. Extraction process of OPEFB's ash

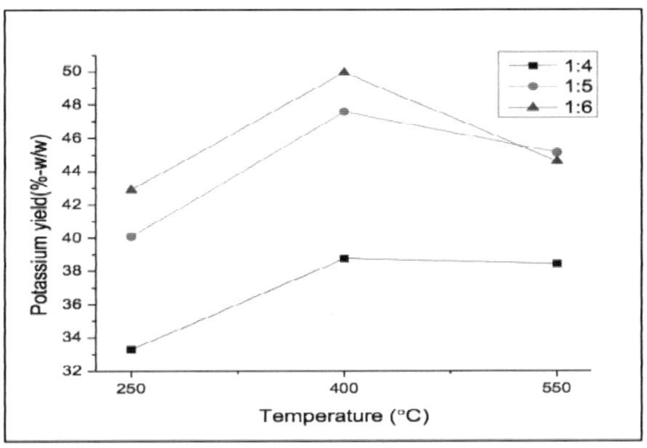

Figure 7. Potassium (K) and K_2O compound in OPEFB's ash after extraction with solid to liquid ratio of 1:4; 1:5, and 1:6.

Figure 8. Photographs of OPEFB ash leaching solution

Figure 7 shows potassium yield resulted from solid-liquid extraction at different ashing temperature, while Figure 8 shows the photographs of OPEFB ash leaching solution. The temperature of extraction is at 80°C. Based on AAS analysis, extracts with the highest potassium concentrations (ppm) were obtained on conditions at ashing temperature variation of 400 °C and variation ratio F:S = 1:4. It can be seen in Figure 7 that the highest potassium yield was produced at temperature variation of 400 °C and variation of F:S = 1:6. From the result, it can also be seen that the more volume of solvent used with the same amount of ash feed, the more amount of potassium is extracted.

One of the raffinates from extraction result was analyzed by XRD for its content. In this study, the raffinate samples that been analyzed were raffinate from variation temperature 550 °C with ratio F:S = 1:5. The result of this analysis can be seen in Figure 9. From this result, its shown that raffinate in the form of OPEFB ash is no longer contained crystalline phase from potassium metal. This shows that most of the potassium has been extracted. The dominant phase in the raffinate is amorphous, which contains quartz, lautite, maghemite, forsterite, and monticellite. The amorphous structure of the raffinates is caused by the leaching process, as the leaching process dilutes all the major crystalline phases in the OPEFB such as potassium carbonate, KCl and K_2SO_4.

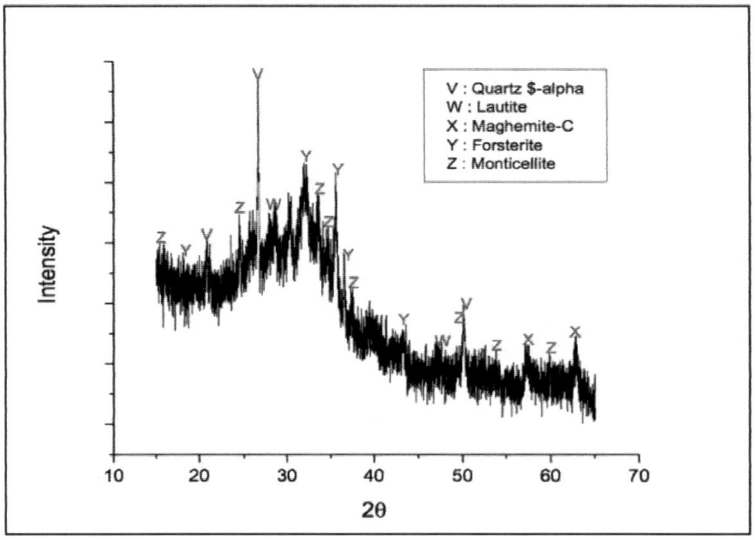

Figure 9. XRD analysis result from example raffinate.

3.3. Crystallization of leaching solution

The process of evaporation of leaching solution is carried out by evaporating the extract solution in beaker glass which is heated on a hotplate. In this study, the evaporation temperature used was maintained at 85 °C. The solid then analyzed using XRD to determine the potassium mineral phase found in the solid phase after evaporation from extract solution. From the XRD result in Figure 10, the solids resulting from the evaporation of extract solution has similar phase to the original ash. The additional information of phase analysis from XRD is shown in Table 2. The mineral phase found in the evaporation solids is KCl, K_2SO_4, and K_2CO_3. However, K_2CO_3 is formed in the hydrate phase ($K_2CO_3.xH_2O$).

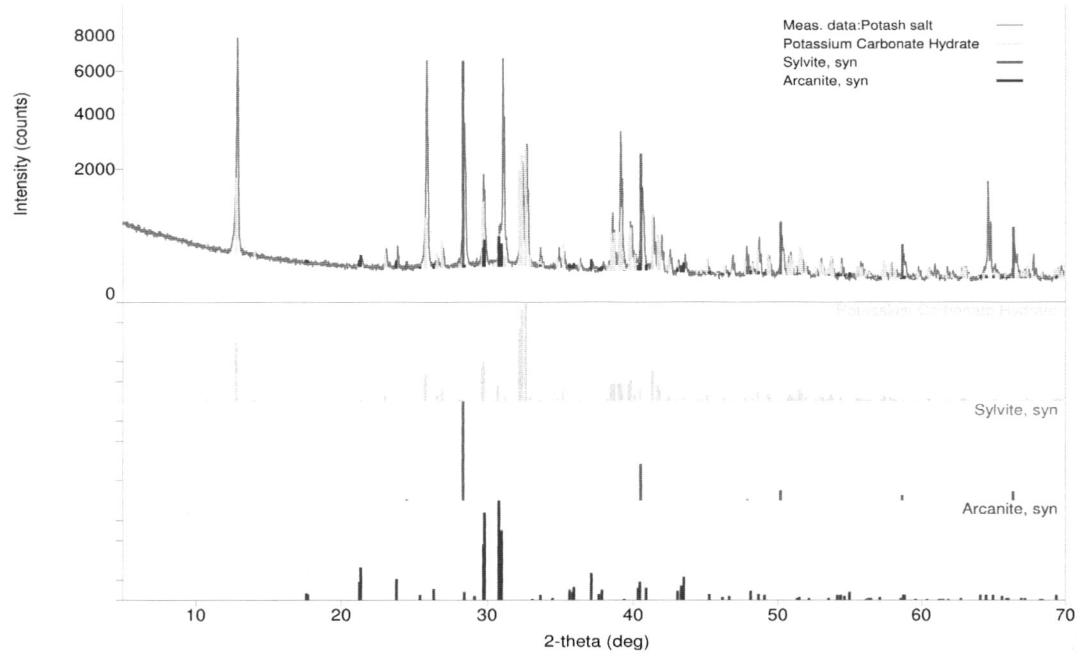

Figure 10. The XRD pattern of the potash salt.

Table 2. The XRD analysis of potash salt.

Phase name	Formula	Figure of merit	Phase reg. detail	DB card number
Potassium Carbonate Hydrate	$K_2CO_3.(H_2O)_{1.5}$	0.777	ICDD (PDF2.DAT)	01-073-0470
Sylvite, syn	KCl	0.895	ICDD (PDF2.DAT)	00-041-1476
Arcanite, syn	K_2SO_4	1.074	ICDD (PDF2.DAT)	00-044-1414

From Figure 11, it is also obtained that from the evaporation result, it can be seen that the highest solid mass is obtained at the varying temperature of 400 °C and variation of F:S = 1:4.

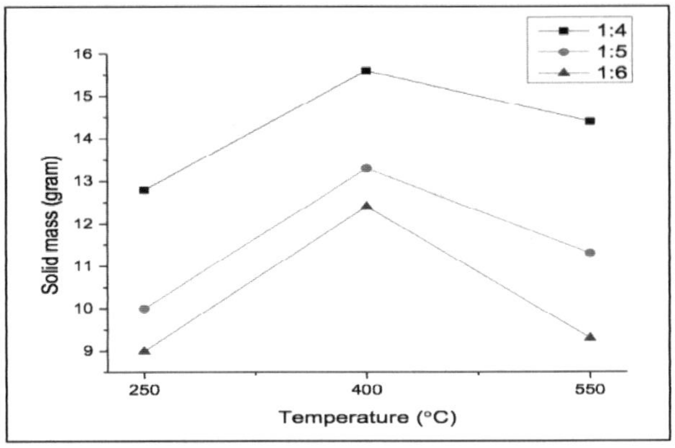

Figure 11. Solid mass from evaporation.

4. Conclusion

Potassium as a raw material for fertilizer can be obtained from OPEFB's ash through an extraction process. The optimum ashing temperature to get the highest amount of potassium is at a temperature of 400 °C. The optimum F:S to obtain the highest yield is at F:S = 1:6 with a yield of 49.94 %. The mass of the most solids is obtained under the temperature conditions of 400 °C filling with F:S = 1:4 with the potassium concentration of 49,94 %-w/w. This value is suitable for liquid fertilizer. Potassium salt resulted from the evaporation of its leaching solution is majorly contains K_2CO_3, KCl, and K_2SO_4.

Acknowledgments

This research is funded by the P3MI grant program 2017 for the Chemical Engineering Product Design and Development Expertise Group at the Faculty of Industrial Technology, Bandung Institute of Technology.

References

[1] Varqa S 2017 Essential palm oil statistics *Palm Oil Analytics* Available from: www.palmoilanalytics.com/files/epos-final-59.pdf (accessed 18.07.2019)
[2] Direktorat Jendral Perkebunan 2017 *Statistik Perkebunan Indonesia 2015-2017* (Jakarta: Kelapa Sawit (Oil Palm))
[3] Hambali E and Rivai M 2017 *IOP Conf. Ser.: Earth Environ. Sci.* **65** 012050
[4] Faizi M K *et al* 2016 *Matec Web of Conferences* **90** 01064
[5] Foo K Y and Hameed B H 2009 *J. Hazard. Mater.* **172** 523-31
[6] Sinsiri T, Kroehong W, Jaturapitakkul C and Chindaprasirt P 2012 *Mater. and Design* **42** 424-33
[7] Niu Y, Zhu Y, Tan H, Wang X, Hui S and Du W 2015 *Proc. Combust Inst.* **35** 2405-13
[8] Teixeira P, Lopes H, Gulyurtlu I, Lapa N and Abelha P 2012 *Biomass and Bioenergy* **39** 192-203
[9] Sadi S 1992 *Proc. 1st Soehadi Reksowardojo Chemical Engineering Seminar* (Bandung, Indonesia)
[10] Bassam N El 1998 *Energy Plant Species* (London: James & James Sci. Publ) p 194
[11] Giorgi C D V 1941 *Malay. Agr. J.* **29** 426-36
[12] Suwandi P P 1991 *Bull. Puslitbun Marihat* **11**(3) 43-6 (in Indonesian)
[13] Samadhi T W, Narcia F and Amril H 2018 *MATEC Web Conferences* **156** 03027
[14] Yin C Y, Kadir S A S A, Lim Y P, Syed-Ariffin S N and Zamzuri Z 2008 *Fuel Processing Technol.* **89** 693-6

Plant extract-assisted biosynthesis of zinc oxide nanoparticles and their antibacterial application

E Rahayu[1], V Wonoputri[1], and T W Samadhi[1]

[1] Departement of Chemical Engineering, Institut Teknologi Bandung, Bandung 40132 Indonesia

E-mail: twsamadhi@che.itb.ac.id

Abstract. ZnO nanoparticles are multi-purposes materials that can be synthesized by several methods, including physical and chemical routes. A novel synthesis method of ZnO nanoparticles is the biological method using plant extracts as reducing and capping agents, such as the fruit extract of *Averrhoa bilimbi*. Plant extracts are superior agents for synthesizing nanoparticles because it provides essential phytochemical substances as reductor, capping agents, and free from toxicants. In this study, the effects of precursor concentrations and the amount of plant extract on the formation and morphology of nanoparticles were investigated. The characteristics of ZnO particles were studied by UV-Vis spectroscopy, XRD, FTIR, TEM, and DLS. The study showed that the formation of ZnO nanoparticles occurred after five hours reaction at 70°C, as indicated by color change of the solution. ZnO nanoparticle formation was confirmed by the maximum absorption at the wavelength of 372 nm and XRD analysis. FTIR analysis showed that the as-synthesized ZnO contained significant organic compounds on its surface, especially compared to commercial ZnO. Reduction reactions using *A.bilimbi* produce nanoparticles in the size from 35.4 to 59.5 nm with round shape and some agglomeration that were observed by TEM. The ZnO antibacterial property was tested against planktonic and biofilm *Escherichia coli*. The result showed that as-synthesized ZnO have comparable antibacterial antibiofilm property as the commercial ZnO nanoparticles at low concentration. Interestingly, this property was diminished when as-synthesized ZnO nanoparticles were used at high concentrations.

1. Introduction

ZnO nanoparticles are multi-purposes material with a wide range of applications, such as photocatalyst and antibacterial agents. These nanoparticles can be synthesized by several methods, such as physical and chemical methods. Physical methods for synthesizing nanoparticles that have been used are ball milling, melt mixing, physical vapour deposition, laser ablation, sputter deposition, electric deposition, and ion implantation. Due to the involvement of specific equipment, physical methods require expensive investment [1,2]. Researchers have developed nanoparticle synthesis processes through a chemical method that is more efficient and able to produce nanoparticles wit goof physicochemical characteristics [1,3]. This method requires the use of an organic and hazardous compound that acts as reductor and capping agent, such as sodium borohydride, ascorbic acid [4], ethanol, NaOH [5], sodium citrate, dimethyl fumarate (DMF), Tollens reagent, and ethyleneglycol [6].

A novel method to synthesize ZnO nanoparticles is biological method that uses plant extracts as the reducing and capping agents. Plant extracts are superior agents for synthesizing nanoparticles because they are free from toxicants and provides essential phytochemical substances as reductor and capping agents. Enzyme, protein, flavonoid, polysaccharide, and phenol in the plant extract play an important role in the reduction process of particle [7]. Other studies also proved that polyphenol is the main substances that influenced the formation and stabilization of nanoparticles [8]. Compared to physical and chemical routes, biological method is an efficient and eco-friendly method because of the relatively mild operating condition [1,9].

Several factors influence the nanoparticles synthesis process and its resulting characteristics. Morphology and particle size are controlled by precursor concentration and volume ratio of precursor-extract which were used in the reaction [10]. Another study also confirmed that the volume ratio of precursor and reductor/capping agent influence the kinetics of reaction [11]. Increasing the extract volume led to an increase in absorbance intensity of the suspension, implying the increase in nanoparticle formation. After reaching a maximum number, absorbance intensity tends to decrease because of metal ions scarcity, which could provoke agglomeration of the smaller particles to form larger particles [12]. In line with the extract addition, the higher precursor volume led to the higher absorbance intensity of suspension [13].

Due to the variety of biomolecules composition and content in the plant extract, nanoparticle synthesis through biological method requires different process conditions from each other producing particle with different characteristics. The fruit of *Averrhoa bilimbi* is a potential plant used in synthesis ZnO nanoparticles. In this part, there are some phytochemical substances such as polyphenol as one of the major constituent, alkaloids, carbohydrate, flavonoid, saponin, and tannin [14]. Another study confirmed that the fruit extract of *A.bilimbi* is a complex matrix that contains bioactive components such as 4H-pyran-4-one, 2,3-dihydro 3,5-dihydroxy-6methyl, hexadecanoic acid, squalene, erucic acid, oleic acid, furfural, boron acid, and mannitol [15]. Therefore, this study intends to identify characteristics of ZnO nanoparticles synthesized by the fruit extract of *A.bilimbi* in different precursor concentration and volume ratio of precursor-extract, and also to explore its potential application as antibacterial/antibiofilm agents.

2. Material and Method

2.1 Materials

Fresh fruit of *A.bilimbi* was collected from Bandung city. Zinc nitrate tetrahydrate, ($Zn(NO3)_2.4H_2O$), and Folin-Ciocalteu reagent were purchases from Merck Indonesia. Commercial ZnO nanoparticles from Sigma-Aldrich was used as a comparison.

2.2 Extraction of phenol from A. bilimbi fruit

10 g fresh fruit were washed and crushed using blender. Fruit pulp was added to 300 mL aqua demineralization (Aqua DM) and shaked in the shaking incubator (200 rpm) at 70°C for 90 minutes. After cooling process, the mixture was filtered using filter cloth and the supernatant was stored at 4°C for total phenol content analysis.

2.3 Total phenolic content analysis

Total phenol content was determined using the Folin-Ciocalteau [16]. 40 μl extract solution was mixed with 1.8 ml of the Folin-Ciocalteau reagent that was previously dissolved in water (1:10). After 5 minutes, 1.2 ml of Na_2CO_3 (7.5% w/v) was added to the mixture. The solution was allowed to stand at room temperature for 1 hour, then absorbance was measured using a UV-Vis spectrophotometer (Thermo Fisher Scientific 4001/4) at 765 nm. Calibration curve was prepared using standard solution of gallic acid at several concentrations (20, 40, 60, 80, and 100 mg/L $r^2 = 0.999$). The measured total phenol was expressed on a fresh weight basis as mg gallic acid equivalents/100 gr sample (mg GAE/100 gr extract). Average measurement of the extract showed that total phenol of fruit extract that is available as reducing and capping agent was 124.285 mg GAE/100 g sample.

2.4 Synthesis of ZnO nanoparticles using A.bilimbi
Experiment was conducted using 2^k full factorial design. The first factor was precursor concentration (0.025 M, 0.050 M, 0.100 M), while the second factor was volume ratio precursor-extract (2:1; 1:1; 1:2). Precursor solution in a certain volume and concentration was heated on hot plate until it reached 70°C. Then, the extract was added using a burette under constant stirring for up to 5 hours. ZnO nanoparticle formation was indicated by a color change of the solution to brownish-yellow. The solution was cooled to 30°C, then it was purified through centrifugation at 14,000 rpm for 5 minutes to get deep white powder. The powder was washed twice using aqua DM, then it was dried for 24 hours at 70°C. The dried solid was crushed with a mortar for analyzing.

2.5 Analytical methods
The characteristics of ZnO nanoparticles were analyzed using UV-Vis spectrophotometer, X-ray diffraction (XRD), Dynamic Light Scattering (DLS), Transmission Electron Microscopy (TEM), and Fourier Transform Infra-red (FT-IR).

2.6 Antibacterial assay

2.6.1 Inhibition of planktonic bacteria growth of Escherichia coli. 1 mL of *E.coli* overnight inoculum was added into 50 mL LB-Broth (10 gr/L tryptone, 5 gr/L yeast extract, 10 gr/L NaCl, pH 7). 0.9 mL of bacterial inoculum was poured into 24-well plates which contained 0.1 mL of ZnO nanoparticles solution in several concentrations (0, 50, 100, and 200 ppm). Then, the plates were incubated for 6 hours, 37°C, and constant agitation at 180 rpm in a shaker incubator. The bacteria population was measured using a spectrophotometer at 600 nm. The result was analysed by ANOVA using statistical software.

2.6.2 Inhibition of biofilm formation of E.coli. 1 mL of E. coli bacterial culture was poured to 50 mL of M9 minimal medium (48 mM Na_2HPO_4, 22 mM KH_2PO_4, 8.6 mM NaCl , 19 mM NH_4Cl, 2 mM $MgSO_4.7H_2O$, 100 μM $CaCl_2$, and 20 mM glucose). 0.9 mL of bacterial inoculum was poured into 24 well-plates containing 0.1 mL of ZnO nanoparticles to reach a certain concentration (0, 50, 100, and 200 ppm) and incubated in a shaker incubator at 37°C and 180 rpm. After incubation for 6 hours, the supernatants were removed from each well and the absorbances were measured at 600 nm to denote the planktonic bacteria concentrations. Then, the wells were washed with phosphate buffer solution (PBS) and stained using crystal violet 0.1 w/v and allowed to stand for 15 minutes. After the incubation process was complete, the leftover crystal violet was removed from the well. The remaining liquid at the well was redissolved using 1 mL ethanol and its absorbance was measured by a spectrophotometer at 540 nm. The result was analysed by ANOVA using statistical software.

3. Result and Discussion

3.1 Formation of ZnO nanoparticles
The synthesis process is carried out at 70°C because it allowed faster water solvent evaporation at high temperatures. Throughout the synthesis process, the solution absorbance was measured to determine the formation of the nanoparticles. The formation of ZnO nanoparticles using plant extract indicated the presence of a peak UV-Vis absorption band at 375 nm [17]. Moreover, the reduction of precursors by plant extracts can also be identified by colour change of the suspension to brownish-yellow and the formation of deep white precipitate.

When 0.05 M Zn precursor was mixed with *A. bilimbi* extract at 1:1 v/v ratio, two peaks at 240 and 300 nm were observed (Figure 1). These peaks were still observable during the first 3 hours. However, when the reaction was continued up to 5 hours, the peak slowly shifted to 372 nm, which indicated the formation of ZnO nanoparticles. Therefore, 5 hours was chosen as the basis for reaction time for the remaining variation.

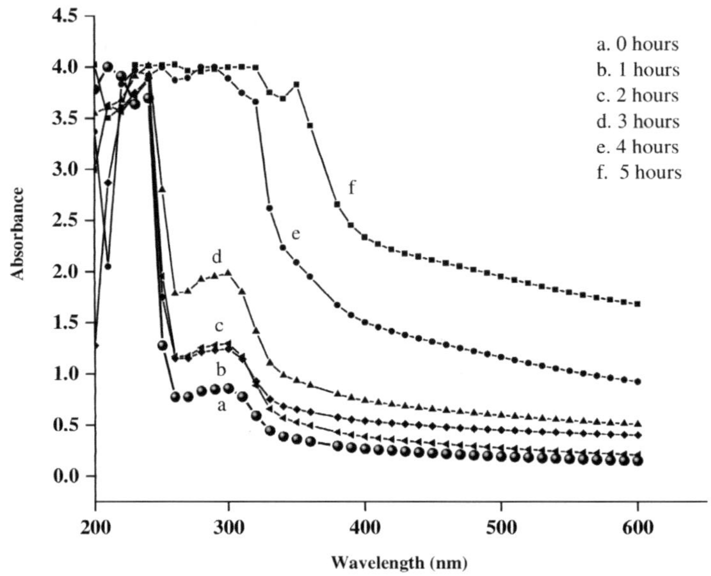

Figure 1. UV-vis spectra recorded as a function of reaction time of 0.05 M $Zn(NO_3)_2.4H_2O$ solution with *A.bilimbi* fruit extract (1:1 v/v)

Table 1. Physical characteristics of solution after 5hours reaction

Precursor concentration (M)	Precursor : Extract (v/v)	Solution color	Precipitate formed	pH Initial	pH Final
0.025	2 : 1	Brownish yellow	No	3,5	2,5
	1 : 1	Brownish yellow	Yes	3,5	1,5
	1 : 2	Brownish yellow	Yes	3,5	1
0.050	2 : 1	Brownish yellow	No	4	2,5
	1 : 1	Brownish yellow	Yes	3	1
	1 : 2	Brownish yellow	Yes	3	1
0.100	2 : 1	Brownish yellow	No	4	2,5
	1 : 1	Brownish yellow	No	3,5	1,5
	1 : 2	Brownish yellow	Yes	3,5	1,5

When the synthesis process variables established previously was tested at other precursor concentration and different precursor:extract volume ratio, some combination failed to yield any precipitate, even after centrifugation (Table 1). This especially happened when the precursor volume used is greater than the extract volume. For instance, the mixture of 0.100 M precursor and 1:1 ratio produced a brown cloudy layer on the solution surface which were suspected to be $Zn(OH)_2$. However, after the centrifugation process, the brown sediment could not be separated from the suspension. This implies that the $[OH^-]$ ions needed to form $Zn(OH)_2$ (and subsequently ZnO) from the extract was not sufficient to bind [Zn^{2+}] ion. On the other hand, when the 1:2 ratio was used, the recoverable precipitate was obtained (Table 1).

3.2 Characterization of as-synthesized ZnO nanoparticles

Confirmation of ZnO formation was analyzed using XRD. The result showed a slight shift in the curve peaks at several diffraction angles on ZnO nanoparticles compared to commercial ZnO. The diffraction peaks of commercial ZnO nanoparticles were found at 31.76°, 34.42°, 36.24°, 47.54°, 56.58°, 62.85°, 63.02°, 66.37°, 67.93°, 68.12°, 69.07°, 72.56° and 76.96°. The reaction using 0.025 M of precursor solution with ratio volume precursor-extract at 1:1 and 1:2 produce ZnO nanoparticle with the diffraction peaks found at 33.68°, 56.35°, 70.65° and 33.71°, 56.36°, 70.69°, respectively. These appearances of new peak diffractions could indicate that the particles contain some amorphous phases with high peak intensity which contributed to the noisy spectra [18].

Formed ZnO nanoparticles had average particle sizes ranging from 35 to 60 nm (Table 2). The size trend of ZnO nanoparticles synthesized from *A.bilimbi* fruit extract was almost close to the size of commercial ZnO nanoparticles (29 nm). However, an inconsistent trend of particle size in the treatment variation of volume ratio of precursor and extract was observed. For reaction using 0.025 M precursor, increasing of extract volume caused the reduction in particle size. However, on the reaction using 0.050 M precursor solution, increasing the ratio of precursor-extract led to increasing particle size from 49.0 nm to 57.1 nm, which have been observed before [19]. Furthermore, Song et al. (2009) argued that this increase in extract volume provides abundant reducing agents that can stimulate particle aggregation because of interaction between capping agent molecules to bind surface particles which cause a secondary reduction to form new nuclei of particle.

Generally, the formed ZnO nanoparticles had a round shape, showed some agglomeration, and had good crystallinity. Although the polyphenols contained in the plant extracts were able to reduce Zn to form nanoparticles (size of ± 50 nm), these compounds can form a mantle that covers the surface of the particles (Figure 3.a). Agglomeration of the nanoparticles was attributed to the nature of the synthesis process (Figure 3.b). Note that throughout the synthesis process, the pH of the solution dropped from 3-4 to 1-2.5 due to the evaporation of water molecules (Table 1). This can affect the size and crystallinity of the nanoparticles. In acidic medium, nanoparticles formation was more difficult due to the limited presence of [OH^-] ions in suspension compared to [H^+] ions, thus limiting the formation of alkoxide ($Zn(OH)_2^-$) which subsequently limit the formation of ZnO [20].

The pH strongly influences the degree of aggregation and particle settling, due to the change of surface charge and zeta potential of particles which allowed for a greater degree of particle interaction [21]. ZnO nanoparticle synthesis processed under acidic or neutral conditions tends to cause particle agglomeration [20]. The optimum pH for synthesizing ZnO nanoparticles and having lower agglomeration is 8 [22,23]. At low pH, the adsorption of metallic particles that were positively charged to the negatively charged organic matter from capping agent was enhanced thus resulting in a high level of aggregation [24]. The acidic process also would change chemical structures and activities of reductant to form alkoxide ions [25]. Other studies confirmed that the size and density of nanoparticles increase as a response to a decrease in the acidity of the medium. When the pH was continuously decreased to certain limits cause re-dissolution of $Zn(OH)_2$ [26]. Good crystallinity of the ZnO nanoparticles was found on synthesized particles with a concentration of 0.100 M precursors at a 1: 2 precursor-extract ratio.

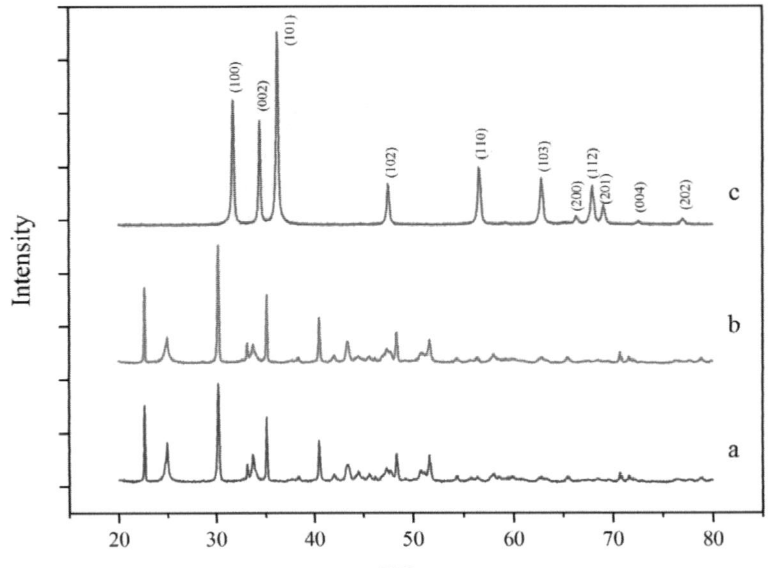

Figure 2. X-ray diffraction pattern of ZnO nanoparticle synthesized using 0.025 M of precursor solution; (a) 1:1 (v/v), (b) 1:2 (v/v), (c) commercial

Table 2. Physical characteristics of ZnO nanoparticle

Precursor concentration (M)	Precursor : extract (v/v)	Average dried weight (g)	Average particle size (nm)
0.025	2 : 1	-	-
	1 : 1	0.065	51.7
	1 : 2	0.104	35.4
0.050	2 : 1	-	-
	1 : 1	0.060	49.0
	1 : 2	0.130	57.1
0.100	2 : 1	-	-
	1 : 1	-	-
	1 : 2	0.105	59.5
ZnO nanoparticle (commercial as control)			28.9

FTIR analyses were performed on the formed nanoparticles. ZnO nanoparticles synthesized using *A.bilimbi* fruit extract transmit significantly more peaks spectrum compared to commercial ZnO nanoparticles, implying the presence of various organic groups on the surface of the sample (Figure 4 and Table 3). The bond between M-O (Zn-O) would be found in areas ranging between 600 cm^{-1} and 400 cm^{-1} [27]. The FTIR spectrum confirms Zn-O bond on the peak area between 445.56 cm^{-1} to 493.78 cm^{-1}. Bending vibrations of C-H from the alkyne compound derived from plant extract were found at 626.87 cm^{-1} to 634.58 cm^{-1}. The peak area between 740.67 cm^{-1} to 821.68 cm^{-1} represented C-H bending vibrations derived from aromatic compounds (alkene groups). The peak area ranging from 1315.45 cm^{-1} to 1363.67 cm^{-1} confirmed vibrations of the C-N bond derived from aromatic amine compounds. Vibration of the C=C from alkene compound with strong to moderate intensity was found at 1633.71 cm^{-1} to 1656.85 cm^{-1}. Stretching vibration of C-H derived from alkane in low intensity was

IOP Conf. Series: Materials Science and Engineering **823** (2020) 012036 doi:10.1088/1757-899X/823/1/012036

found at 2854.65 cm^{-1} to 2926.01 cm^{-1}, while stretching vibration of O-H with strong intensity was found at 3369.64 cm^{-1} to 3379.29 cm^{-1}.

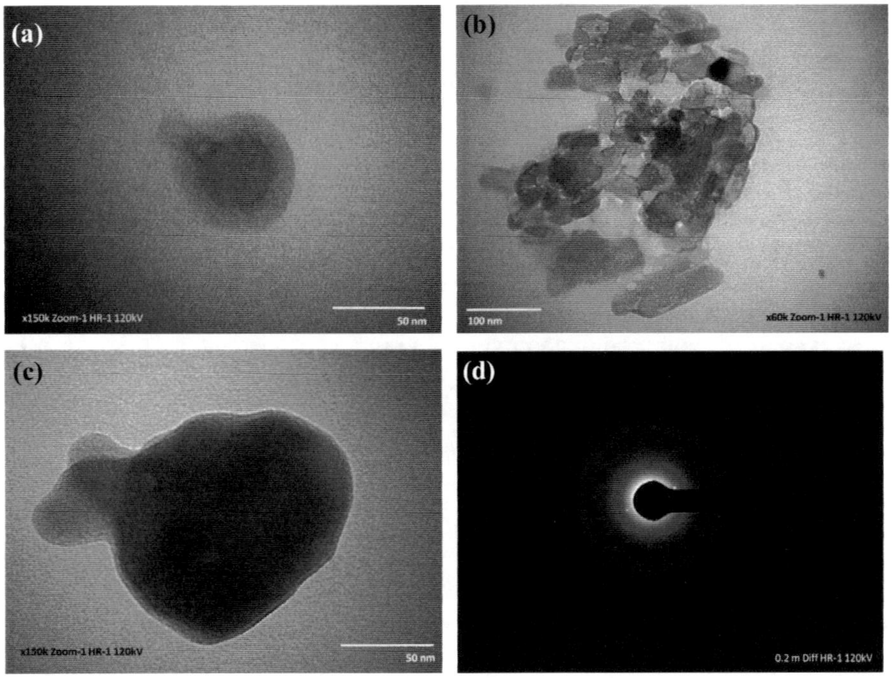

Figure 3. TEM micrograph of ZnO nanoparticles, (a) 0.050 M of precusor solution, 1:1 (v/v), (b) 0.050 M of precursor solution, 1:2 (v/v), (c) 0.010 M of precursor solution, 1:2 (v/v), (d) SAED pattern of ZnO nanoparticles from 0.010 M of precursor solution, 1:2 (v/v)

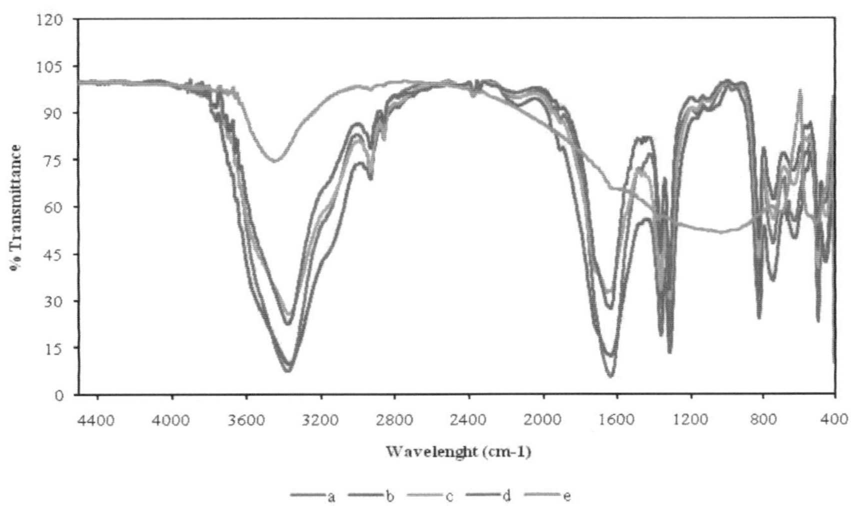

Figure 4. FTIR spectra of ZnO nanoparticles (a) 0.025 M precursor solution, 1:1 (v/v), (b) 0.025 M precursor solution, 1:2 (v/v), (c) 0.050 M precursor solution, 1:2 (v/v), (d) 0.100 M precursor solution, 1:2 (v/v) (e) commercial

IOP Conf. Series: Materials Science and Engineering **823** (2020) 012036 doi:10.1088/1757-899X/823/1/012036

Table 3. Functional groups capping on ZnO nanoparticles surface

Wavenumber (cm^{-1})					Functional groups	Compound
a	b	c	d	e		
453,27	451,34	445,56	453,27	507,28	Zn-O	-
493,78	493,78	493,78	493,78	-	Zn-O	-
626,87	626,87	632,65	634,58	694,37	C-H	Alkyne
742,59	742,59	740,67	740,67	-	C-H	Aromatic
821,68	819,75	821,68	821,68	-	C-H	Aromatic
1317,38	1315,45	1315,45	1317,38	-	C-N	Aromatic amine
1363,67	1363,67	1361,74	1363,67	1381,03	C-N	Aromatic amine
1633,71	1633,71	1654,85	1633,71	1624,06	C=C	Alkene
2854,65	2854,65	2852,72	2854,65	-	C-H	Alkane
2926,01	2926,01	2926,01	2926,01	-	C-H	Alkane
3379,29	3369,64	3373,50	3373,50	3448,72	O-H	Phenol

Note :
a = 0.025 M precursor solution, 1:1 (v/v)
b = 0.025 M precursor solution, 1:2 (v/v)
c = 0.050 M precursor solution, 1:2 (v/v)
d = 0.100 M precursor solution, 1:2 (v/v)
e = Commercial

3.3 Antibacterial and antibiofilm properties of the synthesized ZnO

The antibacterial properties of ZnO nanoparticles were tested against *E.coli* grown in two different life mode, planktonic (free-swimming) form and biofilm form. Three different nanoparticles concentration (50, 100, and 200 ppm) were tested. Again, commercial ZnO was used as a comparison. At 50 ppm, around 30% of planktonic bacteria were inhibited, both for the as-synthesized and commercial ZnO. Increasing the ZnO concentration to 100 ppm saw a slight %inhibition increase for the commercial ZnO (34%), while similar %inhibition was observed for 100 ppm. Interestingly, when the ZnO concentration was further increased to 200%, the as-synthesized ZnO saw a decrease in antibacterial property, as the %inhibition was decreased to 25.5%. On the other hand, commercial ZnO saw a significant increase in %inhibition (37%), which is more commonly observed in other studies, i.e., increasing the ZnO concentration resulted in increasing inhibition efficiency [28].

Similar trend of antibacterial activity of the as-synthesized nanoparticles the formation and morphology of the nanoparticles on biofilm inhibition assay. Initially, 50 ppm of ZnO resulted in 37% and 33% inhibition for as-synthesized and commercial samples, respectively. Increasing the ZnO concentration to 100 ppm significantly decreased the efficiency of as-synthesized ZnO (25% inhibition) while significantly increased the efficiency of commercial ZnO (40% inhibition). Further increment to 200 ppm ZnO saw a further decrease in efficiency for as-synthesized (2.5%) and similar efficiency (40% inhibition) was observed for commercial ZnO.

As mentioned previously, the as-synthesized sample showed a different trend than that normally observed. At this stage, the exact reason is not clear yet. Remembering FTIR result (Figure 4) showed before, it seems that the presence of significant organic compounds on the surface of as-synthesized ZnO could affect the trend observed here. The presence of higher ZnO could result in a higher organic compounds in the system. This organic compound could easily leach out of the ZnO surface and subsequently used by the bacteria as a carbon source for growth, therefore reducing the antibacterial activity of the nanoparticles. Moreover, the presence of organic compound together with metal oxide nanoparticles have been shown to reduce the antibacterial efficiency, especially when compared to its respective metal salts [29]. Meanwhile, TEM results (Figure 3) also suggest the agglomeration of as-synthesized ZnO. It has been argued before that agglomeration of nanoparticles which led to the deposition of the nanoparticles also could contribute to the low antibacterial activity of nanoparticles [30].

IOP Conf. Series: Materials Science and Engineering **823** (2020) 012036 doi:10.1088/1757-899X/823/1/012036

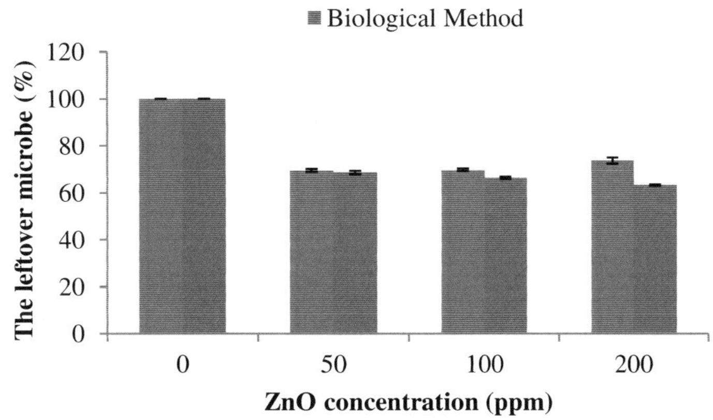

Figure 5. Inhibition activity of ZnO nanoparticles on planktonic bacteria of *E.coli*. Note: Mean of replications ± standard error (n=3).

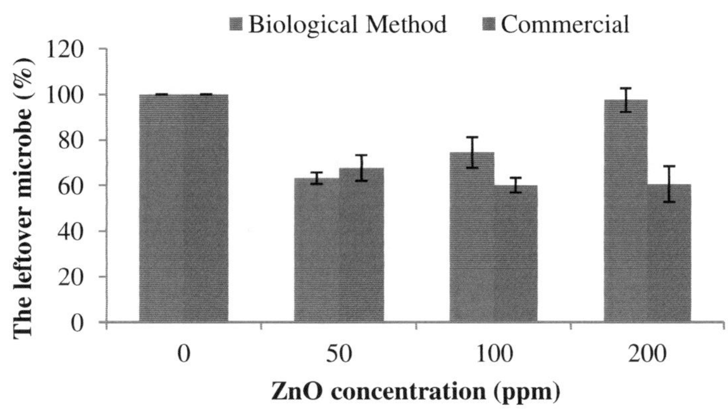

Figure 6. Inhibition activity of ZnO nanoparticles on biofilm of *E.coli*. Note: Mean of replications ± standard error (n=3).

4. Conclusion

ZnO nanoparticles were successfully synthesized using *A. bilimbi* extract as the reducing and capping agent. A preliminary experiment showed that when synthesis was performed at 70°C, the reaction time needed for the formation of ZnO was 5 hours. However, the formation of ZnO was highly dependent on the precursor concentration and volume ratio of precursor-extract. The abundance of reductant in the solution increases non-binding ions accumulation that would bind to the capping agent molecules thus form particles aggregate in a larger size and stimulate a new compound growth that reduces the purity of the ZnO nanoparticles. The formation of ZnO nanoparticles was confirmed by XRD analysis. The as-synthesized ZnO nanoparticles were generally round with good crystallinity and some agglomeration, with sizes ranging from 35.4 to 59.5 nm. Due to the nature of the synthesis process, the as-synthesized ZnO showed a significant presence of organic compounds on its surface, especially when compared to commercial ZnO. Due to agglomeration, ZnO nanoparticles synthesized from *A.bilimbi* fruit extract have lower antibacterial activity compared to commercial ZnO. Antibacterial and antibiofilm testing showed that as-synthesized ZnO had comparable antibacterial/antibiofilm property to commercial ZnO at low concentration (50-100 ppm). However, the antibacterial/antibiofilm property was reduced when high concentration (200 ppm) of as-synthesized ZnO was used.

Acknowledgments
This work was supported by the Departement Indonesian Agency for Agricultural Research and Development (IAARD), Ministry of Agriculture grants and P3MI ITB (*Program Penelitian, Pengabdian kepada Masyarakat, dan Inovasi* Institut Teknologi Bandung/Research Programs, Community Services, and Innovation Institut Teknologi Bandung) funding scheme.

References
[1] Ali A, Zafar H, Zia M, Ul Haq I, Phull A R, Ali J S and Hussain A 2016 Synthesis, characterization, applications, and challenges of iron oxide nanoparticles *Nanotechnology, Science, and Application* **9** 49-67
[2] Ul Haq A N, Nadhman A, Ullah I, Mustafa G, Yasinzai M and Khan I 2017 Synthesis approaches of zinc oxide nanoparticles: the dillema of ecotoxicity *Journal of Nanomaterials* doi: 101155/2017/8510342
[3] Prabhu S and Poulose E K 2012 Silver nanoparticles: mechanism of antimicrobial action, synthesis, medical applications, and toxicity effects *International Nano Letters* **2**(32)
[4] Bogdanovic U, Lazic V, Vodnik V, Budimir M, Markovic Z and Dimitrijevic S 2014 Copper nanoparticles with high antimicrobial activity *Materials Letters* **128** 75-78
[5] Lastovina T, Budnyk A P, Khaishbashev G A, Kudryavtsev E A and Soldatov A V 2014 Copper-based nanoparticles prepared from copper (II) acetate bipyridine complex *Journal of the Serbian Chemical Society* **81**(7) 751-62
[6] Iravani S, Korbekandi H, Mirmohammadi S V and Zolfaghari B 2014 Synthesis of silver nanoparticles: chemical, physical and biological methods *Research in Pharmaceutical Sciences* **9**(6) 385-406
[7] Mittal A K, Chisti Y and Banerjee U C 2013 Synthesis of metallic nanoparticles using plant extracts *Biotechnol Adv* doi: 10.1016/j.biotechadv.2013.01.003
[8] Kharissova O V, Dias H V R, Kharisov B I, Perez B O and Perez V M J 2013 The greener synthesis of nanoparticles *Trends in Biotechnology* **31**(4) 240-48
[9] Parveen K, Banse V and Ledwani L 2016 Green synthesis of nanoparticles: their advantages and disadvantages *2nd International Conference on Emerging Technologies: micro to nano 2015 (ETMN-2015)* doi: 10.1063/1.4945168
[10] Padil V V T and Cernik M 2013 Green synthesis of copper oxide nanoparticles using gum karaya as biotemplate and their antibacterial application *International Journal of Nanomedicine* **8** 889-98
[11] Moloto N, Revaprasadu N, Musetha P L and Moloto M J 2009 The effect of precursor concentration, temperature and capping group on the morphology of CdS nanoparticles *Journal of Nanoscience and Nanotechnology* **9** 4760-66
[12] Bartos-Arrieta J, Florido A, Perez-Rafols C, Serrano N, Fiol N, Poch J and Villaescusa I 2018 Green synthesis of Ag nanoparticles using grape stalk waste extract for the modification of screen-printed electrodes *Nanomaterials* **8** 946
[13] Saranya S, Vijayarani K and Pavithra S 2017 Green synthesis of iron nanoparyicles using aqueous extract of *Musa ornata* flower sheath againts pathogenic bacteria *Indian Journal of Pharmaceutical Sciences* **79**(5) 688-94
[14] Hasanuzzaman M, Ali M R, Hossain M, Kuri S and Islam M S 2013 Evaluation of total phenolic content, free radical scavenging activity and phytochemical screening of different extract of *Averrhoa bilimbi* (fruits) *International Current Phamaceutical Journal* **2**(4) 92-96
[15] Suluvoy J K and Grace V M B 2017 Phytochemical profile and free radical nitric oxide (NO) scavening activity of *Averrhoa bilimbi* L. fruit extract *Biotech* **7**(85)

[16] Alothman M, Bhat R and Karim A A 2009 Antioxidant capacity and phenolic content of selected tropical fruits from Malaysia, extracted with different solvents *Food Chemistry* **115** 785-88

[17] Zhao C, Zhang X and Zheng Y 2018 Biosynthesis of polyphenols functionalized ZnO nanoparticles: Characterization and their effect on human pancreatic cancer cell line *Journal of Photochemistry and Photobiology, B: Biology* **183** 142-46

[18] Pholnak C, Sirisathitkul C, Suwanboon S and Harding D J 2014 Effects of precursor concentration and reaction time on sonochemically synthesized ZnO nanoparticles *Materials Research* **17**(2) 405-11

[19] Song J Y, Jang H and Kim B S 2009 Biological synthesis of nanoparticles of gold nanoparticles using Magnolia kobus dan Diopyros kaki leaf extracts *Process Biochemistry* **44** 1133-38

[20] Alias S S, Ismail A B dan Mohamad A A 2010 Effect of pH on ZnO nanoparticle properties synthesized by sol-gel centrifugation *Journal of Alloys and Compounds* **499** 231-37

[21] Badawy A M E, Luxton T P, Silva R G, Scheckel K G, Suidan M T and Tolaymat T M 2010 Impact of environmental conditions (pH, ionic strength, and electrolyte type) on the surface charge and aggregation of silver nanoparticles suspensions *Environ Sci Technol* **44** 1260-66

[22] Domingos R F, Rafiei Z, Monteiro C E, Khan M A K and Wilkinson K J 2013 Agglomeration and dissolution of zinc oxide nanoparticles: role of pH, ionic strenght and fulvic acid *Environ Chem* **10** 306-12

[23] Zare E, Pourseyedi S, Khatami M and Darezereshki E 2017 Simple biosynthesis of zinc oxide nanoparticles using nature's source, and it's in vitro bio-activity *Journal of Molecular Structure* **1146** 96-103

[24] Illes E and Tombacz E 2006 The effect of humic acid adsorption on pH-dependent surface charging and aggregation of magnetite nanoparticles *Journal of Colloid and Interface Science* **295** 115-23

[25] Sibiya P N and Moloto M J 2014 Effect of precursor concentration and pH on the shape and size of starch capped silver selenide (Ag_2Se) nanoparticles *Chalcogenide Letters* **11**(11) 577-88

[26] Amin G, Asif M H, Zainelabidin A, Zaman S, Nur O and Willander M 2011 Influence of pH, precursor concentration, growth time, and temperature on the morphology of ZnO nanostructures grown by the hidrothermal method *Journal of nanomaterials* 269692

[27] Elumalai K and Velmurugan S 2015 Green synthesis, characterization and antimicrobial activities of zinc oxide nanoparticles from the leaf extract of *Azadirachta indica* (L) *Applied Surface Science* **345** 329-36

[28] Bhattacharyya P, Agarwal B, Goswami M, Maiti D, Baruah S and Tribedi P 2018 Zinc oxide nanoparticle inhibits the biofilm formation of *Streptococcus pneumoniae* Antoine van Leeuwenhoek **111** 89-99

[29] Gunawan C, Teoh W Y, Marquis C P and Amal R 2011 Cytotoxic origin of copper (II) oxide nanoparticles: comparative studies with micron-sized particles, leachate, and metal salts *ACS Nano* **5(9)** 7214-25

[30] Hozyen H F, Ibrahim E S, Khairy E A and El-Dek S I 2019 Enhanced antibacterial activity of capped zinc oxide nanoparticles: a step towards the control of clinical bovine mastitis *Veterinary World* EISSN: 2231-0916

IOP Conf. Series: Materials Science and Engineering **823** (2020) 012037 doi:10.1088/1757-899X/823/1/012037

Identification of plants natural dye by Meto tribe in South Central Timor

Alan Charis Sabuna[1] and Fransina T. Nomleni[1]

[1] Department of Biology Education, Universitas Kristen Artha Wacana, Kupang 85227, Indonesia

E-mail: alan.sabuna@gmail.com

Abstract. Information on plant species, natural dyes and the dying process of the yarn of ikat weaving in South Central Timor is still minimal and not well documented. Therefore, this study aimed to identify natural dye in South Central Timor. The samples were collected by exploration methods. Parts of plants, dyeing processes, and kinds of color were recorded based on observation, interviews with craftsmen and documentation. Our results showed that there are 7 plant species were successfully identified, *Curcuma longa* L., *Indigofera tinctoria* L., *Morinda citrifolia* L., *Phyllanthus reticulates* Poir, *Phaseolus lunatus* L., and *Tectona grandis*. The plant organs used as a source of natural dye are leaves, barks, rhizome, and roots. The method of processing was to be pounded, squeezed, soaked in water and the color produced was red, dark blue, yellow, black, and green.

1. Introduction

Ikat weaving of East Nusa Tenggara has many varieties because each district has a characteristic. East Nusa Tenggara has 21 districts, and each district has multiple motifs. The district the ikat came from could be determined by the dominant color. One of the districts, South Central Timor, has its identity that can be seen from the bright colors of its ikat [1]. This colorful ikat weaving represents the motif of the Meto tribe communities that occupy almost all areas in Timor mainland.

Manufacture of woven fabrics is one of the traditional-home industries using plants' natural dye used for generations. This woven fabric is traditionally made by rural communities and has to be a part of their daily activities. To create woven fabrics, people need dyes that can give patterns and motifs to the fabric [2]. The dye used to stain the fabrics is derived from nature, such as animals, minerals, and plants, either directly or indirectly. This natural dye is obtained by extraction or boiling traditionally. The parts of plants that can be used to dye are the bark, stems, leaves, roots, flowers, seeds and sap. Each plant can be a source of natural dyes because they contain natural pigment. This potential is determined by the intensity of the produced color and highly dependent on the type of dyeing matter. Dyeing matter is a substance that determines the direction of the color of natural dyes and organic compounds contained in sourced natural dyes. One type of plant contains more than one dyeing matter [3].

The advantage of using natural dyes is the color intensity is much lower than the synthetic colors, so that the effect creates an impression in the eyes always supple, soft, and will not produce the same color tone even though using the same recipe. Of course, there are also disadvantages related to their natural properties that are easy to fading, the raw material is uncertain and standards are not guaranteed. However, these disadvantages can be anticipated with special care. The use of natural dyes is more associated with the elements of art, elegant, high taste and reduce environmental pollution so that the consumer target will be upper middle class and overseas. Therefore, the selling price is higher [3].

The use of natural dyes in some countries is still believed to be safer than synthetic dye because it is non-carcinogenic, degradable, manufacture, and relatively use simple technology. It is very suitable for small and medium industries that are currently encouraging the government to support exports. The development of natural dyes for Indonesia as a tropical area is rich in plants that can produce the dye. This situation is expected to revive the image of natural dyes. Therefore, a variety of plants that are capable of producing the dye will have good prospects.

Besides, a sustainable environment is a hot issue that preoccupies Indonesia nowadays; more specifically in South Central Timor, mining could have a negative impact on environmental damage. Indigenous people could lose their jobs from weaving and lost herbal medicine from plants because of forest damage.

Because of information on plant species, natural dyes and ways of processing is still minimal and not well documented, so our goal was to identify the plant species, plant organs, and way of processing organs as a natural dye for the Meto tribe in South Central Timor.

2. Methodology

2.1. Time and location of study

Data were collected from November – January 2015 in Pana Village, Boti Village, Ayotupas Village and Oenasi Village, South Central Timor district, East Nusa Tenggara Province, Indonesia. We chose those villages represented the characteristic of Meto Tribe that are Mollo, Amanatun, and Amanuban Tribe.

2.2. Technique for collecting data

The method used in this study was the interview with Meto tribe communities in South Central Timor Regency. Respondents were determined by using the combination of purposive sampling and snowball sampling techniques [4]. Respondents were elderly and craftsmen. The exploration method was used to observe the plants and which part of them used as a natural dye directly. Flora's book was used to identify the plants [5] and documentation was made on the plant species that have been recognized.

2.3. Data analysis

Data were analyzed by using descriptive qualitative and presented in tabulation form.

3. Results and Discussion

Based on the results of interviews and observations in the field, it is known that the types of coloring plants used by craftsmen from the Meto tribe weaving can be seen in Table 1.

IOP Conf. Series: Materials Science and Engineering **823** (2020) 012037 doi:10.1088/1757-899X/823/1/012037

Table 1. Plant species used as natural coloring agents by the Meto tribe in South Central Timor Regency

No.	Local name / Indonesian name / Scientific name	Plant organs	Processing method	The color produced
1	Huki/ Curcumin / *Curcuma longa* L. 	Rhizome 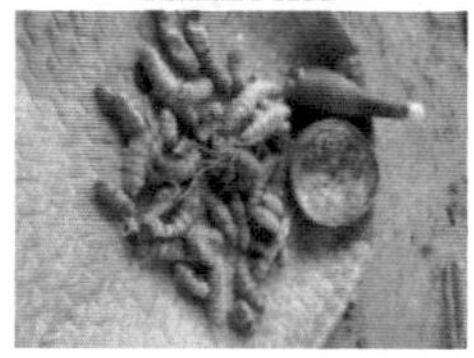	Pounded 	Yellow 
2	Ta'um/ Tarum/ *Indigofera tinctoria* L. 	Leaf 	Soaked in water 	Dark blue/ Black 
3	Bauk ulu/Starvation fruit/ *Morinda citrofolia* L. 	Root 	Pounded 	Red 
4	Meko/ Buah tinta/ *Phylantus reticulatus* Poir. 	Fruit 	Squeezed 	Dark blue 

IOP Conf. Series: Materials Science and Engineering **823** (2020) 012037 doi:10.1088/1757-899X/823/1/012037

5	Koto/ Arbila/ *Phaseolus lunatus* L. 	Leaf 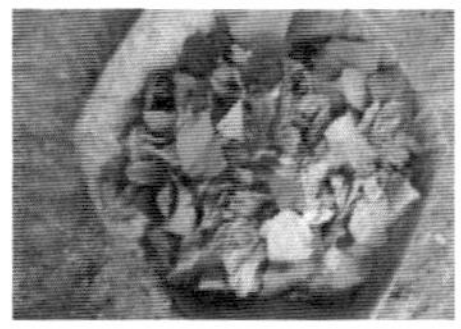	Pounded 	Yellow 
6	Yati/ Jati/ *Tectona grandis* 	Leaf 	Pounded 	Red 

The coloring plants used by the weaving craftsmen of the Meto tribe in Timor Tengah Selatan District were 6 species. The species was identified to consist of 5 families, namely Zingiberaceae, Fabaceae, Rubiaceae, Euphorbiaceae, and Lamiaceae. Even in the coloring of threads, craftsmen still use the soil to produce black, mixed with tarum after that, so the black color becomes thick.

Table 2. Additional ingredients used in the coloring process of the Meto Tribal woven fabric

No.	Local name / Indonesian name / Scientific name	Part used	Function
1	Lime betel	Lime powder	Color amplifier
2	Noa/Coconut/*Cocos nucifera*	Flesh of fruit	Color amplifier
3	Pena/Corn/*Zea mays*	Cooking water	Fastening woven threads
4	Fenu/Candlenut/*Aleurites moluccanus*	Seed	Initial immersion
5	Loba/Noba/*Symplocos* sp.	Bark	Color binder

In addition to the 6 types of coloring plants above, craftsmen also use additional ingredients. Whiting (Ao) and coconut meat (Noa) are used to strengthen the color of the yarn so that it does not easily wear off, and also boiled water from cornflour (pen) to strengthen the white thread. The plants used in the initial immersion process are candlenut (Fenu) and also boiled water from cornflour (pen).

Table 3. Processing of natural coloring plants to produce colors

No.	Color	Processing method
1	Red	- Starvation fruit bark is pounded until smooth and soaked in the water together with yarn and noba. After that, the thread is then squeezed and dried in the sun to dry. - Starvation fruit bark is grounded until smooth and then cooked for 1 hour, along with threads and additional lime.
2	Yellow	- Turmeric rhizome is crushed to pieces, put the thread and pound together again, removed and then squeezed, add water and soak for 3-4 hours, then lifted and

		dried. - Turmeric rhizome is crushed to pieces, put the yarn in and pound together with yarn, then add water and then cook until yellow and then removed and dried.
3	Green	Arbila bean leaves are taken one basket, pounded until smooth, put the yarn and pounded together, and then the thread is removed, squeezed so that the color of the yarn is evenly distributed, then soaked with 1 cup of water for 3-4 hours, lifted and dried in the sun.
4	Dark blue	The fruit of the inked tree is taken as much as 1 winnowing pan then squeezed and then added cornflour and added water soaked for 3 days, then put the thread and soaked for 3-4 hours just removed and dried
5	Black	Tarum leaves are soaked for 4-5 days. Tarum leaves are soaked in the knees to get the marinade. Soaking water is added coconut fruit and lime then the yarn is dipped 3 times

Based on the results of the study, the Meto tribe in the Timor Tengah Selatan district used 6 types of natural coloring plants to be used to dye the weaving yarn. Seran and Hana [6] researched the community of Kaliuda Village, Pahunga Lodu Sub-District, East Sumba Regency, found that there were 9 species of plants. In contrast, Berlin et al. [7] have found 5 types of coloring plants. From the types of plants found, it can be said that the plants used are different because the flora available in the surrounding environment is different. While the same types of plants used are tarum, starvation fruit, and loba specifically in NTT, but when compared to the same in Borneo (Kalimantan), starvation fruit and areca nut are used.

The parts of plants used by the Meto Tribe in Timor Tengah Selatan Regency are leaves, roots, fruit, rhizomes, leaves, and bark (Table 1). According to Winarno [8] each part of the plant has its color pigment characteristics depending on the method of processing. Based on Table 1, the Meto tribe uses more leaves than other parts. This is because the leaves are easier to obtain and easier when processed to become natural coloring agents.

Based on Table 3, several methods for processing natural dyes that have been done by the Meto are obtained by boiling, pounding, kneading and soaking. Kumalasari [9] states that traditional processing is one of the processes carried out by still maintaining its authenticity inherited from generation to generation by elders.

The Meto tribe in Timor Tengah Selatan District mainly processes natural dyes by pounding them compared to other processing methods, and this serves to produce more concentrated colors. Usually, to get a more concentrated color, the Meto add additional ingredients (mordant) in the form of betel lime, loba, corn water, candlenut seeds, and coconut to the thread. For the Meto, adding betel lime, the material to be colored will produce a lighter color and can last longer (Table 2). According to Prabhu and Bhute [10], besides being able to strengthen colors, mordants function to enhance the fixation of the natural color on the fiber by the formation of a complex with the dye.

4. Conclusion

Based on the results and discussion, it was concluded that natural coloring plants used by the people of the Meto tribe in Timor Tengah Selatan regency are *Curcuma longa* L., *Indigofera tinctoria* L., *Morinda citrifolia* L., *Phylantus reticulates* Poir, *Phaseolus lunatus* L., and *Tectona grandis*. The plant organs used as a source of natural dye are leaves, barks, rhizome, and roots. The method of processing was to be crushed, boiled and soaked and the color produced was red, dark blue, yellow, black, and green.

5. References

[1] Noviyati S 2014 *Mengenal rupa-rupa tenun ikat NTT* Available from: http://travel.kompas.com/read/2014/05/22/1357500/mengenal.rupa-rupa.tenun.ikat.ntt

[2] Ati N H, Rahayu P, Notosoedarmo S and Limantara L 2006 The composition and the content of pigments from some dyeing plant for ikat weaving in Timorrese Regency, East Nusa Tenggara *Indo. J. Chem.* **6**(3) 325-31
[3] Sutara P K 2009 Jenis tumbuhan sebagai pewarna alam pada beberapa perusahaan tenun di Gianyar *Jurnal Bumi Lestari* **9**(2) 217–23
[4] Bernard H 2004 *Research Methods in Antropology: Qualitative and Quantitative Methods* 3rd Edition (California: Altamitra Press)
[5] Steenis V 2005 *Flora* (Jakarta: PT. Pradnya Paramita)
[6] Seran W and Hana Y W 2018 Weaving cloth coloring plant identification in Kaliuda Village Pahunga Lodu East Sumba *Jurnal Agribisnis Perikanan* **11**(2) 1-8
[7] Berlin S H, Linda R and Mukarlina 2017 Pemanfaatan tumbuhan sebagai bahan pewarna alami oleh Suku Dayak Bidayuh di Desa Kenaman Kecamatan Sekayam Kabupaten Sanggau *Protobiont* **6**(3) 303–9
[8] Winarno F G 2002 *Kimia Pangan dan Gizi* (Jakarta: Penerbit Gramedia)
[9] Kumalasari 2006 Studi etnobotani tumbuhan obat tradisional oleh masyarakat desa di Kotamadya Banjarbaru *Jurnal Bioscientiae* **2**(1) 25-36
[10] Prabhu K H and Bhute A S 2012 Plant based natural dyes and mordnats: A review *J. Nat. Prod. Plant Resour.* **2**(6) 649-64

Acknowledgments
Authors wishing to acknowledge United Board and Christian Higher Education in Asia for giving a small grant to conduct this project.

IOP Conf. Series: Materials Science and Engineering **823** (2020) 012038 doi:10.1088/1757-899X/823/1/012038

The effect of cationic CTAB surfactants on the performance of graphene electrode for supercapacitor

I Nurdin[1], H R Fitri[1], P Widiatmoko[1], H Devianto[1], T Prakoso[1]

[1]Department of Chemical Engineering, Institut Teknologi Bandung, Jl. Ganesha 10, Bandung, 40132, Indonesia

E-mail: isdi@che.itb.ac.id

Abstract. Graphene is a two-dimensional sp2 bonded carbon nanostructure packed in a honeycomb crystal lattice. Graphene has a high theoretical surface area and electrical conductivity that is suitable for the electrode of supercapacitors. There are many methods to produce graphene, such as mechanical exfoliation using scotch tape, reduction of graphene oxide, and chemical vapor deposition. An alternative and simple method to produce graphene is through pyrolysis process. The previous study shows that the production of graphene from biomass via two-stage pyrolysis process results in an increase of the surface area; however, its capacitance is still low to be applied as the electrode for supercapacitor. This study aims to modification of graphene surface using cetyltrimethylammonium bromide (CTAB) as the surfactant. The graphene was produced from palm kernel shell via two-stage pyrolysis method (the first stage was at 350°C followed by the second stage at 900°C) using $FeCl_3$ as catalyst and $ZnCl_2$ as activating agent, resulting in 16% yield. Graphene was analyzed using Scanning Electron Microscope (SEM), Transmission Electron Microscope (TEM), Raman scattering, X-ray Diffraction (XRD), and Fourier Transform Infrared (FTIR). These analyses show that the two-stage pyrolysis produces multi-layered graphene. The surface properties were analyzed by nitrogen adsorption-desorption measurements, which show some mesoporous graphene product with a surface area of 351.27 m^2/g. The result exhibited more hydrophilic graphene than the unmodified one, but according to the Cyclic Voltammetry (CV) analysis, the specific capacitance of modified graphene (11.91 Fg^{-1}) is lower than unmodified graphene (43.87 Fg^{-1}).

1. Introduction

Supercapacitors have become the most significant energy conversion and storage system in recent renewable and sustainable nanotechnology. Due to its large energy capacity, energy delivery in relatively short time, and longer lifetime, supercapacitors breakthrough in advance energy applications [1]. The energy storage in supercapacitors is based on ion adsorption (electrical double-layer capacitors, EDLCs) or fast and reversible Faradaic reactions (pseudocapacitors) [2]. EDLC store energy through the reversible adsorption of electrolyte ions on the surface of electrode materials by physical processes. Various carbonaceous materials, such as activated carbon and graphene, are used in EDLC [3, 4]. Activated carbon was extensively used as the electrode material in the construction of supercapacitors due to its large surface area and low cost. However, in such structures, there are a lot of carbon atoms that cannot be accessed by the electrolyte ions; consequently, this is a major factor that limits the specific capacitance of activated carbon electrodes. Additionally, it is reported that the low electrical conductivity of activated carbon limits its applications in high power density supercapacitors and results in a low specific capacitance per area unit of material [5].

Therefore, an outstanding candidate for electrode material is graphene. Graphene is a two-dimensional *sp2* bonded carbon nanostructure packed in a honeycomb crystal lattice. Graphene has recently been investigated to exhibit exceptionally high thermal conductivity, electrical conductivity, and strength. Another outstanding characteristic of graphene is its exceptionally high specific surface area up to 2675 m^2/g [6,7]. Unfortunately, since the entire carbon structure is graphene, it is classified as a hydrophobic material [8]. Graphene surface modification that has been performed was modification graphene using splitting induced by UV irradiation [9], dispersed graphene oxide in polysulfone [10], sonicated graphene in water and acetone [11]. Previously, surfactants were used to surface modification of graphene. Furthermore, it has been previously reported that the surfactant can improve the interface property between electrode and electrolyte solution. Surface modification using surfactant was a simple method and can increase the wettability of graphene. In this work, we aim to alter the property of the graphene surface to be hydrophilic by a simple method using a common surfactant to improve the interfacial properties.

Surfactants are organic compounds consisting of two parts: a water-loving (hydrophilic) and a water-hating (hydrophobic) portions. Nowadays, such substances are used for these purposes: improving the wetting ability, emulsifying, increasing the inhibiting properties of the electrodeposition process, etc. Sodium dodecylbenzene sulfonate (SDBS), tetradecyl dimethyl betaine (BS14), and cetylpyridinium chloride (CPC) were used to functionalize graphene, but only tetradecyl dimethyl betaine functionalized graphene achieves a good dispersion in water. Tetradecyl dimethyl betaine is difficult to obtain. Furthermore, it has been previously reported that CTAB surfactant can improve the interface property between electrode and electrolyte solution [12]. Here, cetyltrimethylammonium bromide (CTAB), a positively charged surfactant with quaternary ammonium group, is used for the surface modification of graphene. In this study, we took sonication and dip-coating to study on the effect of mixing method of graphene and CTAB. In our best knowledge, this is the first paper that explains the surface of hydrophilic graphene is not the major requirement that can increase the specific capacitance of graphene. Specific capacitance is affected by the crystallinity and surface area of graphene.

2. Methods

2.1. Graphene synthesis

Palm kernel shell (PKS) was collected from PT Perkebunan Nusantara VIII Bogor, Indonesia. Ferric trichloride ($FeCl_3$, 98%), hydrochloric acid (HCl, 37%), ethanol (Merck, 99,5%), and potassium hydroxide (KOH) (Merck, ≥84%) was used. Zinc chloride ($ZnCl_2$, 99,8%) were bought from PT. Smartlab, Indonesia. Cetyltrimethylammonium bromide (CTAB, 99%) was obtained from Himedia Laboratories Pvt Ltd, Indonesia. Dried palm kernel shell was crushed using a ball mill to obtain coarse granules of 170-200 mesh. The raw material was mixed with urea as a nitrogen source with ratio of 1:1, 1 M $FeCl_3$ as the catalyst, and $ZnCl_2$ as activating agent with ratio of 2:1, all respect to the raw material (w: w). The mixture was stirred and heated on a hot plate (80°C) for 2 hours. Furthermore, the mixed material was dried in the oven for 2 hours. The mixed material then was put into a tubular furnace for carbonization and activation. The tubular furnace was set under nitrogen (99,998%) atmosphere, heated to 350°C and keep at this temperature for 60 minutes. This was the first stage of pyrolysis. Afterward, the temperature was increased to graphitization stage at 900°C and kept the temperature for another 90 minutes under flowing nitrogen. This is the second stage of the pyrolysis. Furthermore, HCl 2 M was added to the solution and stirred for 30 minutes. Consecutively, the solution was washed with demineralized water several times until the rinsed water reached pH of 7. The product was then dried at 105°C in a conventional oven for 24 hours to obtain graphene, denoted as unmodified graphene (G-PKS).

2.2. Surface modification of graphene
Cetyltrimethylammonium bromide (CTAB) was used in our study to modify the surface of graphene. According to reference [13], 500 mg of graphene was diluted to 1 mg/ml by water. The equivalent weight of CTAB was added to the solution, and then the suspension was sonicated for 1 h to form a uniform dispersion. Afterward, the suspension was separated by filtration through a Whatman Filter Paper No. 41 and the powder was dried in oven at 80°C for 24 h, denoted as modified graphene (G-PKS-S). Surface modification of graphene is also performed by dip-coating. Surfactant coatings are prepared by dipping graphene into a CTAB solution. The samples are then pulled out of the solution at a constant speed and were dried in a conventional oven at 80 °C for 10 minutes, denotes as modified graphene (G-PKS-DC).

2.3. Characterization of graphene
Scanning Electron Microscopy (SEM) was performed by using a HITACHI SU3500 instrument operating at 10 kV. Transmission Electron Microscopy (TEM) was performed by using a HITACHI H9500 to study the graphene morphology. Raman spectroscopy analysis was recorded on a Renishaw Raman Microscope. X-ray diffraction (XRD) patterns were obtained using a Bruker D8 advanced diffractometer with monochromatized Cu K radiation. The specific surface area was recorded according to adsorption-desorption isotherms performed by the Surface Area and Porosity Analyzer (NOVA 3200e, Quantachrome Instrument). The specific surface area was calculated according to the Brunauer-Emmett-Teller (BET) equation. Contact angle measurement used for hydrophobic/hydrophilic property or wettability characterization of the graphene surface. Fourier transform infrared spectroscopy (FTIR) was employed to explore the functionalization of graphene.

2.4. Electrochemical analysis
The electrochemical measurements were firstly conducted in a 6 M KOH aqueous solution in a two-electrodes system. The negative and positive electrodes were prepared by mixing the produced graphene with polyvinylidene fluoride (PVDF) at a weight ratio of 90:10. The above materials were coated onto a 1 cm^2 stainless steel mesh current collector and dried at 80°C for 8 h. Filter paper was sandwiched between two electrodes. The test was performed using KOH 6 M aqueous electrolyte solution under ambient conditions using a potentiostat (Gamry V3000). Electrochemical characteristic of samples was studied using Cyclic Voltammetry (CV), Galvanostatic Charge and Discharge (GCD), and Electrochemical Impedance Spectroscopy (EIS) techniques.

3. Results and Discussions
The morphology of graphene resulted by pyrolysis process was analyzed using electron microscopy (SEM and TEM). Figures 1a, b, c show the SEM results of unmodified graphene (G-PKS), modified graphene by sonication (G-PKS-S), and modified graphene by dip-coating (G-PKS-DC), respectively. The G-PKS presents a continuous sheet-like structure of graphene layers that are stacked and folded together, which is beneficial for rapid ion and electron transport during the charge-discharge process of supercapacitors. After modification, the graphene layer is more piled, due to the interaction between nitrogen-doped graphene and CTAB. The interaction was revealed by FTIR analysis.

Morphological analysis of graphene from pyrolysis was also analyzed using TEM and HR-TEM. The TEM image of modified graphene in Figure 1c and d confirm that it consists of several layers with a distance of 0.354 nm between them. This is corresponding to the (002) interatomic distance of the graphitic carbon lattice [14]. The above results indicate that few-layer graphene has been successfully fabricated by two-stages pyrolysis method. The Selected Area Electron Diffraction (SAED) pattern in Figure 1f displays the representative hexagonal crystalline structure of graphene sheets, suggesting its high degree of graphitization.

The graphitic characteristic of G-PKS is confirmed by Raman spectroscopy. Three featured peaks located at 1362 cm^{-1} (D-band), 1569 cm^{-1} (G-band) and 2743 cm^{-1} (2D-band) were obtained from the G-PKS sample (Figure 2a). The D-band is corresponding to the breathing mode of aromatic rings with dangling bonds in planar terminations of disordered graphite and is related to the defects and disorders of structures in carbon materials, while the G-band is associated with the vibration of sp2-bonded carbon atoms in a two-dimensional hexagonal lattice [15]. The ratio between the G-band and D-band intensities (I_G/I_D) is proportional to the degree of crystallinity of carbon materials. The I_G/I_D value is 3.1, suggesting the quite high graphitization degree of G-PKS, which is in accordance with the above SAED results.

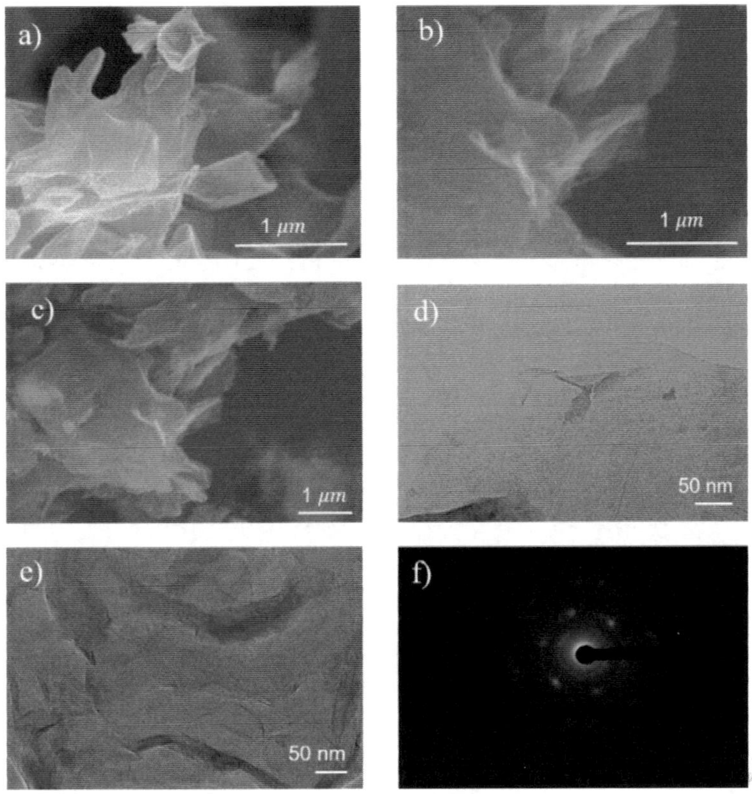

Figure 1. SEM images of graphene (a) G-PKS, (b) G-PKS-S, (c) G-PKS-DC. TEM of graphene (d) G-PKS, (e) G-PKS-S, (f) SAED image of G-PKS.

Figure 2b shows the X-ray Diffraction (XRD) pattern of G-PKS and G-PKS-S. The XRD patterns of both graphene types show a peak at $2\theta \sim 26.5°$, which indicates the atomic diffraction of (002) of graphitic structure. G-PKS and G-PKS-S showed the same intensity at $2\Theta \sim 26.5°$, suggesting that surface modification of graphene did not affect the crystallinity of graphene. XRD pattern of graphene shows impurities (such as cementite) peaks at 2Θ angle of 33.80° and 36.58°. The impurity is produced by the decomposition of the $FeCl_3$ catalyst during the pyrolysis process.

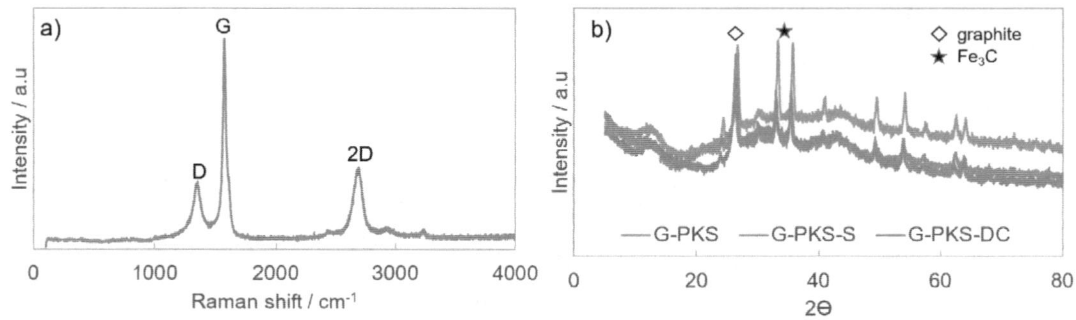

Figure 2. (a) Raman spectrum of G-PKS, (b) XRD patterns of graphene

The hydrophilicity of graphene was measured using the contact angle (CA). As shown in Figure 3a, the CA of the modified graphene is obviously much smaller than that of the unmodified graphene. Water penetration time also demonstrates the wettability of the modified membranes, as shown in Figure 3a. The CA of unmodified graphene decreased slowly and reached 0° within 13 s. However, the CA of modified graphene by sonication and dip-coating decreases rapidly and reached 0° within 3 s and 6 s, which exhibit much higher hydrophilicity than the unmodified graphene. This phenomenon was resulted by the CTAB which is able to reduce surface tension between interfacial phases. In other words, CTAB has the ability to increase the miscibility between polar and non-polar phases. Thus, the adjustment of the modified graphene surface with micelle of CTAB takes advantage of improving the compatibility of graphene and water [16,17].

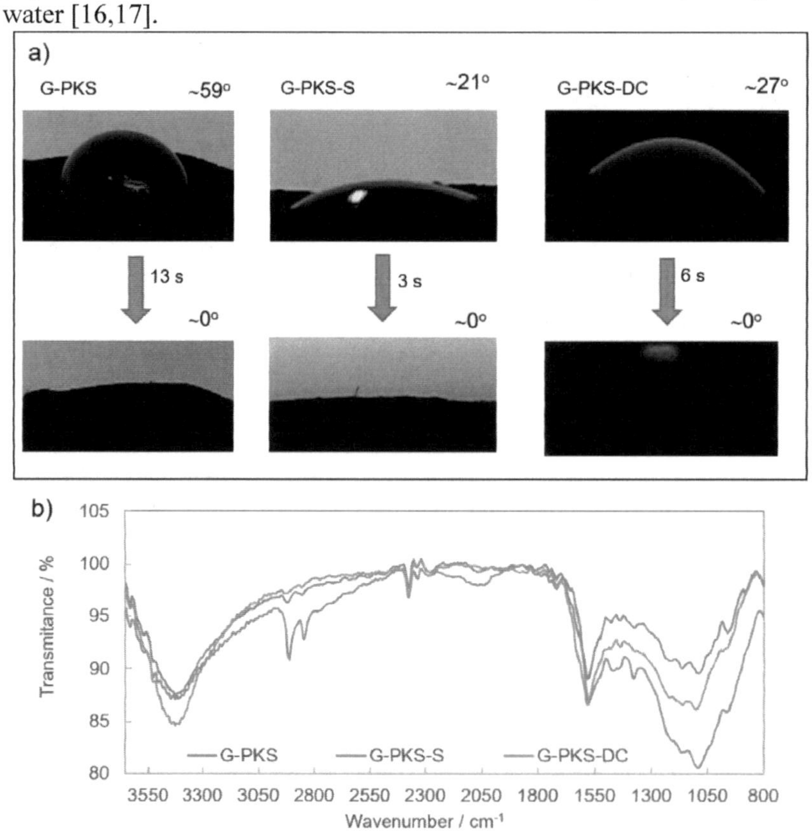

Figure 3. (a) Water contact angle and images of a water droplet on G-PKS, G-PKS-S, and G-PKS-DC (b) G-PKS, G-PKS-S, and G-PKS-DC FTIR spectra.

IOP Conf. Series: Materials Science and Engineering **823** (2020) 012038 doi:10.1088/1757-899X/823/1/012038

FTIR spectrum of produced graphene confirms the success of graphene surface modification. FTIR spectrums of unmodified and modified graphene were shown in Figure 3b. In all curves, there is a broad peak at 3400 cm^{-1}, which is attributed to the N-H stretching vibration of the amine group. The peak at ~1600 cm^{-1} is attributed to the N-H bending vibration of the amine group, and the peak at ~1200 cm^{-1} is attributed to C-N stretching vibration. After the modification process, the transmittance of N-H and C-N bonds were decreased; it showed an increase in N-H and C-N bonds in graphene [18]. These strong evidences demonstrated that the CTAB was attached onto the surface of graphene. The increase in the intensity of the C-N bond is quite significant, the C-N bond of nitrogen-doped graphene is coupled with the C-N bond of the amine bond on CTAB and the C atom bond on CTAB with N on nitrogen-doped graphene. This bonding reduces the distance between the graphene layers and reduces the active surface area of graphene. The peak at 1600 cm^{-1} is attributed to the C = C stretching vibration. The intensity of the C = C bond also increases; this shows the existence of a bond between C in the hydrophobic group ($C_{16}H_{33}$) in CTAB with the C atom in graphene [19]. FTIR peak G-PKS-S is significantly increased while G-PKS-DC is almost similar compared with G-PKS due to sonication cause the CTAB was attached onto the surface of graphene more strongly than dip coating.

The specific surface area of graphene was measured using nitrogen adsorption-desorption isotherms. If this graphene will be used as electrodes of the supercapacitor, it needs a large surface area beside a high degree of crystallinity [20]. Based on the calculation using the BET method, the surface area of G-PKS is 351.26 m^2/g, G-PKS-S is 187.82 m^2/g, G-PKS-DC is 302.52 m^2/g. The surface area of graphene has decreased after CTAB addition, due to the interaction between CTAB and graphene. It stacked the graphene layers and reduced the active surface area of graphene. The surface area of modified graphene by sonication has also decreased since sonication during 30 or 60 minutes breaks the graphene large sheets into rather small pieces. These results are in accordance with the above FTIR results.

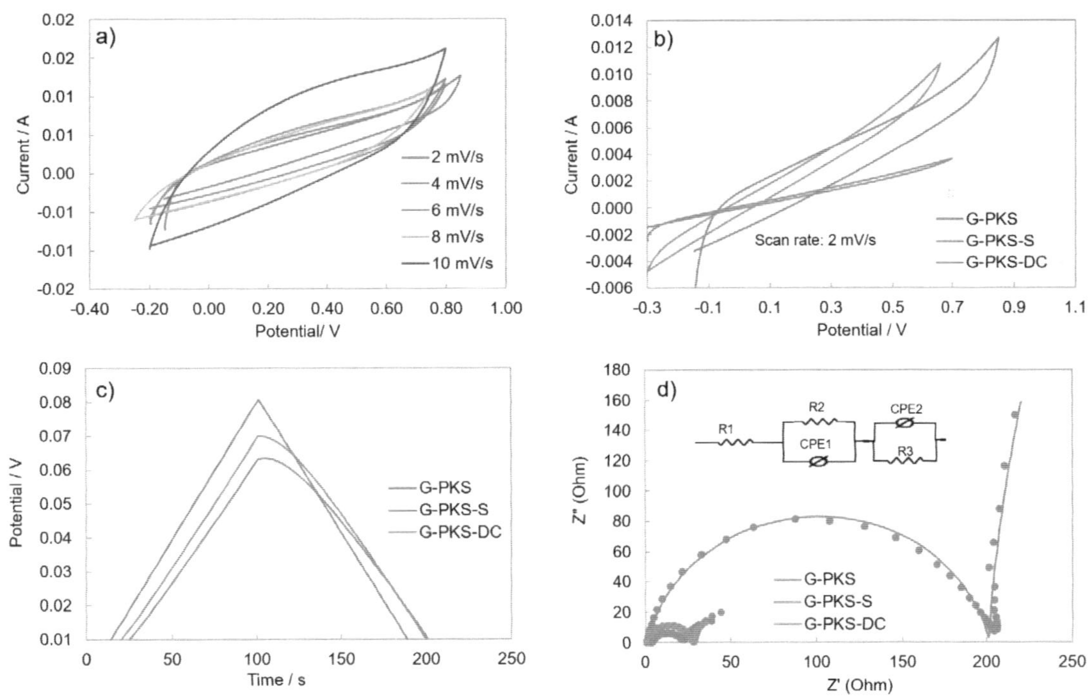

Figure 4. (a) CV curves of unmodified graphene at various scan rates ranging from 2 to 10 mV/s, (b) CV curves of unmodified and modified graphene at scan rate 2 mV/s, (c) Charge discharge curve and (d) Nyquist plot of unmodified and modified graphene.

The electrochemical characteristics of the EDLCs made of unmodified and modified graphene were examined using CV, GCD, and EIS. Cyclic Voltammetry was used to observe the cell characteristics and its capacitance. The measurements were performed at room temperature within the potential range of -0.2 to 0.8 V and over a scan rate of 2 to 10 mV/s. Figure 4a shows the voltammograms of the unmodified graphene symmetrical cell at five different scan rates (2, 4, 6, 8, and 10 mV/s). Figure 4a demonstrates that the current response increased with the scan rate. Table 1 shows the *Csp* values determined using Eq. (1) and the voltammograms of unmodified graphene symmetrical cell obtained at different scan rates. All cells show a common trend of decreasing *Csp* with increasing scan rate. It is well known that for very low scan rates, the *Csp* values are high because the ions have a much longer time to penetrate and reside in all the available electrode pores and form electrical double layers, which are needed to generate higher capacitance [21].

Figure 4b presents the CV curves of unmodified and modified graphene, and Figure 4c presents their GCD curves. The shape of the voltammograms for all the cells is quasi rectangular, representing a typical voltammogram of the Electrical Double-Layer Capacitor [22]. Comparing to the voltammograms of modified graphene by sonication (G-PKS-S), and modified graphene by dip-coating (G-PKS-DC) cells in Figure 4b, it can be observed that the G-PKS cell has a broader voltammogram area. These results indicate that the G-PKS cell has better cycle reversibility and higher electrical double-layer capacitance stability during the charge and discharge processes compared to the other cells. The specific capacitance of the produced graphene was calculated from cyclic voltammogram according to Eq. (1) [23]:

$$C = \sum|I|.\Delta t \ / \ m \ \Delta V \tag{1}$$

The specific capacitance of unmodified graphene, modified graphene with sonication and modified graphene with dip-coating respectively are 43.87, 11.91, and 37.76 F/g. After surface modification, the specific capacitance of graphene decreases. Since the presence of CTAB decreases graphene surface area, therefore it reduces the site of the electrochemical reaction, thus, reduces the occurrence of the electrochemical reaction. The reduction in electrochemical reactions is proportional to the reduction in electric current, so the amount of electric charge stored is also reduced.

Table 1. Specific capacitance of unmodified graphene at various scan rates

Electrode	Specific capacitance (F/g)				
	2 mV/s	4 mV/s	6 mV/s	8 mV/s	10 mV/s
Unmodified graphene	43.87	20.12	14.84	13.88	10.33

The GCD curves of all the graphene cells recorded at a current density of 1 mA cm^{-2} are shown in Figure 4c. As seen in this figure, all cells show a similar symmetrical triangular curve with a nearly linear variation of voltage as a function of time during charge and discharge. This type of curve is typical for carbon-based supercapacitors. However, despite having a similar shape, the curves of the G-PKS cells show higher peak potential than G-PKS-S and G-PKS-DC cells, as shown in Figure 4c. These observations indicate that the unmodified graphene cell has a higher capacitance than modified graphene. These results are in good agreement with the obtained CV results.

Table 2. Impedance parameters of graphene cell

Parameter	Unit	Value		
		Unmodified graphene	Modified graphene by sonication	Modified graphene by dip-coating
R1	Ohm	3.534	1.022	0.845
R2	Ohm	18.61	200	48.97
CPE1	S*s^a	5×10^{-4}	6×10^{-5}	3.95×10^{-1}
a1		7.53×10^{-1}	8.83×10^{-1}	9.15×10^{-1}
CPE2	S*s^a	1.11×10^{-1}	7.02×10^{-1}	6×10^{-5}
a2		8.52×10^{-1}	9.29×10^{-1}	8.78×10^{-1}
R3	Ohm	35.41	41.53	26.95

The EIS of graphene-based supercapacitor cells has been conducted in frequency range of 100 kHz to 100 mHz. Figure 4d shows the Nyquist plots of graphene-based supercapacitor. In the low-frequency region, the Nyquist plot is a straight line for an EDLC supercapacitor. The more vertical the line, the more closely the supercapacitor behaves like an ideal capacitor. Compared to the G-PKS and G-PKS-DC cells, the G-PKS-S cell results in the straightest line with an almost 90° angle, which is the characteristic of better capacitive behavior. However, G-PKS-S has largest semicircular portion that indicates G-PKS-S has high charge-transfer resistance, leading to poor electron transfer through the surface. Adding CTAB impeded the electron transfer due to CTAB has no property of direct improvement of the electrical conductivity. The semicircle diameter of the impedance loop represents the charge transfer resistance in the electrode material. In the high-frequency region, the curve intercept on the real axis is the Equivalent Series Resistance (ESR) [24]. The fitting equivalent circuit model by the coupled nonlinear Schrödinger equation (CNLS) method is displayed in the inset of Figure 4d. The whole capacitor circuit is constituted by R1, R2, CPE1, CPE2, and R3. The R1 is the sum of the contact resistance and material resistance (electrode materials, electrolyte, and separator). The R2 represents the charge transfer resistance at the electrode-electrolyte interface [25]. The R3 is the mass transfer resistance at electrolyte-electrode interface. The charge and mass transfer resistances (R2 and R3) increase after surface modification. It indicates that the CTAB attached on graphene surface hinders mass transfer to the graphene surface. The EIS results further demonstrate that the synergy effect between the large surface area and good electrical conductivity of electrode materials enhances the rate of electron and ion transport.

4. Conclusion

Nitrogen-doped graphene has been synthesized successfully through an easy, low cost and reliable two-stages pyrolysis from palm kernel shell. The graphene morphology shows a 2D structure. Raman studies provide evidence of the presence of graphene. XRD confirmed the formation of graphene. The addition of CTAB surfactant in the surface modification of graphene process enhances the hydrophilicity of graphene because CTAB has the ability to improve the miscibility between polar and non-polar phases. However, the specific surface area of graphene from PKS decreases from 351.26 m^2/g to 187.82 m^2/g. It causes the specific capacitance of graphene to decrease from 43.87 F / g to 11.91 F / g.

5. Acknowledgment

This work is supported by Badan Pengelola Dana Perkebunan Kelapa Sawit, funding year of 2018.

References

[1] Dale A C B and Craig E B 2012 *ChemComm* **48** 1425-27
[2] Sun L, Tian C, Li M, Meng X, Wang L, Wang R, Yin J and Fu H 2013 *J. Mat. Chem. A* **1** 6462-

70
[3] Frackowiak E 2007 *Phys. Chem.* **9** 1774
[4] Zhang S, Zeng M, Li J, Xu J and Wang X 2014 *J. Mat. Chem. A* 2014 **2** 4391-7
[5] Dale A C B and Craig E B 2014 *The Handbook of Graphene Electrochemistry* (London: Springer)
[6] Shahram G and Fatemeh A 2015 *J. Power Sources* **289** 129-37
[7] Chenguang L, Zhenning Y, David N, Aruna Z and Bor Z J 2019 *Nano Lett* **10** 4863-68
[8] Nontapol A, Pranee R, Orawon C and Nadnudda R 2017 *Talanta* **165** 692-701
[9] Zhemi X, Zhimin A, Dewei C, Adnan Y, Chang M L and Sean L 2014 *Sci. Rep* **4** 64650
[10] Ganesh B M, Arun M I and Ismail A F 2013 *Desalination* **313** 199-207
[11] Javad R, Mohammad A R, Zhong Z Y and Nikhil K 2010 *Adv. Mater. Technol* **22** 2151-54
[12] Robab K G, Zahra R and S Mojtaba T 2007 *J. Power Sources* **164** 934-46
[13] Jing M, Jianqiang L, Weihui Z and Wen Q 2018 *Colloids Surf. A* **538** 79-85
[14] Feng C, Juan Y, Tao B, Bo L and Xiangyang Z 2016 *J. Electroanal. Chem* **768** 18-26
[15] Zhongxin S, Mingtao Z, Hang H, Hanwu D, Yeru L, Yong X, Bingfu L and Yinfliang L 2018 *Chem Eng J.* **336** 550-61
[16] Nontapol A, Pranee R, Orawon C and Nadnudda R 2017 *Talanta* **166** 692-701
[17] Songfeng Z, Lili W, Feirong D, Depeng Z, Chao Z and Chaocan Z 2016 *RSC Adv* **6** 71287
[18] Alyaa K, Dayang R A B, Salmiaton A, Shamsul I, Muzab A R, Yusoff H M, Yasin F M and Suryani K 2016 *Int. J. Appl. Chem* **12** 104-8
[19] Dasli N 1986 *Elusidasi Struktur Senyawa Organik* (Bandung: Angkasa)
[20] Wang Z L, Zhang X B, Liu X J, Lv M F, Yang K Y and Meng J 2011 *Carbon* **49** 161
[21] Farma R, Deraman M, Awitdrus, Talib I A, Omar R, Manjunatha J G, Ishak M M, Basri N H and Dolah B N M 2013 *Int. J. Electrochem. Sci* **8** 257-73
[22] Inamdar A I, Kim Y S, Pawar S M, Kim J H, Im H and Kim H 2011 *J. Power Sources* **196** 2393
[23] Yu A, Chabot V and Zhang J 2013 *Electrochemical Supercapacitors for Energy Storage and Delivery* (New York: CRC Press)
[24] Wang Y, Shi Z, Huang Y, Ma Y, Wang C, Chen M and Chen Y 2009 *J. Phys. Chem. C* **113** 13103
[25] Kim Y J, Lee B J, Suezaki H, Chino T, Abe Y, Yanagiura T, Park K C and Endo M 2006 *Carbon* **44** 1592

IOP Conf. Series: Materials Science and Engineering **823** (2020) 012039 doi:10.1088/1757-899X/823/1/012039

Isolation, identification and antimicrobial activity of secondary metabolites of endophytic fungi from annona leaves (*Annona squamosa L.*) growing in dry land

Antonius R B Ola[1,2], Yoseph Sugi[3], Caterina S Lay[1,2]

[1]Chemistry Department, Faculty of Science and Engineering, University of Nusa Cendana, Kupang 85118 Indonesia
[2] Integrated Center Research (Bioscience) Laboratory, University of Nusa Cendana, Kupang 85118 Indonesia
[3]Faculty of Science and Engineering, University of Nusa Cendana, Kupang 85118 Indonesia

E-mail: antonius.ola@staf.undana.ac.id

Abstract. Research on isolation, identification, and antimicrobial activity of secondary metabolites of endophytic fungi from leaves of *Annona squamosa* growing in dryland had been undertaken. The work includes the isolation of endophytic fungi, cultivation, and extraction of fungal extract and identification of chemical metabolites together with the antibacterial test. The pure colony of endophytic fungi was grown on solid rice media using 1 L Erlenmeyer flask. The grown fungi were extracted with ethyl acetate. The ethyl acetate crude extract was then further subjected to chemical analysis and its antibacterial properties. Endophytic fungi species was identified as *Aspergillus niger* based on macroscopic and microscopic analysis. Analysis LC-MS/MS had revealed the presence of five metabolites including ephedradine A, ergosine, Ia, mudanpioside H, and trichosanic acid. The extract showed strong inhibition against *Staphyloccus aureus* with diameter of zone inhibition of 16.1 mm and moderate inhibition against *Escherichia coli* 0175H7 and *Salmonella enteritidis* ATCC 6939 with the observed diameter of zone inhibition of 9.6 mm and 11.3 mm, respectively.

1. Introduction

One of the important sources of bioactive compounds is from microbes including endophytic fungi that grow inside plant tissue such as leaves, flowers, seeds, twigs, stems, and roots without causing tissue damage in the host plant [1]. These fungi are rich in secondary metabolites and have been reported to provide and produce several *lead drugs* and *drugs* to cure deadly diseases such as cancer and bacterial infections [2]. Examples of endophytic fungus bioactive compounds that have been isolated from plants were equisetin, epi equisetin and beauvericin produced by *Fusarium equiseti* isolated from *Piper niger* leaves [3]. Another example was neosartorin, a potent antibiotic compound against multi-resistant bacteria without any significant toxicity observed, together with (-) Palitantin from endophytic fungi *Aspergillus fumigatiafinis* isolated from the leaf of the *Tribulus terrestris* (Zygophyllaceae) [4,5].

Several works regarding the bioactive fungal metabolites isolated from plants growing in Timor Island have been reported. The endophytic fungi *Xylaria* sp isolated from *Curcuma xanthoriza* leaves produced arugosin J, xylarugosin, and resacetophenone [6]. While the endophytic fungi *Diaporthe melonis* isolated from annona twigs produced diaporthemin A and B together with flavomannin dimethyl ether [7]. *Aspergillus flavus* isolated from the medicinal plant *Catharanthus roseus* growing in Timor Island was currently reported to accumulate high content of kojic acid, which has wide application in cosmetics and pharmaceutical industries [8]. In continuation of our work on finding fungal bioactive metabolites from Timor Island, we evaluated antibacterial extract from the endophytic fungi *Aspergillus niger* associated with *Annona* leaves.

2. Material and Methods

2.1 Material
The endophytic strains selected for the obtainment of crude extracts were isolated from the fresh leaf of *Annona squamosa L* located at Timor Island East Nusa Tenggara, Indonesia. Samples of leaves were transported to the laboratory in sterile baggage.

2.2 Isolation of endophytic fungi
The healthy leaf samples were washed several times using sterile distilled water and followed by immersing the tissues in 70% alcohol for 30 s and sterile distilled water for eliminating epiphytic microorganisms. The leaves were cut into 0.5 cm2 pieces and then transferred to Petri dishes containing potato dextrose agar supplemented with chloramphenicol (0.02 g) to suppress bacterial growth. After five days, all fungal colonies were isolated, purified, and maintained in PDA.

2.3 Cultivation and extraction of secondary metabolite
In order to get secondary metabolites for antimicrobial activity screening, the pure culture of every endophytic fungus was grown in 100 g of rice media and incubated for thirty days. After reaching its stationary phase, the fungi were extracted with ethyl acetate. Ethyl acetate was removed by rotary evaporator. The crude extract was analyzed for its chemical profile by using HPLC and LC-MS/MS.

2.4 Antibacterial assay
The paper disk method used to screen the antimicrobial activity of the EtOAc endophytic fungus extracts. The micro-organisms were one Gram-positive bacteria *Staphylococcus aureus camp.* and two Gram-negative bacteria *Escherichia coli* 0175H7 and *Salmonella enteritidis* ATCC 6939. The test microorganisms were cultivated in the test tubes comprising 2 g/100 ml of nutrient broth and incubated for 24 h at 37^0C. Turbidity was adapted to that of a standard for barium sulfate (0.5 McFarland). Paper disks were also inoculated with aquades (10 µl) as the negative control, tetracyclin (30 µg) as the positive control. On the surface of the medium containing bacteria test strain, each 10 µl of extracts were pipetted into 0.66 cm sterile paper disk. All plates were incubated for roughly 24 hours at 37^0C. Zones of inhibition were evaluated and recorded. Antibacterial activity screening was repeated twice.

3. Result and Discussion
A white and black strain of endophytic fungi was isolated from the leaf of *Annona squamosa*. Based on the macroscopic and microscopic characteristics and after comparison with endophytic fungus morphology according to [9], the isolated endophytic fungus was identified as *Aspergillus niger*.

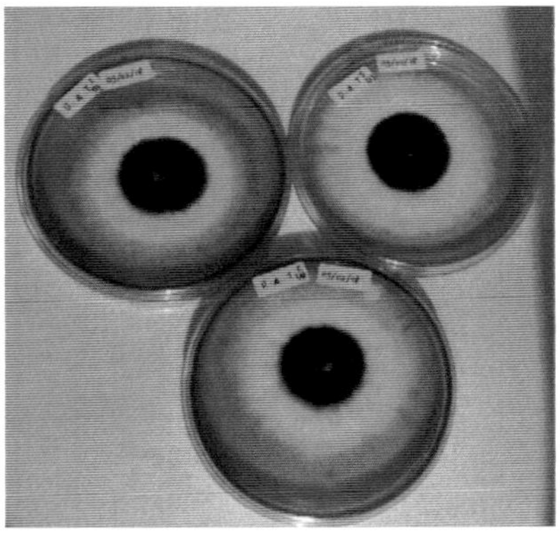

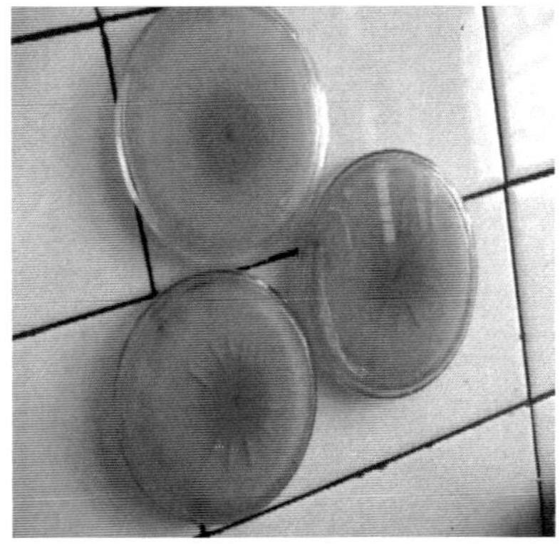

Figure 1. *Aspergillus Niger* from old annona leaves, (a) Front view and (b) Rearview

Figure 1 showed that the black and white endophytic fungi have been successfully isolated and purified from the leaves of the mature annona leaves.

Fungus cultivation was performed for 3 weeks so that the fungus was able to grow until its stationary phase was reached. The endophytic fungi were then extracted with ethyl acetate and left for two nights before filtration. The solvent was removed under vacuum using a rotary evaporator. The crude extract was then analyzed with HPLC and LC-MS for its chemical profile (Figures 2 and 3). In addition, the extract was also evaluated for its antibacterial properties.

The chemical content of fungal extract was performed with HPLC using C-18 analytical column at a flow rate of 0.5 mL/minute for 15 minutes and detection at 204 nm.

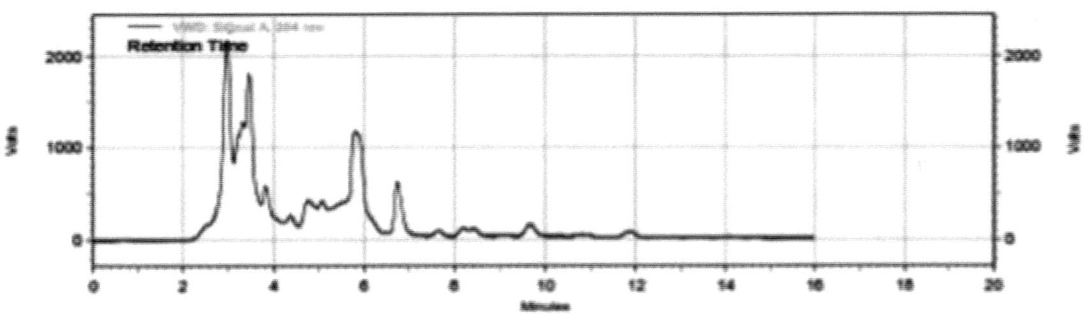

Figure 2. HPLC chromatogram

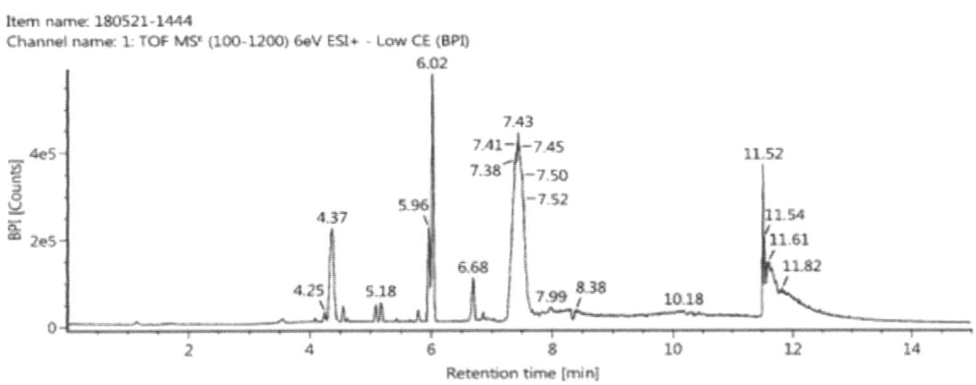

Figure 3. Chromatogram LC-MS / MS injection volume 1.00 µl

Five metabolites were able to be identified using LC-MS/MS method as shown in Table 1.

Table 1. Chemical components of *Aspergillus niger* (black endophytic fungi) ethyl acetate extract

No	Retention time (Minute)	Weight (m/z) [M+H$^+$]	Compound Weight (Dalton)	Compound
1	6,04	493,2814	492,27366	Ephedradine A
2	5,10	548,2873	547,27947	Ergosine
3	6,86	379,2506	378,24062	Ia
4	6,02	639,1695	638,17921	Mudanpioside H
5	6,68	279,2359	278,22458	Trichosanic acid

a. Identification of Ephedradine A

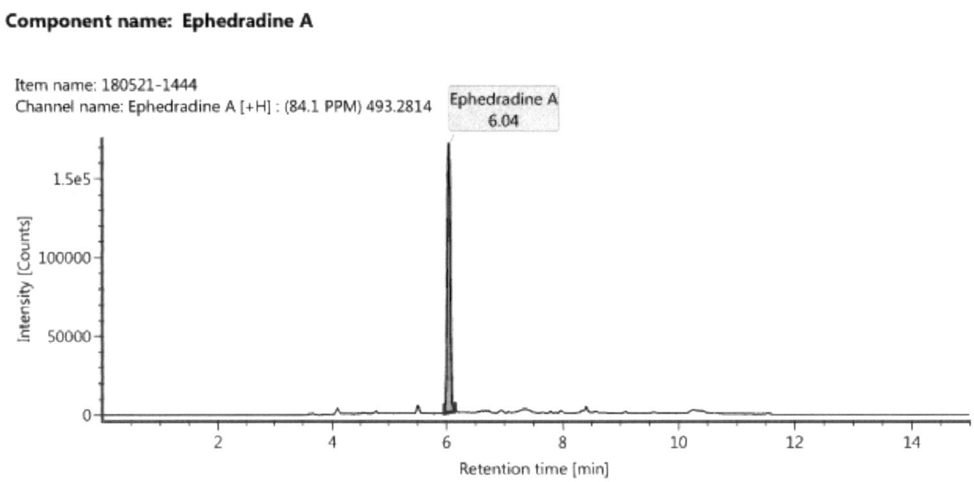

Figure 4. LC-MS / MS chromatogram of Ephedradine A

The structure of ephedradine A was shown in Figure 5.

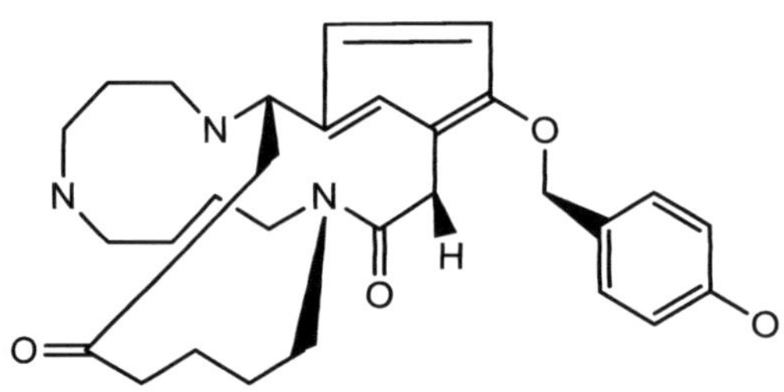

Figure 5. Structure of Ephedradine A compounds.

b. Identification of Ergosine

Component name: Ergosine

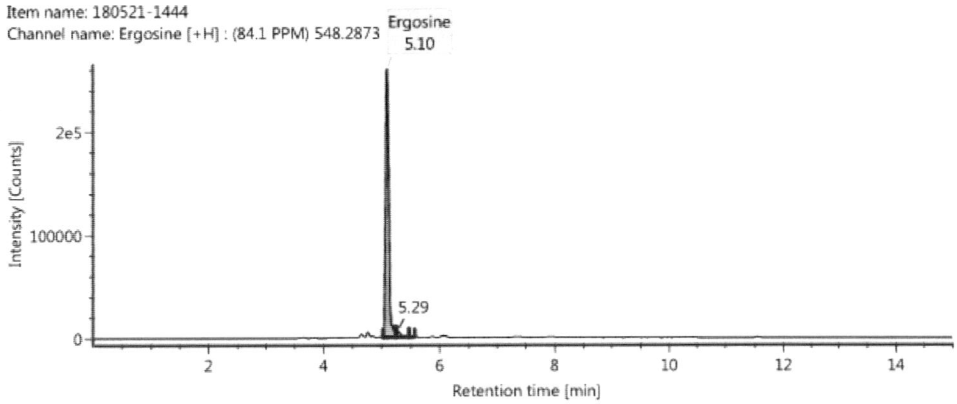

Figure 6. LC-MS / MS chromatogram of Ergosine

Based on the results of the LC-MS / MS chromatogram, the structure of Ergosine was shown in Figure 7.

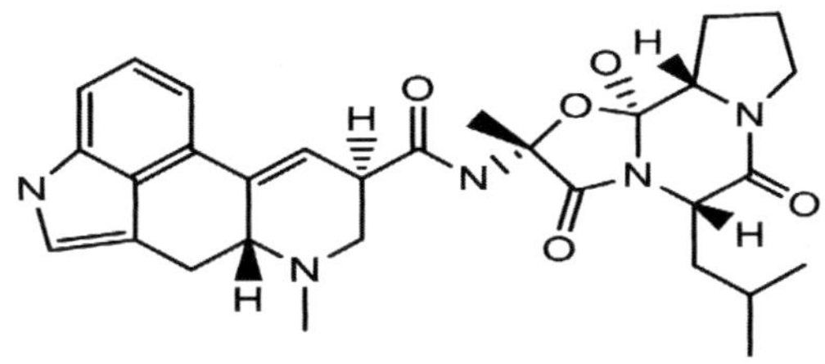

Figure 7. Structure of Ergosine compounds.

c. Identification of Ia

Component name: Ia

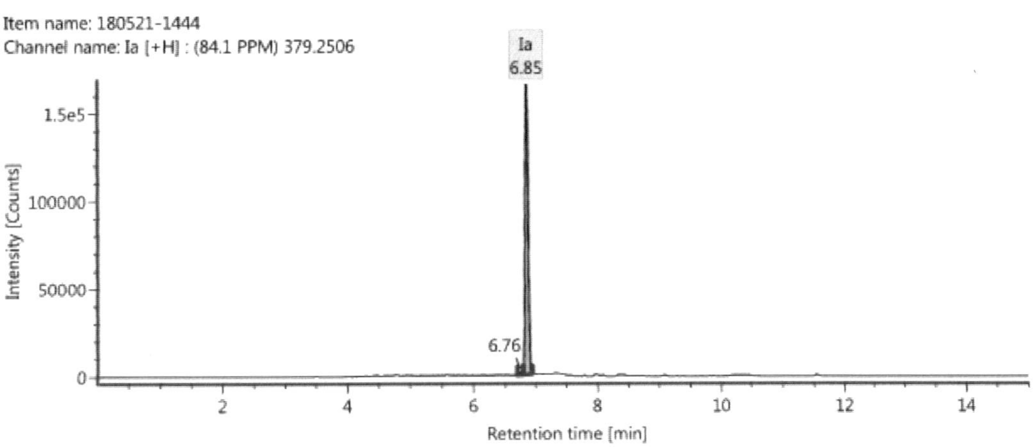

Figure 8. LC-MS / MS chromatogram of Ia

Based on the results of the LC-MS / MS chromatogram, the structure of Ia compounds was shown in Figure 9.

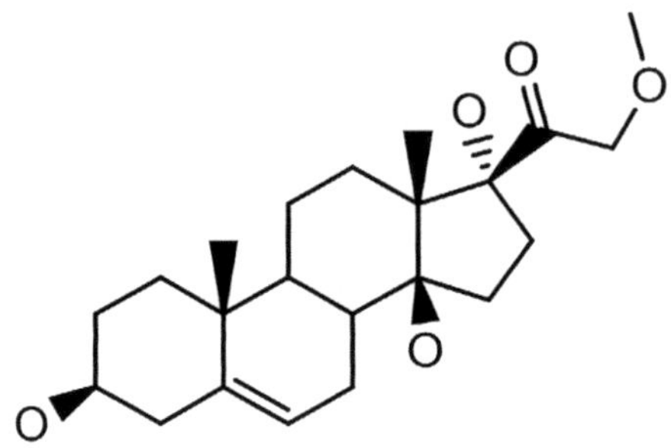

Figure 9. Structure of Ia compounds.

d. Identification of Mudanpioside H

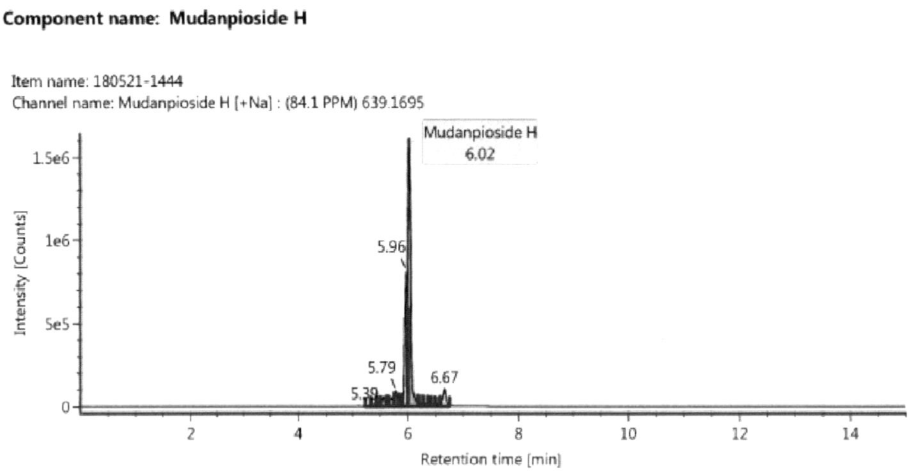

Figure 10. LC-MS / MS chromatogram of Mudanpioside H

Based on the results of the LC-MS / MS chromatogram, the structure of Mudanpioside H was shown in Figure 10.

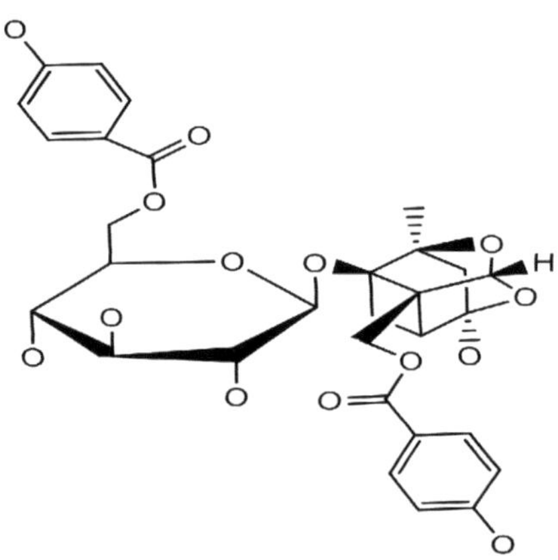

Figure 11. Structure of Mudanpioside H compounds.

e. Identification of Trichosanic acid

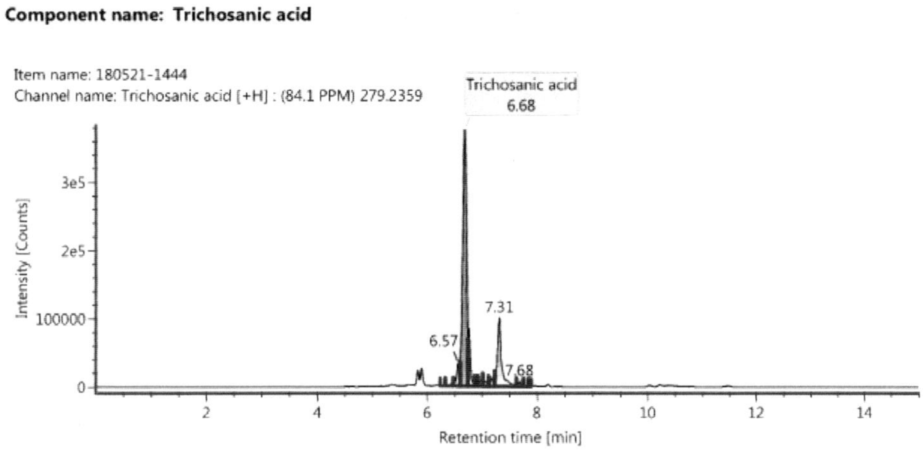

Figure 12. LC-MS / MS chromatogram of Trichosanic acid bioactive compounds

Based on the results of the LC-MS / MS chromatogram, the structure of Trichosanic acid compounds is shown in Figure 13.

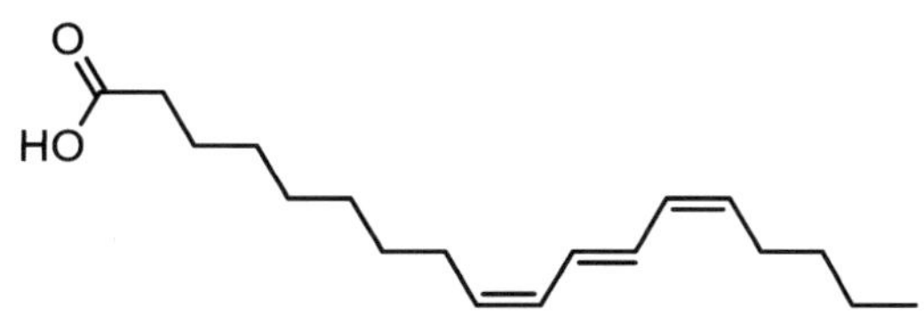

Figure 13. Structure of Trichosanic acid compounds.

The ethyl acetate fungal extract was further evaluated for its antibacterial properties against *S. aureus*, *E. coli* and *Salmonella enteritidis* with tetracycline as positive control and sterile distilled water as the negative control. The result was shown in Table 2.

Table 2. Diametre of zone inhibition of *Aspergillus Niger* ethyl acetate extract

No	Type of Bacteria	Diametre of zone inhibition (mm)		
		Aspergillus Niger ethyl acetate extract (10 μg)	Tetracycline (positive control) (30 μg)	Aquades (negative control) (10 μL)
1	*Staphylococus aureus camp*	16,1	26,8	0,6
2	*Escherichia coli 0175H7*	9,6	28,7	0,6
3	*Salmonella enteritidis ATCC 6939*	11,3	24,5	0,6

Gram-negative and gram-positive bacteria used in this study had a concentration of 10^8 cfu/mL. The fungal extract was found to inhibit bacterial growth at a concentration of 10 μg. Of all the identified components, Mudanpioside H was probably responsible for the antibacterial activity of the extract as it was previously reported to have moderate antibacterial activity [9]. However, further fractionation and isolation of pure metabolite from the fungal extract should be undertaken.

IOP Conf. Series: Materials Science and Engineering **823** (2020) 012039 doi:10.1088/1757-899X/823/1/012039

References

[1] Namasivayam S K R, Swetha R and Srivatsan K V 2014 Evalution of potential biological activities of metabolites from endophytic fungi residing in leaves of *Azadirhacta indica Int. J. ChemTech. Res.* **5** 3116–21

[2] Yu H, Zhang L, Zheng C, Guo L, Li W and Sun L 2010 Recent developments and future prospects of antimicrobial metabolites produced by endophytes *Microbiol Res.* **165**(6) 437–49

[3] Ola A R B, Amal H A, Wen H L, Victor W and Abdessamad D 2014 Absolute configuration and revision of the structure of lateritin *Tetrahedron Letters.* **55** 3147–50

[4] Ola A R B, Amal H A, Ilka Z, Alexandra H, Attila M, Matthias K, Heike B O, Tibor K, Peter, Proksch and Abdessamad D 2014 Absolute configuration and antibiotic activity of neosartorin from the endophytic fungus *Aspergillus fumigatiaffinis Tetrahedron Letters.* **55** 1020–23

[5] Ola A R B, Tawo B D, Belli H L L, Proksch P, Tommy D and Hakim E H 2018 A new antibacterial polyketide from the endophytic fungi *Aspergillus fumigatiaffinis Natural Product Communications* **13**(12) 1934578X1801301202

[6] Hammerschmidt L, Ola A, Werner E G M, WenHan L, Attila M, Tibor K, Peter P and Amal H A 2015 Two new metabolites from the endophytic fungus *Xylaria* sp. isolated from the medicinal plants *Curcuma Xanthorrhiza Tetrahedron Letters.* **56** 1193–97

[7] Ola A R B, Heike B O, Abdessamad D, Peter P and Amal H A 2014 Dihydroanthracenone metabolites from the endophytic fungus *Diaporthe melonis* isolated from *Annona squamosa Tetrahedron Letters.* **55** 3133–36

[8] Ola A R B, Metboki G, Lay C S, Sugi Y, Rozari P D, Darmakusuma D and Hakim E H 2019 Single production of kojic acid by *Aspergillus flavus* and the revision of flufuran *Molecules* **24**(22) 4200

[9] An R.-B, Kim H.-C, Lee S.-H, Jeong G.-S, Sohn D.-H, Park H, Kwon D.-Y, Lee J. H and Kim Y.-C 2006 A new monoterpene glycoside and antibacterial monoterpene glycosides from Paeonia suffruticosa *Archives of Pharmacal Research* **29**(10) 815

Upscaled synthesis of carbon nanotube from palm oil mill effluent using pyrolysis for supercapacitor application

P Widiatmoko, N F Nugraha, H Devianto, I Nurdin, and T Prakoso

Department of Chemical Engineering, Bandung Institute of Technology, 40132, Bandung, Indonesia

E-mail: pramujo@che.itb.ac.id

Abstract. The utilization of Palm Oil Mill Effluent (POME) is advantageous to reduce environmental problems and increase its value as a source of carbon. Carbon nanotube (CNT) was synthesized by pyrolysis of POME and ferrocene mixture at 900°C under a flowing nitrogen atmosphere. CNT is nanosized and has high electrical conductivity; it is compatible as electrodes for energy storage devices such as supercapacitor. The study investigated the process for the upscaled formation of CNT from POME. The synthesis consists of POME polymerization, pyrolysis, and CNT characterization. Additionally, before pyrolysis, the CNT collecting system was applied in tube furnaces. The result showed that this process produced 1.2 g of CNT, a diameter of 47-100 nm, and a surface area of 635.5 $m^2.g^{-1}$. XRD analysis, 2Θ of 26.58° indicated structure of graphite C (002), Raman spectra resulted ratio I_G/I_D of 1.16 indicated higher graphite than disorder. CNT was characterized and applied as electrode material for supercapacitor which exhibits a specific capacitance of 13.98 $F.g^{-1}$.

1. Introduction

Indonesia is the largest palm oil producer in the world with a capacity of 37.8 million tons per year [1]. As a consequence, abundant wastes are generated. Palm oil mill effluent (POME) is the liquid waste that represents 62% of total waste. From each ton of processed palm oil fresh fruit, 20 m^3 POME is produced [2]. Palm oil mill has been regarded as a profit-making industry for the past decades. Besides revenue from the palm oil production itself, the abundance of biomass could generate high economic return to the palm oil mill by converting it to value-added products [2]. POME contains rich organic compounds, so it has potential as a carbon source to produce carbon nanotubes (CNT). Several carbon precursors such as methane, acetylene, benzene have been used as a carbon feedstock to synthesize CNTs [3–5]. These carbon precursors are related to fossil fuels and there may be a crisis for these precursors soon. There are few reports on the synthesis of CNTs from natural precursors, e.g. turpentine oil, neem oil, and even organic wastes, like Bio-oil PCB and waste cooking oil [6-8]. The advantage of using natural precursors as a carbon feedstock for synthesizing CNTs is that they are inexpensive and have no chance of shortage soon.

Many processes have been reported for their syntheses, such as arc discharge, laser ablation, chemical vapor deposition (CVD), and Pyrolysis [9-12]. CVD is the most promising method for synthesis CNT but expensive. We have been using pyrolysis due to simple and inexpensive method [13]. One of them, this method vaporizes carbon precursor at lower temperatures and pyrolyzes at

IOP Conf. Series: Materials Science and Engineering **823** (2020) 012040 doi:10.1088/1757-899X/823/1/012040

higher temperatures in one step furnace while in CVD takes two steps. The pyrolysis method has attracted attention due to the possibility of producing carbon nanotubes on a commercial scale [14].

CNT is a cylindrical allotrope of carbon with nanometric size. CNTs have a high electrical conductivity of 10^7 $S.cm^{-1}$. Application of CNTs in energy storage, such as supercapacitor, produces specific capacitances of 12-120 $F.g^{-1}$. KPEE-ITB Laboratory succeeded in producing CNT from POME by pyrolysis. However, the yield of CNT is still low, and the capacitance of supercapacitor does not meet commercial standards [15]. The study will investigate how to increase the production of CNT by adding three times the amount of raw materials and utilization of the CNT collecting system made by stainless steel mesh at the end of the tube furnace.

2. Methods

2.1. Materials

CNT was produced from Palm oil mill effluent (POME) which was collected from the palm oil mill of PT. Perkebunan Nusantara VIII in Bogor. The POME can also be described as wastewater rich in organic carbon with a colloidal suspension of 95 - 96% water, 0.6 - 0.7% oil, and 4 - 5% total solids, including 2 - 4% suspended solids [6]. Formaldehyde water solution (Merck, purity~37%) used to polymerize POME and ammonium hydroxide (Merck, purity~28%) as catalyst. Pyrolysis of the resin used ferrocene (purity~98%) as a catalyst in nitrogen flow. Ferrocene is frequently used as a precursor to prepare carbon nanostructures because it cannot only act as a carbon source but also give rise to small metal clusters as catalysts.

2.2. CNT Production

Figure 1 represented the process flow diagram of CNT synthesis. POME has to be polymerized with a 37% formaldehyde solution with a weight ratio of 5:6.5, and ammonium hydroxide as catalyst at 95°C for 4 hours. The resin formed is heated at 60°C for 2 hours and at 120°C for 12 hours. After that, the resin was ground to fine powder, mixed with 20-30%-w ferrocene, and homogenized with ethanol at 50°C until the mixture is uniform and all ethanol is evaporated. Afterward, the mixture is loaded in a ceramic crucible and placed in the tube furnace.

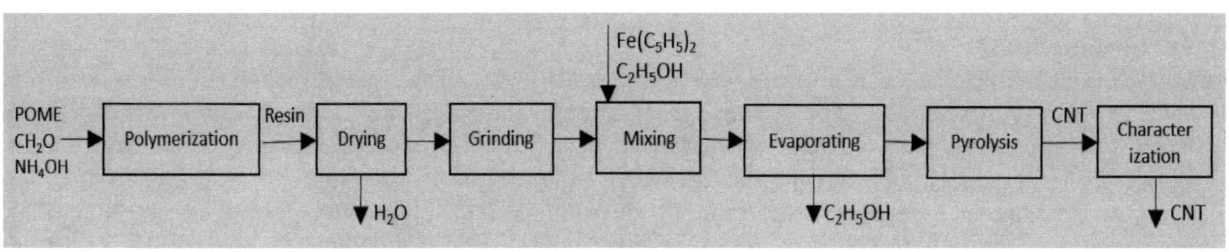

Figure 1. Process flow diagram of CNT synthesis.

Figure 2 represented the utilization of the CNT collecting system made by stainless steel mesh at the end of the tube furnace. The pyrolysis process consists of 2 steps; the first step is conducted at 200°C during 1 hour with heating rate of 5°C/min in natural atmosphere, and the second is run at 900°C for 1 hour with a heating rate of 3°C/min in flowing nitrogen atmosphere.

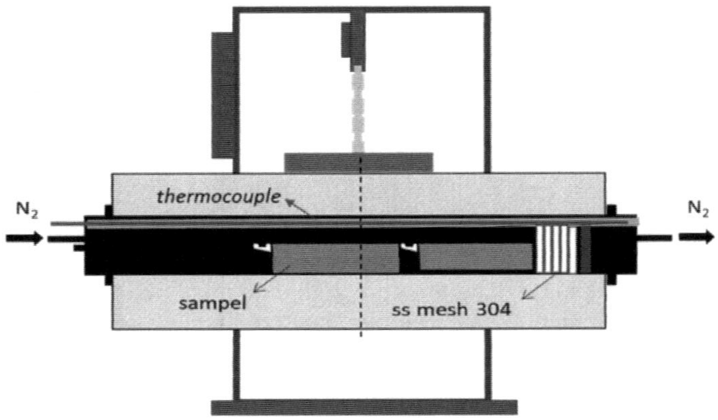

Figure 2. Furnace with CNT collecting system made by stainless steel mesh at the end of the tube furnace.

2.3. *Characterization*

The surface morphology of CNT was investigated by using scanning electron microscopy (SEM) (Hitachi SU3500, Japan) and using a transmission electron microscope (TEM) (Hitachi H9500, Japan). The Crystal structure of CNT was obtained from a powder X-ray diffraction pattern by using a Bruker D8 advanced diffractometer with monochromatized Cu K radiation. Nitrogen sorption isotherm of the sample was measured using a NOVA 3200e, Quantachrome Instruments, USA. Surface areas are obtained from the nitrogen sorption isotherm using multipoint Brunauer-Emmett-Teller (BET). A Raman spectrum of the CNT was measured by Raman spectrometry (Renishaw, model no. H 45517) using an argon ion laser (λ=532 nm).

The electrochemical performances of CNTs were studied by cyclic voltammetry (CV), galvanostatic charge/discharge (GCD) and electrochemical impedance spectroscopy (EIS) methods. Electrochemical measurements were carried out using Gamry V3000 potentiostat–galvanostat, utilizing a symmetrical electrode configurations with 6 M KOH solution as electrolyte at room temperature. Electrode of the symmetrical cell was prepared by mixing CNT with polyvinylidene fluoride (PVDF), which is coated on 1 cm^2 of stainless steel mesh and dried at 80°C for 1 day. Nafion 212 impregnated with the electrolyte, was placed between both electrodes. The specific capacitances of supercapacitors were computed from CV curves by using the following equation (1) [16].

$$C_{sp} = \frac{2}{m|V_2 - V_1|} \int_{t=0(V_1)}^{t(V_2)} i(V)dt \tag{1}$$

Where 'C_{sp}' is the specific capacitance ($F.g^{-1}$), 'm' is the mass of the active electrode (g), 'V' is the cell voltage (V) and integral of i(V)dt is the area under the CV curve. Specific capacitances are also computed from the GCD by using the following equation (2) [17].

$$C_{sp} = \frac{2I}{[(\Delta V/\Delta t)m]} \tag{2}$$

Where 'I' is the current (A), 'ΔV' is the discharge voltage difference between Δt period, and 'm' is the mass of the active electrode (g).

3. Result and Discussion

Increasing of raw material amount and utilization of a stainless steel mesh-based collecting system increased the yield of CNT from 0.46% to 0.96% (the maximum fix carbon yield was 2.94%). The growing mechanism of CNT during the synthesis can be proposed as follows: nanometer iron produced from ferrocene decomposition accumulates carbon on its surface with the formation of the Fe–C solid solution. After a very short period of time, the Fe–C solid solution reached a supersaturated point and carbon atom started to nucleate and grow in a nanotube shape during the slow growth [8]. The CNT that moved from the ceramic boat was caused by the influence of the growth mechanism of nanocarbon at 900°C. When the catalyst is mixed directly with the raw material, CNT growth in the gas phase and is carried by inert gas so that the product covers the inner wall of the tube furnace [9].

The morphology of CNT was identified using SEM and TEM. There are 4 samples of CNT, which are commercial CNT (CNT CO), CNT before scale up (CNT BS), CNT after scale-up at ceramic crucible (CNT CC) and CNT after scale-up at stainless steel mesh (CNT SS). The SEM observation results are shown in Figure 3.

Figure 3. SEM micrographs of (a) CNT CO, (b) CNT BS, (c) CNT CC, and (d) CNT SS.

The SEM images showed that all of CNT were cylindrical. Figure 3a indicated the CNT with a diameter between 5 and 10 nm. CNT with a smaller diameter will give larger capacitance in supercapacitor [6]. CNT BS has a diameter between 30 and 72 nm. Increasing three times amount of raw materials produced CNT with smaller distance diameter of 47-59 nm for CNT CC and 80-100 nm for CNT SS. Transmission electron microscope (TEM) is widely accepted that the carbon nanotube diameter can be determined by directly measuring the distance between two dark lines associated with carbon nanotubes in a TEM image. Representative TEM overviews of the samples investigated are shown in Figure 4. From a qualitative point of view, these samples mostly consist of multiwalled nanotubes with a hollow inside.

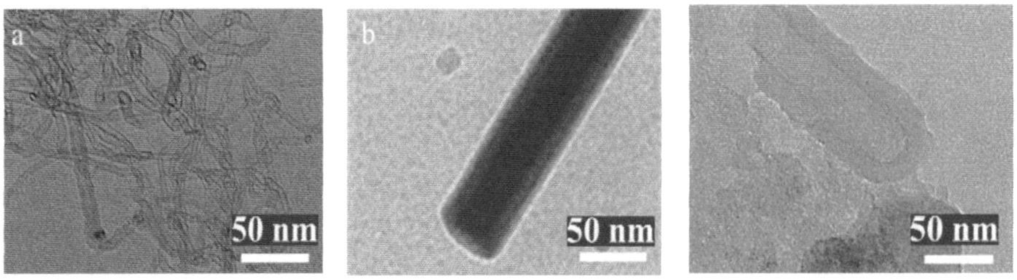

Figure 4. TEM micrographs of (a) CNT CO, (b) CNT BS, and (c) CNT CC.

In particular, the samples labeled CNT CO, BS and CC indicated similar diameter results with SEM. It presents the smaller diameter for CNT CO with the larger capacitance in the supercapacitor. On the contrary in Table 1, CNT before and after scale-up have a similar diameter. CNT has an external diameter of 8.9-57 nm, and multiwalled nanotubes (MWCNT) have an external diameter of 5-100 nm [18]. The large specific surface area includes a relatively small contribution from the surface area of the particles and a major contribution from the pores inside the particles [19].

The internal diameter of CNT or pore of CNT can increase the active surface area for the electrical double layer between electrodes and electrolytes in the supercapacitor [20]. All samples have a diameter of 2-50 nm, which is the size range of the mesopore regime. When the pore size is in the mesopore regime, counter-ions enter mesoporous carbon materials and approach the pore wall to form an electric double-layer capacitor (EDLC) [15]. The pore size under the mesopore will have an impact on the low value of the specific capacitance of the supercapacitor [19].

Table 1. Diameter measurement results from TEM micrographs.

Sample	Average Diameter (nm)	
	Internal	External
CNT CO (a)	4.8	8.9
CNT BS (b)	25	43
CNT CC (c)	26	57

Raman spectroscopy is an important tool for studying CNT samples, which provides information about the structure and the presence of disorder in the sample. Figure 5 shows the representative Raman spectra of CNT CO, CNT BS, CNT CC, and CNT SS. In the Raman-shift range of 500-2000 cm corresponding to D and G bands, respectively. The G band-oriented pyrolytic graphite and suggests the CNT is composed of crystalline graphitic carbon. The higher intensity of the G band peak indicates a higher degree of graphitization/crystallinity. The D band-oriented disorder in the sp2-hybridized carbon and indicates lattice distortions in the curved tube ends. The intensity ratio of G and D peaks ($I_G.I_D^{-1}$) is used to characterize the purity of CNT. Higher $I_G.I_D^{-1}$ ratio indicates a higher degree of graphitization [6].

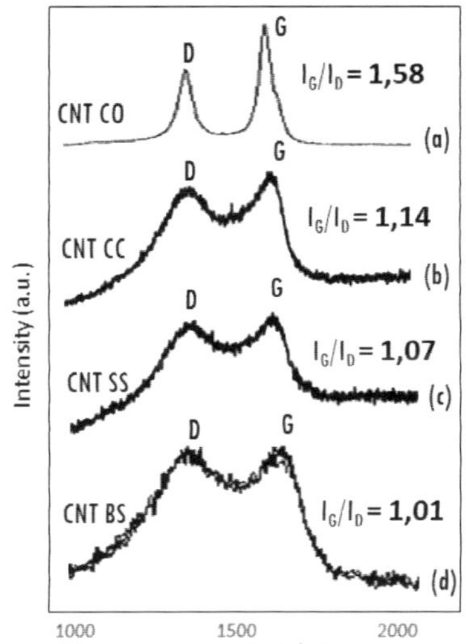

Figure 5. Raman spectra of (a) CNT CO, (b) CNT CC, (c) CNT SS, and (d) CNT BS.

Figure 5 shows crystalline CNTs more dominant than disorder carbon. It can be seen from all CNT samples which have a higher G-band than the D-band. Commercial CNT has the highest ratio of 1.58 while the CNT results of this study have a ratio of 1.01 to 1.14. Although 1.01 to 1.14 is not the best value, it is much higher and encouraging than many recent reports of CNTs by similar thermal decomposition ($I_G.I_D^{-1}$= 0.85–1.3) [21]. Raman's analysis shows CNT CC has a higher $I_G.I_D^{-1}$ ratio than CNT SS. CNT CC is formed with more stable temperature conditions (the location is located right in the heating area) while CNT CC is formed at a lower temperature because it moves away from the heat source of the tube furnace [22].

Figure 6 shows the typical XRD pattern of the CNT sample. The peaks at 2θ of 29.098° with intensity of 3.173 Å, slightly lesser a perfect graphite intensity (3.354 Å) are indexed to be the (002) [8]. The peak on the widened XRD spectrum shows the amorphous carbon structure generated by the destruction process of the initial graphite structure during the pyrolysis process. The resulting carbon structure is an intermediate structure between graphite and amorphous structures called turbostratic structures or random layered lattice structures [23].

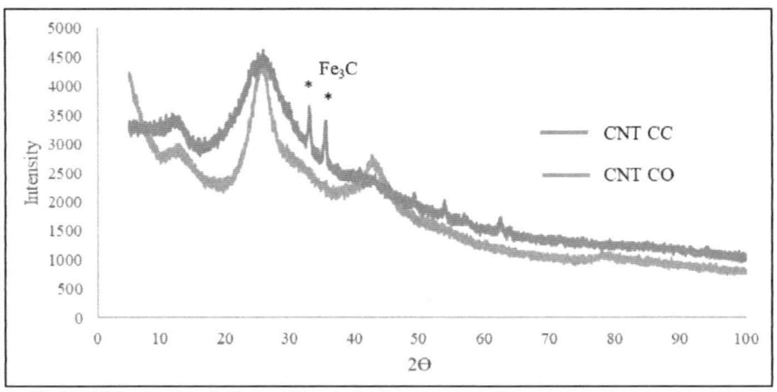

Figure 6. X-ray diffractogram of CNT CC and CNT CO.

Before and after the scale-up, the results of XRD showed the same thing, that the CNT POME had a peak of impurities such as cementite at 33.80° and 36.58°. The impurity is formed due to POME resin decomposition and ferrocene catalyst decomposition. Fe from catalyst decomposition and carbon from resin decomposition react to iron carbide or Fe_3C (cementite) [24].

Analysis of nitrogen absorbed on the surface of the CNT was measured by the method of Brunauer Emmet Teller (BET) to determine the surface area. The surface area of a CNT comes from the surface area of the outer tube (first layer) and inner tube (n-layer). The BET results in Table 2, CNT CC have the highest surface area while CNT SS approaches commercial CNT. The higher the surface area, the less CNT layers are formed. CNT with the SWNT type has a surface area of more than 1200 m^2g^{-1} [18].

Table 2. Surface area measurement results from BET method.

Sample	Spesific surface area (m^2g^{-1})
CNT CO	644.62
CNT CC	669.10
CNT SS	522.68

CNT which produced is aimed for supercapacitor. Therefore, the electrochemical performances of CNT were studied by cyclic voltammetry (CV), galvanostatic charge/discharge (GCD) and electrochemical impedance spectroscopy (EIS) methods. CV study of supercapacitor with CNT electrode was performed in the potential window between -0.4 to 0.4 V with a scan rate from 2 to 10 $mV.s^{-1}$. Figure 7a shows the CV curves of supercapacitors with various scan rates. non-ideal CV curves (not quasi-rectangular) were obtained at all scan rates. The deviation of the rectangular shape in the CV curve ensures the presence of both non-Faradaic reaction / electrical double layer (EDL) and faradaic reaction (pseudo-capacitance) in charge storage.

In the case of supercapacitor with CNT electrode, the pseudo-capacitance associated with redox reactions of surface functional groups is always coupled with the non-Faradaic double-layer capacitance. The specific capacitance will be influenced by functional groups or rather the oxygenated functional groups. Oxygenated functional groups absorbed on the surface of a porous electrode could be oxidized or reduced with the electrode potential as shows on reaction [25].

$$O_2 + H_2O + 2e^- \rightarrow HO_2^- + OH^- \quad (3)$$

The area under the curves increases with the decrease of scan rates, which is mainly due to a higher scan rate, ionic diffusion in pores become faster and induces more ionic collisions. Too large numbers of ionic collisions will produce an irregular electrical double layer, which has a lower capacitance [26-28]. Table 3 shows that the supercapacitor with the CNT electrode offers the highest specific capacitance at a scan rate of 2 mV.s^{-1}.

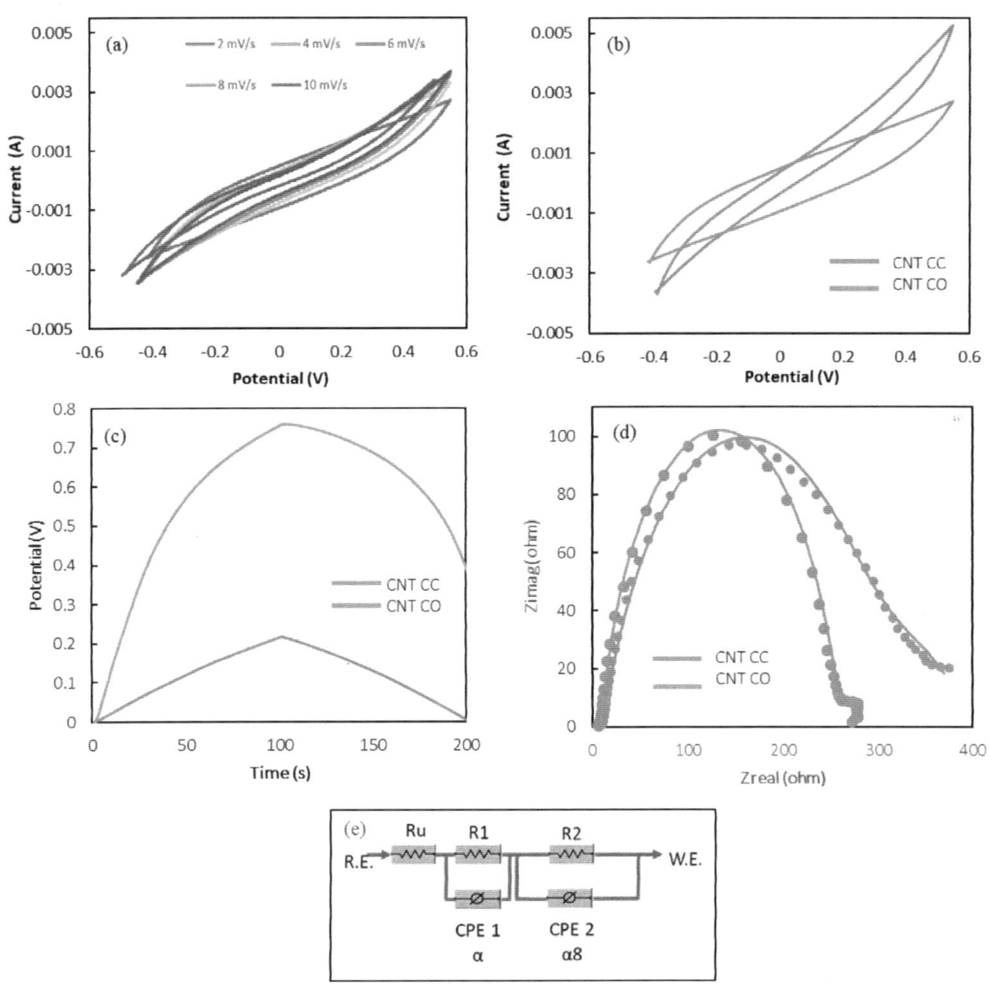

Figure 7. (a) CV curves of supercapacitor with different scan rates, (b) CV curves of supercapacitor with different electrodes, (c) GCD curves of supercapacitor with different electrodes, (d) Nyquist plots of supercapacitor with different electrodes, and (e) equivalent circuit model.

Figure 7b show the CV curves of supercapacitor with various CNT electrode and Table 3 show the specific capacitance. Specific capacitance supercapacitors with electrodes from CNT POME was still low than commercial CNT. Commercial CNT have the highest specific capacitance (23.32 $F.g^{-1}$) due to the higher crystallinity of commercial CNT (Raman spectroscopy results), have smaller pore diameters and have more CNT thereby increasing the formation of electrical double layers (TEM results) [18].

Table 3. Specific capacitance as a function of scan rate with different electrodes.

Sample	Spesific capacitance at scan rate variation ($mV.s^{-1}$)				
	2	4	6	8	10
CNT CO / ($F.g^{-1}$)	23.32	12.63	9.52	7.96	6.36
CNT CC / ($F.g^{-1}$)	14.42	8.41	5.75	4.29	3.43

The GCD curves for the different electrodes recorded at a current density of 1 $mA.cm^{-2}$ are shown in Figure 7c. As seen in this figure, all the cells show a similar symmetrical triangular curve with a nearly linear variation of voltage as a function of time during charge and discharge. This type of curve is typical for carbon-based supercapacitors [17]. However, despite having a similar shape, the curves for the CNT CC didn't return to the initial potential value.

Oxygenated functional groups have initiated faradaic or pseudo-capacitance reactions and affected the GCD curve. These reactions still leave a charge when the discharge time has run out [25]. As seen in Figure 7c, the GCD curve of CNT CO has a GCD curve with lower potential compared to CNT CC. As seen in equation 2, Voltage is inversely proportional to the specific capacitance produced so that the ideal GCD curve is a curve with a low voltage [29]. Based on the calculation of specific capacitance of the CV and GCD curves, both have relatively the same results, and CNT CO is still unable to compete with CNT CO.

Table 4. Specific capacitance with different electrode from CV curves and GCD curve.

Sample	Spesific capacitance ($F.g^{-1}$) at 2 $mV.s^{-1}$	
	(CV)	(GCD)
CNT CO	22.46	23.32
CNT CC	12.14	14.42

Electrochemical Impedance Spectroscopy (EIS) is a non-destructive process to gather the kinetic information of the electrode. The EIS measurements were performed in the frequency range between 100.000 – 0.001 Hz. The obtained data are plotted in the form of the Nyquist plot shown in Figure 7d. This plot reveals three different frequency-dependent regions, which provide extensive information regarding the electrode materials. At the high-frequency region, the starting point of the curve (electrolyte resistance, Ru) followed by a semi-circular loop is present in both curves. The value of Ru does no effect on imaginary Z and follows Ohmic law. Increasing the initial value of Z_{real} induced an increase in the Ohmic resistance generated [16].

The Nyquist curves for all samples have the same starting point, so that the electrode variation has no effect on electrolyte resistance. The diameter of the semicircle represents the charge transfer resistance (R1). Decreasing the diameter of semicircle induced decreases the charge transfer resistance [16]. GCD curve in constant current represented CNT CC when the initial discharge has a higher potential than CNT CO. The greater the resistance, the greater the voltage required ($V=IR$). The higher potential, the smaller capacitance or charge stored in the double layer electrical phenomenon ($C=Q/V$). The Nyquist curve was analyzed by fitting based on equivalent circuit modeling which can be seen in Figure 7e and the results of the resistance values in Table 5.

Table 5. Resistance values of supercapacitor with different electrodes.

Parameter	CNT CO	CNT CC	Unit
Ru	5.2430	9.9140	Ohm
CPE1	0.0538	0.0128	F
α	0.4595	0.2088	-
R1	106.5000	130.2000	Ohm
CPE2	0.0004	0.0004	F
α8	0.8852	0.7722	-
R2	238.1000	256.8000	Ohm

The Ru value between the electrodes is at the same level. The formation of an electrical double layer on the surface of CNT CO is formed more as indicated by the value of α approaching 1. The closer to the value of 1, increase the formation of an electrical double layer [17]. R1 represents the amount of charge transfer resistance in the electrical double layer. CPE1 is comparable in value to double-layer electrical capacitance. Decreasing R1 and increasing CPE1 induced better supercapacitor performance [30].

4. Conclusions

CNT has been successfully prepared on upscaled with the utilization of a stainless steel mesh-based collecting system. It was found that the present technique gives a higher yield of CNT from 0.46% to 0.96% (the maximum fix carbon yield was 2.94%). CNT has relatively similar characteristics to CNT before scale-up. CNT was characterized and applied as electrode material for supercapacitor, which exhibits a specific capacitance of 13.98 $F.g^{-1}$. However, the capacitance of supercapacitor does not meet commercial standards due to still has considerable resistance between electrodes and electrolytes in the EIS analysis.

5. Acknowledgment

This work is supported by Badan Pengelola Dana Perkebunan Kelapa Sawit during the fiscal year 2018.

References

[1] Central Bureau of Statistics 2016 *Indonesian Palm Oil Statistics* (Indonesia: BPS)
[2] Rahayu A S, Karsiwulan D, Yuwono H, Trisnawati I, Mulyasari S, Rahardjo S, Hokermin S and Paramita V 2015 *POME to Biogas - Project Development in Indonesia* (Indonesia: Winrock International)
[3] Awasthi K, Kumar R, Tiwari R and Srivastava O N 2010 *J. Exp. Nanosci.* **5** 498
[4] Vinayan B P, Nagar R, Raman V, Rajalakshmi N, Dhathathreyan K S and Ramaprabhu S 2012 *App. J. Mat. Chem.* **22** 9949
[5] Robati D 2013 *J. Nanostructure. Chem.* **3** 55
[6] Kumar R, Singh R K and Singh D P 2016 *Renew. Sust. Energ. Rev.* **58** 976-1006
[7] Kumar S, Nehra M, Kedia D, Dilbaghi N, Tankeshwar K and Kim K.-H 2018 *Prog. Energy Combust. Sci.* **64** 219-53
[8] Quan C, Li A and Gao N 2010 *J. Hazard. Mater.* **179** 911
[9] Iijima S 1991 *Nat* **354** 56
[10] Yan Y, Miao J, Yang Z, Xiao F.-X, Yang H B, Liu B and Yang Y 2015 *Chem. Soc. Rev.* **44** 3295-346
[11] Saifuddin N, Raziah A Z and Junizah A R 2013 *New. J. Chem* **2013** 18
[12] Nurdin I, Devianto H, Widiatmoko P, Prakoso T, Nugraha N F and Srimurti S 2018 *IOP Conference Series Mater. Sci. Eng.* **543** 012029

[13] Ghosh P, Afre R A, Soga T and Jimbo T 2007 *Mater. Letters* **61** 3768-70
[14] Vivekchand S R C, Cele L M, Deepak F L, Raju A R and Govindaraj A 2004 *Chem. Phys. Lett.* **386** 313-18
[15] Lufrano F, Staiti P and Minutoli M 2004 *J. Electrochem. Soc.* **151** A64
[16] Yu A, Chabot V and Zhang J 2013 *Electrochemical supercapacitors for energy storage and delivery Fund. Appli* (Boca Raton: CRC Press) p 373
[17] Farma R, Deraman M, Awitdrus, Talib I A, Omar R, Manjunatha J G, Ishak M M, Basri N H and Dolah B N M 2013 *Int. J. Electrochem. Sci.* **8** 257-73
[18] Peigney A, Laurent C, Flahaut E, Bacsa R R and Rousset A 2001 *Carbon* **39** 507-14
[19] Huang J, Sumpter B G and Meunier V 2008 *Chem. Eur. J.* **14** 6614
[20] Schindall J 2007 *IEEE Spectrum* **44**(11) 42-6
[21] Kumar M and Ando Y 2003 *Diam. Relat. Mater.* **12** 1845
[22] Kerdnawee K *et al* 2017 *KONA Powder Part. J.* **34** 24-43
[23] Manoj B and Kunjomana A G 2015 *Mater. Sci. Eng.* **73** 012096
[24] Schnitzler M C and Zarbin A J G 2008 *J. Nanopart. Res.* **10** 585
[25] Yeager E 1984 *Electrochim. Acta.* **11** 1527
[26] Taurino I, Carrara S, Giorcelli M, Tagliaferro A and De Micheli G 2012 *Surf. Sci.* **606** 156-60
[27] Zhang Y, Bai W, Cheng X, Ren J, Weng W, Chen P, Fang X, Zhang Z and Peng H 2014 *Angew. Chem. Int.* **53** 14564-8
[28] Termehyousefi A, Bagheri S, Shinji K, Rouhi J, Mahmood M R and Ikeda S 2014 *BioMed Res. Int.* **2014** 691537
[29] Davies A and Yu A 2011 *Can. J. Chem. Eng.* **89** 1342
[30] El-Khodary S A, El-Enany G M, El-Okr M and Ibrahim M 2017 *Synth. Met.* **233** 41-51

IOP Conf. Series: Materials Science and Engineering **823** (2020) 012041 doi:10.1088/1757-899X/823/1/012041

Secondary metabolites of some varieties of *Caulerpa species*

Alfred G.O. Kase[1], Hilconida Calumpong[2] Angreini Rupidara[3]

[1] Fisheries and Marine Sciences Department of Artha Wacana Christian University Kupang
[2] Institute of Environmental and Marine Sciences, Silliman University Philippines
[3] Biology Department of Artha Wacana Christian University Kupang

E-mail: kasealfred@yahoo.com

Abstract. Several species of *Caulerpa* have been investigated and bioactive principles, such as caulerpin and caulerpicine have been isolated from a number of them. Caulerpin produces mild anesthetic action, difficulty in breathing, sedation, and loss of balance. The toxic syndrome has been reported to be somewhat similar to that produced by ciguatera fish poisoning. The neurotrophic activity of caulerpicine is thought to be of clinical value. Tropical green algae and a few of their temperate relatives have yielded a number of bioactive metabolites and some of these are believed to be used by the algae as a chemical defense against herbivorous animals. This research was aimed to determine the secondary metabolites compound of some varieties of *Caulerpa species*. Chromatography and Spectrophotometric methods were used to conduct this research. The result shows that Caulerpin and caulerpicin presence on five species of *Caulerpa* species *Caulerpa racemosa* Siquijor, *Caulerpa lentillifera* wild and farm from Mactan, *Caulerpa clavifera* Mactan and *Caulerpa lentillifera* cultured in the common garden. Higher absorbance of both compounds caulerpin and caulerpicin found in wild *Caulerpa lentillifera* from Mactan while the lowest absorbance of caulerpin found in *Caulerpa racemosa* collected from Siquijor, whereas the lowest absorbance of caulerpicin found in the cultured *Caulerpa lentillifera* in the common garden.

1. Introduction

The most popular edible algae in the Philippines are considered toxic during the rainy season. This toxicity has been attributed to the presence of caulerpicine which has been found to have neurotrophic effects. Its toxicity, presumed to be a chemical defense of the alga.

Several species of *Caulerpa* have been investigated and bioactive principles, such as caulerpin and caulerpicine have been isolated from a number of them. Caulerpin produces mild anaesthetic action, difficulty in breathing, sedation and loss of balance. The toxic syndrome has been reported to be somewhat similar to that produced by ciguatera fish poisoning. The neurotrophic activity of caulerpicine is thought to be of clinical value. Tropical green algae and a few of their temperate relatives have yielded a number of bioactive metabolites and some of these are believed to be used by the algae as a chemical defense against herbivorous animals [1]. The concentration of caulerpenyne varies with water depth and temperature but is always higher in the invasive strain [2]. Water temperatures above 19 °C (spring to autumn) and depths of around 5 m are the optimum conditions for caulerpenyne production [3].

Secondary metabolites produced by *Caulerpa* species, being the toxic terpenoid caulerpenyne Ranielo et al. [4], the most studied, have been investigated to evaluate their ecological role in chemical defense against herbivores while no studies were performed to investigate their possible role as allelochemicals in interspecific competition among marine macrophytes. Dumay et al. [5] provided interesting data about changes in caulerpenyne content in the two invasive species of *Caulerpa, C. taxifolia* and *C. racemosa* var. *cylindracea*. These two algae showed a decrease in caulerpenyne content in the presence of *P. oceanica*, with a concomitant increase in frond length. Further, when growing under the seagrass canopy, the *Caulerpa* species may allocate more energy into frond development to the detriment of other functions such as the production of secondary metabolites.

The marine alga *Caulerpa lamourouxii* is widely distributed in the Phillippines. The upper branches are eaten as a 'salad', despite their peppery and astringent taste. However, the alga is found toxic to some individuals. Chemical investigation of the alga had furnished caulerpicine, caulerpin, cholesterol, taraxerol, β-sitosterol and palmitic acid. Caulerpin had also been isolated from *Caulerpa sertularioides, C. racemosa* var. *clavifera* and caulerpicine *from C. racemosa* [1].

Among the many secondary metabolites produced by algae belonging to the genus *Caulerpa*, caulerpenyne (CYN) is the most abundant cytotoxic sesquiterpenoid to be released. In *Caulerpa taxifolia* (Vahl) C. Agardh, this molecule makes up 0.2% (in spring) to 13% (in summer) of the alga's frond wet weight [5], values that are much higher than those observed in other *Caulerpa* species [6]. In *C. taxifolia* and *C. racemosa* (Forssk°al) J. Agardh, CYN, thought to be the most active substance, is involved either in the chemical defense of the plant against herbivore pressure or within the framework of interspecific competition (antifeedant and antifouling effects) [5,7]. A variety of toxic effects due to CYN has been demonstrated in several organisms at different stages of their growth. The most harmful effect is on sea urchin eggs, which are killed by CYN [7].

The selective force of specialist herbivores is considered an important driver in the evolution of quantitative and qualitative variation in plant chemical defenses [8,9]. For this reason, linking variation in seaweed chemical defenses to the diet specificity of marine herbivores [9,10] is an important precursor to addressing theory derived from specialist associations in terrestrial systems, in particular, Baumgartner et al., [9]. Because most marine herbivores (fishes, sea urchins, and gastropods) are generalists [11], examples of chemically mediated, marine specialist interactions come from a broader phylogenetic pool of species than the traditional plant-herbivore model [9].

However, 2 groups of herbivores (amphipods and sacoglossan opisthobranch mollusks) do contain consumers with consistently narrow host ranges [10,11]. Of these, the sacoglossans are almost exclusively associated with chemically rich seaweeds (siphonous green algae) [12], which provides an opportunity to contrast the efficacy of chemical defenses in closely related seaweeds against multiple specialized herbivores. Feeding preferences could have selected for high levels (an upper threshold) in *Caulerpa* chemical defense, in an attempt to radiate and escape as suggested for terrestrial plants in response to specialization by herbivores [9]. Similarly, any preference for seaweed with intermediate concentrations could manifest as an indirect defense for low concentration *Caulerpa* species (e.g. *C. racemosa* var. *laetevirens*), where such species are common but are avoided by sacoglossans with broader diet niches (e.g. *Elysia* and *Lobiger*) [9].

Raniello et al. [4] demonstrated that the presence of the invasive seaweed in the meadows stimulated the sexual reproduction of seagrasses, reported by many authors to be more frequent under stressful conditions [4,13]. However, the determinants of the observed structural and physiological changes induced in the benthic community have not been fully investigated, although Ceccherelli et al. [14] have speculated on the possible occurrence of an allelopathic interaction by which the release of a phytotoxic compound would affect and damage the competitor physiology. The aims of this study are to determine the secondary metabolites compound of some varieties of *Caulerpa* species using chromatography and spectrophotometry methods.

2. Materials and Methods

2.1. Sample collection
Five species of *Caulerpa* were collected from different sites, *Caulerpa lentillifera wild, Caulerpa lentillifera farm, and Caulerpa clavifera* were collected from Mactan island, *Caulerpa racemosa* were collected from Siquijor, and *Caulerpa lentillifera* was collected from the experiment tanks of Institute of Environment and Marine Sciences (IEMS) Silliman Laboratory (common garden cultured *Caulerpa lentillifera*).

2.2. Sample preparation
Fresh and healthy samples of each species were collected and cleaned from epiphytes, muddy sands and other dirt. Samples were air-dried for 5 to 7 days until they become well dried and was ground to the powder form. Modified method by Doty and Santos [15].

2.3. Sample extraction
The grounded sample was weighed 100 grams and refluxed in 1000ml of 95% ethanol for 3 days at 70°C. A minimum of 10 milliliter ethanol per dry gram of sample was used for each refluxing. After the three days refluxing the ethanol was collected by filtering to separate the extract from the debris.

2.4. Partitioning
Ethanol extracts were concentrated on a Brinkman flash evaporator. Equal volumes of diethyl ether were added to the ethanol and samples were set aside for 24 hours. The samples were filtered and evaporated to dryness.

Equal volumes of ether and water were added to make it 100ml and two layers were separated using a separatory funnel. The bottom part contains water that was discharged and the ether upper layer containing compounds were collected and concentrated to a small volume.

2.5. Chromatography
The separated and purified diethyl ether extract was adsorbed onto 100g neutral alumina in a column (16cm x 4.5cm) and eluted with diethyl ether.

Fractions were concentrated and spotted on silica gel coated thin layer chromatographic plates and were developed using the solvent system made up of hexane: diethyl ether: acetic acid in 30:70:1 ratio.

2.6. Retention factor (Rf) analysis
The spots developed on the TLC plate were marked and measurement of Rf value was done to determine the compounds. The UV light was used to visualized and mark the Rf value of the colorless compound caulerpicin.

2.7. Spectrometric analysis
Spots on the TLC plates scraped and were dissolved in diethyl ether and ethanol. The absorbance spectra for each spot was determined using the wavelength scan mode of the Perkin Elmer Lambda 100 Spectrophotometer. For caulerpin the absorbance spectra were generated using the wavelengths from 200 to 600 nm, while for caulerpicin the wavelength used was from 200 to 350 nm.

3. Results

3.1. Column Chromatography
Five bands were obtained from the column chromatography perform on all species *Caulerpa racemosa* Siquijor, *C. lentillifera* wild and farm from Mactan, *Caulerpa clavifera* Mactan and *Caulerpa lentillifera* cultured in the tank IEMS.

Tabel 1. Number of fraction yielded on column chromatography

No	Species	Number of Fractions
1	*Caulerpa lentillifera* Farmed – Mactan	5
2	*Caulerpa lentillifera* Wild – Mactan	5
3	*Caulerpa lentillifera* Tank	5
4	*Caulerpa clavifera* – Mactan	5
5	*Caulerpa racemosa* – Siquijor	5

The column chromatography test results shown in Table 1, summarizing the five *Caulerpa* species from five different locations, each has five fractions. Which is means that there are five possibilities compounds content in *Caulerpa* species from five species.

Caulerpin found on the first three fractions of column chromatography of each sample, while caulerpicin found at the last fraction column chromatography. The first fraction was dark green color, second fraction light green, third fraction light green (brownness), forth fractions light green to yellow color and the last fraction dark green. All fractions were subjected to thin-layer chromatography (TLC).

3.2. Thin Layer Chromatography

Fractions collected from the column chromatography were subjected to the Thin Layer Chromatography, this technique for chemical separation of mixtures compounds or substances content on the fractions collected from column chromatography. Fourteen to sixteen bonds were generated from the thin layer chromatography for all five *Caulerpa* species extracted,

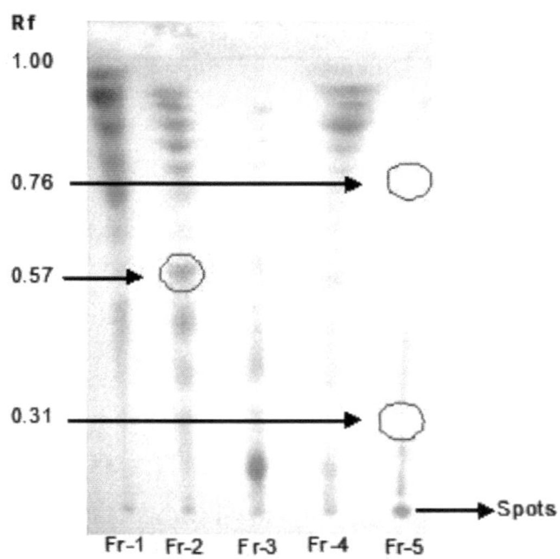

Figure 1. Thin Layer Chromatography result (TLC) of *Caulerpa* species.

The result of Thin Layer Chromatography (TLC) on the second fraction showed the bright orange color in the red circle with Rf value 0.57 which is the characteristic color of caulerpin, while the colorless spot of caulerpicin was obtained from the fifth fraction had two spots with the first spot with Rf value of 0.31 and the second is 0.76. While the yellow color spot with an Rf of 1.00 was identified as α-carotene pigments, chlorophyll pigments as indicated by the green color spots. Some other spots present in the TLC plates were not been identified (Figure 1).

Table 2. Retention factor (Rf) Value of TLC product of *Caulerpa* species

No	Species	Caulerpin		Caulerpicin	
		Presence (+), absence (–)	Rf	Presence (+), absence (–)	Rf
1	*C. lentillifera* Farmed – Mactan	+	0.575	+	0.31
2	*C. lentillifera* Wild – Mactan	+	0.575	+	0.31
3	*C. lentillifera* –Tank	+	0.575	+	0.31
4	*C. clavifera* – Mactan	+	0.566	+	0.31
5	*C. racemosa* – Siquijor	+	0.575	+	0.31

Table 2 shows that caulerpin presents in the 4 *Caulerpa* species extracted with the Retention factor (*Rf*) value of 0.575, except *Caulerpa clavifera*; caulerpin presents at *Rf* 0.566; while caulerpicin presents in all extracted *Caulerpa* species with *Rf* value of 0.31.

3.3. Spectrophotometric analysis

The spectrophotometer on this study was used to measure the intensity of wavelengths in a spectrum of light compared with the intensity of light from a standard source of five *Caulerpa* species.

Table 3. Wavelength and absorbance value of Caulerpin

No	Species	Caulerpin					
		Wavelength (nm)			Absorbance		
1	*Caulerpa lentillifera* Farmed – Mactan	445	427	408	0.559	0.652	0.604
2	*Caulerpa lentillifera* Wild – Mactan	445	428	408	0.707	0.813	0.735
3	*Caulerpa lentillifera* –Tank	445	426	405	0.411	0.497	0.424
4	*Caulerpa clavifera* – Mactan	445	431	408	0.399	0.411	0.386
5	*Caulerpa racemosa* – Siquijor	445	427	408	0.217	0.295	0.304

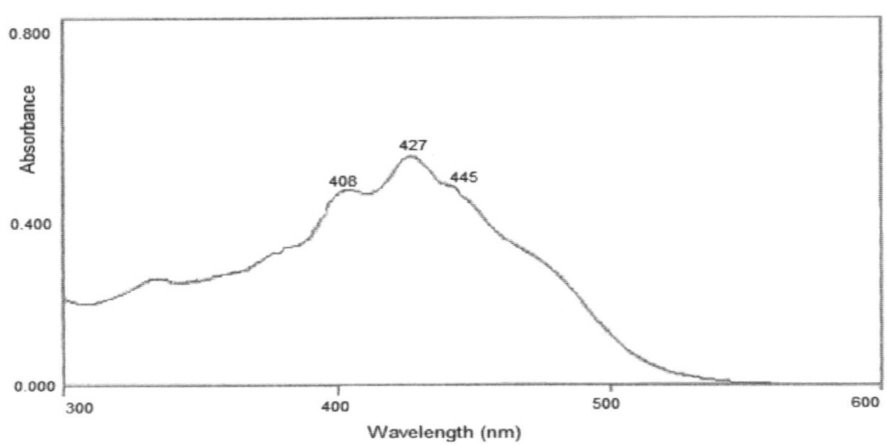

Figure 2. Spectrogram of caulerpin.

Figure 2 showed the spectrogram of caulerpin, which composed of 3 peaks (408 – 427 – 445), while Figure 4 and Table 3 show the wavelength and the corresponding absorbance of caulerpin extracted from five species extracted. The result of spectrophotometer showed that caulerpin absorbed light at wavelengths 408nm to 445nm.

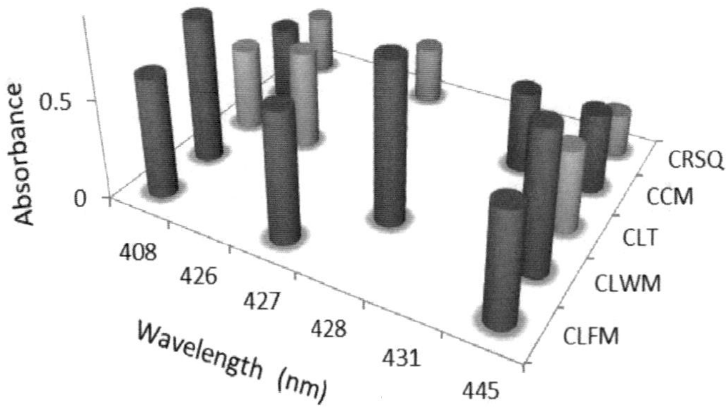

Figure 3. Amount of light absorbed by caulerpin from five *Caulerpa* species at various wavelengths.

Table 3 and Figure 3 showed that *Caulerpa lentillifera* wild taken from Mactan had the highest absorbance of caulerpin, which is (0.707 – 0.813 – 0.735) at the wavelength (445 – 428 – 408) nm, followed by *Caulerpa lentillifera* farm from Mactan with absorbance of caulerpin (0.559 – 0.652 – 0.604) at the wavelength (445 – 427 – 408) nm, while the lowest absorbance found on *Caulerpa racemosa* from Siquijor (0.217 – 0.295 – 0.304) at the wavelength of (445 – 427 – 408) nm.

Table 4. Wavelength and absorbance value of caulerpicin

No	Species	Caulerpicin	
		Wavelength (nm)	Absorbance
1	*Caulerpa lentillifera* Farmed – Mactan	208	0.313
2	*Caulerpa lentillifera* Wild – Mactan	214	0.324
3	*Caulerpa lentillifera* –Tank	216	0.290
4	*Caulerpa clavifera* – Mactan	217	0.304
5	*Caulerpa racemosa* – Siquijor	218	0.295

Table 4 shows the wavelength and the absorbance value of caulerpicin obtained from five species extracted. The result of spectrophotometer showed that Caulerpicin absorbed light at wavelengths 208nm to 218nm.

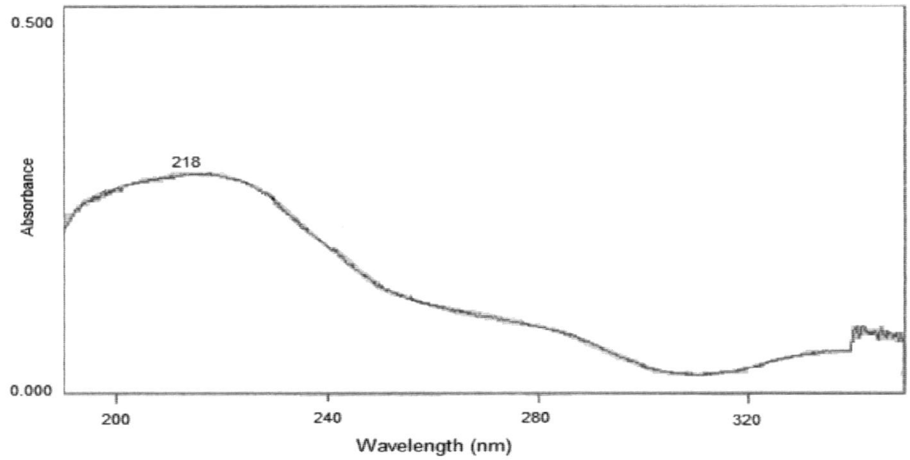

Figure 4. Spectrogram of Caulerpicin.

Figure 4 showed the spectrogram of caulerpicin which appeared one peak located between 200 to 230 nanometer, while Figure 5 showed the amount of light absorbed by caulerpicin from five *Caulerpa*.

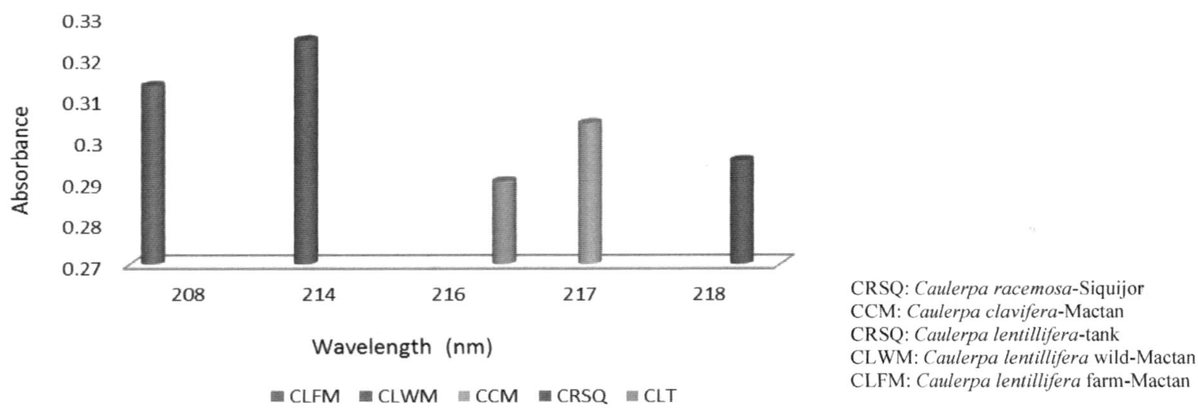

Figure 5. Amount of light absorbed by caulerpicin from five *Caulerpa* species at various wavelengths.

The highest absorbance of caulerpicin found in *Caulerpa lentillifera* wild from Mactan Island with value of 0.324 located in 214 nm of wavelength. The second highest found in *Caulerpa lentillifera* farm from Mactan with absorbance value of 0.313 at the wavelength 208 nm, while the lowest absorbance of caulerpicin found on *Caulerpa lentillifera* cultured on tank with the absorbance value of 0.290 at the wavelength of 216nm.

4. Discussion

Based on the result caulerpin were present in the first to third fraction, while caulerpicin found in the fifth or last fraction, in the study of eight species of green algae of *Caulerpa*, Vest [16] found five to eight bands in the column chromatography, caulerpin were eluted in the first three bands while caulerpicin appeared in the last band or the column wash.

The bright orange color on the Thin Layer Chromatography (TLC) at the Rf of 0.57 was characteristic color of caulerpin, while the colorless spot of caulerpicin at the Rf 0.31 was obtained from the fifth fraction, this result similar with the study of Vest [16], caulerpin appeared as a single bright orange band with an Rf value of 0.58, while caulerpicin appeared as colorless compound with an Rf value of 0.32. Bhakuni and Rawat [1] had been isolated Caulerpin from *Caulerpa sertularioides, C. racemosa* var. *clavifera* and caulerpicin from *C. racemosa*.

According to Valerie [17] Caulerpin is often a minor metabolite in *Caulerpa*. This bright-orange, pigmented compound, probably derived from indole biosynthesis, has been found in over 50% of the *Caulerpa* species [18,19].

Other color spots were also obtained from TLC. A yellow color spot with an Rf of 1.00 was identified as α-carotene pigments, chlorophyll pigments as indicated by the green color spots. Some other spots present in the TLC plates were not been identified. Dawes [20] and Goodwin [21] stated that carotene pigments located in Rf of 1.0 with the range of yellow to red color.

Since in this study, the pure caulerpin and caulerpicin as standard compounds were absent, the compounds were investigated only by comparing the relative abundance by measuring the absorbance in spectrophotometer. *Caulerpa lentillifera* wild taken from Mactan had the highest absorbance of caulerpin, followed by *Caulerpa lentillifera* farm from Mactan while the lowest absorbance found on *Caulerpa racemosa* from Siquijor. Study of Vest et al. [16] in eight species of green algae *Caulerpa*, caulerpin was present in *C. cupressoiedes, C. paspaloides, C. prolifera and C. sertularioides* in amounts varying from less than 1 mg to 5.47 mg per dry gram of algae.

Same as caulerpin, highest absorbance of caulerpicin found in *Caulerpa lentillifera* wild from Mactan, while the lowest absorbance of caulerpicin found on *Caulerpa lentillifera* cultured on tank. Study of Vest [18], caulerpicin was found in *C. ashmeadii, C. paspaloides, C. sertularioides,* and *C racemosa var. uvifera* in lesser concentrations ranging from 0.03 to 0.09 mg per dry gram of algae.

Wild *Caulerpa lentillifera* had the highest caulerpin and caulerpicin, this may be related to environmental factors which *Caulerpa* will produce more caulerpin and caulerpicin as prevention from the predators compare to the farmed and the tank which less in predators. According to Ruensink [22] *Caulerpa* produces secondary metabolites such as caulerpicin that repel predators.

5. Conclusion

Caulerpin and caulerpicin presence on five species of *Caulerpa* species *Caulerpa racemosa* Siquijor, *Caulerpa lentillifera* wild and farm from Mactan, *Caulerpa clavifera* Mactan and *Caulerpa lentillifera* cultured in the common garden.

Higher absorbance of both compounds caulerpin and caulerpicin found in wild *Caulerpa lentillifera* from Mactan while the lowest absorbance of caulerpin found in *Caulerpa racemosa* collected from Siquijor, whereas the lowest absorbance of caulerpicin found in the cultured *Caulerpa lentillifera* in the common garden.

Acknowledgment

The Author would like to gratitude the help of Sir Melchor, Robert Guinoo, Nong Koro, who assisted the field and laboratory work during the research.

References

[1] Bhakuni D S and Rawat D S 2005 *Bioactive Marine Natural Products* (New Delhi: Anamaya Publishers and Springer)

[2] Amade P and Lemee R 1998 Chemical defense of the Mediterranean alga *Caulerpa taxifolia*: variations in caulerpenyne production *Aquatic Toxicology* **43** 287–300

[3] Millar A 2002 The introduction of *Caulerpa taxifolia* in New South Wales, Australia *International Caulerpa taxifolia Conference Proceedings* (San Diego, CA, U.S.A., January 31-February 1, 2002)

[4] Raniello R, Mollo E, Lorenti M, Gavagnin M and Buia M 2007 Phytotoxic activity of

caulerpenyne from the Mediterranean invasive variety of *Caulerpa racemosa*: a potential allelochemical *Bio Invasions* **9** 361–8

[5] Dumay O, Fernandez C and Pergent G 2002 Primary production and vegetative cycle in *Posidonia oceanica* when in competition with the green algae *Caulerpa taxifolia* and *Caulerpa racemosa Journal of Marine Biological Association of the United Kingdom* **82** 379–87

[6] Guerriero A D, Ambrosio M, Guella G, Dini F and Pietra F 1994 Secondary metabolites of the green seaweed *Caulerpa taxifolia* introduced into the Mediterranean sea, and a comparison with ciliates of the genus Euplotes, C. F. Boudouresque, A. Meinesz, and V. Gravez Eds. *First International Workshop on Caulerpa taxifolia* (GIS Posidonie Publication) pp 171–5

[7] Lemee R, Pesando D, Durand-Clement M, Dubreuil A, Meinesz A, Gurriero A and Pietra F 1993 Preliminary survey of toxicity of the green alga *Caulerpa taxifolia* introduced into the Mediterranean *Journal of applied Phycology* **5** 485–93

[8] Verlaque M, Durand C, Huisman J M, Boudouresque C F and Le Parco Y 2003 On the identity and origin of the Mediterranean invasive *Caulerpa racemosa* (Caulerpales, Chlorophyta) *Eur Journal Phycology* **38** 325–39

[9] Baumgartner F A, Cherie A M, Rocky de N and Nicholas A P 2009 Feeding preferences and host associations of specialist marine herbivores align with quantitative variation in seaweed secondary metabolites *Proceeding Department of Marine Ecology, Tjärnö Marine Biological Laboratory* (Strömstad, Sweden: Gothenburg University)

[10] Poore A G B, Hill N A and Sotke E E 2008 Phylogenetic and georaphic variation in host breadth and composition by herbivorous amphipods in the family Ampithoidae *Evolution* **62** 21–38

[11] Hay M E and Steinberg P D 1992 The chemical ecology of plant herbivore interactions in marine versus terrestrial communities *in* Rosenthal G A, Berenbaum M R Eds. *Herbivores: Their Interactions with Secondary Plant Metabolites*: *Ecological and Evolutionary Process* (Sand Diego: Academic Press) CA **2** 371–413

[12] Williams S I and Walker D I 1999 Mesoherbivore-macroalgal interactions: feeding ecology of sacoglossan sea slugs (Mollusca, Opisthobranchia) and their effects on their food algae *Oceanogr. Mar. Biol. Ann. Rev.* **37** 87–128

[13] Phillips R C, Grant W S and Mc Roy C P 1983 Reproductive strategies of eelgrass (*Zostera marina* L) *Aquatic Botany* **16** 1–20

[14] Ceccherelli G, Piazzi L and Balata D 2002 Spread of introduced *Caulerpa* species in macroalgal habitats *Journal Marine Biology Ecology* **280** 1–11

[15] Doty M S and Aquilar-Santos G 1966 Caulerpicin, a toxic constituent of *Caulerpa Nature Lond* **211** 990

[16] Vest S E, Dawes C J and Romeo J T 2009 Distribution of caulerpin and caulerpicine in eight species of the green algae *Caulerpa* (Caulerpales) *Botanica Marina* **26** 313-6

[17] Valerie J P 2002 *Natural Products Chemistry and Chemical Ecology of Caulerpa spp. With an Emphasis on Invasive Caulerpa taxifolia* (University of Guam Marine Laboratory UOG Station vpaul@uog.edu)

[18] Maiti B C, Thomson R H and Mahendran M 1978 The structure of caulerpin, a pigment from *Caulerpa* algae *Journal Chemical Res-S* 126–7

[19] Schwede J G, Cardellina I I J H, Grode S H, James Jr T R and Blackman A J 1987 Distribution of the pigment caulerpin in species of the green alga *Caulerpa*. *Phytochem* **26** 155-8

[20] Dawes C J 1981 *Marine Botany An Introduction* (The United States of America Holt Rinehart and Winston Inc.)

[21] Goodwin T W 1988 *Plant Pigment* (San-Diego: Academic Press Inc.)

[22] Ruesink J and L Vides C 2002 Morphological plasticity and invasive potential of some caulerpa species *Proceedings of the International Caulerpa taxifolia conference* (San Diego, California U.S.A.: National University of México, January 31–February 1, 2002)

IOP Conf. Series: Materials Science and Engineering **823** (2020) 012042 doi:10.1088/1757-899X/823/1/012042

Kinetic studies of turpentine isomerization using hydrochloric acid and acetic acid as catalysts

Muhammad Mufti Azis[1*], Farahfestura Rio Qani'a[1], Suci Indah Pratiwi[1], Jonas Kristanto[1], Suryo Purwono[1] and Antonius Indarto[2]

[1]Department of Chemical Engineering, Faculty of Engineering, Universitas Gadjah Mada, Jl. Grafika No. 2, Kampus UGM Bulaksumur Yogyakarta 55281, Indonesia
[2]Department of Chemical Engineering, Institut Teknologi Bandung, Bandung 40132 Indonesia

E-mail: muhammad.azis@ugm.ac.id

Abstract. Turpentine is a non-wood forest product derived from pine trees. Turpentine oil can be produced from the distillation of pine tree sap (family Pinaceae). The main component of turpentine is α-pinene, and turpentine isomerization may produce various derivative products. The objective of the present study was to investigate the isomerization products of turpentine oil with two acid catalysts: hydrochloric and acetic acid. Here, the influence of acid concentration was investigated by using various concentrations of hydrochloric and acetic acid concentrations between 0.4 to 1 M. Further, the combination of acetic acid and hydrochloric acid as catalyst was also investigated by varying the molar ratio of HCl:Acetic Acid=1 : 0.5 and 1:1. The experiments were carried out in a batch reactor equipped with heater, condenser, and stirrer. The reaction temperatures were maintained at 150°C and reaction time of 6 h. The resulting products from turpentine isomerization were analyzed using GC-MS. The results showed that the largest isomerization product obtained with HCl and combination of HCl:H.Acetic was trans sabinene hydrate. Additional products that could be detected varied among α-terpinolene, α-terpineol, α-terpinene, γ-terpinene and δ-carene. The experimental data showed that α-pinene conversion as high as 62, 63 and 57% were obtained by using HCl concentrations of 0.4, 0.7 and 1 M, respectively. In addition, the use of HCl:H.Acetic=1 :0.5 and 1:1 gave final conversion of 63 and 61%. The modeling results showed the kinetic model for pinene conversion for both catalysts was $\frac{dX}{dt} = 0.5222 \cdot (1 - X)^{2.5210} C_{HCl} - 0.2233 \cdot (1 - X)^{4.5300} C_{HAc}$

1. Introduction

Indonesia is one of the key players in turpentine production in the world by producing more than 13000 tons of turpentine in 2017 [1]. Turpentine is made from pine tree sap which is commonly collected by the local farmers in Java Island. Pine sap processing via distillation may produce turpentine (as top product) and gum rosin (as bottom products). Both turpentine and gum rosin have economic values. Turpentine can be used as raw material for various applications such as paints, solvents, pharmaceuticals and rosin [2,3]. Currently, those products are exported mainly to China and India. In addition to the export market, the demand for gum rosin and turpentine in Indonesia also steadily increased.

The main compound in turpentine is α-pinene which can be converted to various derived products through the isomerization process with acid catalysts [4,5]. Production of various chemicals from turpentine is attractive as it may offer flexibility to cope with the volatile price of turpentine and its derivative products in the market. The price of turpentine derivative products varied greatly from 2000 USD/ton up to 15000 USD/ton.

Production of cineole from turpentine has been reported by Ardiyanto et al. [6] and showed that combination of PTSA and oxalic acid gave cineole yield as high as 21.7 wt.%. Production of α-terpineol from turpentine has also been largely investigated in the present literature. Production of terpineol by hydration of turpentine using low price acid catalyst has been reported by Prakoso et al. [7]. They found that the use of mixed phosphoric acid and acetic acid as catalyst gave terpineol yield as high as 53.5 wt.%. Continuous production of terpineol with reactive distillation has also been investigated in [8].

Kinetic study of turpentine isomerization is useful to predict performance of various catalysts. Utami et al. [9] studied the kinetic of turpentine conversion to terpineol over chloroacetic acid catalyst using a heterogeneous model. In addition, Diana et al [10] reported a kinetic model for acetoxylation of turpentine using a pseudo-homogenous approach. A kinetic model for dual acid catalysts could play an important role to accelerate the screening for combination of acid catalysts.

The objective of present work is to investigate the influence of single catalyst HCl and combination of HCl and Acetic Acid (H.Acetic) as catalysts for isomerization of turpentine. In addition, this study also aimed to investigate product distribution using both acid catalysts. Subsequently, a kinetic model was developed for both systems. The model is expected to serve as a tool to evaluate pinene conversion with the use of mixed HCl and H.Acetic as catalysts.

2. Experimental Method

2.1. Sample preparation

All materials were obtained from Laboratorium Metodika Perancangan dan Pengendalian Proses, Department of Chemical Engineering, Institut Teknologi Bandung which includes: HCl solution, Acetic Acid solution and aquadest. Initially, HCl solutions with concentration of 1 M were prepared in the lab. In addition, the HCl solution was diluted with aquadest to obtain 0.4 and 0.7 M as varied parameters in the present study. In addition, the acetic acid solution of 1 M and 0.5 M were also prepared in the lab. Crude turpentine was received from Perhutani Pine Chemical Industry (PPCI) Pemalang. For the first part of the experiments, we have used a single catalyst of HCl solution to study the turpentine isomerization. In the second part, both HCl and Acetic Acid solutions were used to investigate the influence of dual catalysts to isomerize the turpentine.

2.2. Batch reactor experiment

The experiments of turpentine isomerization were conducted in a three-neck flask reactor equipped with condenser, stirrer, and thermometer. In the first part of the experiment with a single catalyst, 40 ml of HCl solutions with varied concentrations (0.4, 0.7 and 1 M) were added to the reactor. Next, 10 ml of turpentine was also added. The reactor was then heated to 150°C with the total duration of 6 h. The stirrer speed was maintained at 250 rpm or scale 6. The mixed solution forms two phases namely oil phase in the top layer as well as the water phase in the bottom phase. Similarly, for experiment with dual catalyst, 40 ml of acetic acid and 40 ml of HCl solution were added to the reactor. For this part, 20 ml of turpentine was added to form mixed solution of turpentine and acid catalysts.

The sample was collected with small pipette from the reactor after 3 h (for dual catalysts system) and 6 h of experiments. When the experiments were finished, both phases were separated. The resulting oil products of isomerization was then weighed to measure the yield of isomerization products. There are 3 measurements of GC-MS using HCl concentrations of 0.4, 0.7 and 1 M after 6 h of experiment. Additionally, there are 4 measurements at 3 h and 6 h using combination of HCl: H Acetic=1 M: 0.5 M and 1 M: 1 M.

IOP Conf. Series: Materials Science and Engineering **823** (2020) 012042 doi:10.1088/1757-899X/823/1/012042

2.3. Sample analyses
The samples that have been collected in vials were analyzed using Gas Chromatography-Mass Spectrometer (Shimadzu GCMS-QP-2010 with Rtx-5MS capillary column). The α-pinene conversion (X_{pinene}) was evaluated from the following equation:

$$X_{pinene} = \frac{A_{pinene,feed} - A_{pinene,product}}{A_{pinene,feed}} . 100\% \quad (1)$$

A_{pinene} is the chromatographic peak area of α-pinene in the initial and final states of each experiment.

3. Modeling Method
A mathematical model was developed from mole balance of a batch reactor. For a single homogenous catalyst, the lumped kinetic model, as well as the reaction rate which was based on the mol of catalyst, could be presented as follows:

$$V_{tot}\frac{dC_A}{dt} = -r_A \cdot N_{HCl} \quad (2)$$

$$V_{tot}\frac{dC_A}{dt} = -\,k_1 \cdot C_{A0}{}^{n1} \cdot (1-X)^{n1} \cdot N_{HCl} \quad (3)$$

$$V_{tot}C_{A0}\frac{dX}{dt} = k_1 \cdot C_{A0}{}^{n1} \cdot (1-X)^{n1} C_{HCl} . V_{tot} \quad (4)$$

$$\frac{dX}{dt} = k_1{}' \cdot (1-X)^{n1} C_{HCl} \quad (5)$$

Similarly, when we use dual catalysts, the kinetic model was:

$$\frac{dX}{dt} = k_1' \cdot (1-X)^{n1} C_{HCl} + k_2{}' \cdot (1-X)^{n2} C_{HAc} \quad (6)$$

As a result, there are 4 parameters to estimate namely k_1', k_2', n1 and n2 from 7 experimental data of pinene conversion. For parameter estimation, the simulation was conducted in Matlab by using both ode and lsqnonlin solvers.

4. Results and Discussion
Turpentine isomerization was investigated by using HCl and combination of HCl and acetic acid as catalysts. Here, we focused on analysis of product distribution based on GC-MS analyses. A kinetic model was developed to reveal the influence of each catalyst.

4.1. Product distribution based on GC analysis
Figure 1 shows the example of chromatogram from turpentine isomerization by using HCl solution concentrations of 0.4, 0.7 and 1 M as catalysts. As seen here, there are several products that can be detected with various retention times. Several notable peaks that can be detected were 3.7, a range of peaks between 4-4.5 and a peak at 5.2. It could also be observed that there are minor products that could be detected above 5.2 showing the absence of important products.

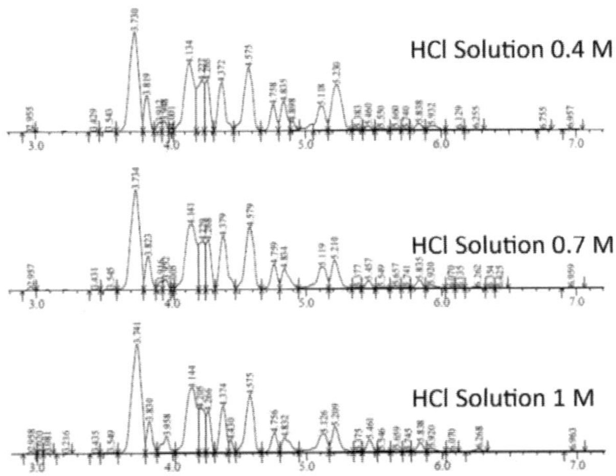

Figure 1. Chromatogram of turpentine isomerization products from GC-MS using various HCl concentration as a catalyst

4.2. *Quantitative analysis of product distribution based on GC-MS analysis*

Analysis of product distribution was conducted using Mass Spectrometer. The analysis here shows percentage area of 11 largest peaks. Figure 2a shows the chromatogram of crude turpentine. As depicted in Figure 2a, the dominance peak is, as expected, α-pinene. It is interesting to note that the crude turpentine also already contained δ-carene. Other minor products that can be detected were β-pinene and δ-limonene.

By using HCl solution of 0.4 M as the catalyst, the three major products include trans sabinene hydrate, α-terpinolene, and α-terpineol (Figure 2b). By using 0.7 and 1 M of HCl solutions (Figure 2c-d), it is interesting to note that trans sabinene hydrate remains as the largest compound. However, the second and third largest compound varied among α-terpinene, α-terpinolene and γ-terpinene.

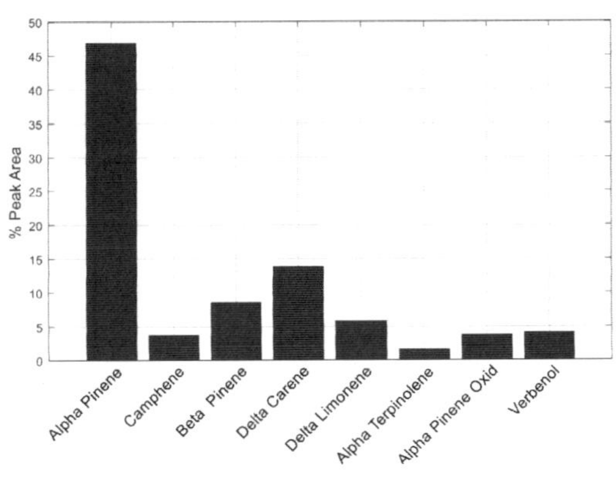

(a)

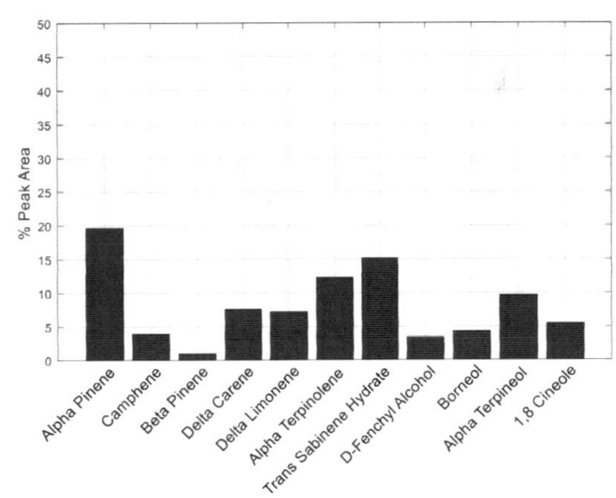

(b)

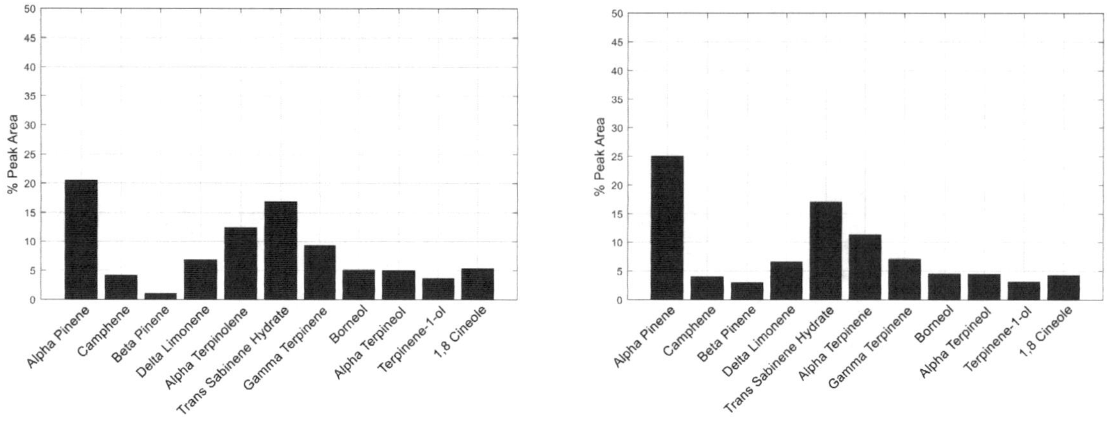

(c) (d)

Figure 2. GC-MS chromatogram of (a) crude turpentine and product distribution of isomerization by using (b) 0.4 M of HCl, (c) 0.7 M of HCl and (d) 1 M of HCl solutions.

The product distribution from the use of dual catalysts system is presented in figure 3. The product distribution appears to be similar to trans sabinene hydrate as the largest product. The next important products varied among α-terpinene, α-terpinolene and δ-carene. For a single catalyst with HCl solutions, the pinene conversions were 62, 63 and 57% by using HCl 0.4 M, 0.7 M and 1 M, respectively. It is worth noting that the highest HCl concentration of 1 M gave a lower conversion of pinene conversion.

With dual catalysts of HCl and H.Acetic, the conversion of pinene increased with the increase of reaction time. By using HCl:H.Acetic=1:0.5 M, it gave conversion of 55% and 63% for 3 h and 6 h, respectively. Similarly, by using HCl:H.Acetic=1:1 M, it gave conversion of 55 and 61% for 3h and 6h, respectively. Hence, it could be inferred that the increase of pinene conversion as a result of H.Acetic addition is relatively small, an increase from 57% by using only HCl 1M to 61 and 63% with addition of H.Acetic.

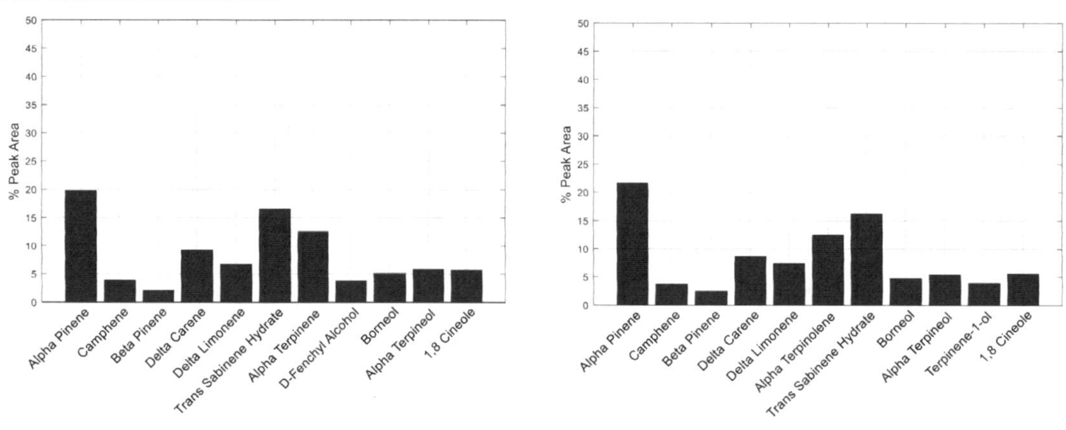

(a) (b)

Figure 3. Product distribution analysis of turpentine isomerization using HCl and CH_3COOH as catalysts with molar ratio of (a) HCl:CH_3COOH=1M:0.5M and (b) HCl:CH_3COOH=1M:1M

4.3. Modeling results

Figure 4 shows the modeling results for both single and dual catalysts system. As illustrated in figure 4a, the model could capture relatively well for HCl concentration of 0.7 M. On the other hand, larger offset was obtained for 0.4 and 1 M. This is due to the increase of HCl concentration from experimental data of 0.7 to 1 M which gave lower conversion. The kinetic parameters of k_1' and n_1 are 0.5222 and 2.5210 which was obtained with HCl as the catalyst.

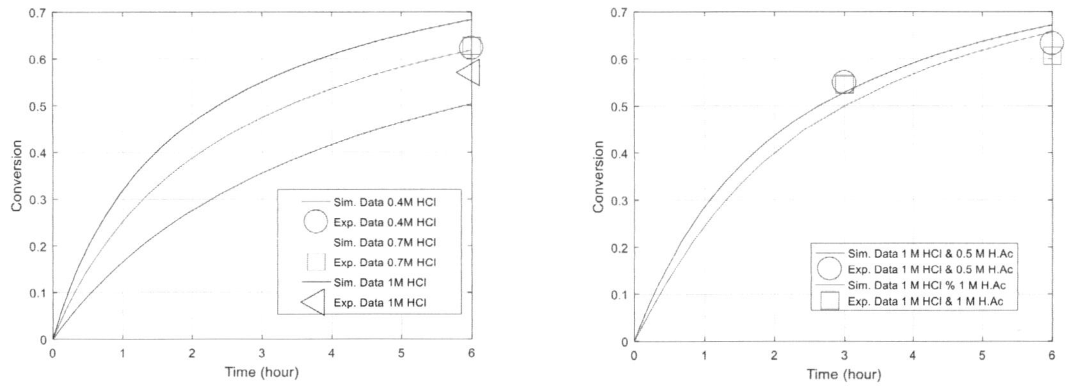

(a) (b)

Figure 4. Modeling result for single catalyst of (a) HCl solutions and (b) dual catalysts of HCl:H.Acetic

Figure 4b shows the modeling results for dual catalysts system. Here, we have two points at 3 h and 6 h. The model could generally capture the data well for 3 h and 6 h. The kinetic model proposed here was developed to accommodate the influence of both acids. In addition to k_1' and n_1, we could also estimate the values of k_2' and n_2 as -0.2233 and 4.5300, respectively. The influence of acetic acid appears to be negative as the model predicted too high conversion for HCl 1 M as 68.4%, higher than that of 57.07% of experimental data (Figure 4a). Hence, the model tried to compensate the calculation by putting negative coefficient of k_2'. The negative value that we obtained here should only be used as an indicator to evaluate the influence of dual catalysts system and can be used as a tool for extensive screening of any combination of acid catalysts. From this work, the final kinetic equation for dual catalyst system of HCl and H. Acetic is:

$$\frac{dX}{dt} = 0.5222 \cdot (1 - X)^{2.5210} C_{HCl} - 0.2233 \cdot (1 - X)^{4.5300} C_{HAc} \tag{7}$$

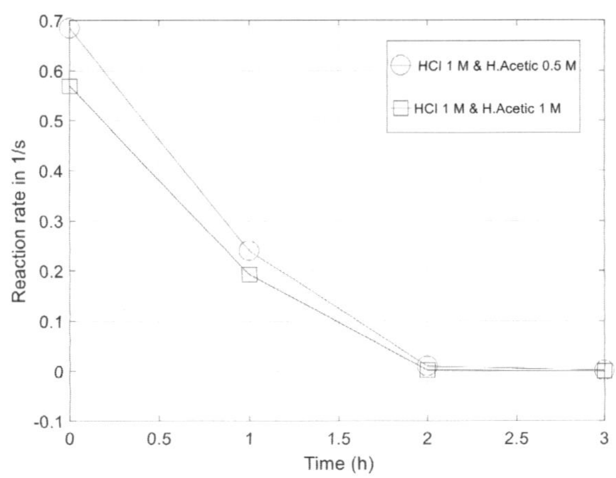

Figure 5. Modeling result for quantification of reaction rate by using dual catalysts using equation (2)

Quantification of reaction rate by using equation (2) is presented in Figure 5. As indicated by equation (7), addition of H.Acetic tends to lower the reaction rate and thus found to be disadvantageous. It is also possible that other kinetic models such as the heterogeneous kinetic model as in [9] or pseudo homogenous model [10] could probably improve the present model.

In addition, the present work also indicated that HCl as well as combination of HCl:H.Acetic did not selectively produce a specific product. Instead, it produced a rather large variety of 11 possible products. Hence, further work is needed to improve the selectivity of turpentine isomerization to a specific product which may improve the economic value of turpentine.

5. Conclusions

Turpentine isomerization has been investigated by using two acid catalysts: hydrochloric and acetic acid. Here, the influence of acid concentration was investigated by using various concentrations of hydrochloric and acetic acid concentrations between 0.4 to 1 M. The results showed that the largest isomerization product obtained with HCl and combination of HCl:H. Acetic was trans sabinene hydrate. Additional products that could be detected varied among α-terpinolene, α-terpineol, α-terpinene, γ-terpinene and δ-carene.

The experimental data of α-pinene conversion showed that the pinene conversions were 62, 63 and 57% by using 0.4, 0.7 and 1 M of HCl solutions. In addition, the use of HCl:H. Acetic=1 M :0.5 M gave the final conversion of 63%. Similarly, by using HCl:H. Acetic=1 M:1 M gave the final conversion of 61%. Hence, it is apparent that the increase of acetic acid concentration lowered the pinene conversion. The modeling results showed the kinetic model for pinene conversion using both catalysts was :

$$\frac{dX}{dt} = 0.5222 \cdot (1 - X)^{2.5210} C_{HCl} - 0.2233 \cdot (1 - X)^{4.5300} C_{HAc}$$

This model is useful to predict pinene conversion by using various concentrations of HCl and acetic acid as catalysts.

References

[1] Perum Perhutani 2019 *Buku Statistik Perum Perhutani 2013-2017* (Jakarta: Perum Perhutani Kantor Pusat)
[2] Gscheidmeier M and Fleig H 2000 *Ullmann's Encyclopedia of Industrial Chemistry*
[3] Monteiro J L F and Veloso C O 2004 *Top. Catal.* **27**
[4] Román-Aguirre M, Torre-Sáenz L D, Flores W A, Robau-Sánchez A and Elguézabal A A 2005 *Catal. Today* **107–108**
[5] Handojo L, Putra I A, Azis M M, Prakoso T, Soerawidjaja T H and Indarto A 2019 *AIP Conference Proceedings* **2085** 020056
[6] Putra I A, Azis M M, Soerawidjaja T H and Indarto A 2019 *AIP Conference Proceedings* **2085** 020057
[7] Prakoso T, Hanley J, Soebianta M N, Soerawidjaja T H and Indarto A 2018 *Catal. Letters* **148** 725–31
[8] Budiman A, Arifta T, Diana, D and Sutijan 2015 *Mod. Appl. Sci.* **9**
[9] Utami H, Budiman A, Sutijan, Roto and Sediawan W B 2011 *World Acad. Sci. Eng. Technol.* **54**
[10] Diana, Sutijan, Rochmadi and Budiman A 2016 *AIP Conference Proceedings* **1755** 050003

Acknowledgments

We would like to thank the financial support from the Department of Chemical Engineering, UGM which enabled us to conduct this study in Bandung. Secondly, we would also like to thank both the Department of Chemical Engineering UGM and ITB that allows the realization of the present study. We would also like to thank lham Ardiyanto Putra, Yestria Yaswari, and Fira for their invaluable help during the experimental works as well as result interpretations.

IOP Conf. Series: Materials Science and Engineering **823** (2020) 012043 doi:10.1088/1757-899X/823/1/012043

Performance efficiency of sand media amended with biochar for phosphorus removal using column filtration

Philiphi de Rozari

Department of Chemistry Faculty of Science and Engineering Nusa Cendana University, Kupang, Indonesia

E-mail: phderozari@yahoo.com or p_derozari@staf.undana.ac.id

Abstract. Increase of phosphorus concentrations from wastewater could cause a significant impact on water quality deterioration. Thus, it is crucial to regulate nutrient concentrations from wastewater effluents before the effluents are discharged to the environment. Establishment of wastewater treatment is one of the solutions. This study investigated the efficiency of column filtration using sand media amended with biochar to remove phosphorus in the laboratory scale. The experimental design consisted of 6 packed columns based on the proportion of biochar in the sand media (0 – 25%). The columns were fed with 100 mL phosphate solution (10 mg/L) at the flow rate of 66.6 mL/h. The same steps were also performed using a combination of PO_4^{3-}-P (10 mg/L) and NH_4^+-N (10 mg/L) solution. The results showed that phosphate concentration in the outflow decreased smoothly and in line with the increase of biochar percentage in the sand media. Removal efficiency of PO_4^{3-}-P ranged from 60.9 – 80.9 % and 35.9 - 44.2 % for media loaded PO_4^{3-}-P and combination of PO_4^{3-} -P and NH_4^+-N (1:1). It means that the availability of biochar in the media influenced the amount of phosphorus adsorbed into media.

1. Introduction

Currently, East Nusa Tenggara Province (ENTP) is facing water shortage problems due to polluted water from both point and non-point sources (discharge from drainage, run-off, unsealed septic tanks, domestic wastewater effluent, etc.). Decline in the water quality in some water resources in ENTP is mainly caused by the increase of population and urbanization. Domestic wastewater contributed approximately 60% of water pollution [1]. Discharging untreated domestic wastewater could increase concentrations of nutrients (N and P) into the water body. This condition could cause pollution of water bodies leading to environmental and human health problems.

Phosphorus is a limiting nutrient in most surface water and excess of phosphorus leads to acceleration of eutrophication processes [2]. Awual and Jyo [3] reported that the concentrations of phosphorus in water body should be less than 10 μg/L to preserve surface water from eutrophication. Thus, it is important to find a proper treatment to reduce phosphorus concentrations from domestic wastewater to meet admissible level of water body systems.

Conventional methods for treating wastewater, such as activated sludge and membrane bioreactors, cannot be applied in ENTP because of high operational and investment costs. Filtration is a potential alternative method due to its cost-effectiveness. In this technology, the media plays a crucial role in phosphorus removals [4]. Phosphorus removal takes place via several mechanisms i.e. sorption on the media, plant and microbial uptake and sedimentation [5, 6]. The selection of media is based on the capacity of media to bind phosphorus since sorption is one of the most important mechanisms of phosphorus removals [7].

Materials used as a media for phosphorus removal are classified into three categories; natural materials, industrial by-products, and man-made products [6]. In general, natural materials such as a variety of sands, gravels and soils which are locally available have been utilized as a filter media. In the sand media, phosphorus compounds are primarily bound with calcium (Ca), aluminum (Al) and iron (Fe), depending on the pH value [8]. Adsorption of phosphorus into iron and aluminum oxides took place at pH level higher than 6. In this pH condition, precipitation could occur when phosphate ions react with calcium ions. The precipitation of iron phosphates and aluminum phosphates mainly take place at pH levels lower than 6 [6].

Biochar, the porous solid carbon obtained from pyrolysis of biomass, is an attractive option of media for phosphorus removal due to its large surface area and microporosity [9]. Therefore, sand media should be amended with other material such as biochar that could improve phosphorus removal in filtration system. Based on laboratory experiments, Kizito, et al. [10] found that biochar from corn cob and wood was significantly higher phosphate removal (>71%) than gravel media. Trazzi, et al. [11] reported that biochar prepared from *Miscanthus giganteus* biomass had higher P adsorption (15.5 mg g^{-1}) compared to sugar cane biomass (12.8 mg g^{-1}). Wang et al. [12] investigated P adsorption from biochar blended with calcium with various mass ratios and concluded that the 2:1 mass ration had the highest adsorption capacity (314 mg g^{-1}). Since the findings showed a better performance of biochar in removing phosphorus, it is necessary to utilize biochar as a media amendment in filtration systems. The objective of this study was to investigate the efficiency of biochar amended in the sand media filters in removing phosphorus.

2. Methods

2.1. Media characterization

The sand was bought from local store and biochar was made with slow pyrolysis method from Eucalyptus plants. The physical and chemical parameters of the media measured included porosity, pH, cation exchange capacity (CEC) and organic matter. Displacement methods were used to determine porosity and bulk density [13]. The pH was determined in 0.01 M $CaCl_2$ solution using pH meter based on procedure number C2B/2 and CEC was determined using the silver thiourea method [14]. The organic matter (OM) was determined by weighing after igniting the media at 550°C for 4 hours in the furnace.

2.2. Column experiment

The column experiment was carried out in the laboratory using glass columns with inner diameter = 2.5 cm and height = 50 cm (Figure 1). There were six columns (BCR00% - BCR25%) representing the six types of media based on biochar content amended in the sand media with 5% increment of biochar (Table 1). The columns were filled with 1 cm layer of gravel (10 – 20 cm) at the bottom. The sand media or sand media amended with biochar were added into following layer to the height of 38 cm. To compact the sand layer, 50 g iron weight was dropped 10 times from 15 cm height. Finally, 1 cm of gravel was placed to ensure that all water was equally distributed and to maintain biochar floatation. The media introduced in the columns were in wet conditions.

Table 1. Characterisation of media in the column filtration

Parameters	Media					
	BCR00%	BCR05%	BCR10%	BCR15%	BCR20%	BCR25%
Porosity (%)	42.61±0.13	40.19±1.51	39.65±1.18	38.07±0.55	36.95±1.84	36.63±1.58
pH	6.79±0.02	6.81±0.01	6.88±0.01	6.95±0.01	7.06±0.02	7.19±0.02
OM (%)	0.59±0.04	1.47±0.09	2.32±0.12	3.56±0.17	4.85±0.16	5.67±0.56
CEC meq/100g)	5.73±0.54	6.85±0.54	7.55±0.30	8.37±0.27	9.23±0.35	10.21±0.13

CEC = cation exchange capacity
OM = organic matter

2.3. Chemicals

All chemicals were analytical grades. Phosphate solution (PO_4^{3-}-P= 10 mg/L) was made up by dissolving 43.9 mg KH_2PO_4 into 1 L volumetric flask. Ammonium solution 10 (NH_4^+- N =10 mg/L) was made up by 38.19 mg NH_4Cl into 1 L volumetric flask. The sample stocks were stored at 4°C in the fridge.

2.4. Sample collection and analysis

This column experiment was divided into two parts according to the effluents used: (1) the experiment using phosphate (PO_4^{3-}-P) solution and (2) the experiment using a combination of phosphate (PO_4^{3-}-P) and ammonium (NH_4^+- N) solution. Measurements of phosphate concentrations were carried out on inflow and outflow. For the first five days, each column was loaded daily with 50 ml deionized (DI) water at a flow rate of 16.6 mL/h. The outflow was collected daily in 100 ml beaker glass and stored in fridge at 4°C until analyzed. The outflow collected from the column was determined phosphate concentration using a molybdenum blue method and measured using a spectrophotometer at a wavelength of 880 nm. This step intended to obtain the information of leachable phosphorus in the media using DI water.

After feeding with 50 ml DI water for the first five days, 50 ml of phosphate solution 10 mg/L was allowed to percolate through the column at the same as the previous flow rate for the second five days. Each day, the outflow was collected in 100 ml beaker glass for further phosphate determination. Then, the concentrations of phosphate from both inflows and outflows were determined using a spectrophotometer with a wavelength of 880 nm. This step aimed to determine phosphate accumulated in the media after percolation with 50 ml of PO_4^{-3}-P (10 mg/L) during 5 days loading. The accumulation of phosphate in the media was determined by calculating the difference between the concentration of phosphorus before and after percolation. The same steps of column experiment as above were also performed using a combination of PO_4^{3-}-P (10 mg/L) and NH_4^+-N (10 mg/L) solution. This step intended to enlighten the influence of NH_4^+-N ion in phosphate adsorption. Spectrophotometer (molybdenum blue method) was used to measure phosphorus concentration at 880 nm. All the column experiments were conducted in duplicate.

2.5. Data Analysis

Data were analysed using the SPSS 21 to perform all statistical tests. The ability of each treatment to remove PO_4^{-3}-P was conducted by determining inflow and outflow concentrations of PO_4^{-3}-P. Percentage removal efficiency was calculated as $\%R = \frac{C_{in}-C_{ef}}{C_{in}} x100\%$ where C_{in} and C_{ef} are inflow and outflow of the parameters measured. The mean and standard deviation for each dataset were determined. One-way ANOVA followed by Tukey HSD posthoc tests were applied to identify the differences among treatments for removing PO_4^{-3}-P as well as to determine which treatments were significantly different. We used a significant level ($\alpha = 0.05$) in all cases.

IOP Conf. Series: Materials Science and Engineering **823** (2020) 012043 doi:10.1088/1757-899X/823/1/012043

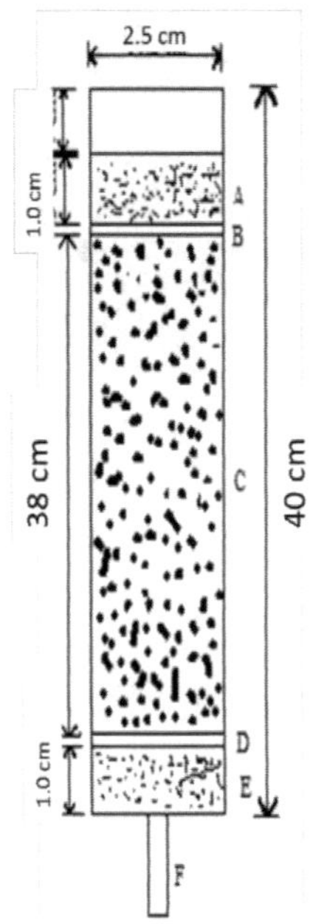

Figure 1. Column experiment design.

3. Results and Discussion

The results are demonstrated in three sections. First, the outflow concentrations of PO_4^{-3}-P when loaded with DI water are introduced. Second, the efficiencies of the experimental column containing sand media amended with various proportions of biochar for phosphorus removal are presented. Finally, the effect of NH_4^+-N on phosphorus removal in experimental column containing sand media amended with biochar is explained. Discussion is presented in each relevant section.

3.1. Concentration of PO_4-P loaded with DI water

The results showed that outflow concentration of PO_4^{-3}-P in six types of column filtration media using 50 ml of DI water for five days ranged from 0.014 mg/L to 0.112 mg/L (Figure 2). The lowest concentration of PO_4^{-3}-P was found in the media without addition of biochar (BCR00%) meanwhile the highest concentration was in the media with 25% of biochar (BCR25%). This is noted that increasing percentage of biochar amended in the sand media led to increasing PO_4^{-3}-P concentrations in the outflow. The availability of phosphorus in the biochar varies depending on the feedstock [15]. This suggested that biochar used in this study influenced the availability of loosely bound phosphorous in the media. The biochar utilized in this experiment was produced from hardwood and this type may contain small amount of loosely bound phosphorus compound. In addition, the outflow concentrations of PO_4^{-3}-P decreased over time because the leachable phosphorus has been released from the media.

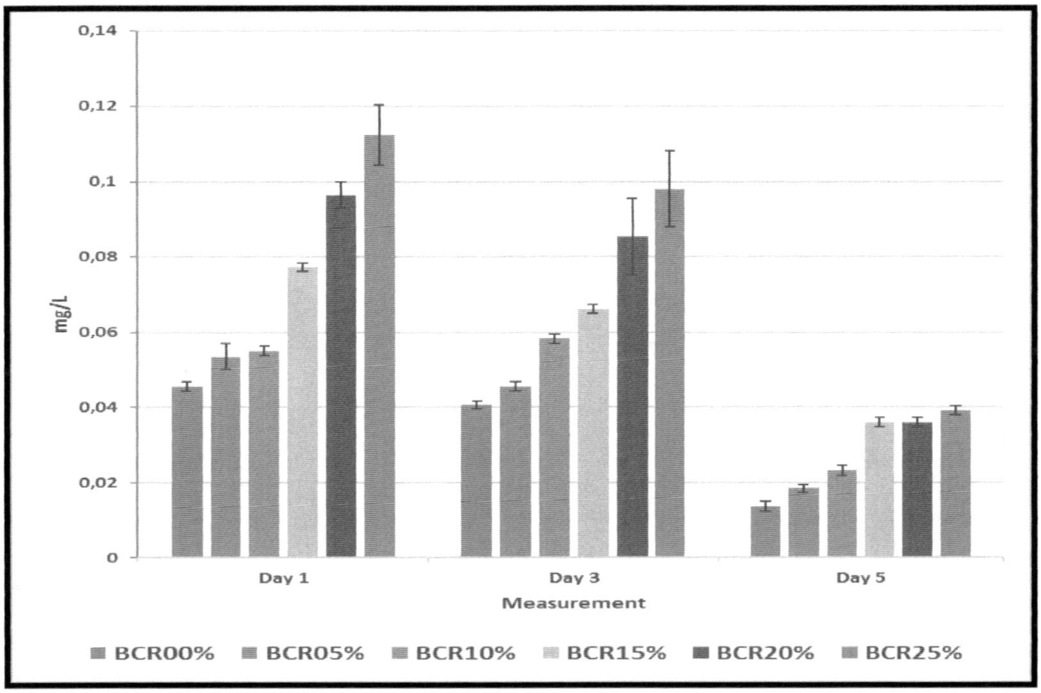

Figure 2. PO_4^{-3}-P concentrations in six types of media loaded with 50 ml of DI water for five days.

3.2. Efficiencies of sand media amended with biochar in column experiment loaded with PO_4^{-3}-P

When the column filtrations were loaded with PO_4^{-3}-P 10 mg/L, the PO_4^{-3}-P concentration in the outflow decreased ranging from 1.50 mg/L – 3.56 mg/L (Figure 3). The decrease of outflow phosphate concentration was in line with the increase of biochar percentage in the sand media. Figure 3 also shows that the removal efficiencies of PO_4^{-3}-P ranged from 60.9 – 80.9 %. Sand media amended with 25% of biochar (BCR25%) was the highest removal efficiency; meanwhile, pure sand media (BCR00%) performed the poorest efficiency. This means that the availability of biochar in the media increased the amount of phosphorus adsorbed into media. This result indicated that the higher percentage of biochar on the media, the higher the phosphate absorbed by the media, leading to an increase of phosphorus removal. This was in line with the results obtained by Yao, et al. [16] stating that biochar was able to adsorb PO_4^{3-}-P from aqueous solution. Yao, et al. [16] pointed out that the abundance of colloidal and nano-sized periclase (MgO) on biochar's surface may influence the increased phosphate removal since this periclase was able to bind phosphate strongly in aqueous solution.

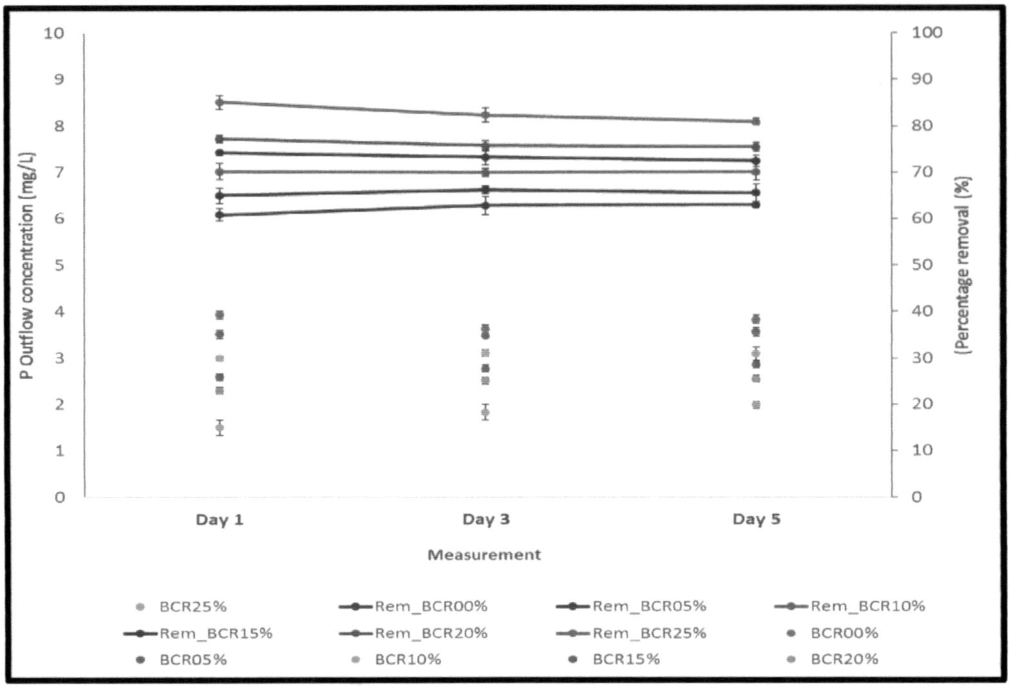

Figure 3. PO_4^{3-}-P concentration in six types of media loaded with the 50 ml of PO_4^{-3}-P (10 mg/L) for five days

3.3. Effect of NH_4^+-N on PO_4^{-3}-P removal in experimental column

This section used a combination of a phosphate (PO_4^{3-}-P) and ammonium (NH_4^+-N) solution. The purpose of this section was to explain the influence of NH_4^+-N ions in the adsorption processes of PO_4^{3}-P in the sand media amended with biochar. The result showed that the concentration of phosphate in the outflow varied, ranging from 5.71 mg/L to 6.55 mg/L (Figure 4). The highest outflow concentration was in the pure sand media (BCR00%) and the lowest concentration was in the media with 25% of biochar (BCR25%). This is noted that the removal percentage of PO_4^{-3}-P in the outflow increased smoothly from 35.9 % in the media without addition of biochar to 44.2% in the sand media amended with 25% of biochar.

Compared to the previous column experiment that utilized only phosphate solution, the adsorption of phosphate in this part tended to be lower. It can be inferred that there was a competition among the ions particularly PO_4^{3-} and NH_4^+ to the surface of the media. The ability of media to adsorb nutrient elements relies on either media and nutrient types [16]. Since most biochar surfaces have negatively charged functional groups, this is obvious that biochar preferred to bind with cationic species from solution provided, resulting in the effectiveness in removing cationic species [17, 18]. In this experiment, the sand media amended with biochar were probably bound with NH_4^+ which had positive charges in comparison with PO_4^{3-} which had negative charges.

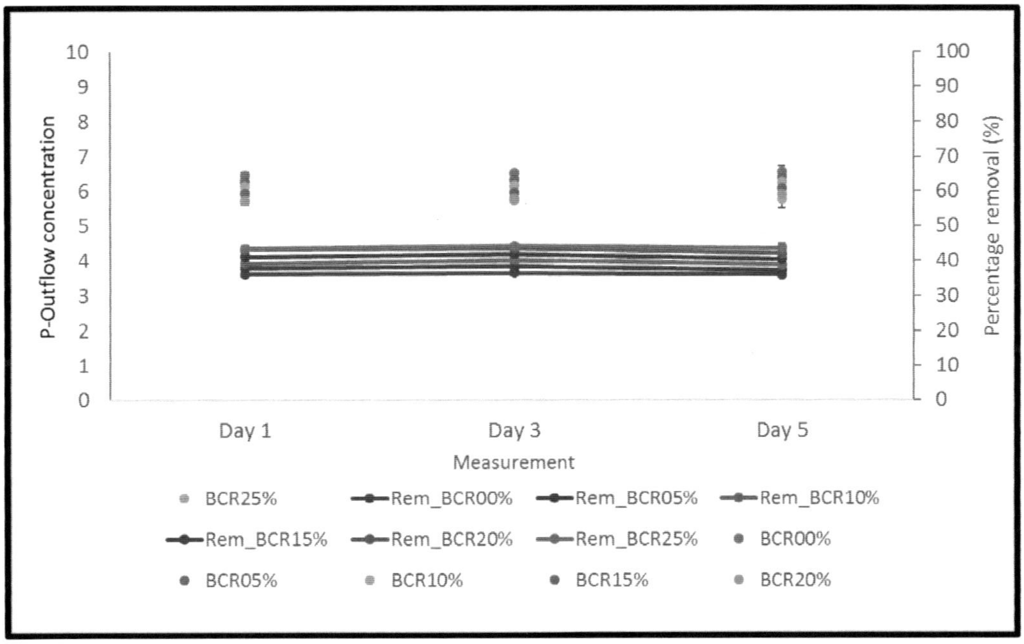

Figure 4. PO_4^{3-}-P concentration in six types of media loaded with the 50 ml of combination PO_4^{3}-P (10 mg/L) and NH_4-N 1(0 mg/L) for five days

4. Conclusion

From this experiment, it can be concluded that removal efficiencies of PO_4^{3-} -P loaded with PO_4^{3-} -P solution were in the range of 60.9 – 80.9 % meanwhile PO_4^{3-} -P removal efficiency for the media loaded with PO_4^{3-} -P and NH_4^+-N solution (1:1) ranged from 35.9 – 44.2 %. It means that the use of biochar from Eucalyptus plants amended in sand media in filtration system could be used for an alternative option to treat PO_4^{3}-P from domestic wastewater in ENTP. The adsorption of phosphate onto surface areas of biochar depends on the types of solution poured in the sand media amended with biochar. The results showed that higher percentage of biochar in sand media provided more phosphate adsorbed into the media, resulting in the increase of phosphorus removal. However, if the solution introduced to the sand media amended with biochar contained a combination of phosphate and ammonium ions, the adsorption of phosphate was influenced by competition processes between these ions. The sand media amended with biochar could have an opportunity to bind with NH_4^+ which had a positive charge and also with PO_4^{3-} which had a negative charge. Since domestic wastewater contains many types of pollutants, it would be better if the application of this technology considered hydraulic loading rate (HLR) and could be combined with other eco-technology such as constructed wetlands.

Acknowledgement

This research is funded by the Directorate of Research and Community Service of the Directorate General of Strengthening Research and Development at the Ministry of Research, Technology and Higher Education through contract No. 47/UN15.19/PL/2019 The authors would like to acknowledge the Ministry of Research, Technology and Higher Education for funding this research project.

IOP Conf. Series: Materials Science and Engineering **823** (2020) 012043 doi:10.1088/1757-899X/823/1/012043

References

[1] Hendrawan D, Widarnako S, Moersidik S S and Triweko R W 2013 Evaluation of centralized WWTP and the need of communal WWTP in supporting community-based sanitation in Indonesia *European Scientific Journal* **9**(17)

[2] Park J H, Ok Y S, Kim S H, Cho J S, Heo J S, Delaune R D and Seo D C 2015 Evaluation of phosphorus adsorption capacity of sesame straw biochar on aqueous solution: influence of activation methods and pyrolysis temperatures *Environmental Geochemistry and Health* **37**(6) 969-83

[3] Awual M R and Jyo A 2011 Assessing of phosphorus removal by polymeric anion exchangers *Desalination* **281** 111-7

[4] El Hanandeh A, Bhuvaneswaran A, de Rozari P, editors 2017 Removal of nitrate, ammonia and phosphate from aqueous solutions in packed bed filter using biochar augmented sand media *MATEC Web of Conf.* **120** 05004 (EDP Sciences)

[5] Ayaz S Ç, Aktaş Ö, Findik N and Akça L 2012 Phosphorus removal and effect of adsorbent type in a constructed wetland system *Desalination and Water Treatment* **37**(1-3) 152-9

[6] Vohla C, Kõiv M, Bavor H J, Chazarenc F and Mander Ü 2011 Filter materials for phosphorus removal from wastewater in treatment wetlands—A review *Ecological Engineering* **37**(1) 70-89

[7] Rustige H, Tomac I and Höner G 2003 Investigations on phosphorus retention in subsurface flow constructed wetlands *Water Science and Technology* **48**(5) 67-74

[8] Yeoman S, Stephenson T, Lester J and Perry R 1988 The removal of phosphorus during wastewater treatment: a review *Environmental Pollution* **49**(3) 183-233

[9] Ahmad M, Rajapaksha A U, Lim J E, Zhang M, Bolan N, Mohan D, Vithanage M, Lee S S and Ok Y S 2014 Biochar as a sorbent for contaminant management in soil and water: a review *Chemosphere* **99** 19-33

[10] Kizito S, Lv T, Wu S, Ajmal Z, Luo H and Dong R 2017 Treatment of anaerobic digested effluent in biochar-packed vertical flow constructed wetland columns: role of media and tidal operation *Science of the Total Environment* **592** 197-205

[11] Trazzi P, Leahy J J, Hayes M H and Kwapinski W 2016 Adsorption and desorption of phosphate on biochars *Journal of Environmental Chemical Engineering* **4**(1) 37-46

[12] Wang S, Kong L, Long J, Su M, Diao Z, Chang X, Chen D, Song G and Shih K 2018 Adsorption of phosphorus by calcium-flour biochar: Isotherm, kinetic and transformation studies *Chemosphere* **195** 666-72

[13] Bashour I I and Sayegh A H 2007 *Methods of Analysis for Soils of Arid and Semi-arid Regions* (Rome: FAO)

[14] Rayment G and Higginson F R 1992 *Australian Laboratory Handbook of Soil and Water Chemical Methods* (Port Melbourne: Inkata Press Pty Ltd)

[15] Chan K Y and Xu Z 2009 Biochar: nutrient properties and their enhancement *Biochar for Environmental Management: Science and Technology* 5th ed pp 67-84

[16] Yao Y, Gao B, Zhang M, Inyang M and Zimmerman A R 2012 Effect of biochar amendment on sorption and leaching of nitrate, ammonium, and phosphate in a sandy soil *Chemosphere* **89**(11) 1467-71

[17] Lehmann J, Rillig M C, Thies J, Masiello C A, Hockaday W C and Crowley D 2011 Biochar effects on soil biota–a review *Soil Biology and Biochemistry* **43**(9) 1812-36

[18] Mukherjee A, Zimmerman A and Harris W 2011 Surface chemistry variations among a series of laboratory-produced biochars *Geoderma* **163**(3-4) 247-55

IOP Conf. Series: Materials Science and Engineering **823** (2020) 012044 doi:10.1088/1757-899X/823/1/012044

Calcium soap from Palm Fatty Acid Distillate (PFAD) for ruminant feed: The effect of CaO quality on reaction temperature

Lienda A. Handojo[1], Antonius Indarto[1], Dian Shofinita[1], Nicholas George Stanley[1], Karen[1], Ignatius David[1] and Jorgensen[1]

[1]Department of Chemical Engineering, Faculty of Industrial Technology, Institut Teknologi Bandung, Bandung, Indonesia

E-mail: lienda@che.itb.ac.id

Abstract. Indonesia is the largest palm oil producer in the world and contributes to about 58% of the world's total palm oil production in 2018. Palm fatty acid distillate (PFAD) is a palm oil refining by-product that contains lots of free fatty acids. Previous research related to the production of calcium soap has successfully found the right stoichiometry and operating conditions at the laboratory and bench scale. This research examines the influence of CaO active concentration on the reaction temperature of calcium soap production on the pilot scale. First, molten PFAD, CaO solids, and water are feed into a CSTR. The mixture is discharged from the CSTR to the screw reactor at a certain temperature, which is then released to the belt conveyor. The experimental result shows that the rate of temperature increment of the mixture using CaO with an active concentration of 79% is greater than CaO of 72%. This may be caused by the different amounts of heat released during the reaction, viscous mixture properties which then influence the mass transfer process between the reaction components. The result also shows that there is a 14% - 19% difference between the calculated and measured temperature, which is predicted due to heat loss, particularly caused by water evaporation during the reaction.

1. Introduction

Indonesia is the world's largest producer of palm oil with the amount of 43 million tons, which is equivalent to 58% of the total world's palm oil production in 2018 [1]. Besides palm oil, crude palm oil (CPO) refining process also produces a by-product called palm fatty acid distillate (PFAD) [2]. PFAD can be used as raw material for oleochemical and bioactive industries such as vitamin E, phytosterols, and squalene. In other countries, PFAD can be processed into animal feed supplements that can increase energy supply [3].

The calcium soap, as a form of protected fat used for ruminant feed supplement, is preferable than the free fatty acid. Unlike directly feeding free fatty acid to a ruminant, the protected fat does not interfere with fermentation in cow's rumen [4]. Protected fat has a high energy concentration which gives benefits to dairy cows, for instance, an increased milk production [5] as well as pregnancy rate [6].

Research related to the production of calcium soap by saponifying PFAD with various CaO active levels has been done on a laboratory scale. It was found that the reaction with a lower concentration of active calcium of CaO resulted in products with a higher acid value, which indicates a lower reaction conversion [7]. Saponification is known as a spontaneous and exothermic reaction that produces heat during the reaction [7]. Meanwhile, there is also heat produced by the presence of calcium oxide and water as the result of dissolved calcium in the water. The heat involved in this reaction can be seen in the increase of the mixture temperature [8].

This study aims to determine the effect of CaO active concentration on the mixture temperature in the production of calcium soap through the saponification reaction in a pilot scale. The CaO active concentrations that are used in the experiment are 79% and 72%. The calcium soap product will be analyzed its acid number as it can indicate the amount of free fatty acid that is converted to product. The favorable calcium soap for ruminant feed is the one that has a low acid number.

The hypothesis of this research is that the saponification reaction with a higher active concentration of CaO will result in a higher rate of temperature increment of the mixture where the reaction occurs. This is because the reaction emits more heat due to the higher CaO concentration which reacts with PFAD.

2. Material and Methods

2.1. Materials

The PFAD used in this study was obtained from the crude palm oil refining plant in PT Tunas Baru Lampung Tbk. The calcium oxide came from the limestone calcination process in Padalarang. Water was also used in the reaction to accelerate the saponification rate.

2.2. Experimental set-up

The equipment used in this experiment includes a CSTR, a screw reactor, and a belt conveyor shown in Figure 1.

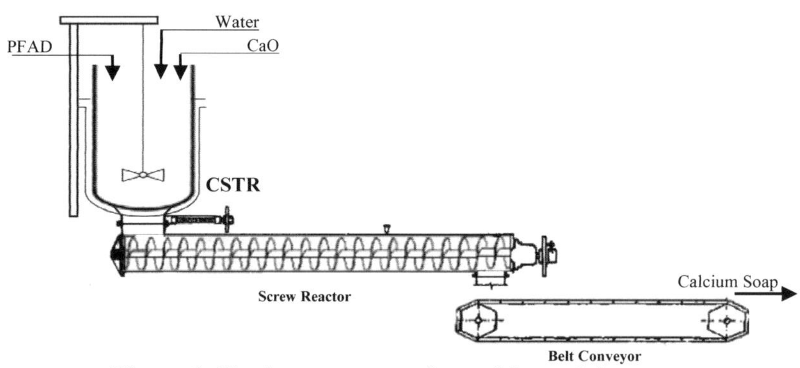

Figure 1. Equipment set-up for making calcium soap

First, the size of the calcium oxide chunks is reduced using a disc mill; then it is sieved on several layers to get 100 mesh CaO powder. As for the calcium soap making, PFAD is melted in the CSTR at a certain temperature. CaO is reacted with the molten PFAD with a certain stoichiometric ratio [6]. Water is then added to the homogeneous mixture. After a certain reaction temperature is reached, the mixture is released from the CSTR to the screw reactor and led to belt conveyor.

The products are tested for the acid number using ISO 6060:2009 and evaluated to estimate the fatty acid conversion. The sample of calcium soaps (1 g) is dissolved in ethanol, heated to 60°C, and stirred for approximately 10 minutes. Phenolphthalein is added as the indicator to the mixture and then titrated by 0,01 N potassium hydroxide (KOH). The acid number is calculated as the number of KOH in milligrams required to neutralize the free fatty acids in 1 g of the sample. The CaO active concentration is determined by using ASTM 1984 from The Schumacher Centre for Technology & Development, UK. The sample of CaO (2,5 g) is added by free-carbon-dioxide water (35 - 45 mL) and sugar solution of 50% concentration. The mixture is then stirred for around 30 minutes. Water is added to the mixture until the mixture reaches 250 mL volume in the Erlenmeyer glass. The mixture is filtrated subsequently. The first 15 – 30 mL filtrate is separated from the glass. 25 mL filtrate is titrated by 0,357 N sulphuric acid (H_2SO_4) with phenolphthalein as the indicator.

2.3. *Variations*

There are 2 procedures done in the experiments, making calcium soap with CaO A, which contains 79% active CaO, and CaO B, which contains 72% active CaO. The difference between the procedures lies in the variation of the CSTR discharge temperature. The mixture in CSTR using CaO grade A was released from CSTR to screw reactor at various temperatures, i.e. 84°C, 86°C, and 90°C. On the other hand, the mixture in CSTR using CaO grade B was released at temperature 72°C, 76°C, 80°C, and 86°C. This range was chosen because the saponification reaction temperature to form calcium soap starts around 78°C. From the experiments, it is observed that the 72% active CaO content produced the viscous reaction mixture. As a result, the mixture should be released at a lower temperature from the CSTR, since the reaction mixture will become more viscous at higher temperatures, so that it cannot be discharged from the CSTR into the screw reactor properly. Another variation in experiments was the use or omission of the screw reactor. The list of experiments is shown in Table 1.

Table 1. The list of experiments

Run	CaO Level (%)	Use of Screw Reactor	T_{outlet} CSTR (°C)
1	72	No	80
2	72	No	76
3	72	No	78
4	72	Yes	72
5	79	No	86
6	79	No	90
7	79	No	84
8	79	Yes	90
9	72	Yes	84
10	72	Yes	78
11	72	No	72
12	72	No	86

3. Results and Discussion

The temperature profile in the CSTR versus reaction time was recorded for each experiment, shown in Figure 2.

The reactions during the process can be divided into 2 steps, i.e. the reaction between CaO and water, and the saponification of fatty acid with hydroxide of calcium, as can be seen in Equations 1 and 2.

$$CaO + H_2O \rightarrow Ca(OH)_2 \quad (1)$$

$$2\, C_{16}H_{32}O_2 + Ca(OH)_2 \rightarrow C_{32}H_{62}CaO_4 + 2\, H_2O \quad (2)$$

As the saponification reaction occurs at the temperature about 78°C, the temperature profile can be divided into two zones. The first zone is the temperature below 78°C, where the temperature gradient results from the heat produced by reaction 1. The second zone is above 78°C, where the temperature gradient results from the combination of heat released by reaction 1 and reaction 2. In this work, only the temperature gradient in zone 1 is discussed.

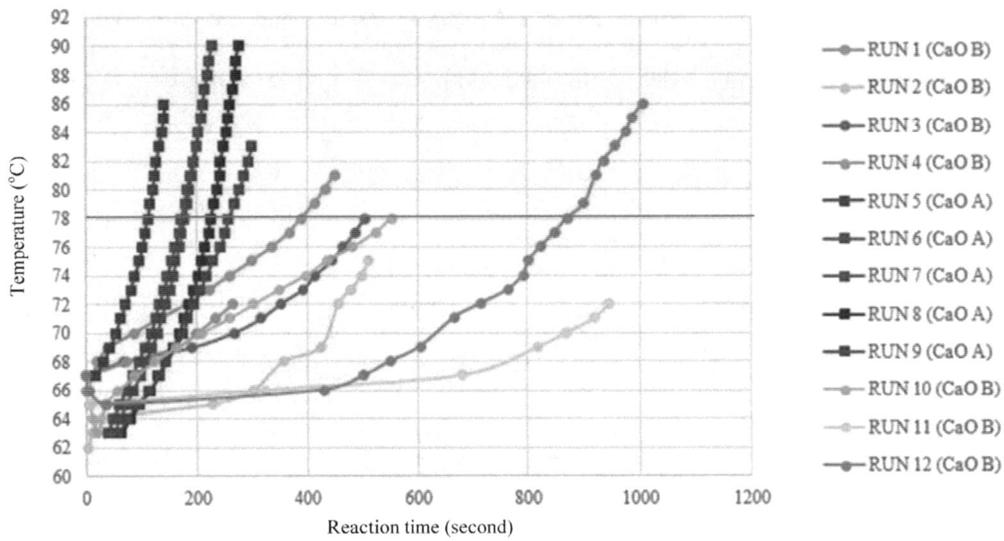

Figure 2. Graphic of the relationship between temperature to the reaction time of making calcium soap experiments

Based on the calculation results, the enthalpy reaction between CaO and water forming $Ca(OH)_2$ at 78°C is -67.44 kJ/mol while the enthalpy reaction between $Ca(OH)_2$ and PFAD forming calcium soap and water at 78°C is -660.47 kJ/mol. This causes the saponification reaction to produce greater heat, thus the raise in the mixture temperature becomes faster after 78°C. The temperature gradient is only investigated from the experiments whose mixture temperature exceeded 78°C. Those, which did not reach 78°C show very different physical properties. The mixture became so viscous that the reaction could not take place further. From the observation during the experiments, the reaction between CaO and water took place immediately after the water was poured into the mixture, at which the initial temperature was set to 63°C for CaO A and 67 °C for CaO B. Hence, the rise of mixture temperature from initial temperature to 78°C was solely caused by the heat released from the reaction of CaO and water. The temperature increase after 78°C is assumed to result from the heat released by a combination of the CaO-water reaction as well as the saponification reaction of PFAD and $Ca(OH)_2$.

Figure 2 shows that differences in the active CaO concentration affect the rate of temperature increase, as the temperature raise with CaO A is faster than with CaO B. The heat released by the reaction can be calculated using Equation 3.

$$Q = n \, . \, \Delta H \quad (3)$$

With Q = reaction heat (kJ), n = reaction mole (mol), and ΔH = reaction enthalpy (kJ/mol). It is assumed that all of the CaO active concentration perfectly reacts with water. For CaO A, the initial active mole of CaO is 5.5 mole, while CaO B is 5 mole. The reaction heat is used to calculate the rate of temperature increment using the transient heat transfer equation shown in Equation 4.

$$Q = m \times Cp \times \Delta T \quad (4)$$

The calculated temperature was then compared with the measurement to determine the effect of CaO levels on the mole that reacts. The temperature calculation of the experiments using CaO B is

IOP Conf. Series: Materials Science and Engineering **823** (2020) 012044 doi:10.1088/1757-899X/823/1/012044

based on run 1, 3, 10, and 12 of the experiments, where the temperature exceeded 78°C. The differences between the calculated and measured temperatures from run 1 and 3 are shown in Figure 3.

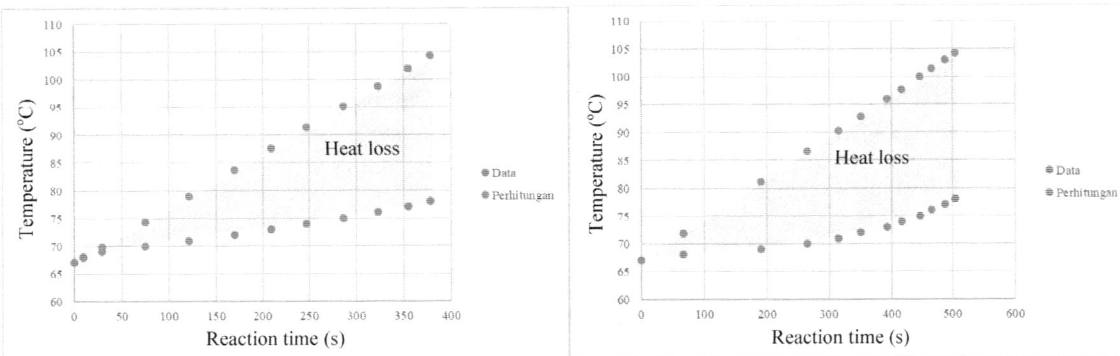

Figure 3. Comparison between the calculated temperature and the experimental temperature (1) run 1; (2) run

The rate of moles that reacted and % temperature differences for the whole experiments are shown in Table 2 below.

Table 2. The rate of reaction and percentage of temperature differences

Run	CaO Level	Rate of temperature rise (°C/sec)	Mole rate of reaction (mole/sec)	Difference (%)
1	72%	0.03	0.013	13.4
3	72%	0.02	0.010	19.3
10	72%	0.02	0.011	14.3
12	72%	0.03	0.014	17.3
5	79%	0.16	0.059	19.1
6	79%	0.11	0.042	16.9
7	79%	0.12	0.044	17.6
8	79%	0.09	0.034	18.6
9	79%	0.07	0.028	17.1

From Table 2, the differences between the calculated temperature with the measured temperature are around 14% - 19%. It can be caused due to heat loss from the system, particularly due to water evaporation during the reaction. There is heat loss to the surroundings as well, as the water in the jacket of CSTR at the beginning of the operation was 63°C or 67°C. A higher mixture temperature also caused the heat loss rate to be larger.

From the Table, it can be seen that the rate of temperature increment using CaO A is greater than with CaO B. This is due to the lower CaO active concentration in CaO B compared to CaO A, so that less heat was released from the reaction. On the other hand, the reaction mixture using CaO B was much more viscous than using CaO A. The high viscosity of mixture affected the mass transfer rate, so that the contact between the reactants was slower or more difficult compared to the system using CaO A.

From Figure 2, it can be seen that the time needed to reach mixture temperature of 78 °C using CaO A was between 2.5 – 5 minutes. Meanwhile, the runs using CaO B took between 7 – 9 minutes to

IOP Conf. Series: Materials Science and Engineering **823** (2020) 012044 doi:10.1088/1757-899X/823/1/012044

reach that temperature. From this experiment, it can be concluded that the temperature of the mixture of PFAD with CaO A raised faster than with CaO B.

The parameter that determines the quality of calcium soap produced is the acid number. In this experiment, the acid numbers obtained are between 0.4 to 5.8. This result shows that there is a certain amount of free fatty acids from PFAD that did not react with CaO, particularly in the system with CaO B, where the active CaO concentration is lower than in CaO A. During the saponification reaction, water was produced and a fraction of it vaporized due to the mixture temperature that increased with time. The amount of water vaporized affected the extent of the reaction or the degree of conversion of PFAD. This is consistent with the *Le Chatelier's Principle* [10], where the reaction will shift toward the product as more water vaporizes.

4. Conclusion

The experimental result shows that the higher concentration of active CaO will produce a higher rate of temperature increment of the mixture. This could be due to differences of amount of heat released by reaction, viscous mixture properties, and mass transfer between the reaction components.

References

[1] Indexmundi.com 2019 *Palm Oil Production by Country in 1000 MT - Country Rankings* Available from: https://www.indexmundi.com/agriculture/?commodity=palm-oil (Accessed 10 Jun. 2019)

[2] Top A G 2010 Production and utilization of Palm Fatty Acid Distillate (PFAD) *Lipid Technology* **22** 11

[3] Gunes F E 2013 Medical use of squalene as a natural antioxidant *Journal of Mamara University Institute of Health Sciences* **3**(4) 220-28

[4] Chalupa W, Rickabaugh B, Kronfeld D and Sklan D 1984 Rumen fermentation in vitro as influenced by long-chain fatty acids *J. Dairy Sci* **67** 1439-44

[5] Purushothaman S, Kumar A and Tiwari D P 2008 Effect of feeding calcium salts of palm fatty acids on performance of lactating crossbred cows *Asian-Aust. J. Anim. Sci.* **21**(3) 376-85

[6] Suksombat W 2009 Improve the productivity of lactating dairy cows through supplementation *International Dairy Topics* **8** 7-11

[7] Handojo L A, Indarto A, Shofinita D, Meitha A, Nabila R, Triharyogi H, Hakim M G A and Saadi M R 2018 Calcium soap from palm fatty acid distillate (PFAD) for ruminant feed: quality of calcium source *MATEC Web of Conferences* **156** 02062

[8] Handojo L A, Indarto A, Shofinita D, Meitha A, Nabila R, Triharyogi H and Kevin L 2018 Calcium soap from palm fatty acid distillate for ruminant feed: the influence of water temperature *IOP Conf. Series: Earth and Environment Science* **141** 012010

[9] Handojo L A, Indarto A, Shofinita D, Meitha A, Nabila R and Triharyogi H 2017 Calcium soap from palm fatty acid distillate for ruminant feed: reaction method *International Journal of Engineering and Advanced Technology* **8** 422-25

[10] Colina C 2007 *Stability, Displacement and Moderation of Chemical Equilibrium: Rediscovering Le Chatelier's Principle* (Venezuela: Simon Bolivar University Venezuela)

IOP Conf. Series: Materials Science and Engineering **823** (2020) 012045 doi:10.1088/1757-899X/823/1/012045

Production of valuable chemical compounds isolated from plants by endophytic fungi

Antonius R B Ola[1,2]

[1]Chemistry Department, Faculty of Science and Engineering, Nusa Cendana University, Kupang 85118, Indonesia
[2]Integrated Center Research (LRT Biosains) Laboratory, Nusa Cendana University, Kupang 85118, Indonesia

E-mail: antonius.ola@staf.undana.ac.id

Abstract. Natural products have always been an integral part of medical therapy. Many drugs have been developed from natural products mostly derived from plants. For example, the most effective anticancer compounds such as anthracyclines, vinblastine, vincristine, paclitaxel and camptothecins are produced by plants. However, productions of these highly valuable compounds have threatened the existence of the plants as it requires a large number of materials from plants. This paper highlights the potential and importance of endophytic fungi as the producers of highly valuable compounds as it needs only a small amount of material from plants. For example, anticancer drugs like paclitaxel, vincristine and vinblastine were more commonly found in endophytic fungi than in plants.

1. Introduction

Plants have been widely known as the major sources of bioactive agents for treatment of cancers. Some of the important anticancer drugs from plants include vinca alkaloids, vinblastine and vincristine produced by *Catharanthus roseus* [1, 2]. Etoposide and teniposide are also clinical drugs used for the treatment of cancer which was developed from natural product epipodophyllotoxin [3]. Epipodophyllotoxin is an isomer of podophyllotoxin found in the roots of various species of the genus *Podophyllum* [4]. These plants have been long traditionally used for the treatment of skin cancers and warts in the American and Asian communities [5].

The most valuable anti-cancer drug, paclitaxel (Taxol®), was originally from the bark of Pacific Yew (*Taxus brevifolia*) [6]. Paclitaxel became the most exciting plant-derived anticancer drug with average sales approximately over $1 billion [7]. The annual demand for paclitaxel was estimated at around 100-200 kg per year. On the other hand, paclitaxel is present in limited quantities from its natural sources as 1 kg of paclitaxel requires 10,000 kg of *Taxus* bark [7]. Thus, it is estimated thousands of Pacific Yew trees will cut down and this will bring serious consequences on the ecological perspective of this valuable and slow-growing tree. Therefore, alternatives production of drugs should be developed. Endophytic fungi have attracted great attention for drug development [8-10] since the discovery of endophytic fungus *Taxomyces andreanae* from *Taxus brevifolia* which was

IOP Conf. Series: Materials Science and Engineering **823** (2020) 012045 doi:10.1088/1757-899X/823/1/012045

also able to produce paclitaxel [11]. This paper highlights and discusses the production of drugs originally from plants by endophytic fungi.

2. Endophytic Fungi as Producer of Valuable Chemical Compounds Originally from Plants

The co-evolution of endophytic fungi with their host plant has been recognized to shape the production of chemical metabolites by the endophytic fungi. Host plant provides nutrition and habitat for endophytic fungi while endophytic fungi produce bioactive metabolites to help the adaptation of their host plant against biotic and abiotic stress. Several reports have shown the biosyntheses of valuable chemical compounds derived from plants by endophytic fungi especially the anticancer drugs and drug leads from plants.

2.1 Paclitaxel from Endophytic Fungi

Paclitaxel produced by Yew tree has broadband of antitumor activities including ovarian, neck, breast, lung and head cancers as well as Kaposi's sarcoma. The first report of the production paclitaxel from endophytes was from the endophytic fungus *Taxomyces andreanae* [11] isolated from the plant Pacific yew *Taxus brevifolia* (Taxaceae). Many other endophytic fungi have been known to produce paclitaxel. These endophytic fungi were isolated not only from the plant of *Taxus* species but also from other species such as hazelnut producing plants [12], *Citrus medica* (Rutaceae), *Podocarpus* sp. (Podocarpaceae), *Torreya grandifolia* (Taxaceae), *Cupressus* sp. (Cupressaceae), *Hibiscus rosa-sinensis* (Malvaceae), *Taxodium distichum* (Taxodiaceae), *Cardiospermum halicacabum* (Sapindaceae), *Terminalia arjuna* (Combretaceae), *Ginkgo biloba* (Ginkgoaceae) and *Wollemia nobilis* (Araucariaceae) [13].

There are more than 20 genera of endophytic fungi that are able to produce paclitaxel and its analogs including *Botryodiplodia*, *Tubercularia*, *Alternaria*, *Botrytis*, *Cladosporium*, *Ectostroma*, *Fusarium*, *Metarhizium*, *Monochaetia*, *Pithomyces*, *Pestalotiopsis*, *Phyllosticta*, *Pestalotia*, *Taxonomyces* and *Aspergillus*) [14]. Among these fungi, fungal strains *Metarhizium anisopliae* was able to produce 846.1μg/L of paclitaxel [15] while fungal strain *Cladosporium cladosporioides* isolated from *Taxus media* produced around 800 μg/L [16].

Of special interest, there are several interesting reports regarding the production of paclitaxel from fungi inhabiting the plants other than *Taxus* species such as the fungal strain *Phyllossticta citricarpa* residing in *Citrus medica* and *Phyllossticta dioscoreae* isolated from *Hibiscus rosa-sinensis* [17] produced 265 and 298 μg/L taxol, respectively.

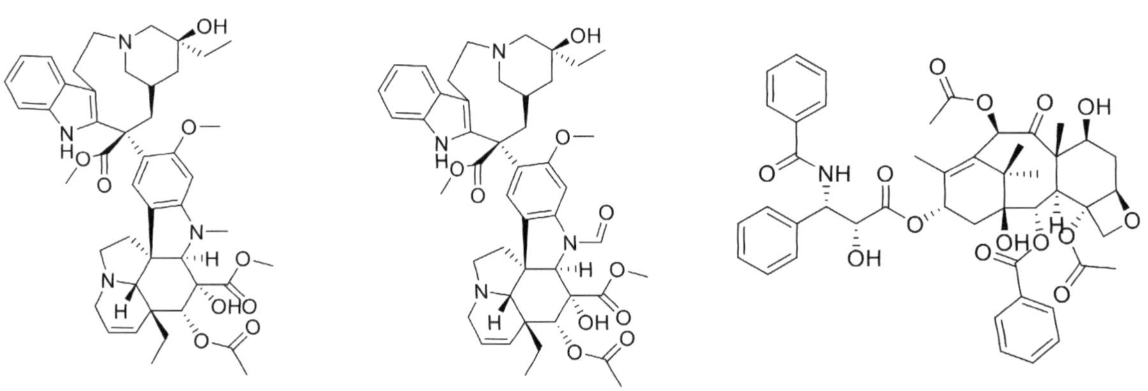

Figure 1. Chemical structure of left: vinblastine; middle: vincristine; right: paclitaxel

2.2 Vinca alkaloid: Vincristine and vinblastine from Endophytic Fungi

Vinblastine and vincristine are dimeric indole alkaloids produced by the plant *Catharanthus roseus*. They are clinically used for the treatment of breast, lung, leukemia and lymphoma cancers. *Alternaria sp* from *C. roseus* was first reported to produce vinblastine [18]. Moreover, endophytic fungi

Fusarium oxysporum has been known to produce both drugs. Endophytic fungus *F. oxysporum* isolated from *C. roseus* was reported to produce vincristine [19]. In 2013, *F. oxysporum* isolated from *C.roseus* collected in India was known to produce both vinblastine and vincristine around 76 µg/L and 67 µg/L, respectively. In addition, endophytic fungus *Talaromyces radicus* isolated from *Catharanthus roseus* was also observed to produce both vincristine and vinblastine [20].

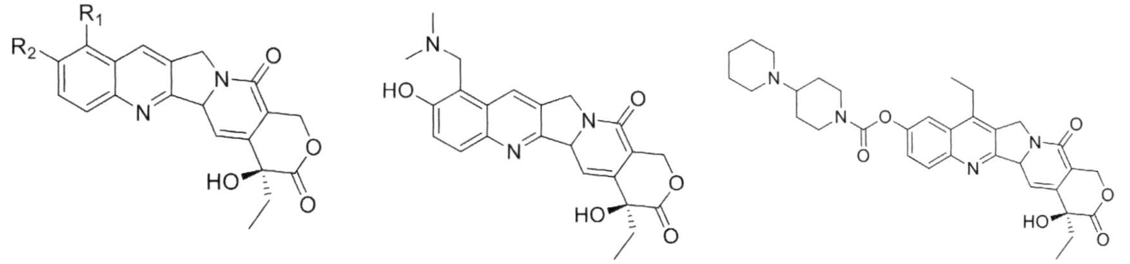

Figure 2. Chemical structure of left: camptothecin (R_1=R_2=H) or 9-methoxycamptothecin (R_1=OMe, R_2=H) or 10-hydroxycamptohecin (R_1=R_2=OH); middle: topotecan; right: irinotecan

2.3 Campthotecin and it's analogue from Endophytic Fungi

Camptothecin was originally isolated from the plant *Campthoteca acuminata* in 1966 [21]. It has become the lead compound for drugs irinotecan and topotecan that were clinically used for the treatment of small lung, ovarian and refractory ovarian cancers [22]. Camptothecin was still supplied from the wild tree of *Camptotheca acuminate* (Nyssaceae) and *Nothapodytes nimmoniana* (Icacinaceae). Another source of camptothecin was first reported from an endophytic fungus *Entrophospora infrequens* associated with the plant *Nothapodytes foetida* [23]. The endophytic fungus *Neurospora* sp isolated from the seeds of *Nothapodytes foetida* was also reported to produce camptothecin [24]. From the plant of *Campthoteca acuminata*, the endophytic fungus *Fusarium solani* produced camptothecin and its anolugues 9-methoxy camptothecin together with 10-hydroxy camptothecin [25]. Two fungal strain endophytic *Fusarium solani* MTCC9667 and MTCC9668 obtained from the plant *Apodytes dimidiate* (*Icacinaceae*) were also known to produce camptothecin, 9-methoxy camptothecin and 10-hydroxy camptothecin [26]. Shweta et al in 2013 [27] reported the production of camptothecin from three endophytic fungi *Alternaria alternata*, and *Phomopsis* sp isolated from the seeds and fruits of plant *Miqueliadentata* Bedd which is known to accumulate very high concentrations of camptothecin in its seeds.

2.4 Podophyllotoxins from Endophytic Fungi

Podophyllotoxin is the lead compound of clinically anti-cancer drugs etoposide and teniposide which can inhibit the topoisomerase by preventing the formation of mitotic-spindle of microtubules [28]. Podophyllotoxin is currently supplied from the plant *Podophyllum* species [14]. The species of *podophyllum* plants have been categorized as endangered species due to overexploitation.

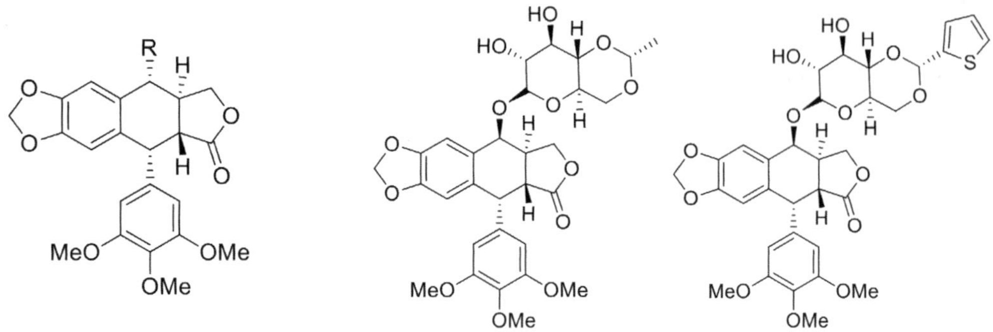

Figure 3. Chemical structure of left: podophyllotoxin (R=OH) or deoxypodophyllotoxin (R=H); middle: etoposide; right: teniposide

Other sources of podophyllotoxins have been discovered from several endophytic fungi such as *Phialocephala fortiniii* and *Trametes hirsuta* obtained from the plants of *Sinopodophyllum peltatum* and *S. hexandrum* [29, 30], *Alternaria* sp from *Sinopodophyllum hexandrum* [31] and *Fusarium oxysporum* obtained from *Sabina recurva* (Kour *et al.,* 2008) and *Fusarium solani* isolated from the root of *S. hexandrum* [32] together with *Aspergillus fumigatus* obtained from *Juniperus communis*. Ran et al. in 2017 [33] has also reported the accumulation of camptothecin from the endophytic *Fusarium solani* obtained from the host *Campthoteca acuminate*.

3. Conclusion

Recent development has shown that the occurrence of anticancer drugs originally derived from plants was found more common in endophytic fungi. Of special interest, the endophytic fungi producing the valuable drugs were also isolated from the host plants which were not known to produce the drugs. As isolation of endophytic fungi required only a small material of plants, this provides an alternative source for future direction in drug discovery programs.

References

[1] Johnson I S, Armstrong J G, Gorman M and Burnett J P 1963 The vinca alkaloids: a new class of oncolytic agents *Cancer Res.* **23** 1390-427

[2] Cragg G M and Newman D J 2013 Natural products: A continuing source of novel drug leads *Biochim. Biophys. Acta* **1830**(6) 3670-95

[3] Mishra B B and Tiwari V K 2011 Natural products: An evolving role in future drug discovery *Eur. J. Med. Chem.* **46**(10) 4769-807

[4] Liu C L and Jiao B H 2006 LC determination of Podophyllum lignans and flavonoids in Podophyllum emodi Wall. var. chinesis Sprague *Chromatographia* **64**(9-10) 603-7

[5] Liu Y-Q, Yang L and Tian X 2007 Podophyllotoxin: current perspectives *Curr. Bioact. Compd.* **3**(1) 37-66

[6] Wani M C, Taylor H L, Wall M E, Coggon P and McPhail A T 1971 Plant antitumor agents. VI. Isolation and structure of taxol, a novel antileukemic and antitumor agent from Taxus brevifolia *J. Am. Chem. Soc.* **93**(9) 2325-7

[7] Dias D A, Urban S and Roessner U 2012 A historical overview of natural products in drug discovery *Metabolites* **2**(2) 303-36

[8] Ola A R B, Aly A H, Lin W, Wray V and Debbab A 2014 Structural revision and absolute configuration of lateritin *Tetrahedron Letters* **55**(45) 6184-7

[9] Ola A R B, Tawo B D, Belli H L L, Proksch P, Tommy D and Hakim E H 2018 A new antibacterial polyketide from the endophytic fungi aspergillus fumigatiaffinis *Natural Product Communications* **13**(12)

[10] Ola A R B, Metboki G, Lay C S, Sugi Y, Rozari P D, Darmakusuma D and Hakim E H 2019

Single production of kojic acid by aspergillus flavus and the revision of flufuran *Molecules* **24**(22) 4200

[11] Stierle A, Strobel G and Stierle D 1993 Taxol and taxane production by Taxomyces andreanae, an endophytic fungus of Pacific yew *Science* **260**(5105) 214–6

[12] Service R F 2000 Hazel trees offer new source of cancer drug *Science* **288** 1609–10

[13] Newman D J and Cragg G M 2015 Endophytic and epiphytic microbes as "sources" of bioactive agents *Frontiers in Chemistry* **3** 34-34

[14] Zhao J, Shan T, Mou Y and Zhou L 2011 Plant-derived bioactive compounds produced by endophytic fungi *Mini Rev. Med. Chem.* **11**(2) 159-68

[15] Liu K, Ding X, Deng B and Chen W 2009 Isolation and characterization of endophytic taxol-producing fungi from Taxus chinensis *J. Ind. Microbiol. Biotechnol.* **36**(9) 1171-7

[16] Zhang P, Zhou P P and Yu L J 2009 An endophytic taxol-producing fungus from Taxus media Cladosporium cladosporioides MD2 *Curr. Microbiol.* **59**(3) 227-32

[17] Kumaran R S, Muthumary J, Kim E-K and Hur B-K 2009 Production of taxol from Phyllosticta dioscoreae, a leaf spot fungus isolated from Hibiscus rosa-sinensis *Biotechnol. Bioprocess Eng.* **14**(1) 76-83

[18] Guo B, Li H and Zhang L 1998 Isolation of a fungus producing vinblastine *Yunnan Daxue Xuebao, Ziran Kexueban* **20**(3) 214-5

[19] Zhang L, Guo B, Li H, Zeng S, Shao H, Gu S and Wei R 2000 Isolation of endophytic fungus of Catharanthus roseus and its fermentation to produce products of therapeutic *Zhongcaoyao* **31**(11) 805-7

[20] Palem P P C, Kuriakose G C and Jayabaskaran C 2015 An endophytic fungus, Talaromyces radicus, isolated from Catharanthus roseus, produces Vincristine and Vinblastine, which induce apoptotic cell death *PLOS ONE* **10**(12)

[21] Wall M E, Wani M C, Cook C E, Palmer, Keith H, McPhail A T and Sim G A 1966 Plant antitumor agents. I. The isolation and structure of camptothecin, a novel alkaloidal leukemia and tumor inhibitor from Camptotheca acuminata *J. Am. Chem. Soc.* **88**(16) 3888-90

[22] Srivastava V, Negi A S, Kumar J K, Gupta M M and Khanuja S P 2005 Plant-based anticancer molecules: A chemical and biological profile of some important leads *Bioorg. Med. Chem.* **13**(21) 5892-908

[23] Puri S C, Verma V, Amna T, Qazi G N, Spiteller M 2005 An endophytic fungus from Nothapodytes foetida that produces camptothecin *J. Nat. Prod.* **68**(12) 1717-9

[24] Rehman S, Shawl A S, Verma V, Kour A, Athar M, Andrabi R, Sultan P and Qazi G N 2008 An endophytic Neurospora sp. from Nothapodytes foetida producing camptothecin *Prikl. Biokhim. Mikrobiol.* **44**(2) 225-31

[25] Kusari S, Zühlke S and Spiteller M 2009 An endophytic fungus from Camptotheca acuminata that produces camptothecin and analogues *J. Nat. Prod.* **72**(1) 2-7

[26] Shweta S, Zuehlke S, Ramesha B T, Priti V, Mohana, Kumar P, Ravikanth G, Spiteller M, Vasudeva R and Uma S R 2010 Endophytic fungal strains of Fusarium solani, from Apodytes dimidiata E. Mey. ex Arn (Icacinaceae) produce camptothecin, 10-hydroxycamptothecin and 9-methoxycamptothecin *Phytochemistry* **71**(1) 117-22

[27] Shweta S, Gurumurthy B R, Ravikanth G, Ramanan U S and Shivanna M B 2013 Endophytic fungi from Miquelia dentata Bedd., produce the anti-cancer alkaloid, camptothecine *Phytomedicine* **20**(3) 337-42

[28] Guerram M, Jiang Z-Z and Zhang L-Y 2012 Podophyllotoxin, a medicinal agent of plant origin: past, present and future *Zhongguo Tianran Yaowu* **10**(3) 161-9

[29] Eyberger A L, Dondapati R and Porter J R 2006 Endophyte fungal isolates from Podophyllum peltatum produce Podophyllotoxin *J. Nat. Prod.* **69**(8) 1121-4

[30] Puri S C, Nazir A, Chawla R, Arora R, Riyaz-ul-Hasan S, Amna T, Ahmed B, Verma V, Singh S, Sagar R, Sharma A, Kumar R, Sharma R K and Qazi G N 2006 The endophytic fungus Trametes hirsuta as a novel alternative source of podophyllotoxin and related aryl tetralin

lignans *J. Biotechnol.* **122**(4) 494-510

[31] Cao L, Huang J and Li J 2007 Fermentation conditions of Sinopodophyllum hexandrum endophytic fungus on production of podophyllotoxin *Shipin Yu Fajiao Gongye* **33**(9) 28-32

[32] Nadeem M, Ram M, Alam P, Ahmad M M, Mohammad A, Al-Qurainy F, Khan S and Abdin M Z 2012 Fusarium solani, P1, a new endophytic podophyllotoxin-producing fungus from roots of Podophyllum hexandrum *Afr. J. Microbiol. Res.* **6**(10) 2493-9

[33] Ran X, Zhang G, Li S and Wang J 2017 Characterization and antitumor activity of camptothecin from endophytic fungus Fusarium solani isolated from Camptotheca acuminate *Afr Health Sci* **17**(2) 566-74

IOP Conf. Series: Materials Science and Engineering **823** (2020) 012046 doi:10.1088/1757-899X/823/1/012046

Secondary metabolites compound of sea plumes gorgonian (*Rumphella, Isis and Ellisella*) from Maumere Water- East Nusa Tenggara Indonesia

Y H Teffu [1], A G O Kase[1], A Trianto[2], D P Wijayanti[2]

[1]Faculty of Fisheries and Marine Sciences, Universitas Kristen Artha Wacana Kupang University, NTT
[2]Department of Aquatic Resources, Faculty of Fisheries and Marine Sciences, Jl. Prof. H. Soedharto, SH, Tembalang, Semarang, Indonesia 50275

E-mail: hapynesnaldi@yahoo.com; kasealfred@yahoo.com; agustrianto.undip@gmail.com; diah_permata@mail.com

Abstract. The abundance and diversity of gorgonian sea plumes in the aquatic of Maumere - East Nusa Tenggara lead us to undertake research of their secondary metabolites. The objective of this study was to identify the species sea plumes gorgonian and to characterize its secondary metabolites that had biological activity. Identification of genus was based on morphology characteristics: sclerite forms; while phytochemical tests was undertaken on gorgonian extract of sea plumes. The phytochemical test showed that secondary metabolites of gorgonian extract contained alkaloids, flavonoids, quinone, steroid/triterpenoid, tannin and saponin. The identification results were found three gorgonian genera, namely Genus *Rumphella*, *Isis* and *Ellisella*. The highest yield of gorgonian extract was *Rumphella* sp. 7.84%. Components of secondary metabolites identified were alkaloids, flavonoids, quinone, steroids, triterpenoids, tannins/ phenolics and saponin.

1. Introduction

One of the richest and most complex ecosystems is marine environment. The extreme chemical and physical conditions in environment have stimulated the production of various important molecules with unique structural features. Molecules derived from the sea display various types of biological activities [1]. Discovery of novel compounds from marine ecosystem is increasingly diverse and rapidly growing. The exploration of new compounds from marine organisms, especially invertebrates has been widely reported. Study on *Rumphella antipathies, Carijoa* sp., Octocoral *Carijoa multiflora* gorgonian showed potential biological properties as antibacterial [2,3,4,5], as antiprotozoal [6]. Octocoral *Dendronephthya griffin* and Soft Coral *Scleronephthya gracillimum* have anti-inflammatory properties [7,8]; gorgonian *Rumphella antipathies* exhibited inhibition of superoxide anion and the release of elastase by human neutrophils [9,10]. Gorgonian *Isis hipirus*, *Rumphella* sp. and *Hicksonella* sp. were toxic against *Artemia salina* Leach [11,12]. The abundance of sea plumes gorgonian in the aquatic of East Nusa Tenggara is one of the potentials that should be further researched. Tuti [13] reported on sea plumes gorgonian identified 16 genera from 8 families

IOP Conf. Series: Materials Science and Engineering **823** (2020) 012046 doi:10.1088/1757-899X/823/1/012046

of gorgonian in the water of Maumere East Nusa Tenggara. This research was limited to identification of gorgonian. Study on secondary metabolites in the gorgonian *Rumphella* sp. and *Hicksonella* sp. by Teffu *et al.* [14] revealed that simplicia and sea plumes extract contain several active compounds namely saponin, triterpenoids, steroids, alkaloids, and flavonoids as anti-inflammatory and antioxidant. Early research on secondary metabolites compound needs to be done due to the potential existence of gorgonian diversity dispersed in the aquatic of East Nusa Tenggara. The diversity of aquatic characteristics in East Nusa Tenggara allows sea plumes gorgonian has different secondary metabolites compounds from the same sea plumes in other aquatic. The objective of this research was to identify sea plumes gorgonian and its secondary metabolite compound having potential biological activities as anti-bacterial, anti-fouling, anti-viral, anti-cancer, anti-inflammatory and anti-tumor.

2. Material and Method

2.1. Material

Raw material of this research was sea plumes gorgonian, ethyl acetate, phytochemical reagents and gorgonian identification books gorgonian in Indonesian waters [15]. The tools used such as scuba diving, pH meter, salinometer, flow/current meter, sechi disk, *Global Positioning System* (GPS Garmin etrex), C23 binocular microscope (Olympus), micropipette (corning), rotary evaporator (Eyela N-1200BS-W).

2.2. Method

2.2.1. Sampling. Sampling was carried out in the water of Wailiti Maumere East Nusa Tenggara at 3-20 m depths using diving equipment. Gorgonian than dried by aerated for 6-7 days. Identification of the gorgonian morphology characteristics: colony form, sclerite needles, location of needle in the body, tentacle by observing the colony form and sclerite needles morphological shape located in the body or tentacles according to Tuti [13] method. Samples were taken approximately 15 cm at the tip, middle and base of the sample cut into 1 cm. Furthermore, separated sample tissue by 5-7 drops of bleach. Waited for 5-10 minutes. Colony central axis would separate with sclerite forming flour, then washed of sclerite by clean water. Observed the morphological of clean sclerite with a microscope. The result of sclerite photo was compared with references to determine the form of polyps, surfaces and subsurface using Fabricius and Alderslade identification books [16].

2.2.2. Extraction. Extraction was carried out using maceration method based on Wikanta *et al.* [17] modified using a single solvent. Dried sea plumes separated into skin and axial than mushed up. Weighed 600 g of skin sample for *Rumphella* sp., 130 g of *Isis* sp., ad 95 g of *Ellisella* sp. The solvent used was ethyl acetate with a ratio between sample and solvent 1:3. Shaken and stored the mixture in the dark room for 24 H. Thereafter, filtered using Whatman paper no 42. The first filtrate saved into refrigerator. Extraction was done until the solution becomes clear. The filtrate was collected and combined then evaporated using a rotary evaporator with a temperature of 28°C till become concentrate. The concentrated extract was transferred into a vial bottle. The results of crude extract were weighed to determine the yield.

2.2.3. Qualitative phytochemical analysis. Qualitative phytochemical analysis was carried out according to Harborne [18] consist of alkaloid, steroid / triterpenoid, flavonoid, saponin, phenol hydroquinone and tannin tests. Alkaloid test was 1 mg of extract from of each solvent was dissolved in a few drops of sulfuric acid 2 N, after that will be tested with several alkaloid reagents including Dragendorff, Meyer and Wagner. The presence of alkaloid indicated by the appearance of brown precipitate for Wagner solution, cream-colored precipitate for Meyer solution and red to orange colored precipitate for Dragondorff solution. Meyer's reagent was made by adding 1 mL of $HgCl_2$

with 0.5 g of KI then dissolved and diluted with distilled water to 100 mL with a volumetric flask. This reagent was colorless. Wagner reagent was made of 10 mL distilled water added 2.5 g of iodine and 2 g of potassium iodide, then dissolved and diluted with distilled water to 200 mL in a volumetric flask. This reagent was brown color. Dragendorff reagents were prepared by dissolving 0.8 g of bismuth subnitrate in 10 mL of acetic acid and 40 mL of water. This solution mixed with a solvent made from 8 g of potassium iodide in 20 mL of water. Before used, 1 volume of this mixture was diluted with 2.3 volume mixture of glacial acetic acid 20 mL and 100 mL of water. This reagent was orange color. Steroid/triterpenoid test was 1 mg of extract was dissolved into 2 mL of chloroform in the test tube. Then, added 10 drops of anhydrous acetate and 3 drops of concentrated sulfuric acid into the mixture. The appearance of red color solution and turns into blue-green indicates the presence of steroid/triterpenoid. Flavonoid test was 1 mg extract of each solvent, added 0.1 mg of magnesium, 0.4 mL of amyl alcohol (a mixture of 37% hydrochloric acid and 95% ethanol in the same volume) and 4 mL of alcohol. The mixture was shaken. The appearance of red, yellow or orange color in the lining of amyl alcohol indicates the presence of flavonoid. Saponin (foam test) was 1 mg of extract dissolved in hot water and shaken. The formation of stable foam for 30 minutes and did not disappear if added 1 drop of HCl 2N indicates the presence of saponin. Phenol hydroquinone test ($FeCl_3$ reagent) was 1 mg of sample extracted with 20 mL ethanol 70%. Taken 1 mL of the resulting solution, then added 2 drops of 5% FeCl solution. The appearance of green and green-blue color solution indicates the presence of phenol hydroquinone. Tannin test was 1 mg of sample added with $FeCl_3$ then, the mixture was homogenized. The appearance of green-black color indicates the presence of tannin

3. Result

3.1. Identification of Sea Plumes Gorgonian

The results of identification of sea plumes gorgonian have led to identification of *Rumphella* sp. *Isis* sp. and *Ellisella* sp. genus gorgonian. Identification based on sclerite needle shape commonly used to identify the type of gorgonian. The sclerite shape shown in the samples visually similar.*Rumhella* sp. genus sclerite had sharper tip. *Isis* sp. sclerite had round shape, the sclerite form of *Ellisella* sp. was similar to *Rumphella* sp. Similar forms of sclerite, often difficult to identify. According to Teffu *et al.* (2015) described sea plumes gorgonian *Rumphella* sp. genus from Raijua Island-East Nusa Tenggara found bulges gada-like and spindle shape, similar sclerite form with the results of the research conducted. Visualization of the sclerite forms and sea plumes gorgonian shown in Figure 1.

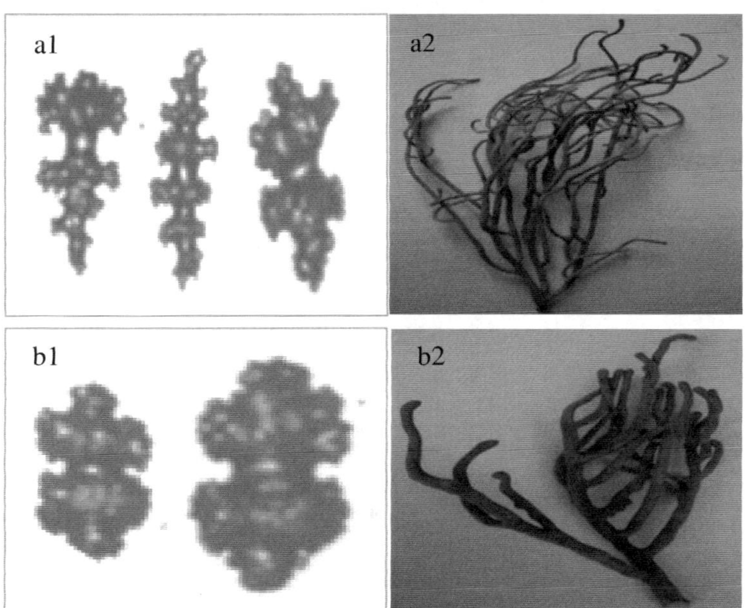

IOP Conf. Series: Materials Science and Engineering **823** (2020) 012046 doi:10.1088/1757-899X/823/1/012046

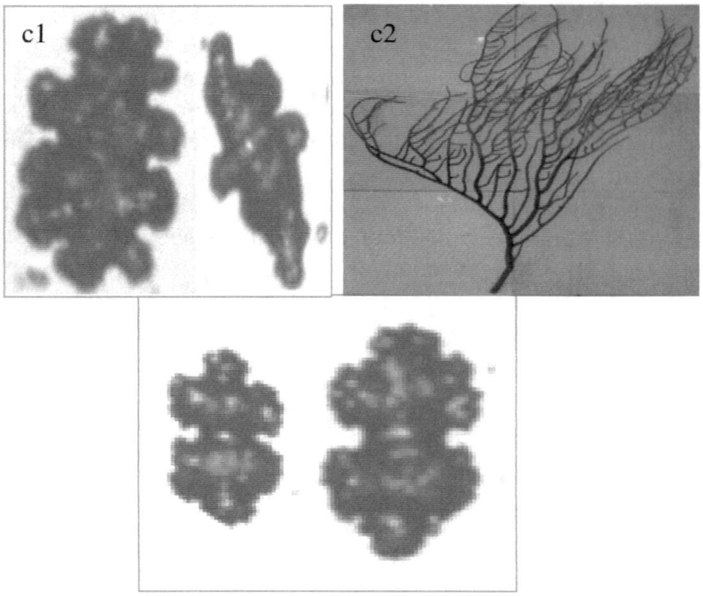

Figure 1. Sclerite form of sea plumes gorgonian. a1) Sclerite of *Rumphella*; a2) *Rumphella* sp; b1) Sclerite of *Isis*; b2) *Isis* sp; c1) Sclerite of *Ellisella*; c2) *Ellisella* sp

Morphologically, the axis of *Rumphella* sp. and *Isis* sp. genus looks different. *Rumphella* sp. had black and hard axis, while axis on *Isis* sp. looks like bamboo segments, bounded by black color between the segments. The morphology of *Ellisella* sp. was different from the two gorgonian genus. The axis had red color, like wood and easy to cut. Visualization of sclerite and axis shape from 3 genus of gorgonian shown in Figure 2.

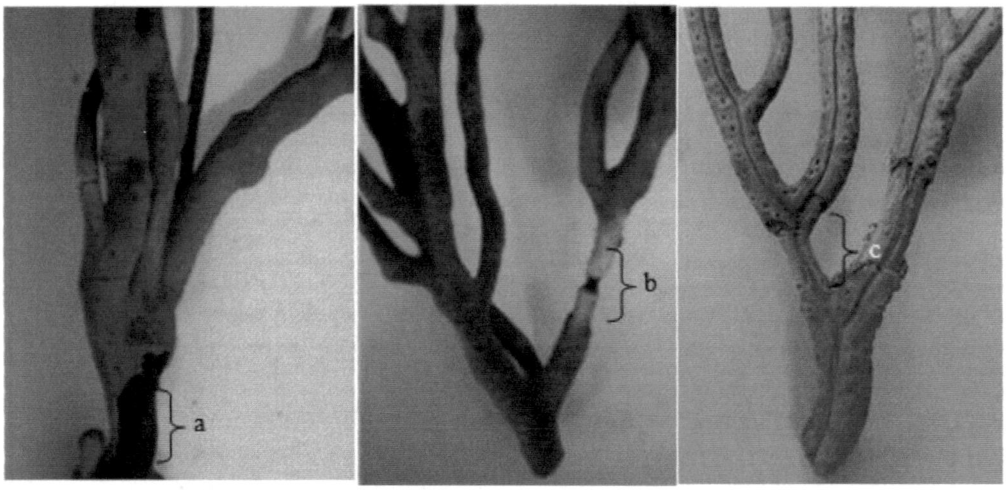

Figure 2. Axis form of sea plumes a) *Rumphella* sp.; b) *Isis* sp.; c) *Ellisella* sp.

3.2. Ecstrac yield

The extraction result was carried out on sea plumes gorgonian using ethyl acetate solvent. The yield of three gorgonian was presented in Table 1.

Table 1. The yield of sea plumes gorgonian extract

Gorgonian	Extract yield (%)
Rumphella sp.	7,84
Isis sp.	6,95
Ellisella sp.	1,93

There were significantly different between three extraction yield. The highest yield obtained from *Rumphella* sp. Different result gained from Teffu *et al.* [12] study, with the same sample and solvent, the yield result for both *Rumphella* sp. and *Hicksonella* sp. gorgonian sample was 0.15% and 0.74% respectively. Ethyl acetate solvent used to take out the same characteristic which the same semi polar compound. The difference in extract yield results was caused by many factors. Factors that affecting include sample size, sample number, extraction methods used. Solvent used affected the yield result. Higher yield commonly used methanol due to the polar properties of solvent, able to take out the same secondary metabolite characteristic or semi-polar. Different from ethyl acetate had semi-polar properties.

3.3. Phytochemical of Sea Plumes Gorgonian

Phytochemical analysis obtained the three gorgonian genera used showed different outcomes from previous studies. Phytochemical properties of three gorgonian genera were summarized in Table 2.

Table 2. Phytochemical characteristic of three gorgonian genera

Phytochemical test	Sea plumes gorgonian genera			
	Rumphella sp	*Isis* sp	*Ellisella* sp	Standard (color)
Alkaloid	+	-	-	Present of brown precipitate
Flavonoid	+	+	-	Present of yellow to red color
Quinone	-	-	+	Present of yellow
Steroid	-	-	-	Presentof red
Triterpenoid	+	+	+	Presentof blue
Tannin /phenolik	+	-	-	Presentof green
Saponin	+	+	-	Foam formed on the filtrate

Note: (+ indicates present, - indicates absent)

The results of phytochemical analysis on three gorgonian, terpenoid secondary metabolite components present in all gorgonian extract samples. *Ellisella* sp. genera only detected secondary metabolites of quinone and triterpenoid. Shen *et al.* and Qi *et al.* [19,20] reported that *Junceella fragilis* and *Junceella juncea* gorgonian from the Elliselidae family contained terpenoid compounds and diterpenoids as potential anti-inflammatory and antifouling properties. Sheu *et al.* [21] found secondary metabolites of terpenoids as antitumor drugs in *Isis hippuris* gorgonian. Sesquiterpenoid compounds also found in *Rumphella antipathies* Gorgonian coral potentially inhibited superoxide anion and the release of elastase by human neutrophils [9,10]. The secondary metabolite component of steroids was not detected, this was in contrast with Teffu *et al.* [14] detected the presence of a secondary metabolic component of steroids from both *Rumphella* sp. and *Hicksonella* sp. extract in ethyl acetate solvents. Steroids were also found in *Isis hippuris* gorgonian as antitumor drugs [22,23,24]. The secondary metabolite component of saponin was not detected in Teffu *et al.* [14] on the combined extracts of *Rumphella* sp. and *Hicksonella* sp., however saponins were detected in this research using ethyl acetate solvent in *Rumphella* sp. and *Isis* sp. by forming foam in the extract. Flavonoids compound were detected in this research and Teffu *et al.* [14] on combined *Rumphella* sp. and *Hicksonella* sp. extract. The secondary metabolites of alkaloids detected in gorgonian *Rumphella* sp. in ethyl acetate solvents, in contrast to Teffu *et al.* [14] secondary metabolite of alkaloids were detected using methanol and n-hexane as solvents. The difference between secondary metabolites compounds may be influenced by environmental factors where the sea plumes gorgonian live. Sea plumes gorgonian that lives in conservation environments had dissimilar type and a number of secondary metabolites from the secondary metabolite compound of sea plumes live in open ocean environment condition. The difference in amount and type of secondary metabolite compounds determines the potential of these to be utilized compounds. Manuputty [25] stated active compounds in the body tissues of octocoral animals were used as a defense, the rapid formation and expansion of the colony or in an effort to compete for a place to live for their extension. Another characteristic of animal life was *allelopathic* used as a strategy to seize land from new corals.

4. Conclusion

Sea plumes gorgonian sample used in the research was identified as Gorgonian Genus *Rumphella* sp., *Isis* sp., and *Ellisella* sp. based on its sclerite form. The highest yield of extracts was obtained from gorgonian *Rumphella* sp. phytochemical test showed the presence of secondary metabolites of alkaloids, flavonoids, quinones, steroids, triterpenoids, tannins/phenolics and saponins.

References

[1] Jain R, Sonawane S and Mandrekar N 2008 Marine organisms: potential source for drug discovery *Current Science* **94** 292

[2] [a]Sung P J, Chuang L G, Kuo J, Chen J J, Fan T Y, Li J J, Fang L S and Wang W H 2007 Rhumphellolides A-F, six new caryophyllane-related derivatives from the formosan gorgonian coral *Rumphellaantiphaties. Chemicals and Pharmaceutical Bulletin* **55** 1296

[3] [b]Sung P J, Chuang L G, Kuo J, Fan T Y and Hu W P 2007 Rhumphellatin A, the first chloride-containing caryophyllane-type norsesquiterpenoid from *Rumphella antiphaties Tetrahedron Letters* **48** 3987

[4] Zhao H Y, Shao C L, Li ZY, Han L, Cao F and Wang C Y 2013 Bioactive pregnane steroids from a south china sea gorgonian *Carijoa*sp. *Molecule* **18** 3458

[5] Diaz-Marrero A R, Porras G, Aragon Z, de la Rosa J M, Dorta E, Cueto M, D'Croz L, Mate J and Darias J 2011 Carijodienone from the octocoral *Carijoamultiflora* aspiropregnane-based steroid *Journal Naturals Products* **74** 292

[6] Reimao J Q, Migotto A E, Kossuga M H, Berlinck R G S and Tempone A G 2008 Antiprotozoan activity of Brazilian marine cnidarian extracts and of a modified steroid from the octocoral *Carijoa riisei Parasitoly Reseach* **103** 1445

[7] Chao C H, Wen Z H, Su J H, Chen I M, Huang H C, Dai C F and Sheu J H 2008 Further study on anti-inflammatory oxygenated steroids from the octocoral *Dendronephthya griffin Steroids* **73** 1353

[8] Fang H Y, Liaw C C, Chao C H, Wen Z H, Wu Y C, Hsu C H, Dai C F and Sheu J H 2011 Bioactive pregnane-type steroids from the soft coral *Scleronephthyagracillimum. Tetrahedron* **68** 9694

[9] Chung H M *et al.* 2013 Rumphellclovanes C-E, new clovane-type sesquiterpenoids from the gorgonian coral *Rumphella antipathies Tetrahedron* **69** 2740

[10] Chung H M, Wang W H, Hwang T L, Li J J, Fang L S, Wu Y C and Sung P J 2014 Rumphellaones B and C, New 4,5-*seco*-caryophyllane sesquiterpenoids from *Rumphella antipathies Molecules* **19** 12320

[11] Trianto A, Has Y Y, Ambariyanto and Murwani R 2004 Uji toksisitas ekstrak gorgonian *Isis hippuris* terhadap nauplius *Artemia salina Ilmu Kelautan* **9** 61

[12] Teffu Y H, Suwandi R and Nurjanah 2016 Aktivitas antioksdan dan toksisitas ekstrak gorgonian akar bahar *Rumphella* sp dan *Hicksonella* sp. *Prosiding Seminar Nasional Sains Ke-3* (Kupang: Universitas Nusa Cendana)

[13] Tuti H M I Y 2014 *Keberadaan Gorgonian (Akar Bahar) di Perairan Maumere, Kabupaten Sikka, Monitoring Kesehatan Terumbu Karang dan Ekosistem Terkait, Perairan*

Kabupaten Sikka 2013 (Jakarta: Pusat Penelitian Oseanografi Lembaga Pengetahuan Indonesia)

[14] Teffu Y H, Suwandi R dan Nurjanah 2015 Komponen kimia dan bioaktif akar bahar gorgonian (Genus *Rumphella* Dan *Hicksonella*) dari pulau Raijua-Nusa Tenggara Timur *Jurnal Pengolahan Hasil Perikanan Indonesia* **18** 84

[15] Tuti Y dan van Ofwegen L P 2018 *Gorgonians in Water Indonesian, Program Coremap-CTI* (Pusat Penelitian Oseanografi, Lembaga Ilmu Pengetahuan Indonesia)

[16] Fabricus K and Alderslade P 2001 *Soft Corals and Sea Fans: Acomprehensive Guide to The Thropical Shallow-Water Genera of The Central West Pacific, The Indian Ocean and The Red Sea* Includex index (Australia: The Australian Intitute of Marine Science PMB)

[17] Wikanta T, Hedi, Januar I and Nursid M 2005 Uji aktivitas antioksidan, toksisitas dan sitotoksisitas ekstrak alga merah *Rhodymenia palmata Jurnal Penelitian Perikanan Indonesia* **11** 41

[18] Harborne J B 1987 *Metode Fitokimia* Translator Padmawinata K and Soediro I, ed. Mansur S (Bandung: Institut Teknologi Bandung) Terjemahan dari: Phytochemical Methods 1st edition

[19] Shen Y C, Chen Y H, Hwang T L, Guh, J H and Khalil A T 2007 Four new briaranediterpenoids from the gorgonian coral *Junceella fragilis Helv. Chim. Acta* **90** 139

[20] Qi S H, Zhang S, Qian P Y and Xu H H 2009 Antifeedant and antifouling briaranes from the South China Sea gorgonian *Junceella juncea. Chem. Nat. Compd* **45** 49

[21] Sheu J H, Chao C H, Wang G H, Hung K C, Duh C Y, Chiang M Y, Wu Y C and Wu C C 2004 The first A-nor-hippuristanol and two novel 4,5-secosuberosanoids from the gorgonian *Isis hippuris Tetrahedron Lett* **45** 6413

[22] González N, Barral M A, Rodríguez J and Jiménez C 2001 New cytotoxic steroids from the gorgonian *Isis hippuris* Structure-activity studies *Tetrahedron* **57** 3487

[23] Sheu J H, Hung K C, Wang G H and Duh C Y 2000 New Cytotoxic Sesquiterpenes from the *Gorgonian Isis hippuris J. Nat. Prod* **63** 1603

[24] Chao C H, Huang L F, Yang Y L, Su J H, Wang G H, Chiang M Y, Wu Y C, Dai C F and Sheu J H 2005 Polyoxygenated steroids from the gorgonian *Isis hippuris. J. Nat. Prod.* **68** 880

[25] Manuputty A E W 2008 *Isis hippuris* Linnaeus 1758 : oktokoral penghasil anti virus *Oseana* **33** 19

Acknowledgement

The author thanks to the Ministry of Research, Technology and Higher Education for all of financial support in the research activities through a research grant scheme Higher Education Research Cooperation with the contract number: 0918/L8/KM/2019.

IOP Conf. Series: Materials Science and Engineering **823** (2020) 012047 doi:10.1088/1757-899X/823/1/012047

Innovative utilization of waste chocolate condensate as cocoa powder

Lienda Handojo[1,2], Judy Retti B. Witono[3], Irene Natasya Sulaiman[3], Sardenianto[1], and Antonius Indarto[1,2]

[1]Department of Chemical Engineering, Institut Teknologi Bandung, ITB Ganesha Campus, Bandung, Indonesia
[2]Department of Food Engineering, Institut Teknologi Bandung, ITB Jatinangor Campus, Indonesia
[3]Department of Chemical Engineering, Universitas Katolik Parahyangan, Bandung, Indonesia

E-mail: lienda@che.itb.ac.id

Abstract. The chocolate condensate, a side product of the roasting process in chocolate production, contains mostly volatiles and liquid chemical compounds similar to the cocoa beans. Currently, it is only considered as waste, although some valuable chemicals could be obtained by extraction. This study purposes of extracting valuables chemicals, such as pyrazine that has a strong aroma to chocolate, from chocolate condensate using three different solvents, i.e., n-hexane, toluene, and ethanol. Later, cocoa powder production from chocolate condensate was developed comparing two different drying methods (spray dryer and tray dryer) and analyzed using FTNIR, FTIR, and GC-MS.

1. Introduction

Cocoa plants are vastly cultivated in many tropical countries, such as Asia region (Indonesia, Malaysia, and Papua New Guinea), Western Africa region (Cote d'Ivoire, Ghana, Nigeria, and Cameroon), South America (Brazil and Ecuador). In 2017, the world production of cocoa reached 4.7 million tonnes, increased from 4.0 million tonnes in 2016 [1]. Cote d'Ivoire, Ghana, and Indonesia showed to be the three largest country producers of cocoa with percentage of 42%, 17%, and 7%, respectively. For every single tree, the production could reach 20 to 50 of cocoa pulp/fruit and harvesting time could be done through the year. The cocoa pulp consists of fruit peel, pulp, seed shell, and cocoa beans [2]. The most important part of the cocoa fruit is cocoa beans itself, while the other parts are only considered as waste.

Chocolate condensates were obtained by condensing the evaporated cocoa beans roasting process. Roasting involves complex chemical transformations, attributed to Maillard reactions, the caramelization of sugars, protein degradation, and other chemical reactions. During roasting, the volatile fraction is released from the bean and resulted in a liquid condensate when it is cooled. Currently, chocolate condensates were usually being dumped as it is considered as waste with no benefit to the production line. Several studies [3–9,13,15] mentioned that pyrazines had been identified in chocolate condensate that plays an essential role in cocoa flavor or additives. It also has a rich content of ester groups that are classified as chocolate fats. Besides those two major components, the other compounds are aldehydes, ketones, alcohol, ethers, esters, and organic acids [10].

This study aims to thoroughly analyze the content of condensate and the possibility to utilize this side product as an additional side income for the chocolate industry. Based on the preliminary identification, pyrazine, one of the essential components responsible for strong cocoa flavor, was tried to be extracted. Then, pyrazines-free condensate was dried using spray dryer and tray dryer to produce commercial condensate powder similar to cocoa powder.

2. Experimental Setup

The raw material of chocolate condensate was obtained directly from PT. Papandayan, Bandung, an industrial chocolate plant located in West Java, Indonesia. The feed material consisted of liquid condensate and solid matter. At first, analysis of chocolate condensate by using GC-MS (Shimadzu GCMS-QP-2010 with Rtx-5MS capillary column) analyzer was conducted to obtain the composition and concentration of species in liquid and solid matter. The extraction process then was carried out using various solvents, i.e., n-hexane, toluene, and ethanol. The product from each extraction process was then analyzed using GC-MS as well as the acid number and saponification number. The pretreatment of the sample before GC-MS analysis consisted of dilution of sample in dichloromethane for 100 times. The condensate was also pretreated by filter and centrifuge to separate possible presence of solid.

After the extraction, chocolate condensate residue was dried to produce cocoa powder by evaporating the water or liquid using spray dryer and tray dryer at temperatures of 80°C, 90°C, 100°C, and 110°C. The analysis of fat content and moisture content of each sample was cone by FTNIR. FTIR was used to compare the spectrum of the resulted cocoa powder from the standard commercial cocoa powder.

2.1. Extraction

25 mL of chocolate condensate was mixed with a solvent at a volume ratio of 1 to 1 in separating funnel. The funnel was then shaken for 10 minutes and settled down for 5 minutes with the funnel was opened to release the accumulated vapor. The condensate and solvent mixture was then treated again for another 10 minutes of shaking and 5 minutes of settling down for two cycles to obtain the optimum extraction. The extract and raffinate then separated and contained in different bottles.

The obtained extract then diluted using the same solvent by 1:20 volume. The diluted extract then filtered using filter paper and injected to centrifuge for 10 minutes by 500 rpm. The extract from centrifuge then re-filtered using filter paper before injected to GC-MS for the analysis.

2.2. Condensate Drying

Evaporation of water from chocolate condensate was carried out to produce cocoa powder by utilizing two different processes, i.e., spray dryer (EYELA Spray Dryers SD-1) and tray dryer, at temperatures of 80°C, 90°C, 100°C, and 110°C. In every run, 200 mL of condensate was used, and 5%-wt dextrin was added to the feed. The drying process carried out at an airflow rate of 1.67 mL/min for 2 hours.

In the case of the tray drying process, the experiments were done at a similar temperature used in the previous cases of spray drying. The drying residence times in the tray dyer were varied for 30, 60, 120, and 180 minutes. The weight fluctuation was recorded online, and the drying process was stopped at desired drying period. In the end, the moisture content of the samples at various drying time was measured.

3. Result and Discussions

3.1. Analysis of raw condensate

In all experiment works, the obtained raw condensates from industrial chocolate plants were used directly without any further pretreatment. The GC-MS result of raw chocolate condensate shows that some chemical groups are present in the sample, as shown in Figure 1.

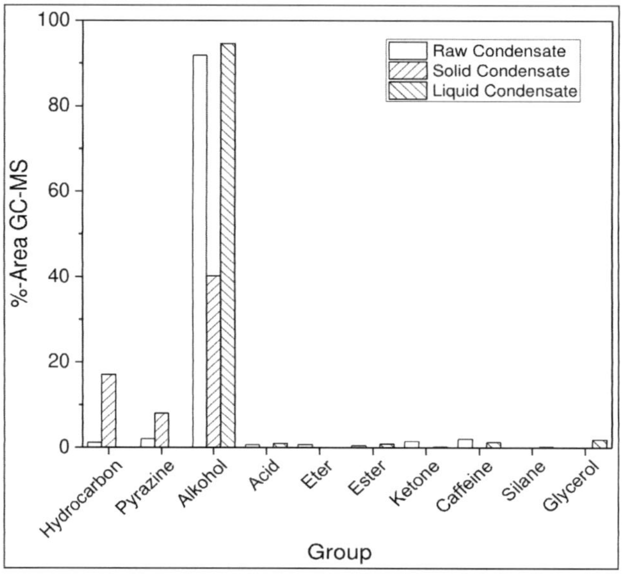

Figure 1. The GC-MS analysis of chocolate condensate.

The obtained raw condensate has similar compound variation and distribution, mentioned in the references [9,10], such as alcohols, aldehydes, ketones, esters, organic acids, and pyrazine compounds. Pyrazine, the compound that gives chocolate flavor could be found both in raw and stable condensates. It was expected that the pyrazine would be most likely to be found in stable condensate (9%) due to its melting point of 53°C. While the solubility of pyrazine in water is relatively high (ca. 300 g/L), it is also expected to be found in the liquid condensate. Nevertheless, the value was too low to be detected by our analytical instrument. In general, it did not require high concentration of pyrazine to give a flavor of chocolate as food additive. Among other compounds, alcohol group was found to be dominant, with the concentration 40-95%.

3.2. Extraction method

Pyrazine is a valuable compound that should be extracted from chocolate condensate using the extraction method. The extraction process was carried out by varying solvents (n-hexane, toluene, and ethanol). The variation of solvent was considered as aimed to get the most effective one in extracting pyrazine from condensate.

The extraction process using ethanol resulted in a single-phase mixture. It indicates that all compounds were dissolved in ethanol as a polar solvent, including the pyrazine. On the other hand, two-phase liquid occurred when n-hexane and toluene were employed as solvent. The use of ethanol as solvent was applied to examine whether the formed solid in chocolate condensate could be dissolved in a high polarity solvent. The extraction process by n-hexane and toluene were then analyzed using GC-MS, and the results are shown in figure 2. The extraction using ethanol was not analyzed further as all compounds were dissolved in the solvent, it means the aim of extraction was not achieved as pyrazine was still in the same liquid phase with others.

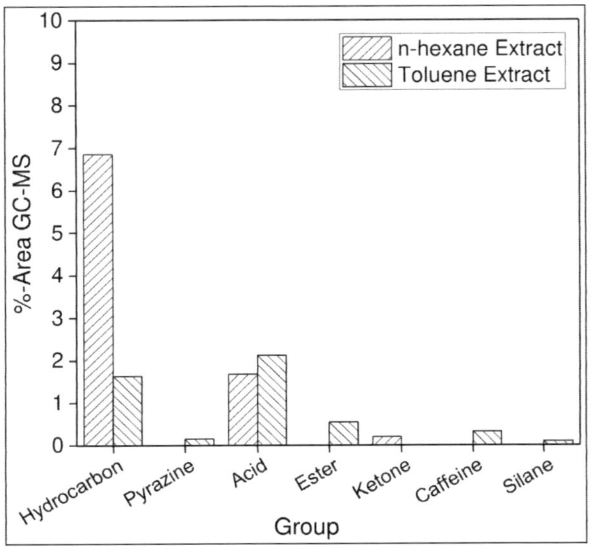

Figure 2. GC-MS extraction analysis

As shown in Figure 2, the pyrazine could be extracted using n-hexane solvent but not in toluene. Pyrazines should be dissolved in polar solvent as it was completely dissolved in water or ethanol as mentioned above. Toluene has a higher polarity compared to n-hexane. Therefore, the pyrazine was predicted to be more favorable in toluene than in n-hexane solvent. The analysis using GC-MS shows different results, there is no pyrazine identified in extracted products by toluene, but it was found in n-hexane solvent. This phenomenon could be caused by the extreme dilution of toluene extract sample before it was injected to GC-MS to obtain a clear visibility sample, and it could lead GC-MS unable to detect the pyrazine in the sample. On the other hand, it shows that the polarity of pyrazine is low or semi-polar, that it could be dissolved in n-hexane, a non-polar solvent.

In addition to GC-MS analysis, the obtained samples, which are raw condensate, liquid condensate, stable condensate, the mixture of condensate and ethanol, the extract and raffinate of each n-hexane and toluene were analyzed for their acid number and saponification number. The acid number, which was converted to propionate acid content and saponification number, is shown in Figure 3 and Figure 4.

The free fatty acid content of cocoa butter is of interest to both producers and chocolate manufacturers since higher acid percentage leads to the reduction of quality in fermented cocoa beans as well as decrease in hardness of the cocoa butter [11]. The free fatty acid considered in this analysis is fatty acid that was not attached as form of triglyceride. The acid number shown in Figure 3 is fatty acid numbers counted as propionate acid in the sample. The determination of fatty acid number considered as propionate acid was the acid that found in all of the analyzed samples by GC-MS. The total summation of acid number from liquid condensate and stable condensate are expected to have similar content to acid number of initial condensate, but in this study, the acid number from both liquid and stable condensate has more significant acid number than initial condensate. This could be caused by a large number of acids in liquid condensate, as supported by the GC-MS analysis result for liquid condensate that has a lot of acids.

The smallest acid number is from the sample of n-hexane extract, due to its property as a non-polar solvent. The acids dissolved in n-hexane are few. On the other hand, n-hexane raffinate has a higher acid number, as it was mainly comprised of chocolate condensate, and the amount of acid number from extract and raffinate of n-hexane has an equal amount of acid number content in the first condenser. The result of GC-MS analysis shows that acid in initial condensate has high amount of acids.

A different result of acid number was observed for the toluene extract and raffinate. The toluene raffinate has a higher acid number compared to raw condensate. The raffinate has the highest acid number among all of the samples. The high acid content in the toluene raffinate is suspected due to the occurrence of fatty acid degradation in the raffinate. Therefore, the acid number is increased.

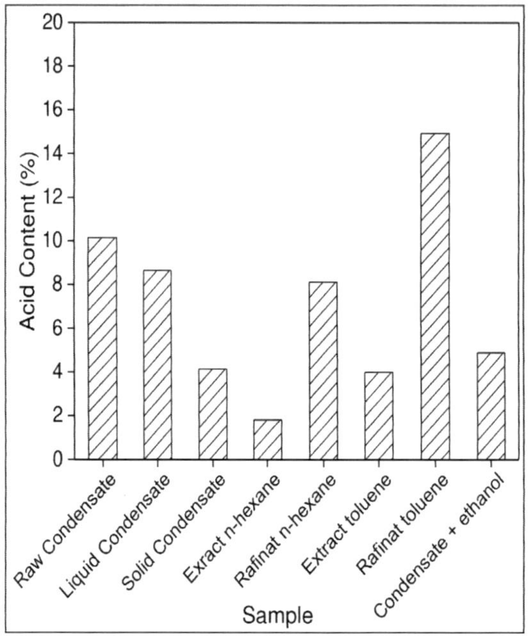

Figure 3. Analysis of propionate acid content.

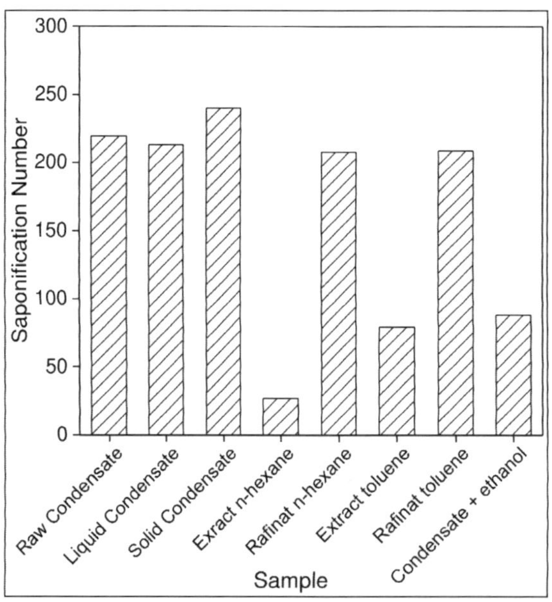

Figure 4. Analysis of saponification number.

The saponification number form initial condensate should be equal to the summation of the saponification number from both liquid and stable condensate. Saponification number measures the mean molecular weight of the fatty acids present in the oil or fat [12]. In this study, it was observed that stable condensate has the highest saponification number. It indicates that the triglyceride in the condensate is the majority contained in the solid condensate. The saponification number of each extract has lower number compared to its respective raffinate. This could be affected by the volume used for extraction. The volume ratio of condensate and solvent is 1:1. Therefore the triglyceride that could be extracted from the condensate is only a few.

3.3. Drying method

As known from previous analyzes there are no dangerous compounds in the chocolate condensate, and the pyrazine is challenging to be analyzed and extracted. Besides, the resulted FTIR spectrum of dried condensate at 100°C oven has similarities with commercial cocoa powder. Therefore, the chocolate condensate is then being dried to a product that could be consumed as cocoa powder.

The drying process was conducted by using a spray dryer and a tray dryer. This variation in drying method was carried out to obtain which method is more effective in making the cocoa powder from chocolate condensate. The resulted powder from spray and tray dryer then analyzed using FTNIR, FTIR, and CG-MS to get data about the contained components, as well the moisture content analysis by weighing the condensate at before and after the drying. Drying time in try dyer was varied by 30, 60, 120, 280, and 240 minutes. The cocoa powder that was taken for further analysis are powders that have similar moisture content to cocoa powder from spray dryer at same temperature. The comparison of moisture content of cocoa powder from spray and tray dryer shown in Figure 5.

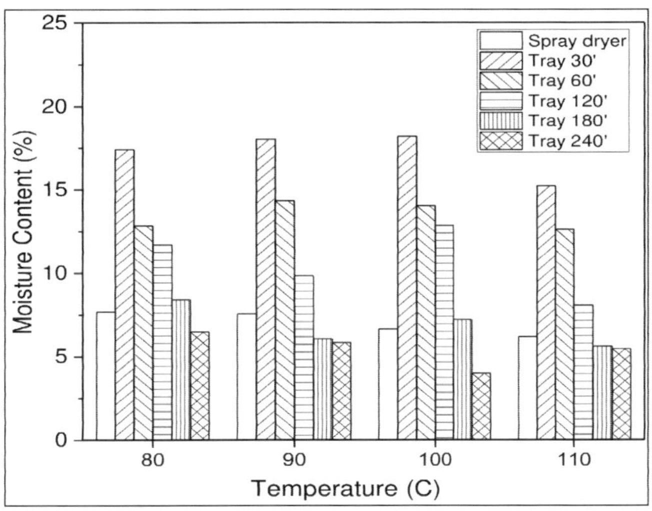

Figure 5. The moisture content of cocoa powder from spray dryer and tray dryer.

As shown in Figure 5 above, the moisture content of cocoa powder from 180 minutes of drying is comparable to powder from the spray dryer. It can be deducted that samples of 180 minutes drying in a tray dryer would have similar cocoa powder with spray dryer; therefore, it was chosen for further analysis. The other way to measure the moisture content in the powder is by analyzing it using FTNIR as well as its fat content analysis. The analysis result of FTNIR for cocoa powder from spray and tray dryer is shown in Figure 6.

From the graphic of FTNIR results in Figure 6, fat and moisture content of cocoa powder are significantly different compared to the commercial cocoa powder. The lower fat content in the condensate can be caused by the injected sample to spray dryer has been separated from its stable, as it is required that the sample entering the spray dryer should be solid-free. The same sample is also injected into the tray dryer. It was expected that most of fat in the condensate is contained in solid phase rather than in liquid condensate. Due to its high melting point, the fat will be in solid phase while stored at room temperature. Therefore the cocoa powder from liquid condensate will have lower fat content than in the commercial cocoa powder, while higher moisture content in cocoa powder from condensate could be caused by an incomplete drying process.

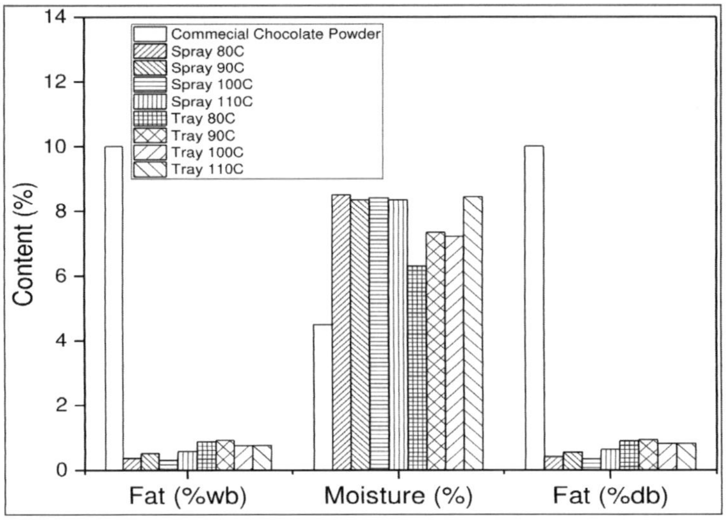

Figure 6. FTNIR analysis results.

3.4. FTIR

Condensate cocoa powder resulted from spray and tray dryer were analyzed using FTIR to compare the similarity of its spectrum to commercial cocoa powder. The FTIR analysis result for cocoa powder at 80°C, 90°C, 100°C, and 110°C was shown in Figure 7, Figure 8, Figure 9, and Figure 10.

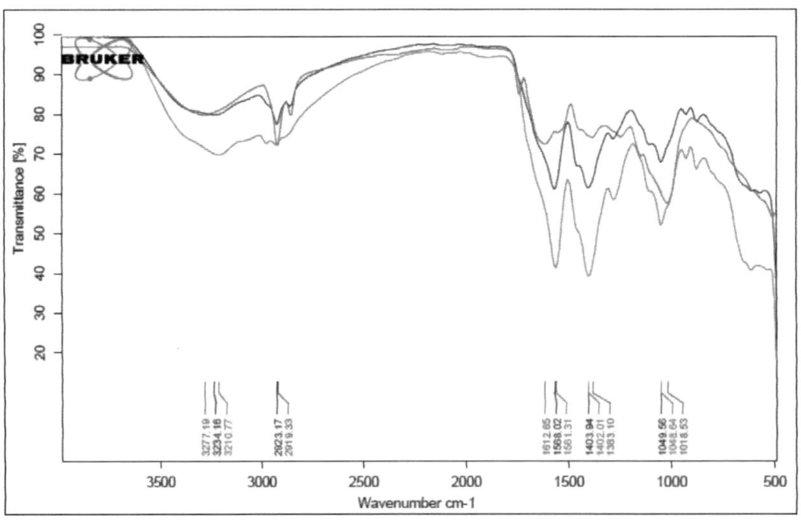

Figure 7. FTIR spectrums for commercial cocoa powder (red), spray dryer (violet), and tray dryer (blue) at 80°C.

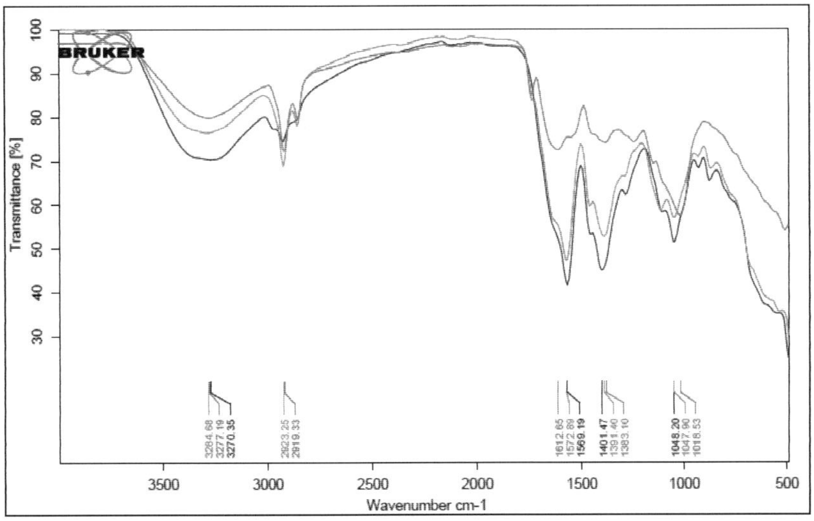

Figure 8. FTIR spectrums for commercial cocoa powder (red), spray dryer (violet), and tray dryer (blue) at 90°C.

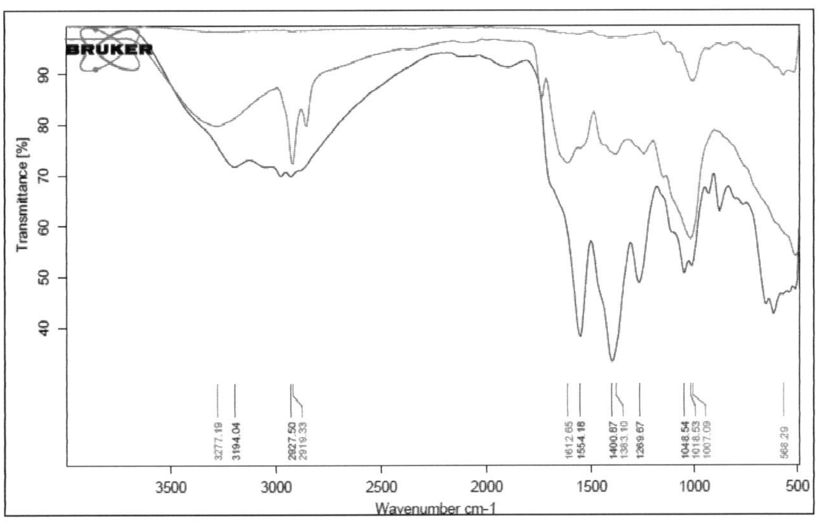

Figure 9. FTIR spectrums for commercial cocoa powder (red), spray dryer (violet), and tray dryer (blue) at 100°C.

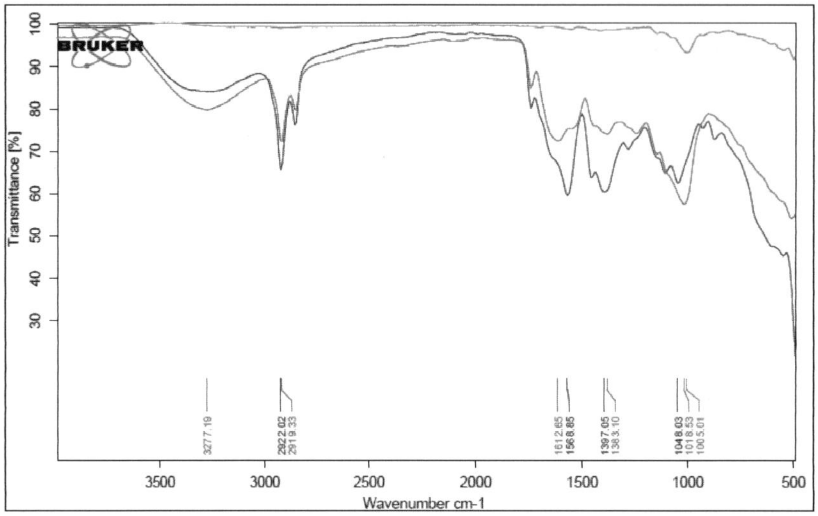

Figure 10. FTIR spectrums for commercial cocoa powder (red), spray dryer (violet), and tray dryer (blue) at 110°C.

Among the FTIR results for various temperatures (80 to 110°C), the spectrum that has the most similarity to commercial cocoa powder is spectrum cocoa powder from 90°C condensate drying for both tray and spray drying. Therefore, a suitable temperature for spray drying is 90°C.

The first peak at 3200 cm^{-1} in Figure 8 is corresponding to the alcohol bond [17,18]. Among all three samples, the cocoa powder that has the highest alcohol content qualitatively is powder obtained from tray dyer method. Theoretically, powder that will have highest content of alcohol should be sample from spray dryer due to its drying time using spray drying method is the shortest one. The less strong peak of alcohol in powder from spray dryer could be caused by the addition of dextrin to the spray dryer sample, and then it would cover up the alcohol bond in the sample. The second peak is indicating the aldehyde bound in powder sample. Based on analysis using GC-MS, the highest aldehyde is contained in cocoa powder from spray dryer.

The third peak at 1500 – 1600 cm^{-1} shows that there are alkene-aromatic compounds [17,18]. It is suspected that the absorbed spectrum in this region is corresponding to pyrazine. The smallest %transmittance peak among the three samples is powder from tray dryer, although it was not significantly different from spray dryer powder. Therefore, the pyrazines are still contained in the condensate sample, although it has been dried became to cocoa powder.

The presence of ester, ether, and carbocyclic acid was indicated by the absorbance in the 1300-1400 cm^{-1} spectrum [17,18]. These compounds can be found most in powder from tray dryer, even it is only slightly higher than spray dryer powder, while the content in commercial cocoa powder is the lowest one. The difference of compounds in each powder could be caused by the source of the cocoa powder each itself. Commercial cocoa powder is obtained from liquor chocolate, while the others came from chocolate condensates, which are the by-product of cocoa bean roasting.

Figure 11 shows the comparison of several components from GC-MS analysis in each powder. Cocoa powder condensates are chosen by comparing the samples obtained from the tray and spray drying, which has the closest FTIR spectrum to commercial cocoa powder, i.e., 90°C.

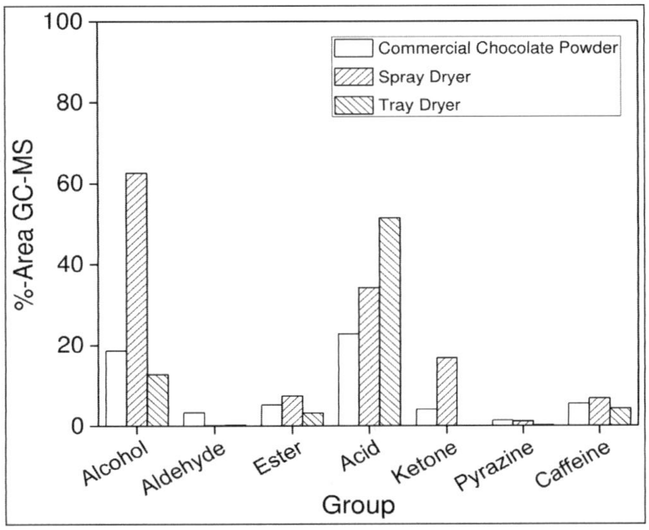

Figure 11. GC-MS components analysis comparison.

Some components such as alcohol, ester, and ketone are found less in tray dryer cocoa powder than in the powder from spray dryer. It was affected by the longer drying time in the tray drying method compared to the spray dryer. These volatile components would be evaporated more in tray drying than in spray drying. Another factor that could affect these components is the addition of dextrin in spray drying method. The added dextrin would reduce or prevent additional evaporation of volatile components. Moreover, the analyzed component in powder from the spray dryer has a close similarity to commercial cocoa powder. Therefore the suitable method to produce cocoa powder from chocolate condensate is spray dryer.

4. Conclusions

The production of cocoa powder from chocolate condensate has been conducted. Chocolate condensate is containing pyrazines, valuable components that provide chocolate flavor. The results showed that pyrazine compounds could be extracted using a non-polar solvent because the pyrazine compounds contained in the chocolate condensate are non-polar. The most suitable cocoa powder production is by using spray drying method with the temperature of the production from chocolate condensates is at 90°C.

References

[1] International Cocoa Organization August 2017 *Cocoa Market Review* (Cote d'Ivoire: International Cocoa Organization)

[2] Afoakwa E O, Quao J, Takrama J, Budu A and Saalia F 2011 Chemical composition and physical quality characteristics of Ghanian cocoa bean as affected by pulp pre-conditioning and fermentation *Journal of Food Sci Tecnology* **50**(6) 1097-105 doi: 10.1007/s13197-011-0446-5

[3] Margaret O, Mikael A P and Hanne H 2011 Effect of fermentation method, roasting and conching conditions on the aroma volatiles of dark chocolate *Journal of Food Processing and Preservation* **36**(5) 446-56 doi: 10.1111/j.1745-4549.2011.00602.x

[4] Brunetto M R, Cayama Y D, Gutiérrez L, Roa S C, Mendez Y C, Gallig-nani, Zambrano M A, Gomez A and Ramos G 2009 Headspace has chromatography-mass spectrometry determination of alkylpyrazines in cocoa liquor samples *Food chemistry* **112** 253-57 doi: 10.1016/j.foodchem.2008.05.082

[5] Frauendorfer F and P Schieberle 2008 Changes in key aroma compounds of criollo cocoa beans

during roasting *Journal of Agricultural and Food Chemistry* **56** 10244-51 doi: 10.1021/jf802098f

[6] Krings U, Zelena K, Wu S and Berger R G 2006 Thin-layer high vacuum distillation to isolate volatile flavor compounds of cocoa powder *European Food Research Technology* **223** 675-81 doi: 10.1007/s00217-006-0252-x

[7] Bonvehi J S and Coll F C 2002 Factors affecting the formation of alkyl-pyrazines during roasting treatment in natural and alkalized cocoa powder *Journal of Agricultural and Food Chemistry* **50** 3743–50 doi: 10.1021/jf011597k

[8] Counet C, Callemien D, Ouwerx C and Collin S 2002 Use of gas chromatography-olfactometry to identify key odorant compounds in dark chocolate, comparison of samples before and after conching *Journal of Agricultural and Food Chemistry* **50** 2385-91 doi: 10.1021/jf0114177

[9] Jinap S, Ishak W R W, Rahman R A and Nordin L M 1998 Effect of roasting time and temperature on volatile components profile during nib roasting of cocoa beans (Theobroma cacao) *Journal of the Science and Food Agriculture* **77** 441–48

[10] Rodriguez-Campos J, Escalona-Bendia H B, Contreras-Ramos S M, Orozco-Avila I, Jeramilo-Flores E and Lugo-Cervantes E 2012 Effect of fermentation time and drying temperature on volatile compounds in cocoa *Food Chemistry* **132**(1) 277-88 doi: 10.1016/j.foodchem.2011.10.078

[11] Afoakwa E O, Ofosu-Ansah E, Takrama J F, Budu A D and Mensah-Brown H 2014 Changes in chemical quality of cocoa butter during roasting of pulp pre- conditioned and fermented cocoa (Theobroma cacao) beans *International Food Research Journal* **21**(6) 2221-27

[12] Krysiak W 2011 Effects of convective and microwave roasting on the physicochemical properties of cocoa beans and cocoa butter extracted from this material *Grasasy Aceites* **62**(4) 467-78 doi: 10.3989/gya.114910

[13] Liu M, Liu J, He C, Song H, Liu Y, ZhangY and Su X 2016 Characterization and comparison of key aroma-active compounds of cocoa liquors from five different areas *International Journal of Food Properties* **20**(10) 2396–408 doi: 10.1080/10942912.2016.1238929

[14] Suparman, Wiranti S H, Elza S and Septiana D S 2015 The use of Fourier Transform Infrared Spectroscopy (FTIR) and Chromatography-Mass Spectroscopy (GCMS) for halal authentication in imported chocolate with various variants *Journal of Food and Pharmaceutical Sciences* **2** 6-11

[15] Lee K G and Shibamoto T 2002 Analysis of volatile components isolated from Hawaiian green coffee beans (Coffee Arabica L) *Flavour and Fragrance Journal* **17**(5) 349-51 doi: 10.1002/ffj.1067

[16] Vas G and Vekeyz K 2004 Solid-phase microextraction: A powerful sample preparation tool prior to mass spectrometric analysis *Journal of Mass Spectrometry* **39**(3) 233-54 doi: 10.1002/jms.606

IOP Conf. Series: Materials Science and Engineering **823** (2020) 012048 doi:10.1088/1757-899X/823/1/012048

Ethnobotany study of lontar tree (*Borassus flabellifer* L.) at Raijua Island

D L Age[1], A Ch Hendrik[1], and A D N Rupidara[1]

[1]Biology Education Study Program, Faculty of Teacher Training and Education, Universitas Kristen Artha Wacana. Kupang, Jl. Adisucipto Oesapa, Indonesia

E-mail: adn.rupidara@gmail.com

Abstract. The aims of the study were to determine the utilization of Lontar Plant (*Borassus flabellifer* L.), its organ used in relation to the culture of the Savu community. Data were collected using semi-structured techniques interviews, field observations, questionnaires, preference and direct matrix ranking with an index of cultural significance. Data were analyzed to calculate the index value of cultural significance which was included 3 aspects, namely the use-value (q), the intensity value (i), and the exclusivity value (e) of the plant and were described qualitatively. The results showed that the utilization of plants in the cultural life of the Savu community was very high. The utilization of plant parts was carried out by cutting, slicing, tapping and weaving. The parts mostly used were the roots, stems, stalks, leaves, flowers, and fruit.

1. Introduction

Ethnobotany, in its use in the community, relates to certain cultural and plant customs that are used in the culture of the community. The utilization assessment covers ecological, cultural and economic factors that are closely related to daily life [1, 2].

The people of East Nusa Tenggara are one of the most culturally and culturally active in the community. The Savu community on the island of Savu Raijua includes a community of strong cultural traditions and practicing ethnobotany in its cultural culture. Origin customs in their community always use both plants and animals for primary needs are also used in traditional rituals, medicine, literature, traditional textiles/weaving, cosmetics, coloring, construction, tools, and social life [3, 4].

Lontar is a kind of palm tree that grows in many areas in Indonesia, including in East Nusa Tenggara. Distribution of palm plants can be found on the coast of North to South Flores islands and Timor island, East and South coast of Sumba island and on small islands. The largest dispersal of the plant in Kupang Regency (West Timor Island, Rote Island) and *Kapuwe Duwe* (Savu Island) is known as *Tuak*, East Sumba Regency (Rindi Umalulu District and Pahungalodu District) are called *Manggita*, *Manggitu* (Sumba), South Central Timor Regency, South and North of Belu are called *Tua* (Timorese) and East Flores [5, 6].

Lontar (*Borassus flabellifer* L.) is one of the plants that dominate the Savu and Raijua land because of suitable geographical and climatic conditions. Generally, these trees grow wildly and spread so widely that planting is carried out in certain areas to facilitate processing, especially tapping sap [7].

IOP Conf. Series: Materials Science and Engineering **823** (2020) 012048 doi:10.1088/1757-899X/823/1/012048

Therefore, Lontar is often applied in various needs of daily life such as food - drinks from roomie, building materials from leaves and stems as well as in traditional ceremonies such as *Peoke* (disaster prevention ceremony), *Pehere jara* (ceremony of horse race event) uses palm leaves as horse bracelets in racial ceremonies, in the cultural or customs of Savu-Raijua community.

The aims of this study were: 1. to find out the relationship between the lontar plant (*Borassus flabellifer* L.) and the culture of the people of Raijua Island; 2. to know how the utilization of the plant that used in the culture of the people of Raijua Island; 3. to know the lontar plant organ used by the people of Raijua Island.

2. Method

This research has been conducted at Raijua Island, Raijua Subdistrict, Savu Raijua District in January 2019.

Figure 1. Geographical of research site at Savu Raijua

2.1. *Research subject*

The area of Savu Raijua is 36 km^2, consisting of 2 villages which are Ledeunu, and Ledeke, and 3 Villages which are Bolua, and Ballu. Raijua Island has the following territorial boundaries: the Northern part bordering the Savu Sea, the Eastern part bordering the Savu Sea, the Southern part bordering the Indian Ocean, and the Western part bordering the Savu Sea [8].

The subject of research is the Savu community on Raijua Island, Raijua District, Savu Raijua Regency, which is considered to represent the research objectives. Respondents were selected based on certain considerations that were considered to be the most knowledgeable about the use of Lontar plants in the cultural customs of the Savu community which consisted of: the traditional leader or tribal chief (10 people) representing all the tribes on Raijua Island, the Village Head (5 people) representing five existing villages, community leaders (5 people) and people who use palm trees (10 people) representing five villages in Raijua Island to be interviewed. In this study also carried out documentation of plants and plant parts used, processed by Lontar plants using a Nikon D800 camera.

2.2. *Research method*

Ethnobotanical data were collected using semi-structured interviews, field observations, preference and direct matrix ranking with traditional index of cultural significance [9, 10]. The determination technique of respondents in this study was using purposive sampling. The data that have been collected was then analyzed descriptively qualitatively. The description is done by describing the palm plants found at the location and their uses, namely:

1) Plant parts which are utilized.
2) What is it used for?
3) Local wisdom of the people of Raijua Island.

2.3. Analisis data

Data were analyzed to calculate cultural significance index using the formula of Turner [11] and Yuniati [12]. Cultural importance index is the result of quantitative ethnobotany analysis which shows the importance of plant species based on the needs of the community which includes 3 aspects namely use-value (q), intensity value (i), and exclusivity value (e). Index of cultural significance is calculated using the formula:

$$\text{ICS} = \sum_{i}^{n}(\text{q1 x i1 x e1})\text{n1} + (\text{q1 x i1 x e1})n2 \ldots + (\text{q1 x i1 x e1})n$$

Which are:

ICS : The sum of the calculation of the utilization of a plant species from 1 to n, where n indicates the umpteenth utilization (last)

I : Intensity value.

Q : The quality value is calculated by using a score or scale to the quality value of a plant species. The scoring are 5 = staple food; 4 = secondary food + primary material; 3 = other food + secondary material + medicine; 2 = ritual, mythology, recreation, etc., and 1 = mere recognition

e : Exclusivity value

3. Results and Discussion

The findings of the research showed that the utilization of Lontar (*Borassus flabellifer* L.) plants in the Savu community at the island of Raijua was very high. Society uses all parts of palm plants such as roots, stems, leaves, stems, flowers and fruit varieties.

All parts of the lontar plant have high cultural significance to the people of Raijua Island as follows: there are 21 kinds of utilization for leaves, 4 kinds of stems, 2 kinds of roots, 2 kinds of uses of stems, 2 kinds for fruits and 1 flower as well as daily necessities and to the cultural traditions uses of the people of Raijua Island.

Table 1. The utilization of Lontar Tree (*Borassus flabellifer* L.) parts in the culture of Savu tribes community at Raijua Island

Plant organ	Utilization
Root	1. Drill (*Dere*)
	2. Water shelter (*Jiab Kerabba Jiu*)
Stem	1. Building material (*Adjju ammu*)
	2. Pig flood plates (*Keraba wawi*)
Stalk	1. Robe (*Pipe*)
	2. Fence (*Lau*)
	3. Bearer (*Dpa dui*)
	4. Fishing gear (*Kenaha*)
Leaves	1. Roof of a traditional house (*Rau Ammu Ada*)
	2. Dwelling house (*Rau Ammu Pe*)
	3. Mat (*Dappi*)
	4. Ketupat *Pedo'a* dance (*Kedu'e Do'a*)
	5. Big water container or *Haik besar* (*Haba Dau*)
	6. Small *Haik* (*Haba Kao*),
	7. Traditional hat (*Heguddu*),
	8. Betel nut plate (*Pai Kenana*),
	9. Borse eye protector (*Ro Wiri*),
	10. Lontar flower wrapping (*Kebiha*)
	11. Small *ketupat* (*Kedu'e*)

IOP Conf. Series: Materials Science and Engineering **823** (2020) 012048 doi:10.1088/1757-899X/823/1/012048

	12. Place of agricultural products (*Hope*) 13. Rice winnowing or *Nyiru* (*Kerigi dai*), 14. Plate to eat (*Kerigi nga'a*) 15. *Kerigi wore* 16. *Oko* 17. Bag (*Beka*) 18. *Kehedi* 19. Cigarette paper (*Roko rokalli*) 20. Knife sheath (*Hope tudi*) 21. Horse leg jewelry (*Lale Kae*) 22. Dry leaves for burning animals (*Tunu Bada*).
Flower	1. Producer of lontar sap (*Ai due*) which is then cooked into sugar. The process of tapping the palm sap by the community is done through the ritual which is called *Tu'i Taga*. 2. The last ritual related to tell the community that the results of palm juice can be consumed by the public is called *Nga'a Ma Rao*.
Fruit	1. Animal food (*Wokalli*) 2. Appetizer (*Wohiru* - young fruit) during the cutting palm leaves activities.

Tambunan [6] reported that the by-product of lontar parts that been used were root, stem, leaves, fonds, fuits including young fruit either in South Sulawesi and Kupang. The by-products of leaves are mostly used, consist of handicraft business products, such as baskets or baskets, hats, carpets, brushes, mat and traditional musical instruments.

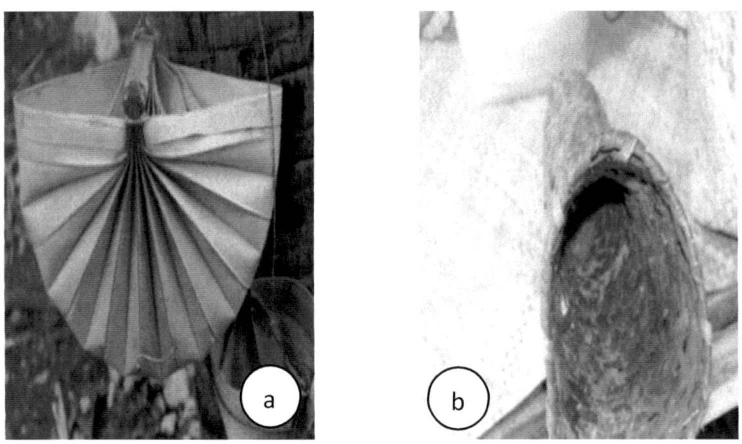

IOP Conf. Series: Materials Science and Engineering **823** (2020) 012048 doi:10.1088/1757-899X/823/1/012048

Figure 2. Utilization of Lontar (*Borassus flabellifer* L.) plant organs in the cultural traditions of the Savu people. a. Big water container or *Haik Besar* (*Haba Dau*) from leaves, b. Fishing Gear (*Kenaha*) from the stalk, c. The roof for building material (*Ajju Ammu*) from the stem, d. Drum or Tambur (*Dere*) from the root.

From measuring the indices of the culture importance value of lontar in the life of the people of Raijua Island, data shown in the following table.

Table 2. Measurement of Cultural Importance Index of community lontar plants on Raijua Island.

NO	UTILITY	Q	I	E	ICS
1	Staple food	5	5	2	50
2	Ingredient food from flower	4	5	2	40
3	Wood building material, wood for container	4	4	2	32
4	Beverage ingredients	4	4	2	32
5	Cigarette material	3	5	2	30
6	Mat material	3	5	2	30
7	Traditional handicraft or traditional technology	4	3	2	24
8	Environment Indicators	2	5	2	20
9	Valuable plant for medicinal	2	5	2	20
10	Medicinal for internal pain	3	3	2	18
11	Birth Ritual	2	4	2	16
12	Rope material, waterproof material	3	4	1	12
13	Edible fruits	4	3	1	12
14	Agriculture activity	2	3	2	12
15	Food flavor enhancer	3	3	1	9
16	Plant types play a role in myth	2	4	1	8
17	Suplemen food ingredient a mixture in food menu, food packaging	3	2	1	6
18	Livestock feed or animal food	3	2	1	6
19	Ritual of burial, heroism	2	2	1	4

20	Fuel wood	4	2	0.5	4
21	Plant species play a role in myths that are magical and religious	2	2	1	4
22	Medicine for animal diseases	3	1	1	3
23	Plant species have natural role in myths or history	2	1	1	2
24	Burns medicinal ingredient	3	1	0.5	1.5
	Total Indeks of cultural significance (ICS)				395.5

The way of using lontar trees (*Borassus flabellifer* L.) in the culture of the people on Raijua Island, namely cutting, cutting and tapping as well as weaving in daily life and during traditional ceremonies. Based on the results of research on Raijua Island, the results of the index of cultural importance of the trees showed that there were 9 uses with a high category (T) that has ICS values above 20, there are 5 uses with a medium category (SD) with a range of ICS values of 10-19.9 and 10 uses with a low (R) category or have an ICS category of 1 - 9.9 [13]. The use of lontar in the life of Savu Raijua community is as a staple food with ICS value was 50, and the lowest is used as an ingredient in burns with ICS value was 1.5.

Table 3. Range of *Index Cultural Significance* categories

ICS Value	Category	Code
≥ 20	High	T
10 - 19.9	Moderate	SD
1 - 9.9	Low	R

Adapted from [13]

4. Conclusion

Based on the results of research on the ethnobotany of lontar plants (*Borassus flabellifer* L.) on the island of Raijua can be concluded as follows: The relationship between lontar plants (*Borassus flabellifer* L.) with the culture of the people on Raijua Island was very high in utilization value (T). The way of using the trees in their culture, namely cutting, slicing and tapping as well as weaving in daily life and during traditional ceremonies. Lontar plant organs that are used by people on Raijua Island are leaves, stems, roots, stems, fruit, and flowers.

5. Acknowledgement

The author would like to gratitude Mr. Stefanus Leo Age and Mrs. Tiatira Weo as motivators. Special thanks to Apriliana Ballo, S.Si., M.Si., and Yanti Daud, S.Pd, M.Si., as mentors who improve the final writing. Savu Village head and community who contributed and help data gathering in the fulfillment of this research.

6. Conflict Of Interest

The author states there is no Conflict of Interest in this study.

7. References

[1] Ginting J S 2012 Perubahan Sosial Budaya Merdang Merdem sebagai Tradisi pada Masyarakat Karo (Medan: Bartong Jaya)

[2] Soekarman and Riswan S 1992 Status pengetahuan etnobotani di Indonesia Prosiding Seminar dan Lokakarya Nasional Etnobotani (Cisarua, Bogor)

[3] Supriono 1997 Tumbuhan Obat Indonesia: Penggunaan dan Khasiatnya (Jakarta: Pustaka Populer Obor)

[4] Acharya D and Anshu S 2008 Indigenous Herbal Medicines: Tribal Formulations and Traditional Herbal Practices (Jaipur: Aavishkar Publishers Distributor)
[5] Heyne K 1987 Tumbuhan Berguna Indonesia (Jakarta: Penerbit Jakarta Sarana Wana Jaya)
[6] Tambunan P 2010 Potensi dan kebijakan pengembangan lontar untuk menambah pendapatan penduduk (The Potential and Policy for Lontar Development to Increase the People Income) Jurnal Analisis Kebijakan Kehutanan **7**(1) 27–4
[7] Jayusman 2010 Pengembangan budidaya lontar di Pulau Sawu Nusa Tenggara Timur Jurnal Paramita **20**(1) 67–71
[8] Statistic of Kupang Regency 2018 Sabu Raijua Regency in Figures (Kupang: Statistic of Kupang Regency)
[9] Esterberg K G 2002 Qualitative Methods in Social Research (New York: Mc Graw Hill)
[10] Sing A and Sing P K 2009 An ethnobotanical study of medicinal plants in Chandauli District of Uttar Pradesh India J. Ethnopharmacol **121**(2) 324–9 doi: 10.1016/j.jep.2008.10.018. Epub 2008 Oct 30
[11] Turner N J 1988 The Importance of a Rose: Evaluation the Cultural Significance of Plants in Thompson and Lillooet Interior Salish *American Anthropologist* **90**(2) 272–90
[12] Yuniati E M 2004 Pengaruh faktor sosial budaya dan ekonomi terhadap keanekaragaman jenis tumbuhan pekarangan pada perkampungan yang dihuni oleh masyarakat Sunda dan Jawa di Kabupaten Brebes *Tesis* (Bogor: Pasca Sarjana IPB)
[13] Setyaningrum 2009 Struktur Tumbuhan dan Bioprospeksi Jenis di Taman Hutan Raya Raden Soerjo Wilayah Kecamatan Jombang Jawa Timur Tesis (Bogor: Institut Pertanian Bogor)

Preliminary study of wet-free CO_2 absorption through membrane diffuser

I G Wenten[1,2], A K Wardani[1], K Khoiruddin[1] and D Ariono[1]

[1] Department of Chemical Engineering, Institut Teknologi Bandung, Jl. Ganesha No.10, Bandung, 40132, Indonesia
[2] Research Center for Nanosciences and Nanotechnology, Institut Teknologi Bandung, Jl. Ganesha No.10, Bandung, 40132, Indonesia

E-mail : igw@che.itb.ac.id

Abstract. Wetting phenomenon is major drawback of membrane contactor in CO_2 absorption which limits the performance stability. In this study, hollow fiber membrane diffuser was proposed for providing wet-free gas absorption. The performance of membrane diffuser as function of operating pressure (0.5-3.0 bar), membrane hydrophobicity (hydrophobic and superhydrophobic polypropylene membrane), and membrane type (polypropylene and polysulfone) was investigated. Result shows that CO_2 concentration in absorbent increases with operating pressure. Superhydrophobic polypropylene membrane exhibits higher CO_2 transfer than the hydrophobic membrane. Meanwhile, the CO_2 absorption by PSf membrane is relatively low.

1. Introduction

The capture of carbon dioxide (CO_2) has gained great interest due to the increased concerns of global warming and greenhouse gas emissions as well as the exhaustion of easily accessible fossil fuel resources [1]. The capture of CO_2 from natural gas is important in order to meet the specifications required by consumers, while CO_2 capture from flue gas is needed to mitigate the increasing concentration of CO_2 in the atmosphere. Currently, various CO_2 capture technologies have been developed, including absorption, solid adsorption, cryogenic distillation, and membrane-based processes. Among them, absorption is the most mature and widely used CO_2 capture technology. Absorption involves the use of a liquid sorbent (e.g. monoethanolamine (MEA), diethanolamine (DEA), methyldiethanolamine (MDEA), potassium carbonate, and water) to separate CO_2 from natural/flue gases [2]. This process is limited by the requirement of solvent regeneration that leads to high energy consumption. Cryogenic distillation is also extremely energy-intensive and is not considered economical for industrial applications [3, 4]. Meanwhile, solid adsorbents (e.g. zeolites, silica, metal-organic frameworks, activated carbons, and graphene) that chemically adsorb CO_2 in active sites can lose efficiency very quickly with high temperatures [3, 5].

Membrane as a relatively new CO_2 capture technology offers advantages of low cost, low energy, flexible in operation, and modularity [6-10]. The membrane can also be integrated with another process such as gas absorption, called membrane contactor. Membrane contactor combines the benefit of high selectivity from chemical solvent and modularity and compactness of membrane permeation [11-14]. The use of porous membrane to replace conventional packing provides an extra area for contact between the gas and liquid phases and able to avoid the dispersion between them [15]. In addition, the flow of gas and liquid phases can be controlled independently which provides better optimization of both phases. However, the operation of membrane contactor is still limited by wetting phenomenon which affects its performance stability. Wetting leads to the increased membrane resistance that may result in a significant drop of CO_2 absorption efficiency [16]. It has been proved by Wang et al. [17] where the CO_2 absorption in non-wetted membrane contactor is six times higher than that in the wetted one.

To prevent the wetting in membrane-based CO_2 capture, various researches and developments have been conducted, including membrane material design and membrane system engineering [2]. Membrane material design focuses on synthesize or modification of membrane to obtain desirable membrane hydrophobicity [18-20]. Meanwhile, membrane system engineering aims to develop membrane-based CO_2 capture processes with optimal configurations to achieve the separation target. Membrane diffuser can be a promising approach in membrane system engineering to obtain wet free CO_2 absorption. In membrane diffuser, CO_2 is transferred through the membrane pores and partitions into a liquid [21, 22]. This method allows high interaction of gas-liquid and possible to operate at lower gas pressures, as there is no need to counterbalance hydrostatic heads [23]. Since the gas phase is operated under dead-end mode, it is possible to keep the pore filled by the gas. This configuration is expected to avoid pore membrane wetting.

Kumar et al. [21] investigated the performance of a hollow fiber membrane diffuser for CO_2 sequestration from combustion gas coupled with wastewater treatment. The membrane was a composite laminated hollow fiber with a thin dense polyurethane layer sandwiched between two microporous polyolefin layers. The results showed that the use of hollow fiber membrane diffuser was able to obtain high CO_2 mass transfer efficiencies with minimal loss of CO_2 to the atmosphere. The same configuration was used by Jana et al. [24] using ceramic hydrophobic membrane for providing CO_2 in algal photobioreactor during cultivation of *Arthrospira* sp. An enhanced overall mass transfer coefficient of about 38×10^{-4} m/s was obtained using 19.5% CO_2 gas mixture and 0.23 m/s gas velocity. Jana et al. also showed that the membrane was efficient in removal of dissolved oxygen produced during photosynthesis, resulting in higher algal growth and higher CO_2 sequestration. Meanwhile, Nordin et al. [25] compared the performance of a polyvinylidene fluoride (PVDF) membrane diffuser and a bubble diffuser. The size of the bubbles produced from the membrane diffuser was much smaller compared to the bubble diffuser, thus the total amount of bubbles produced and the total contact area provided per minute by the membrane diffuser was quite higher than the bubble diffuser.

In this work, a preliminary study was conducted to investigate the performance of wet-free CO_2 absorption through polypropylene (PP) and polysulfone (PSf) hollow fiber membrane diffuser. The influence of operating pressure on CO_2 concentration was investigated. In addition, the effect of membrane hydrophobicity and material was also examined.

2. Materials and Methods

2.1. Materials

PP and PSf hollow fiber membranes were provided by GDP Filter, Indonesia. The diameter and pore size of the membranes are summarized in Table 1. CO_2 gas (purity 99.5%) and demineralized water were used for filtration tests.

Table 1. Membrane characteristic

Membrane	PP	PSf
Outlet diameter (mm)	0.4	2.2
Inlet diameter (mm)	0.3	1.8
Average pore size (nm)	50	10

2.2. CO_2 absorption and analysis

Filtration test was conducted by flowing the CO_2 gas from inside out through the hollow fiber membranes that have soaked in demineralized water, as shown in Figure 1. The operating pressure of the CO_2 gas was varied from 0.5 to 3 bar. The flowrate of inlet gas was constant at 100 mL/minute and the ratio of water volume to membrane area was 30.5 m^3/m^2. The pH of demineralized water was measured every minute by electrode pH analyzer (LUTRON PE-03). The pH value was then converted to the CO_2 concentration by carbonate equilibria approach using "Bjerrum plot" [26, 27]. "Bjerrum plot" portrays the relation of pH value to the concentration of CO_2, HCO_3^-, and CO_3^{2-}. Meanwhile, to measure the water contact angle, a droplet of demineralized water was placed on the membrane surface using a micro-syringe 10μL. The water droplet was recorded by a digital camera and the contact angle was determined from the digital image of the water drop on the membrane surface. The water contact angle was measured at 5 different locations of individual membrane surface to minimize experimental errors.

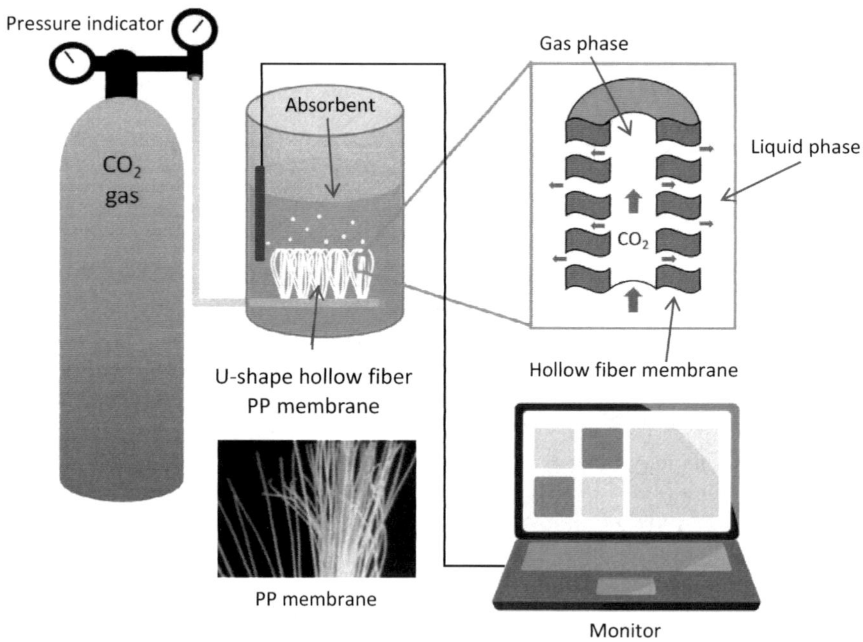

Figure 1. Experimental set-up of membrane diffuser

3. Results and Discussion

3.1. The effect of operating pressure

In this work, the operating pressure was varied from 0.5 to 3 bar to study its effect on the CO_2 concentration. Figure 2 shows that the CO_2 concentration increases by increasing the operating pressure. In 0.5 bar, the CO_2 concentration is almost constant for 13 minutes of measurement. The CO_2 concentration then slowly increases for operating pressure of 1-2.5 bar. Meanwhile, there is a huge jump of CO_2 concentration when operated in 3 bars. These results are due to the decrease of the CO_2 gas volume at higher pressure, thus increases the absorption of CO_2 molecules in water. In addition, the CO_2 is able to be absorbed in water by physical absorption which the solvent absorption capacity increases nearly linearly with pressure following Henry's law [28]. Therefore, the CO_2 solubility in water is enhanced at higher pressure, resulting in an increase of CO_2 concentration. Meanwhile, the operating time has an almost similar effect on all operating pressure, where the CO_2

IOP Conf. Series: Materials Science and Engineering **823** (2020) 012049 doi:10.1088/1757-899X/823/1/012049

concentration increases as the increase of operating time. It is due to the accumulation of CO_2 in the solution.

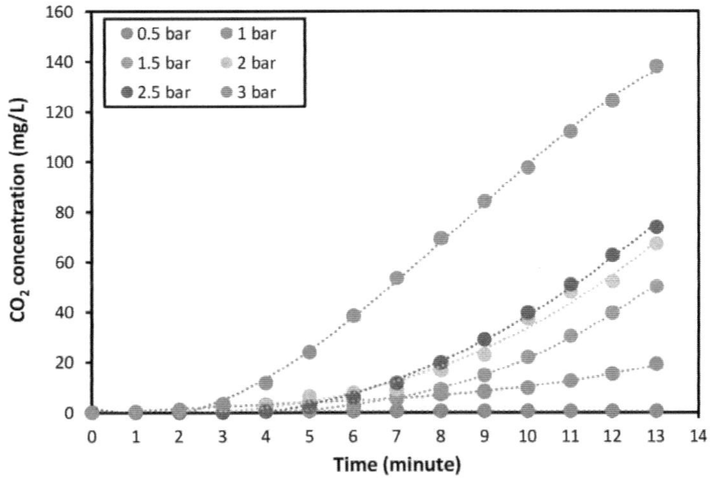

Figure 2. Effect of operating pressure on CO_2 concentration (PP hydrophobic membrane)

3.2. The effect of membrane hydrophobicity

To study the effect of membrane hydrophobicity on CO_2 concentration, two types of PP membranes were used in this work. The first PP membrane has a water contact angle (WCA) of 118.4±1.5°, namely as a hydrophobic PP membrane. The second PP membrane has WCA more than 150°, namely as a superhydrophobic PP membrane. The two PP membranes were operated at 1.5 bar and the results are depicted in Figure 3. It can be seen that the increase of hydrophobicity of the membrane leads to the increase of CO_2 concentration. The superhydrophobic PP membrane results in higher CO_2 concentration than the hydrophobic PP membrane. It is due to the lower surface energy and polarity of superhydrophobic membrane, resulting in a weak covalent interaction between water and membrane surface [29]. This weak interaction between water and membrane surface may reduce the resistance on membrane surface, thus CO_2 is able to easily permeated through the membrane pores. Similar to the previous results, the operating time has similar effect on both hydrophobic and superhydrophobic PP membrane, where the CO_2 concentration increases as the increase of operating time due to the accumulation of CO_2 in the solution.

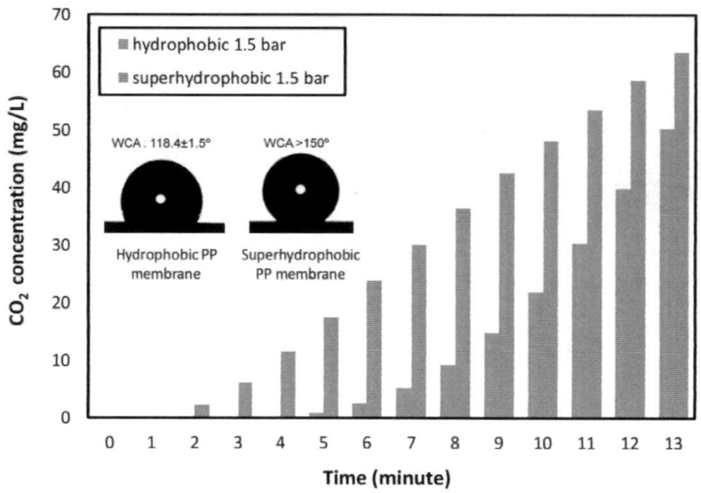

Figure 3. Effect of membrane hydrophobicity on CO_2 concentration

3.3. The effect of membrane material

In this work, the performance of PP and PSf ultrafiltration hollow fiber membranes as membrane diffuser was studied. Figure 4 shows the comparison of CO_2 concentration by PP and PSf membranes. In PP membrane, the CO_2 concentration significantly increases after 13 minutes. Meanwhile, the CO_2 absorption by PSf membrane is relatively low. It is due to the different pore structure between PP and PSf membrane. PP membrane was produced by stretching method, thus resulted in symmetric pore structure. Meanwhile, PSf membrane has asymmetric pore structure due to the solvent/non-solvent exchange during its production by phase inversion method.

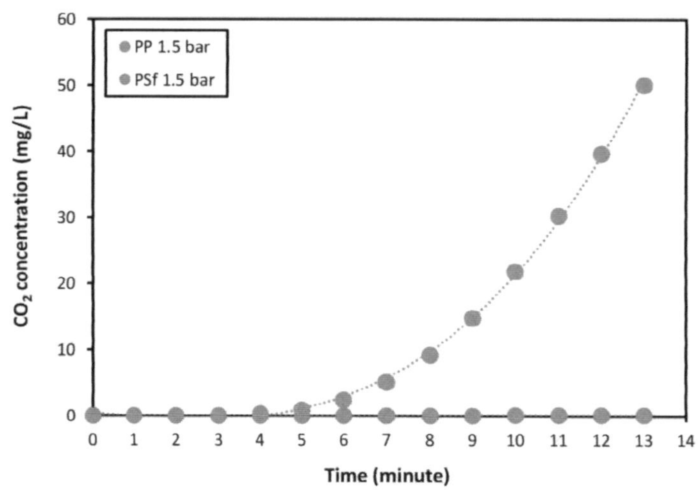

Figure 4. Effect of membrane material on CO_2 concentration

4. Conclusion

In this work, PP and PSf ultrafiltration hollow fiber membranes were used as membrane diffuser. The results showed that membrane diffuser has great potential for wet-free CO_2 absorption. The CO_2 concentration increased by increasing the operating pressure. Superhydrophobic polypropylene membrane exhibits higher CO_2 transfer than the hydrophobic membrane. Meanwhile, the CO_2 absorption by PSf membrane is relatively low.

Acknowledgment
This research is funded by the Indonesian Ministry of Research, Technology, and Higher Education under "Penelitian Dasar Unggulan Perguruan Tinggi 2019".

References

[1] Znad H, Naderi G, Ang H M and Tade M O 2012 CO_2 biomitigation and biofuel production using microalgae: photobioreactors developments and future directions *Adv. Chem. Eng.* (InTech) 230-44
[2] Khalilpour R, Mumford K, Zhai H, Abbas A, Stevens G and Rubin E S 2015 *J. Cleaner Prod.* **103** 286-300
[3] Goeppert A, Czaun M, Surya Prakash G K and Olah G A 2012 *Enegy Env. Sci.* **5** 7833-53
[4] Belmabkhout Y, Guillerm V and Eddaoudi M 2016 *Chem. Eng. J.* **296** 386-97
[5] Song C, Liu Q, Ji N, Deng S, Zhao J and Kitamura Y 2017 *Appl. Therm. Eng.* **114** 887-95
[6] Wenten I G, Khoiruddin K, Hakim A N and Himma N F 2017 The bubble gas transport method *Membrane Characterization* N Hilal *et al.* Ed. (Elsevier) 199-218
[7] Himma N F, Prasetya N, Anisah S and Wenten I G 2019 *Rev. Chem. Eng.* **35** 211-38
[8] Himma N F, Wardani A K, Prasetya N, Aryanti P T P and Wenten I G 2019 *Rev. Chem. Eng.* **35** 591-625
[9] Widodo S, Khoiruddin, Ariono D, Subagjo and Wenten I G 2018 *IOP Conf. Series Mater. Sci. Eng.* **285** 012008
[10] Wardani A K, Ivan I, Darmawan I R, Khoiruddin K and Wenten I G *J. Teknol.* **81**
[11] Zhao S, Feron P H M, Deng L, Favre E, Chabanon E, Yan S, Hou J, Chen V and Qi H 2016 *J. Membr. Sci.* **511** 180-206
[12] Siagian U W R, Raksajati A, Himma N F, Khoiruddin K and Wenten I G 2019 *J. Nat. Gas Sci. Eng.* **67** 172-95
[13] Comite A, Costa C, Demartini M, Di Felice R and Rotondi M 2016 *Int. J. Greenhouse Gas Control* **52** 378-86
[14] Zhang Z 2016 *J. Nat. Gas Sci. Eng.* **31** 589-95
[15] Cui Z and deMontigny D 2017 *J. Membr. Sci.* **540** 297-306
[16] Wenten I G and Himma N F 2017 *Adv. Sci. Letters* **23** 5768-71
[17] Wang R, Zhang H Y, Feron P H M and Liang D T 2005 *Sep. Purif. Technol.* **46** 33-40
[18] Himma N F and Wenten I G 2017 *AIP Conf. Proceed.* **1840** 090005
[19] Himma N F and Wenten I G 2017 *J. Phys. Conf. Series* **877** 012010
[20] Himma N F, Wardani A K and Wenten I G 2017 *Mater. Res. Express* **4** 054001
[21] Kumar A, Yuan X, Sahu A K, Dewulf J, Ergas S J and Van Langenhove H 2010 *J. Chem. Technol. Biotechnol.* **85** 387-94
[22] Ariono D, Chandranegara A S, Widodo S, Khoiruddin and Wenten I G 2018 *IOP Conf. Series Mater. Sci. Eng.* **285** 012010
[23] Carvalho A P and Malcata F X 2001 *Biotechnol. Prog.* **17** 265-72
[24] Jana A, Bhattacharya P, Guha S, Ghosh S and Majumdar S 2017 *Algal Research* **27** 223-234
[25] Lutfi U A M I, Hizam S M, Lau K S, Nordin N A H M, Bilad M R, Putra Z A and Wirzal M D H 2018 *J. Appl. Membr. Sci. Technol.* **22**
[26] Deffeyes K S 1965 *Limnology Oceanography* **10** 412-26
[27] Lower S K 1999 *Carbonate Equilibria In Natural Waters* (Canada: Simon Fraser University)
[28] Plaza M G, Pevida C and Rubiera F 2016 Ongoing Activity on CO_2 Capture in the Power Sector: Review of the Demonstration Projects Worldwide *Reference Module in Chemistry, Molecular Sciences and Chemical Engineering* (Elsevier)
[29] Wardani A K, Ariono D, Subagjo and Wenten I G 2019 *Polym. Adv. Technol.* **30** 1148-55

IOP Conf. Series: Materials Science and Engineering **823** (2020) 012050 doi:10.1088/1757-899X/823/1/012050

Insect diversity profile of mangrove ecosystem in Menipo Nature Tourism Park, East Amarasi, East Nusa Tenggara

Chatarina Gradict Semiun, Eufrasia R. A. Lengur, Gaudensius U. U. Boli Duhan

Biology Study Program Faculty of Mathematics and Science
Widya Mandira Catholic University, Kupang-East Nusa Tenggara, Indonesia

E-mail: gr4dict@gmail.com

Abstract. Insects are one of the biotic components that make up the mangrove ecosystem. The diversity of insects is believed to be used as one of the bio-indicators of the condition of the mangrove ecosystem. This study aimed at determining the profile of insect diversity in the mangrove ecosystem in Menipo Nature Tourism Park, East Amarasi, East Nusa Tenggara. The study was conducted in February 2018 to June 2018. Insect capture was carried out using the yellow pan trap method. Data analysis was done quantitatively to determine the profile of mangrove insects. The profile of mangrove insect diversity was determined based on species richness index, Shannon-Wienner diversity index, evenness index, and Jaccard similarity index. The results showed that there were 14 species, 11 families, and 5 orders with a total number of 56 individuals in mangrove ecosystem of Menipo Nature Tourism Park. Genetic wealth of mangrove buffer was still relatively low (R1 <3.5). The diversity of mangrove insects was classified as moderate (H '= 1- 3). All mangrove insects had almost the same level of evenness (E <1). The similarity index of mangrove insects in Menipo Nature Tourism Park, East Amarasi ranges from 0.18 to 0.5, meaning that the similarity of mangrove insect species among the three stations was different from each other.

1. Introduction

Indonesia has the largest mangrove and the largest biodiversity in the world. Mangroves are a tidal habitat comprised of salt-tolerant trees and shrubs. Compare to rainforests, mangroves have a mixture of plant types. Sometimes the habitat is called a tidal forest or a mangrove forest to distinguish it from the trees that are also called mangroves [6]. Mangrove forest is a type of forest that grows in tidal areas which are flooded during high tide and free from inundation at low tide with the plant community tolerating salt [10]. Mangrove plants have a special ability to adapt to extreme environmental conditions, such as inundated soil conditions, high salinity and less stable soil conditions [20].

In Indonesia, the area of mangrove forest is reduced, along with the rapid needs of the community, as well as the increasing development in the coastal area. The conservation of mangroves in the coastal area of Menipo Nature Tourism Park is very important because of the ecological and economic functions provided by the mangrove ecosystem. Ecologically mangroves have very important functions i.e. as a food chain in water, as the place where various types of fish, shrimps and mollusks live [10]. In addition, mangrove ecosystem also acts as a protector against waves, serves as sinks for carbon, and provides a nutrient sink [21].

Insects are one of the biotic components that make up the mangrove ecosystem. The study of insect diversity can help in preserving and managing mangroves well and sustainably. Insects, as a component of biodiversity, also have an important role in the food chain as herbivores, carnivores, and detriors [9]. In addition insects also function as a bio-indicator environment [8]. The diversity of insects is believed to be used as one of the bio-indicators of the condition of a mangrove ecosystem. Insect diversity provides a unique response to the level of environmental damage so that it has the potential as an indicator species to detect changes and environmental health [14].

The important role of insects in the ecosystem and the many types of insects that have not been identified become the reason why it is worth doing to examine the profile of insect diversity in the mangrove ecosystem. On the other hand, research on insects is still rarely done especially in the mangrove area; as a result, the information gathered from this field is still limited.

2. Material and Method

2.1 Time and location of the study
This research was conducted in the mangrove area in Menipo Nature Tourism Park (Figure 1). The location for observation was determined based on the type of mangrove stand. Separation and identification of insects were carried out in the Biology Laboratory Widya Mandira Catholic University. The study was conducted in February 2018 until June 2018.

2.2 Materials and tools
The materials used in this study were mangrove stands, insects that were caught by yellow traps, detergent, plastic bags, label papers, and 70% alcohol. The measurement of environmental factors was done by using GPS and air thermometer.

In collecting samples, the tools used were yellow traps, tweezers, plastic bags, rubbers and writing instruments. In identifying insect samples a microscope was used with a magnification of 20 times, and a digital camera (Sony DSC-S730) was used for documentation.

2.3 Description of the study area
Menipo Nature Tourism Park is composed of savanna ecosystems, coastal forests, and mangrove forests [1]. The savanna ecosystem is dominated by *lontar* (*Borassus flabellifer*), acid (*Tamarindus indica*), *kesambi* (*Schleichera oleosa*), and *waru* (*Hibiscus tiliaceus*).

Mangrove forests are dominated by *Rhizophora mucronata*, *Rhizophora stylosa*, *Ceripos tagal*, *Bruguiera conjugate* (*Bruguiera gymnorrhiza* (L.) Lamk) and *Bruguiera exaristata*, and additional components in the form of peanut leaves (*Ipomea pes-caprae*) and rolls (*Spinifex littoreus*). This is in line with the research findings (Table 1). In addition, the people in Enoraen and Pakubaun villages have local wisdom, namely the prohibition on logging of mangroves and the destruction of coral reefs.

Table 1. Geographical condition and types of mangrove trees in the study area

No	Station	Mangrove Species
1	I	*Ceriops tagal; Ceriops decandra; Rhizopora mucronata; Bruguiera gymnorrhiza* (L.) Lamk
2	II	*Ceriops tagal; Ceriops decandra; Rhizopora mucronata*
3	III	*Bruguiera exaristata; Ceriops tagal; Bruguiera gymnorrhiza* (L.) Lamk *; Rhizopora apiculata*

Figure 1. Sampling location

Mangrove plants characteristically range from trees (like species of *Avicennia*), *Rhizophora*, to shrubs (species of *Aegiceras*, *Aegialitis*, *Pemphis*, and *Conocarpus*), to the trunkless palm (*Nypa fruticans*) and ground fern (*Acrostichum*). Further, trees and shrubs vary where they might be columnar and erect (*Pelliciera rhizophorae*, *Bruguiera parviflora*), to spreading, sprawling (*Acanthus spp*., *Scyphiphora hydrophylacea*), and multiple-stemmed (*Ceriops decandra*). Growth form might also vary with in the same species (*Lumnitzera littorea* and *Rhizophora*), having both an erect tree form, and low tangled thicket forms. In general, edge plants (both waterfront and land-ward) have lower limbs and foliage, and their stems are laterally sprawling and sinuous, rather than erect and straight. Some species typically form combined closed canopies (*Avicennia marina*, *Rhizophora apiculata*, *Bruguiera parviflora*, *Bruguiera gymnorhiza*, *Camptostemon schultzii*, *Xylocarpus spp*.), while others are commonly found as under canopy plants beneath the closed canopy (like species of *Aegiceras, Cynometra*, *Acanthus*, *Acrostichum*, and *Ceriops decandra*) [6].

2.4 Work procedures

The line method was used as the sampling plot for data retrieval. At each location one path was created. The path length was 100 m and the width was 10 m with the direction parallel to the coastline. In each lane, a plot of 10 m x 10 m was made with a distance between plots on the lane was 20 m. Each type of stand was made in four plots. Insect capture was done using the yellow pan trap, a yellow tray with a length of 12 cm, a width of 20 cm, and a height of 6 cm. The yellow pan trap was placed in a 10 m × 10 m plot and filled with detergent liquid so that insects trapped would not fly and die. Yellow pan trap was installed for 24 hours started at 06.00 PM. Five pieces of yellow pan trap with a diagonal position were placed in each plot. Insect capture by using yellow pan trap was carried out for two days at each observation area. The measurement of insects' environmental factors was done by measuring ecological temperature. Temperature was measured using an air thermometer by placing the tool in the middle of the sampling plot.

The captured insects were separated and identified based on morphospecies in the Biology Laboratory Widya Mandira Catholic University. Insects' identification process was carried out by using an identification source in the form of an insectarium and identification guide books that already exist. The books used in insect identification are: A Field Guide to Insects in 1970, by Donald J. Borror and Richard E. White; Introduction of Insect Lessons, in 1996, by Donald J. Borror, Charles A. Triplehorn, and Norman F. Johnson translated by Partosoedjono; and Hymenoptera of the World: An

IOP Conf. Series: Materials Science and Engineering **823** (2020) 012050 doi:10.1088/1757-899X/823/1/012050

Identification Guide to Families, 1993, by Henry Goulet and John T. Huber.

The data obtained in this study were variable quality data on insects, which was tabulated using Microsoft Excel 2010 Data analysis was done quantitatively to determine the profile of mangrove insects. The profile of mangrove insect diversity was determined based on the species richness index (equation (2)), the Shannon-Wienner diversity index (equation (1)), evenness index (equation (4)), and Jaccard similarity index (equation (3)). The equation of the index is presented as follows:

$$H' = -\Sigma pi \ln pi;\ pi = ni/N \quad (1)$$

$$R1 = \frac{S-1}{\ln N} \quad (2)$$

$$Cj = \frac{J}{(a+b-j)} \quad (3)$$

$$E = \frac{H'}{\ln(S).} \quad (4)$$

Information:

H ' = diversity index
R1 = species richness
E = evenness type
Cj = the same type of jaccard
Pi = The proportion of the number of individuals with the number of individuals in all types
ni = Number of species individuals i
N = Total number of all individuals
S = Number of types
J = Number of species found in habitat a and b
a = Number of species found in habitat a
b = Number of species found in habitat b

3. Results and Discussion

3.1 Types of mangrove ecosystem insects in Menipo Nature Tourism Park

In mangrove ecosystem of Menipo Nature Tourism Park there was found 14 species, 11 families, and 5 ordos with a total number of 56 individuals (Table 2). The findings of each station were described as follows: there were 8 families at Station I, there were 3 families at station II, and there were 4 families at station III. In identifying mangrove insect specimens, it was limited to the family level. This was because the damaged road infrastructure has caused some specimens to suffer from damage due to shocks on the way to the laboratory. Besides that, some traps were lost (especially at station II) allegedly due to high tide at night.

Table 2. Types of mangrove insects caught in *yellow pan trap*

No	Ordo	Family	Species	Station		
				I	II	III
1	Diptera	Ptychopteridae		4		
		Dolichopodidae	*Tentanocara vicina*	5		
		Syrphidae		3		
		Xylomidae				2
		Sciaridae		2		
		Culicidae	Big black mosquito	5		5
2	Hymenoptera	Formicidae	Big black ant	2		1

			Big black ant with wings			2
			Big red ant	4	3	1
3	Coleoptera	Colydiidae		2		8
		Buprestidae				1
4	Hemiptera	Aphididae		1	1	
5	Lepidoptera	Lyonetiidae			1	

In all three stations, mangrove insects were found from the family of Formicidae. Formicidae (ants) is one of the families of the Hymenoptera ordo whose existence is ubiquitous, and the number exceeds most other land animals. These animals also play a role in the translocation of organic matter from the surface into the soil and are able to adapt to paddy habitats following changes in land conditions and plant age. Naturally, ants could be preferred as an indicator of environmental condition changes; it also roles to maintain the nutrition cycle and structure of soils. Ants can be used for habitats indicator status [25].

Tentanocara vicina was only found at station I. This species belonged to Dolichopodidae family. The Dolichopodidae family is usually small to medium -sized and can live in various conditions. But it is generally found in marshy areas and grasslands [5].

At stations I and III large black mosquitoes were found. These insects are members of the Culicidae family with extensive distribution. There are about 3,523 species of mosquitoes (Culicidae) described throughout the world [12]. Male mosquitoes do not bite, while female mosquitoes bite so that they are considered pests. Mosquitoes can also act as vectors carrying diseases such as malaria, filariasis, encephalitis, and others. The Culicidae family is divided into 3 subfamilies i.e. Anophelinae, Culicinae, and Toxorhynchitinae [5]. Mosquitoes are always found where the air is relatively cool and the humidity is high. Many species live within a few meters of the ground whereas many sylvan species occur primarily in the forest canopy. Vertical distribution is largely dependent on feeding preferences. All males and the females of many species feed exclusively on plant liquids, including nectar, honeydew, fruit juices and exudates [11].

Colydiidae and Buprestidae are families from Coleoptera order found in station I and III. The Colydiidae are widely distributed and diverse family of beetles found throughout the world and on many oceanic islands. Many species are saproxylic, feeding on fungi associated with decomposing wood, or else are predaceous on the larvae of scolytin [17]. The beetle family Buprestidae (jewel beetles) comprises about 15000 species and has a world-wide distribution and most of the species are distributed in the humid tropics and semi desert areas of the planet [2].

Aphididae was found at stations I and II. The About 4,700 species of aphids popularly known as plant-lice or ant-cows, constitute a group of small-sized (0.7 to 7.0 mm in length) insects that suck the phloem fluid of plants and about 1000 species are injurious to crops throughout the world [23]. Aphididae family in Indonesia is commonly called *Kutu daun*. *Kutu daun* can be identified with a typical peach-like shape with a pair of cornicles at the posterior end of the abdomen. Cornelike cornicles are tube-like structures arising from the fifth and sixth segments of the dorsal abdomen (Borror et al. 1996). *Kutu daun* has different body sizes, shapes, and colors. These differences can be influenced by host and environmental factors [13]. The population density influences physiological changes in plants such as withered, death of shoots, deciduous leaves, leaf shape changes, and plant death.

3.2 Richness index

Richness of mangrove insects refers to the number of species found in an ecosystem. The total number of species in a community depends on the sample size and time. Some indices are recommended to measure Richness species that are not dependent on sample size but based on the relationship between a community and the total number of individuals observed which increases by increasing sample size [16].

Based on the criteria of Ludwig and Reynolds (1988), the richness index of mangrove insects in Menipo Nature Tourism Park was low (R1 <3.5). The observation data showed that the highest richness index value was at station I (R1 = 2.40), followed by station III (R1 = 2.00), and station II (R1 = 1.44) (Figure 2).

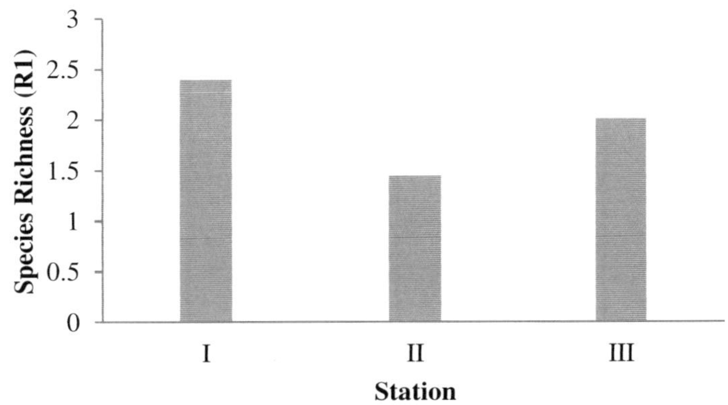

Figure 2. Mangrove insect species richness index in Menipo Nature Tourism Park

Wealth index value is strongly influenced by the number of mangrove species found. Based on the results of the data analysis, it was revealed that the index value of mangrove insect species in Menipo Nature Tourism Park had a low value of wealth. High and low levels of mangrove insects were influenced by the presence of food sources and climate. A stable climate conditions caused insect species wealth to be high [22].

3.3 Diversity index

The observation results (Figure 3) indicated that the diversity index of mangrove insects in Menipo Nature Tourism Park was classified as medium (H '= 1- 3). Moderate species diversity indicated the number of individuals was not diverse [7].

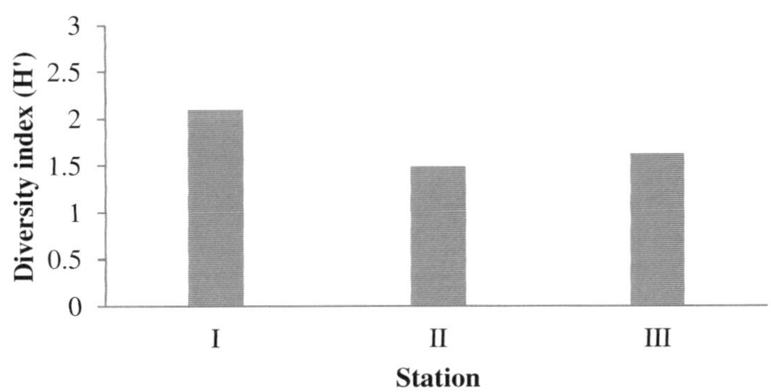

Figure 3. Mangrove insect diversity index in Menipo Nature Tourism Park

Based on Figure 3, the value of the diversity index from high to low sequentially was station I (2.10), Station III (1.62), and station II (1.49). The diversity index depends on the food chain. The more complex the food chain then the higher the value of the diversity index in an ecosystem. Within an ecosystem that has an abundance and diverse high-flying insects, there were complex interactions in the food chain that can support ecosystem stability. In addition, the differences in mangrove insects encountered in Menipo could be attributed to the fact that there are more desirable mangrove trees that could attract insect and create a conducive environment for insect habitats.

3.4 Evenness index

The analysis results of the evenness index (E) of mangrove insects in Menipo Nature Tourism Park (Figure 4) approached 1 (<1). In concerns to the value of E which approaches 1, all types of insects have similar levels of evenness [16]. The evenness index in succession from high to low was station I (0.95), station III (0.72), and station II (0.72).

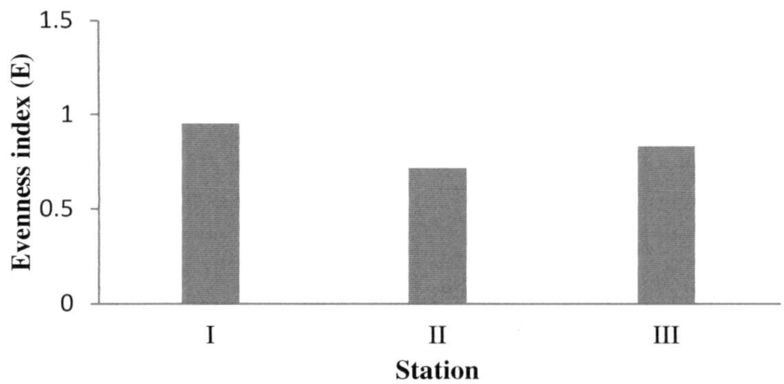

Figure 4. Evenness index of mangrove insects in Menipo Nature Tourism Park

According to Odum (1993) [24] if the evenness index is > 0.5, then evenness is high. It is because the soil physic-chemistry of land of research station is not much different. The high index of evenness of mangrove insects is caused by these animals that have almost the same ability to utilize various environmental conditions to sustain their lives [7].

The Jaccard similarity index (Cj) ranges of 0-1, when approaching the value of 1 it shows the level of similarity between habitat is high, and when approaching the value of 0 it indicates the level of similarity between habitat is low [18]. Based on Figure 5 the similarity index of mangrove insect species in Menipo Nature Tourism Park ranged from 0.18 - 0.5. The range of values tends to approach the value of 0, meaning that the similarity of mangrove insect species among the three stations was different from each other. It is due to the different constituent components of mangrove vegetation among three stations so that only certain species can live and adjust according to the environmental conditions in each location.

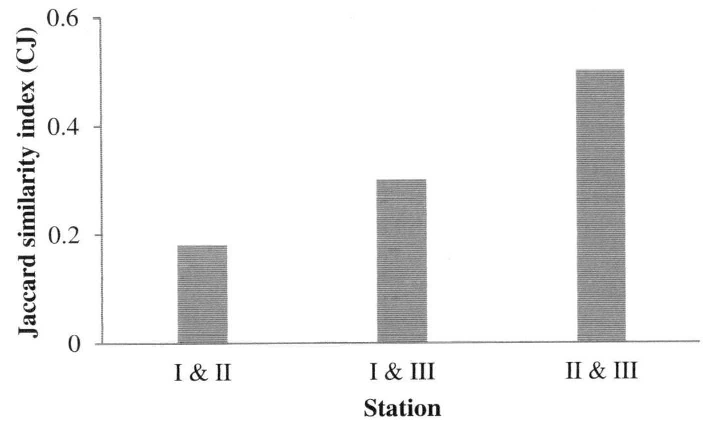

Figure 5. Jaccard similarity index of mangrove insects in Menipo Nature Tourism Park

The abiotic factor measured in this study was only the ecological temperature. The ecological temperature at station I was 27 °C, station II, and station III had the same ecological temperature value of 27.5 °C. Insects can live in a certain range of effective temperatures (i.e., minimum temperature of 15 °C, optimum temperature of 25 °C, and maximum temperature of 45 °C) [15]. Insects have a certain temperature range where they can live. Outside the range, insects will die of cold and heat, which affects the process of insect physiology.

4. Conclusion

Based on the findings and discussion, it can be concluded that the types of mangrove insect families found in the mangrove ecosystem in Menipo Nature Tourism Park were Ptychopteridae, Dolichopodidae, Syrphidae, Xylomidae, Sciaridae, Culicidae, Formicidae, Colydiidae, Buprestidae, Aphididae, and Lyonetiidae. The richness index of mangrove insects found in mangrove ecosystems in Menipo Nature Tourism Park was still relatively low (R1 <3.5). The data observation showed that the highest richness index value was at station I (R1 = 2.40), followed by station III (R1 = 2.00), and station II (R1 = 1.44).

The diversity of mangrove insects found in the mangrove ecosystem in Menipo Nature Tourism Park was classified as moderate (H '= 1-3). The value of the diversity index sequentially from high to low was station I (2.10), station III (1.62), and station II (1,49). All mangrove insects found in the mangrove ecosystem of Menipo Nature Tourism Park had almost the same level of evenness (E <1). Evenness index of all types in succession from high to low was station I (0.95), station III (0.72) and station II (0.72). The similarity index of mangrove insect species in Menipo Nature Tourism Park ranged from 0.18 to 0.5. The range of values tends to approach the value of 0, meaning that the similarity of mangrove insect species among the three stations was different from each other.

Acknowledgments

Thank you to the Chancellor of Widya Mandira Catholic University through LPPM, who fully funded this research in 2018. The team that had helped and had supported the implementation of this research.

References

[1] BBKSDA NTT 2017 *Informasi Kawasan Konservasi Lingkup Wilayah Kerja Balai Besar KSDA* (Nusa Tenggara Timur, Kupang)
[2] Bílý S, Kubáň V, Volkovitsh M G and Kalashian M Y 2011 *Arthropod fauna of the UAE* **4** 168223

[3] Borror, D J, Triplehorn C A, Johnson N F 1992 *Pengenalan Pelajaran Serangga Edisi Keenam* (Yogyakarta: Universitas Gadjah Mada)

[4] Borror, D J, Triplehorn C A, Johnson N F 1996 *An Introduction to the Study of Insects* (Yogyakarta (ID): UGM Press)

[5] Borror D and White R 1970 *A Field Guide to Insects America North of Mexico* (Boston New York: Houghton Mifflin Company)

[6] Duke C N 2017 *Springer International Publishing* 1753

[7] Febrita R, Suwondo and Mayrita E 2008 *J. Pilar Sains*, **7**(1) 37-45

[8] Gogoi A and Guha B 2015 *IJSR* **6** 9

[9] Grampurohit B and Karkhanis H 2013 *National Conference on Biodiversity: Status and Challenges un Conservation-FAVEO* 108-15

[10] Haneda N F, Kusmana C and Kusuma F 2013 *Journal of Tropical Silviculture* **4**(1) 42-6

[11] Harbach E R 2007 *Zootaxa* **1668** 591638

[12] Hutchings R S G, Hutchings R W and Sallum M A M 2013 *Zoologia* **30**(1) 1-14

[13] Irsan C 2004 Tumbuhan inang, parasitoid, dan hiperparasitoid kutu daun Myzus persicae (Sulzer) (Homoptera: Aphididae) di sekitar Bogor dan Cianjur, Jawa Barat *Disertasi* (Bogor: Institut Pertanian Bogor)

[14] Jones T J and Eggleton P 2000 *Journal of Applied Ecology* **37** 191203

[15] Jumar 2000 *Entomologi Pertanian* (Jakarta: Rineka Cipta)

[16] Ludwig J A and Reynolds J F 1988 *Statistical Ecology: A Primer methods and computing* (New York: JohnWiley & Sons)

[17] Majka G C and Cline A R 2006 *The Coleopterists Bulletin* **60**(3) 105-11

[18] Magurran A E 1988 *Ecological Diversity and Its Measurement* (London: Croom Helm Ltd)

[19] Magurran A E 2004 *Measuring Biological Diversity and Its Measurement* (UK: Blackwell Science Ltd)

[20] Noor Y S, Khazali M and Suryadiputra I N N 2006 *Panduan Pengenalan Mangrove di Indonesia* (Bogor: Wetlands International Indonesia Programme)

[21] Romañach S S, DeAngelis D L, Koh H L, Li Y, Teh S Y, Raja Barizan R S and Zhai L 2018 *Ocean and Coastal Management* **154** 72-82

[22] Semiun G C and Stanis S 2016 *BioWallacea Jurnal Ilmiah Ilmu Biologi* **2**(3)

[23] Patel S and Singh R 2016 *International Journal of Research Studies in Zoology (IJRSZ)* **2**(4) 55-67

[24] Odum E P 1993 *Dasar-dasar Ekologi edisi ketiga* ed Tjahjono Samingan (Yogyakarta: Gadjah Mada University Press)

[25] Zayadi H, Hakim L and SetyoLaksono A 2013 *The Journal of Tropical Life Science* **3**(3) 166-71

IOP Conf. Series: Materials Science and Engineering **823** (2020) 012051 doi:10.1088/1757-899X/823/1/012051

Analysis of mineral content in Hotspring Water samples in Tubbe and Aramaba Village, Pantar Tengah District, Alor-NTT

Martasiana Karbeka, Loth Botahala, Pither Duka

Department of Chemistry, Universitas Tribuana Kalabahi, Alor, 85812 Indonesia

E-mail: martasianakarbeka@untribkalabahi.ac.id

Abstract. Clean water standards are tasteless, colorless and odorless. Hot spring water in Tubbe and Aramaba villages is used by the community for drinking, cooking, washing, bathing and other needs. However, the physical parameters of the water has no longer met the standard because it is slightly colored and has distinctive odor of sulfur. Low quality of water affects the health of people who consume the water. The main problem is clearly seen from the yellow teeth of the people in Tubbe and Aramaba villages. Therefore, it is important to analyzed water quality to determine the levels of iron (Fe), phosphate $(PO_4)^{3+}$ and sulfate $(SO_4)^{2-}$ in hot spring water in Tubbe Village and Aramaba Village, Pantar Tengah District. Iron content testing has been carried out using an atomic absorption spectrophotometer. UV-Vis spectrophotometer has been used to test the concentration of phosphate and sulfate. The measurement results show that the hot spring water samples in Tubbe Village contain 0.04mg / L iron (Fe), 0.09 mg/L phosphate $(PO_4)^{3+}$ and 11,31mg/L sulfate $(SO_4)^{2}$. This value is below the maximum threshold, according to Minister of Health Regulation No. 149 of 2010. The results of the Aramaba Village are 0.04 mg/L; 0.09 mg/L and 11.31 mg/L for iron (Fe), phosphate $(PO_4)^{3+}$ and sulfate $(SO_4)^{2-}$. This value is above the threshold of calcium content according to Minister of Health Regulation No. 149 of 2010.

1. Introduction

Water plays a very important role in our life, especially for drinking, cooking and other needs. The quality of drinking water have to fulfill three requirements according to the regulation of the Ministry of Health Republic of Indonesia No. 492/MENKES/PER/IV/2010 [1,2]. Physical requirements, consumable water has to be clear having normal temperature equals to the ambient temperature (20 °C-26 °C). Drinking water does not give flavor, color and smell. Chemical requirements, water acidity (pH) has to be neutral (i.e., pH = 7). Drinking water does not contain toxic chemicals, salt or metal ions. Microbiological requirements, water does not contain pathogenic bacteria such as *Coli*-type, *Salmonella*-type, *Vibrio-cholera*; it also does not contain nonpathogenic bacteria such as *Actinomycetes*, *Phytoplanktoncoliform* and *Dadocera*.

Local people in Tubbe and Aramaba villages in Alor regency have long used hot spring water for their daily consumption due to the lack of clean water. Based on the physical appearance, the hot spring water is not clear and produces a distinctive odor of sulfur that might be originated from Mount Sirung, an active volcano. The same condition of high surfur content is also found in hot springs at the area of Samboja, Kutai Kartanegara district because it is near the volcano area [**3**].

Water with high-temperature can dissolve mineral ions contained in the rocks, such as iron (Fe^{3+}), phosphate $(PO_4)^{3+}$ and Sulfate $(SO_4)^{2-}$, Na^+, K^+, Ca^{2+}, S, and Cl^- [4,5,6]. According to regulation of the Minister of Health of Republic of Indonesia No. 492/MENKES/PER/IV/2010, the maximum limit of water content containing mineral iron (Fe^{3+}) of < 0.3 mg/L, mineral phosphate $(PO_4)^{3+}$ of 0.2 mg/L, and mineral sulfate $(SO_4)^{2-}$ amounting to 250 mg/L [2]. The hot spring water consumed is slightly yellowish in colour indicating that it contains high concentration of sulfur. The use of sulfur contented hot spring water in a long period by the local people caused tooth decay (yellowish tooth) [9]. One of the potential development of hot spring water is to be used a hot spring for health therapy. Hot springs is a very important aspect to relate their balneotherapeutic property and enhance the Medical Tourism of the state. Besides sulfur and sulfate, other nutrients and minerals in hot springs water are also important for therapeutic properties [7,8]. Therefore, in this study focused on the iron (Fe^{3+}), phosphate $(PO_4)^{3+}$ and Sulfate $(SO_4)^{2-}$ the cotained in Tubbe and Aramaba hot spring water.

2. Materials and Methods

2.1 Sampling

Samples of hot springs water were collected according to its purpose. Station I, water samples were taken from drinking and cooking water in Tubbe village (± 12 Km from Mount Sirung). Station II, water samples were taken from bathing and washing water in Tubbe village. Station III, water samples were taken in Aramaba village, located on the roadside of the main road (there is a reservoir) which is ± 4 Km from Mount Sirung. The hot spring water from Station III is not used for daily consumption by the community. Hot spring water samples were are collected in polyethylene containers and brought to the laboratory for analyses. Samples of hot springs water were chemically preserved by adding oxidizing agents (HNO_3, 6 M) to inhibit biological activity and any loss of ions fractions by adsorptions onto the wall of containers. The parameters of hot spring water quality were analyzed using standard methods according to the Indonesian National Standard (SNI).

2.2 Measurement of iron (Fe), (SNI 6989.4:2009)

A total of 50 mL hot spring water sample was inserted into the Erlenmeyer 100 mL and then added 5 mL of HNO_3 concentrated and then closed and homogenized. Then heated the sample water slowly until the remaining volume of 15-20 mL. This process is done repeatedly until all the dissolved metals. This is evident from the color of the deposits in the hot spring water samples that will turn clear. Enter the hot spring water sample into the cup cups and then the hot spring water samples are moved into the 50 mL measuring flask and added mineral-free water to the precise marking of the impressions and homogenized then measured by the atomic absorption spectrophotometer. Calculated iron levels with the following formula: Iron content (Fe) (mg/L) = C x F_P, C is concentration Fe of the measuring result (mg/L), F_P is the dilution factor.

2.3 Measurement of phosphate $(PO_4)^{3+}$, (SNI 06-6989.31-2005)

A total of 50 mL sample solution was inserted into the Erlenmeyer and then 1 drop of phenolphthalein indicator was added to the Erlenmeyer. If the color changes to pink, sulphuric acid solution was added drops by drop until the color disappears and then added 8 mL of a mixed solution consisting of potassium antimonyl tartrate solution, ammonium of molybdate solution and ascorbic acid solution was then homogenized. Furthermore, the sample solution and blanko solution were analyzed using UV-Vis at 880 nm wavelength. Calculated phosphate levels with the following formula: Phosphate $(PO_4)^{3+}$ (mg/L) = C x F_P, C is concentration $(PO_4)^{3+}$ of the measuring result (mg/L), F_P is the dilution factor.

2.4 Measurement of sulfate $(SO_4)^{2-}$, (SNI 06-6989.20-2004)

A total of 100 mL water samples was inserted into the Erlenmeyer 250 mL; then 20 mL buffer solution was added to the Erlenmeyer and homogenized by stirring by using magnetic stirrer at a fixed speed for 60 sec. While stirring, added 0,3 g barium chloride ($BaCl_2$). Measurements are performed with UV-Vis spectrophotometers and are calculated sulfate levels $(SO_4)^{2-}$ with the following formula; Sulfate $(SO_4)^{2}$, (mg/L) = C x F_P; C is concentration $(SO_4)^{2-}$ of the measuring result (mg/L) and F_P is the dilution factor.

3. Results and Discussion

3.1 Analysis of hot spring water at station I

Hot spring water at the station I is used by the local community for dringking and cooking consumption. Based on the results of the analysis shows that iron (Fe), phosphate $(PO_4)^{3+}$ and sulfate $(SO_4)^{2-}$ can be seen in Figure 1.

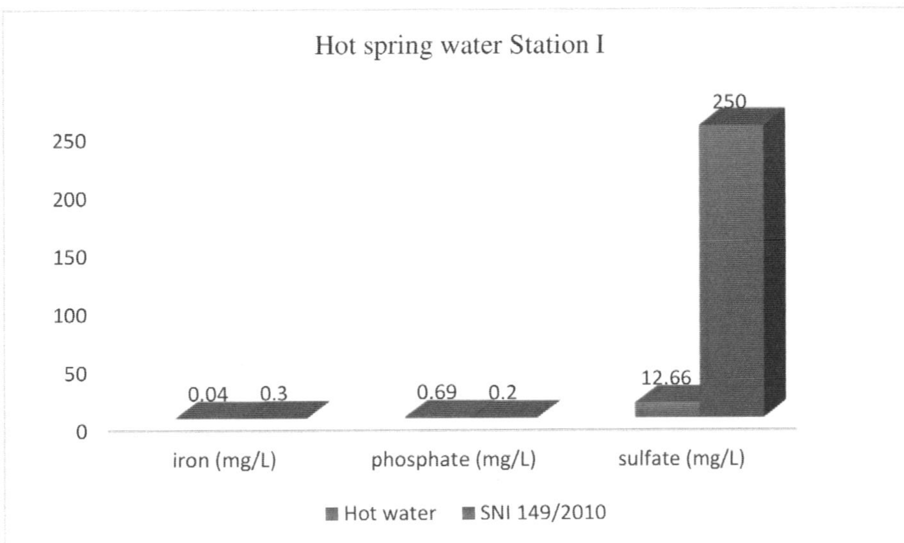

Figure 1. Hot spring water test results at station I Tubbe Village

Referring to the regulation of the Minister of Health Republic of Indonesia No. 149/MENKES/PER/IV/2010 then the hot spring water can be consumed but the conversion for a long time period can cause side effects. The use of hot spring water for long time period can cause mineral accumulations in the body which is evident in the dental corrosion experienced the community around both children and adults. Phosphorus is the second largest mineral stored in the body as a calcium phosphate salt, which is part of the hydroxyapatite crystal $Ca_{10}(PO_4)_6(OH)_2$, in the bones and teeth [9]. Dental corrosion (indicated by yellowish tooth) experienced by the local community can be caused by the high content of sulfate in hot spring water. Sulfate $(SO_4)^{2-}$ is replacing $(PO_4)^{3+}$ the hydroxyapatite crystal $Ca_{10}(PO_4)_6(OH)_2$, causing white teeth turns yellow, which is a typical sulfur/sulfur color. Compared to the villagers in Tubbe and Aramaba, the surrounding communities, who utilize well water to meet their daily needs, experience less cases of tooth decay.

3.2 Analysis of hot spring water at station II

Hot spring water at station II is used by the community for bathing and washing. Iron (Fe), and phosphate $(PO_4)^{3+}$ levels are very small, while the sulfate $(SO_4)^{2-}$ is 11.98 mg/L.

Table 1. Hot spring water testing result at a point II of Tubbe village.

Parameter	Result (mg/L)	Minister of health 149/2010
Iron	<MDL	0,3
Phosphate	<MDL	0,2
Sulfate	11,98	250

The hot spring of Tubbe village at station II is a pond located on the shore and there is a water channel to the sea. If there is a tide, water will cover the surface of hot springs and there is a mixing dilution between seawater and hot spring water. There is a process of dilution by seawater due to the tidal process that allows the mineral levels of phosphate and iron is very small. Because the two types of water had mixed up. Sulfate level $(SO_4)^{2-}$ is possible high enough so that with the presence of dilution by seawater does not lower the sulfate rate significantly. When compared to sulfate at station I, there is no significant difference in station II and station III; therefore, it can be concluded that the dilution factor by seawater at station II does not give a meaningful influence. Based on its chemical composition, the hot spring water in Tubbe and Aramaba village has the potential to be used as a bath for health therapy purposes [7,8].

3.3 Analysis of hot spring water Mineral levels III

Hot spring water at point III located in Aramaba Village is not used by the community to fulfill daily needs. Based on the results of measuring iron (Fe), phosphate $(PO_4)^{3+}$ and sulfate $(SO_4)^{2-}$ in hot spring water at station III are above the threshold that has been determined by regulation of the Minister of Health 149/2010, about the limit of drinking water content that is worth consuming.

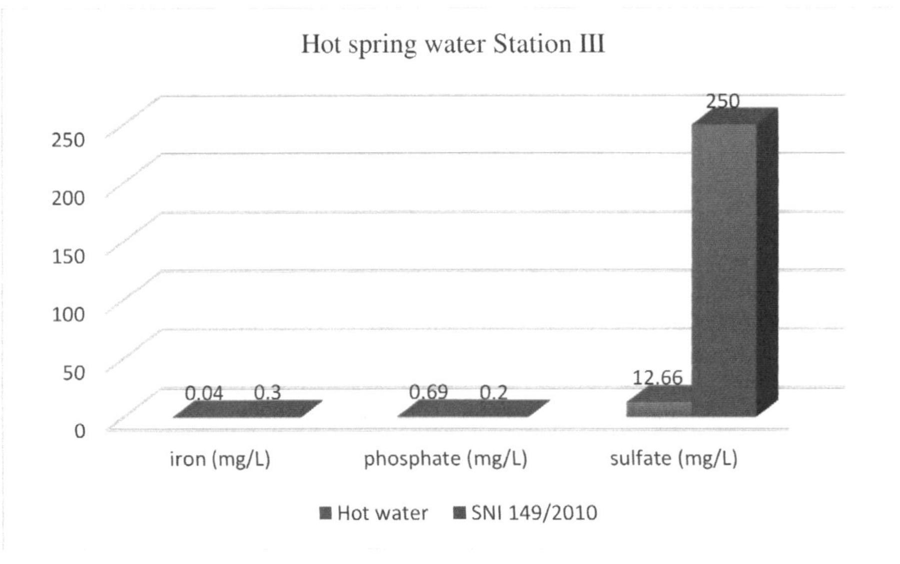

Figure 2. Results of hot spring water test III Aramaba Village

This high level of the dissolved mineral is possible because hot spring water at station III is near Mount Sirung with a distance of ± 4 Km. High rates are possible due to the flow of suspension from ash, stone, pebble, and sand in a mass of hot volcanic gas that comes out of Mount Sirung and flows through its slope into hot spring water at station III in Aramaba village.

4. Conclusion

Hot spring water used for daily consumption does not meet the physical requirements of drinking water.

According to the regulation of the Ministery of Health, Republic of Indonesia, the dissolved mineral contents in hot spring water at the station I and II is below the threshold, while at station III is above the threshold. Although the mineral level is below the threshold, hot spring water at the station I is not worth consuming.

Reference

[1] Effendi H 2003 *Telaah Kualitas Air bagi Pengolahan Sumber Daya Lingkungan* (Yogyakarta: Kanisius)
[2] Menteri Kesehatan Republik Indonesia 2010 Persyaratan kualitas air minum *Peraturan Menteri Kesehatan Republik Indonesia Nomor 492/MENKES/PER/IV/2010* (Jakarta: Departemen Kesehatan Republik Indonesia)
[3] Marzani Y 2013 Kimia mata air panas bumi untuk pengembangan pariwisata di daerah Samboja Kabupaten Kutai Kartanegara Provinsi Kalimantan Timur *Seminar Nasional ke 8 Tahun 2013: Rekayasa Teknologi Industri dan Informasi* (Depok)
[4] Torres-Ceron D A, Acosta-Medina C D and Restrepo-Parra E 2019 Geothermal and mineralogic analysis of hot springs in the Purace-La Mina Sector in Cauca, Colombia *Geofluids* 1-21
[5] Sahoo M, Maharana R M and Seth P 2016 Physico-chemical analysis of surface and groundwater around talcher coal field, District Angul, Odisha, India *Journal Of Geoscience And Environment Protection* 26-37
[6] Maseke S A and Vegi M R A 2019 Comparative study of water quality between hot spring and borehole waters of Mara, Shinyanga and Manyara Regions of Tanzania *Science Journal of Chemistry* **7**(4) 82-9
[7] Simon N, Unjah T, Yusry M and Dzulkafli M A 2019 Phsyco-chemical characterisation and potential health benefit of Hulu Langat hot spring in Selangor, Malaysia *Sains Malaysiana* **48**(11) 2451-62
[8] Hamzah Z, Rani N L A, Saat A and Wood A K 2013 Determination of hot springs physico-chemical water quality potentially use for balneotherapy *Malaysian Journal of Analytical Sciences* **17**(3) 436-44
[9] Parinduri F K, Rahfiludin M Z and Pradigdo S F 2017 Hubungan asupan kalsium, vitamin D, fosfor, kafien, aktivitas fisik dengan kepadatan tulang wanita dewasa muda *Jurnal Kesehatan Masyarakat* **5**(4) 664-74

IOP Conf. Series: Materials Science and Engineering **823** (2020) 012052 doi:10.1088/1757-899X/823/1/012052

Conceptual design on the integrated CO_2 mineralization process with a sugar plant

A. Murnandari[1], **P. Victor**[1], **J.P. Sitompul**[1]

[1]Department of Chemical Engineering, Institute of Technology Bandung, Faculty of Industrial Technology, Bandung 40132, Indonesia

Email: sitompul@che.itb.ac.id

Abstract. Sustainable development goals have set challenges to be achieved in the next ten years. All the stakeholders, including industries, have been forced to take serious action in preserving the environment. Therefore, recycle of waste into energy and valuable chemicals has become attention in industries throughout the world. This paper focused on the integrated-waste utilization in a sugar industry by capturing the CO_2 from the boiler and absorbing it into the mineral solution extracted from filter cake (mud) for producing mineral carbonates with the capacity of 3000 tonnes sugarcane per day. The mud first undergoes the extraction using Ammonium Chloride (NH_4Cl) to extract the minerals. The alkali extracted from the mud then will be converted into Calcium Carbonate and Magnesium Carbonate via mineralization. Simulation using Aspen® plus was carried out using stoichiometric (R_{Stoic}) and yield (R_{Yield}) reactor block to find out the heat duty, the thermodynamic feasibility, and the yield distribution. $CaCO_3$ was produced from both methods with the yield of R_{Stoic} and R_{Yield} 0.718 and 0.676 mol $CaCO_3$/mol CaO, respectively. Based on the capacity of 3000 tonnes sugarcane per day, the process could produce the desired $CaCO_3$, approximately 210 kg/hr. The CO_2 to mineralize 1 tonne of filter cake was 74.2 kg CO_2, while the heat duty is 76.3 kWh for 2.8 tonnes filter cake/hr. The conceptual design is thermodynamically favorable and would be potential to be applied in sugar mills.

1. Introduction

The demand of sugar is increasing along with the growth of population. In 1943, Indonesia was used to be world rank number 2 for world sugar production. However, the production of sugar from sugarcane nowadays ranks number 12 worldwide due to most of sugar companies suffering from low productivity. The demand for sugar for food and beverages is increasing by 6% in demand for 2018, with the consumption of sugar in 2017 is counted as 6.32 million tonnes. With the annual population growth of 1.3%, sugar consumption has escalated by 4.3% per year[1]. Indonesia currently has 63 sugar mills owned by 18 companies with total production capacity reach 245,900 tonnes cane per day (TCD) with each factory produce 3900 TCD on average [2]. With the demand for sugar keeps increasing, the production of sugar will have to increase eventually.

Sugar is produced along with solid, gas, and liquid waste. The solid waste is produced in the form of sugar bagasse, filter cake, litter, boiler ash, and sludge. The gas waste is produced from the boiler as smoke, and the liquid by-product is the molasses that can be utilized as the raw material for ethanol production. Sugar bagasse and litter are used as boiler fuel, while the waste with higher moisture content needs to be treated in other ways. The filter cake is commonly used as organic fertilizer through composting.

Filter cake contains various valuable chemicals that can be extracted such as calcium, wax, and protein [3]. The lipids that can be converted into wax come from the sugarcane plant dispersed in the juice and bagasse while the calcium comes from the use of lime milk in the liming process. The common composition of filter cake could be found as stated in Table 1 below. In Indonesia, the filter cake is called *blotong* and has a similar composition as of common filter cake.

The hazard of untreated filter cake from the sugarcane purification is the depleting oxygen content in the water, resulted in bad odor and darkened water [4]. Waste utilization allows emission reduction by up to 25% [5]. The smoke from the boiler typically contained 10-14% of CO_2. In 2013, Ramadhan [6] studied Karangsuwung Sugar Mills, with the CO_2 emission of boiler using bagasse as the boiler fuel was 0.485 tonnes/MWH and equivalent with 9872 tonnes CO_2/month.

Therefore, this study proposed the complete utilization by integrating the waste as raw material to produce higher value-added material, mineral carbonates. The conceptual designs are to use the filter cake containing calcium and magnesium together with boiler flue gas containing the CO_2 to be converted into the mineral carbonates. Process of removing the wax and protein has already been established by using toluene [7], the liquid that has been separated will contain the needed material to be converted into Calcium Carbonate ($CaCO_3$) and Magnesium Carbonate ($MgCO_3$) using mineralization method. Simulation using the Aspen Plus® and calculation using the black-box method of mass balance will be compared and evaluated using this study. The feasibility of the thermodynamically favored process is expected, and the yield of the $CaCO_3$ will be presented in this paper.

Table 1. Typical composition of filter cake from sugar industry

Components	Content (%)	
	Common filter cake [8,9]	*Blotong* [10]
C	33.73	26.51
N	2.36	1.04
CaO	1-4	5.785
P_2O_5	1-3	6.142
MgO	0.5-1.5	0.419
Wax, lipids, resin	4-14	-
Fiber	15-30	-
Protein	5-15	-
Total ash	9-20	-

2. Materials and Methods

While the integration of the whole waste utilization is complicated using only one simulation tool, the aspen simulation will include the main process of alkali extraction and mineralization. Figure 1 shows the simplified block flow diagram for this study. Alkali extraction is basically separating the mineral content in the filter cake by diluting it using the ammonium chloride (equation 1). There is no built-in model for the mineralization process; therefore, the calculation will be carried out using the available mode of reactor in comparison to stoichiometric reaction and to yield reactor. Mineralization of the extracted solution is conducted by flowing CO_2 into solution which contains mineral salts in order to be converted into the solid mineral carbonates (equation 2). For the complete process of waste utilization, calculation using black-box mass balance was provided in the Aspen file. The Aspen® Plus V9 is then used to calculate the mass balance for the process as schemed in Figure 1.

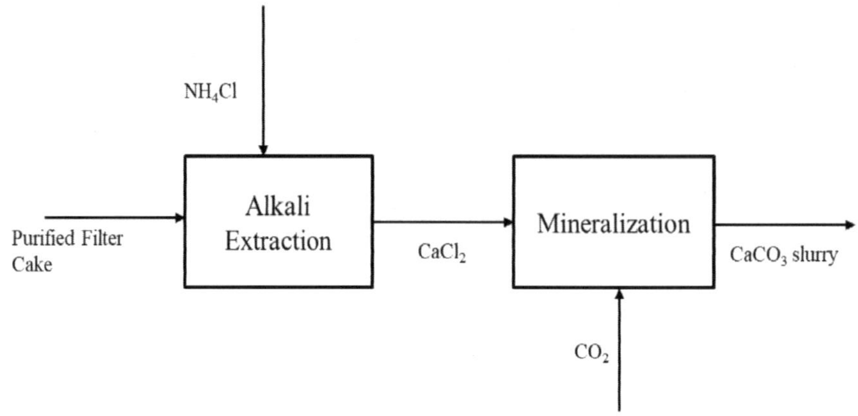

Figure 1. Block Flow Diagram for integrated sugar waste utilization simulated using Aspen®.

The two main reactions are involved in the alkali extraction and mineralization as follows

$$\text{Alkali Extraction: } 2NH_4Cl + CaO \longrightarrow CaCl_2 + 2NH_3 + H_2O \quad (1)$$
$$\text{Mineralization: } CaCl_2 + 2NH_3 + H_2O + CO_2 \longrightarrow CaCO_3 + 2NH_4Cl \quad (2)$$

The schematic model and process parameters used in the simulation are presented in figure 2 and Table 2, respectively. Note that the operating parameter for process and yield were collected from the available literature. Some assumptions have to be made in order to allow the simulation running. Note that the fluid package in both types of reactors is NRTL which is the best result for predicting the thermodynamic properties in this study. In stoichiometric reactor calculation, the base yield is 0.76 mol $CaCl_2$/mol CaO for alkali extraction and 0.95 mol $CaCO_3$/mol $CaCl_2$. In the yield reactor calculation, the same yield was applied for alkali extraction, while in the mineralization, the yields for CaO and $CaCl_2$ has to be determined to obtain convergence result. Therefore in this case, the conversion for mineralization from CaO 50%, while from $CaCl_2$ in the extract 85%[11]. The resulted calcium oxide (CaO) and calcium carbonate ($CaCO_3$) are considered as slurry in the model. Particle size distribution and pure component data were all provided by NIST (National Institute of Standards and Technology) through the Aspen Plus analysis.

In order to complete the simulation, a set of properties of pure components and binary interactions are determined using the NRTL method. All the material type is conventional and available in Aspen® Plus. NIST data is provided to make assumptions in the NRTL method calculation as provided in Table 3. The heat duty is calculated using the standard free energy of formation (DHFORM) and the standard Gibbs energy of reaction (DGFORM). Heat duty for each block is the result of outlet heat streams subtracted from outlet heat stream. Binary interaction included in the calculation is Ammonia (NH_3) and water with the property unit on-hand by APV90 VLE-IG. Heat capacity of each component in each phase, density, surface tension, boiling temperature, and vapor pressure also has to be defined and assumed using NRTL method calculation in each pure component state before executing the R_{Yield} calculation.

IOP Conf. Series: Materials Science and Engineering **823** (2020) 012052 doi:10.1088/1757-899X/823/1/012052

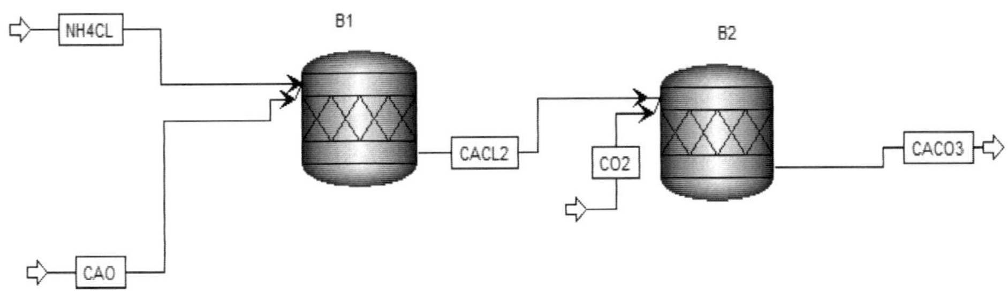

(a)

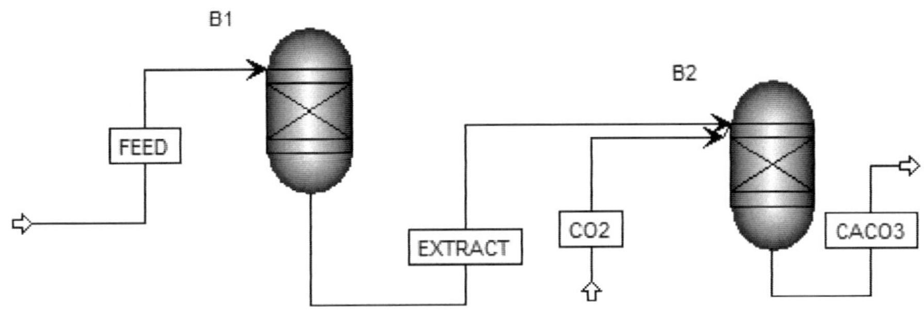

(b)

Figure 2. Typical flow chart for filter cake mineralization; (a) Stoichiometric reactor (b) yield reactor

Table 2. Operating parameter for model simulation using Aspen Plus

Operating Parameter	Value (unit)
NH_4Cl Concentration	2 M [11]
Extraction Temperature and Pressure	25°C, 1.013 bar [11]
Extraction Efficiency	76% [11]
Mineralization Mass Yield	0.858 $tCaCO_3/tCaCl_2$ [12,13]
Mineralization Mol Yield	0.95 mol $CaCO_3$/mol $CaCl_2$ [12,13]
Mineralization Temperature and Pressure	40°C,1.013 bar [12,13]
CO_2 mol fraction inlet	15% CO_2, 85% N_2
Filter cake feed	2.8 tonnes/hr
Mol ratio Ca^{2+}:NH_4Cl	1:2

Table 3. NIST data for heat duty calculation

Parameters	Units	Data set	Component	Component	Component	Component
			NH_4CL	$CACO_3$	CAO	$CACL_2$
VB	cc/mol	1	542.318	140.903		
RKTZRA		1	0.291852		0.291766	0.291767
VLSTD	cc/mol	1	434.571		866.928	46.1659
TB	C	1		186.82		
DHFORM	cal/mol	1		-135299		
DGFORM	cal/mol	1		-124962		

3. Results and Discussion

3.1 Aspen simulation of alkali extraction and mineralization

The molar balance (in fractions) of the process using stoichiometric reactor, so-called R_{Stoic} model, is presented in Table 4. The total yield obtained from the mass balance calculation was 0.718 mol $CaCO_3$/ mol CaO. Note that CaO was presented in 5.785% of Calcium in the filter cake. Hence, 74.2 kg $CaCO_3$ was produced per tonne of filter cake using the assumption of stoichiometric reaction occur in the system. The needs of CO_2 to mineralize 1 tonne of filter cake was 74.2 kg CO_2 per tonne filter cake with the yield of $CaCO_3$ 2.27 $tCaCO_3/tCO_2$. The kinetics of the process is not yet known; thus, using the stoichiometric reactor the error was found in molar balance 12%. The error in the stoichiometric calculation may be caused by the valid phase of the reactants and products, which was only available in vapor-liquid phase, without considering solid phase as the reaction naturally occurred in the presence of solid. However, the simulation tool was limited to calculate the liquid and vapor phase under condition of unknown solid data such as particle size distribution and phase equilibrium. The heat duty needed to process the reaction was calculated based on the reaction stoichiometry as much as 274.5 MJ/h or 76.3 kWh with an outlet temperature of 40°C as needed for the mineralization reaction.

Table 4. Molar fraction of stream in the Aspen using Stoichiometric Reactor (R_{Stoic} Model)

Component	Stream names				
	CaO	NH_4Cl	CaCl	CO_2	$CaCO_3$
NH_4Cl	0	0.734797	0.280918	0	0.327351
$CaCl_2$	0	0	0.104073	0	0.003956
H_2O	0.405531	0.265203	0.374	0	0.188997
CaO	0.594469	0	0.032865	0	0.02253
NH_3	0	0	0.208145	0	0.007912
CO_2	0	0	0	0.15	0
N_2	0	0	0	0.85	0.381865
$CaCO_3$	0	0	0	0	0.067388

R_{Yield} Model is the block in the Aspen Plus tool used for calculation based on the yield of reaction. The block does not require exact information of the reaction stoichiometry and kinetics. The inert component could be defined using the condition if the unreacted component is known. Also, there is no need to separate the feed stream as in using the R_{Stoic} block. The model is closer to the actual process using the R_{Yield} block. However, the heat duty could not be calculated based solely on the yield data. Therefore, the use of both calculations represents the model complementary to each other and both could be compared and analyzed. Table 5 shows the molar fraction balance using the R_{Yield} model.

The yield of $CaCO_3$ production using R_{Yield} Model was 0.676 mol $CaCO_3$/mol CaO. Lower mass yield of $CaCO_3$ using this motel was found around 69.83 kg $CaCO_3$ per tonne of filter cake. However, using the R_{Yield} model, the consumption of CO_2 was found to be lower, around 44%. Calculation using R_{Yield} Model distributed the yield based on the known yield composition. The error calculated using the R_{Yield} block was significantly lower than R_{Stoic} model which is only 1%. Calculation using R_{Yield} shows equilibrium between the ammonium salts, water, and ammonium ions present in the system. In terms to satisfy the condition of all the components inside the system, R_{Yield} block calculation assumes that there must be the same amount of element N, Ca, and CO_2. Hence, the error for the mol balance is significantly low. The weakness of the process is that clearly it does not take into account that $CaCO_3$ is produced in solid-state despite its solubility in the aqueous state, and thus the lower yield than R_{Stoic} is obtained.

Table 5. Molar fraction of streams in the Aspen Model using Yield Reactor (R_{Yield} Model)

Component	Stream Name			
	FEED	EXTRACT	CO_2	CACO$_3$
NH_4Cl	0.67	0.2359	0	0.0948
NH_3	0	0.2359	0	0.0948
CaO	0.33	0.0566	0	0.0415
H_2O	0	0.2359	0	0.0948
$CaCl_2$	0	0.2359	0	0.0124
N_2	0	0	0.85	0.4958
$CaCO_3$	0	0	0	0.0828
CO_2	0	0	0.15	0.0829

3.2 Pretreatment and Purification

As stated above, the utilization of sugar waste using the integrated concept is to produce pure $CaCO_3$. Magnesium and Calcium are both found in the filter cake because it is difficult to separate during the extraction and mineralization process as they possess quite similar properties. The method proposed to separate the $CaCO_3$ from the $MgCO_3$ produced from the mineralization is by floatation with addition of Sodium Naphthenate ($C_{10}H_{17}NaO_2$) in the suspension of the solution. After the mixing, $CaCO_3$ will float and be easily separated from $MgCO_3$ [14]. In addition to the high value-added products, the environmentally friendly process supports sustainable development goals. Taken into account the purification method, the mass balance was calculated based on the yield data using a black-box method, $CaCO_3$ was produced using the method 105.90 kg/hr and $MgCO_3$13.07 kg/hr.

4. Conclusion

The conceptual design for accommodating the zero waste principle has been applied in the sugar mills industries for producing valuable minerals. Using mineralization with low pressure and temperature, the capital cost is quite low. Mineralization also allows the recycle of ammonium salts in the process. The simulation showed that design was thermodynamically favorable and would be potential to be applied in sugar mills. Calcium carbonate was produced in the ambient pressure and low temperature with reactor stoichiometric model and also using the reactor yield model. Based on the capacity of 3000 tonnes of sugarcane per day, the process could produce the desired $CaCO_3$, approximately 210 kg/hr. The CO_2 to mineralize 1 tonne of filter cake was 74.2 kg/hr with heat duty 76.3 kWh for 2.8 tonnes filter cake/hr.

5. References

[1] Sulaiman A A, Sulaeman Y, Mustikasari N, Nursyamsi D and Syakir A M 2019 *Land Suitability Analysis and Sugar* pp 1–17
[2] Toharisman A and Triantarti 2016 An overview of sugar sector in Indonesia *Sugar Tech* **18**
[3] Rabelo S C, Eduardo C and Rossel V 2015 Learn more about filter cake industrial waste recovery environmental concerns in soybean processing *Sugarcane* 4–14

[4] Supari, Pabrik D and Trangkil G 2015 Analisa kandungan kimia pupuk organik dari blotong tebu limbah dari pabrik gula trangkil *Prosiding Seminar Nasional Sains Dan Teknologi* **1** 1

[5] Gunawan, Bantacut T, Romli M and Noor E 2019 Life cycle assessment of cane-sugar in Indonesian sugar mill : Energy use and GHG emissions *IOP Conf. Ser. Mater. Sci. Eng* **536**

[6] Ramadhan A F 2014 Analisis Emisi Gas CO_2 Dalam Penilaian Daur Hidup Industri Gula pada Pg. Karangsuwung PT. Rajawali II Cirebon *Skripsi* (Bogor: Departemen Teknologi Industri Pertanian IPB)

[7] Partha N and Velmurugan S 2006 Recovery of chemicals from pressmud – A sugar industry waste *Indian Chem. Engr* **48**

[8] Kenneth R H 1971 By-products of the Cane Sugar Industry : J. M. Paturao *Carbohydr. Res.* **16** 251

[9] Gupta N, Tripathi S and Balomajumder C 2011 Characterization of pressmud : A sugar industry waste *Fuel* **90** 389–94

[10] Fadjari T 2009 *Memanfaatkan Blotong Limbah Pabrik Gula* Available from: https://lordbroken.wordpress.com/2010/01/14/pemanfaatan-limbah-pabrik-gula/

[11] Moon S, Hyun S, Kwan S, Hye M, Duc D, Woong S and Su S 2017 Calcium extraction from steelmaking slag and production of precipitated calcium carbonate from calcium oxide for carbon dioxide fi xation *J. Ind. Eng. Chem.* **53** 233–40

[12] Murnandari A, Youn M H, Park K T, Kim H J, Kim Y E and Jeong S K 2017 Single process for CO_2 capture and mineralization in various alkanolamines using calcium chloride *Energy & Fuels* **31** 763-9

[13] Murnandari A, Kang J, Youn M H, Park K T, Kim H J, Kang S-P and Jeong S K 2017 Effect of process parameters on the CaCO3 production in the single process for carbon capture and mineralization *Korean J. Chem. Eng.* **34** 935–41

[14] Schoenlaub and Robert A 1943 *United States Patent Office 2433297*

IOP Conf. Series: Materials Science and Engineering **823** (2020) 012053 doi:10.1088/1757-899X/823/1/012053

Electrochemical reduction of CO_2 to Formic Acid on Pb-Sn Alloy Cathode

Pramujo Widiatmoko, Isdiriayani Nurdin, Hary Devianto, Budi Prakarsa, and Henry Hudoyo

Department of Chemical Engineering, Institut Teknologi Bandung, Bandung 40132 Indonesia

E-mail: pramujo@che.itb.ac.id

Abstract. Carbon dioxide released to the atmosphere due to fossil combustion needs to be reduced to decrease the global warming effect. Electrochemical reduction of CO_2 is one of promising technology to convert CO_2 to various valuable compounds, reducing the amount of released CO_2 to the atmosphere. Metal alloys are interesting material for cathode. In this preliminary work, a CO_2 electrochemical reduction was performed on a Pb-Sn alloy cathode using catholyte of $KHCO_3$ and $NaHCO_3$ at various concentrations. An anode from Pt-Ir and H_2SO_4 0.1 M as anolyte were used in the experiment. The CO_2 was bubbled into 400 mL catholyte at rate of 75 mL min^{-1}. It was found that Pb-Sn alloy in $KHCO_3$ electrolyte provides higher productivity and faradaic efficiency than in $NaHCO_3$. Increasing overpotential from 0.1 V to 1 V increases both formic acid productivity and efficiency. The optimum amount of formic acid produced was 9400 μmol in catholyte $KHCO_3$ 0.5 M with faradaic efficiency of 67.19% and conversion of 0.326%.

1. Introduction

Extensive consumption of fossil fuel increases the concentration of CO_2 in the atmosphere which contributes to the greenhouse effect. Indonesia, as the 4th populated country, fulfill almost 95% of the energy consumption by fossil fuel [1], with increasing rate of approximately 6-8% per annum. Therefore, technologies for CO_2 mitigation should be developed.

Various metal electrodes have been examined in electrochemical reduction of CO_2 to formic acids such as Pb, Pd, Pt, and Sn [2,3]. Use of metal alloy, however, is still limited. Kwon et al. have reported a selective electrochemical reduction of carbon dioxide to formic acid using indium-zinc bimetallic nanocrystals [4]. A composition of In : Zn = 0.05 shows high catalytic activity with 95% faradaic efficiency for HCOOH production. Li et al. also reported Sn-In co-deposited on Cu alloy as cathode for formic acid synthesis with faradaic efficiency of 88% [5]. Knaeco et al. used copper alloy in methanol-based electrolyte and showed methane-dominant products which proportional to copper content [6]. Choi et al. used Sn-Pb alloys powder on carbon paper as new electrocatalyst [7]. The alloys show higher electrocatalytic activity than its single metal electrode.

IOP Conf. Series: Materials Science and Engineering **823** (2020) 012053 doi:10.1088/1757-899X/823/1/012053

In this study, electrochemical reduction of CO_2 was conducted using Pb-Sn alloy wire. High electronic conductivity of the Pb-Sn alloy is expected to increase reaction efficiency. Effect of electrolyte types, concentration, and overpotential toward amount of formic acid produced, conversion, and current efficiency was studied. It was found that electrochemical reaction using the Pb-Sn alloy in $KHCO_3$ catholyte shows higher performance than ones in $NaHCO_3$ catholyte.

2. Methodology

Experiment was conducted using H-cell with Nafion® 212 as separator. A Pt-Ir wire anode was used as counter electrode in H_2SO_4 0.1 M anolyte. A Pb-Sn alloy wire with surface area of 12.26 cm^2 was used in the catholyte of Sodium bicarbonate ($NaHCO_3$) and potassium bicarbonate ($KHCO_3$) with various concentration of 0.1 M; 0.2 M; 0.5 M. All chemicals were obtained from Merck without prior treatment before used. Electrolysis is operated under atmospheric pressure and temperature with CO_2 bubbling rate of 75 mL/min for 3 h. A Stability test was conducted for 5 – 7 h. The operation voltage was set at overpotential of 0.5 V, 0.75 V, and 1 V. A saturated calomel electrode (SCE) was used as reference electrode. Figure 1 shows the experimental set-up.

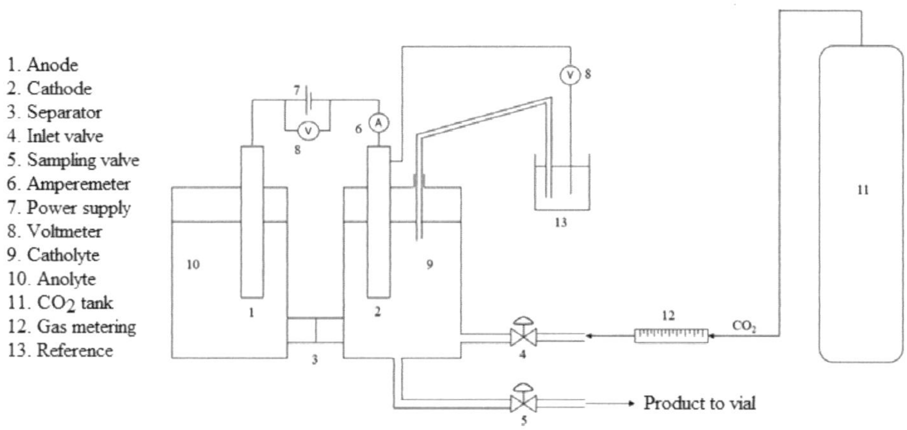

Figure 1. Experiment set-up

3. Result and Discussion

3.1. Working potential of reaction

Equilibrium potential of CO_2 reduction to formic acid is – 0.20 V/RHE [8]. Due to activation overpotential, the potential where reaction starts observable will have onset to the equilibrium ones. Here we define the summation of equilibrium potential and activation overpotential as initial voltage. The increasing voltage will increase the electric current in the ohmic overpotential region and then reach the concentration overpotential. In this study, working overpotential was determined based on the initial voltage instead of the equilibrium ones. A current-voltage curve was prepared to calculate the initial potential, as shown in Figure 2. Table 1 shows the calculated initial voltage vs SCE at 10 mA of current.

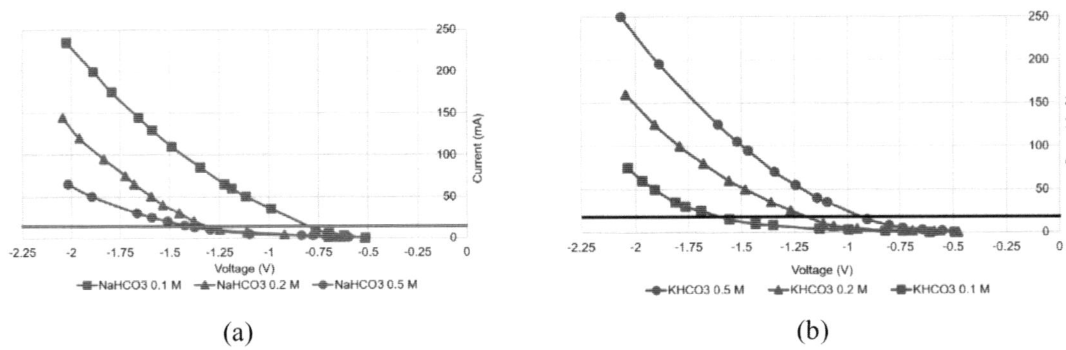

Figure 2. I-V curves for (a) $NaHCO_3$ and (b) $KHCO_3$ at various concentration

Table 1. Initial potential data for various types and concentrations of electrolytes

Electrolytes	Initial potential (V vs SCE) in various concentration of catholyte		
	0.1 M	0.2 M	0.5 M
$NaHCO_3$	-0.766	-1.250	-1.298
$KHCO_3$	-1.351	-1.067	-0.802

The increase of electrolyte concentration will decrease cell equilibrium potential [9]. However, the initial voltage for reaction in $NaHCO_3$ seem requires higher voltage as the concentration increase. To confirm the trend, an experiment using potentiostat was conducted to get Tafel curve. The equilibrium potential E_0 then was determined for each curve, as shown in Figure 4. Based on the Figure 3, reaction equilibrium potential for NaHCO3 0.5 M, 0.2 M, and 0.1 M are 1.24 V, 1.04 V, and 0.99 V vs SCE. Those values are equivalent to 1.48 V, 1.28 V, and 1.23 V against SHE, respectively. Therefore, the trend of $NaHCO_3$ at various concentrations is in the correct path.

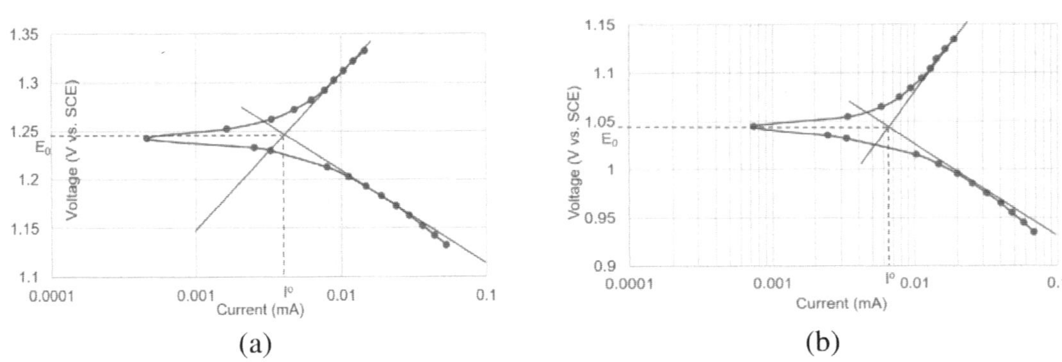

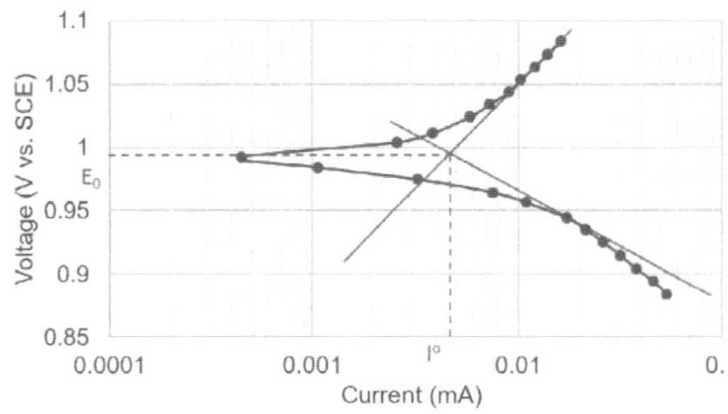

(c)

Figure 3. Tafel curve for electrolysis cells with electrolyte $NaHCO_3$ (a) 0.5 M; (b) 0.2 M; (c) 0.1 M

3.2. Electrochemical reduction performances

Performance of electrochemical reduction of CO_2 for formic acid synthesis is measured based on produced formic acid, reaction conversion, and current efficiency, shown in Table 2. Reaction in $KHCO_3$ electrolyte provides higher productivity than in $NaHCO_3$ electrolyte. Higher electrolyte conductivity of $KHCO_3$ than $NaHCO_3$ in the same concentration causes the reaction can go faster. It is also reported that $KHCO_3$ can dissolve more CO_2 than $NaHCO_3$, providing more reactants in the electrolyte [10].

Table 2. Electrochemical reduction parameters

Catholyte type	Over potential (V)	Produced formic acid (μmol)			Reaction conversion of formic acid formation (%)			Current efficiency (%)		
		0.1 M	0.2 M	0.5 M	0.1 M	0.2 M	0.5 M	0.1 M	0.2 M	0.5 M
$NaHCO_3$	0.5	120	39	0	0.004	0.001	0.000	9.47	9.52	0.00
	0.75	192	760	4480	0.007	0.026	0.155	11.93	26.43	58.94
	1	272	5480	7680	0.009	0.190	0.266	13.09	61.63	61.84
$KHCO_3$	0.5	40	320	4000	0.001	0.011	0.139	10.05	23.53	61.93
	0.75	1000	2200	6000	0.035	0.076	0.208	24.46	41.02	64.77
	1	5040	5640	9400	0.175	0.196	0.326	61.18	62.08	67.19

Concentration of electrolytes has a proportional effect to performance of electrolysis. Higher concentration increases the conductivity of electrolytes [11, 12]. Moreover, CO_2 binding capability increase with concentration. Both conditions increase reaction efficiency and productivity. As an overpotential increases from 0.5 to 1 V, the reaction rate increases. Within the overpotential range, no side reaction and limiting current condition were observed. Faradaic efficiency however still lower than reaction on Sn-In reported by Kwon et al. (88%) or on Sn-Pb alloys reported by Choi et al. (79.8%). Low soluble CO_2 and contact of gas – electrode due to lack of CO_2 bubbling method are supposed to be the cause.

3.3. Stability of electrolysis reaction

Continuous electrolysis may cause degradation on the electrode material, as well as their performance. Therefore, a stability test was conducted on Pb-Sn electrode using $KHCO_3$ 0.5 M and 1 V overpotential for 5 hours continuous operation. Current as function of time and produced formic acid were measured as shown in Figure 4. The result show stable current and production rate, which indicates stable cathode material within 5 hours test.

Observation on the cathode before and after reaction was performed under scanning electron microscope. The cathode was used for 7 h continuous reaction prior to the examination. Figure 5 shows significant different morphology of the cathodes before and after electrochemical reaction. After reaction, a rough deposit layer was formed on the surface of cathodes. An EDX analysis revealed oxide compound of K, Pb, and Sn which increases along with electrolysis duration. The layer seems not to affect to electrolysis performance as reaction rate keep a linear trend with time. Further research is required to examine effect of the oxide layer on cell performance.

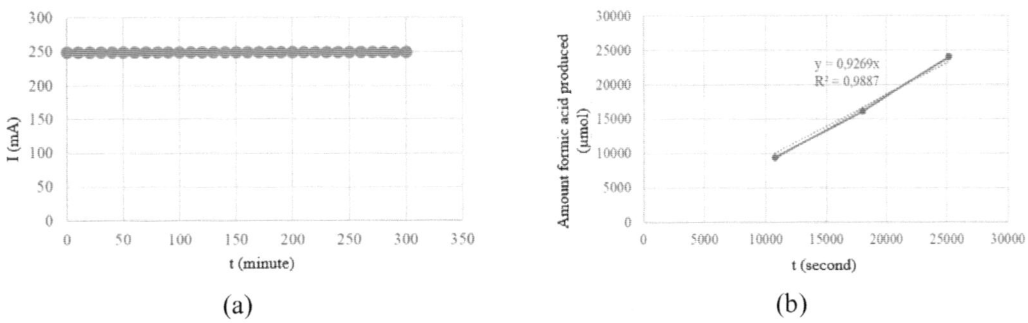

Figure 4. (a) Current – time and (b) Formic acid production – time curves

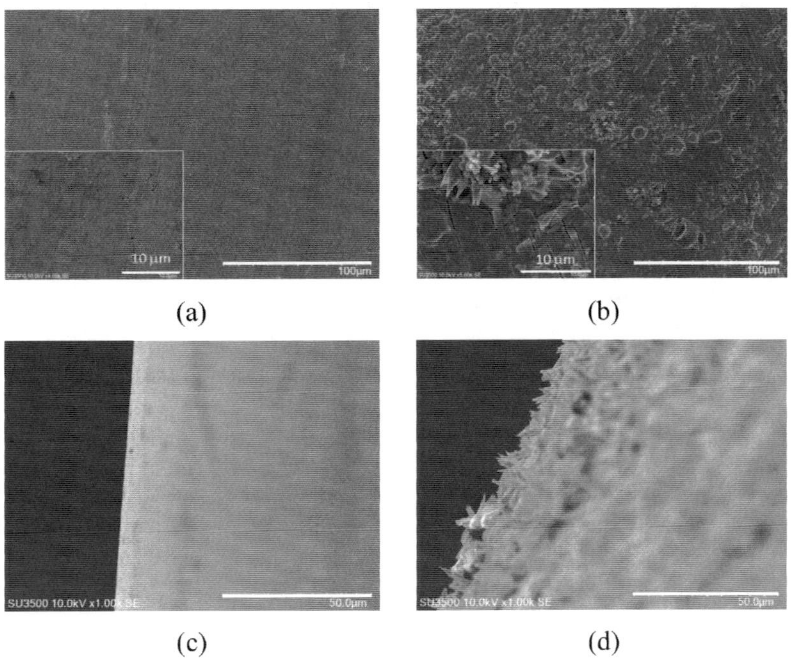

Figure 5. The SEM results of cathode before (a and c) and after (b and d) used in electrochemical reduction of CO_2

4. Conclusion

We successfully synthesize formic acid from CO_2 via electrochemical reduction on Pb-Sn alloys. Use of $KHCO_3$ as a catholyte can produce higher amounts of formic acid, current efficiency, and conversion than $NaHCO_3$ ones. The electrolyte concentration is directly proportional to produced formic acid formed, current efficiency, and conversion. In the overpotential range of 0.5 – 1 V, the performance of electrolysis is increase without any side reactions were observed. The Pb – Sn alloys electrode is stable within electrolysis duration of 7 h.

IOP Conf. Series: Materials Science and Engineering **823** (2020) 012053 doi:10.1088/1757-899X/823/1/012053

References

[1] Pusat Data dan Teknologi Informasi Kementrian ESDM 2016 *Data Inventory Emisi GRK Sektor Energi* ISBN:978-602-0836-22-5 (Jakarta)
[2] Jitaru M, Lowy D A, Toma M, Toma B C and Oniciu L 1997 *J. Appl. Electrochem.* **27** 875-89
[3] Jitaru M 2007 *J. Univ. Chem. Technol. Metallurgy* **42**(4) 333
[4] Kwon I K, Debela T T, Kwak I H, Seo H W, Park K, Kim D, Yoo S J, Kim J.-G, Park J and Kang H S 2019 *J. Mater. Chem. A* **7** 22879-83
[5] Li F, Zhang H, Ji S, Liu W, Zhang D, Zhang C, Yang J, Yang F and Lei L 2019 *Int. J. Electrochem. Sci.* **14** 4161-72
[6] Kaneco S, Yamauchi H, Katsumata H, Suzuki T and Ohta K 2004 *Stud Surf Sci Catal* **153**(04) 80264-7
[7] Choi S Y, Jeong S K, Kim H J, Baek I.-H and Park K T 2016 *ACS Sustainable Chem. Eng.* **4**(3) 1311-18
[8] Chen C, Kotyk J F K and Sheehan S W 2018 *Chem* **4** 2571
[9] Pletcher D 2015 *Electrochem. Commun.* **61** 97
[10] Salazar-Villalpando M D 2011 *J Electrochem. Soc.* **33** 77
[11] Weast R C and Lide D R 1989 *CRC Handbook of Chemistry and Physics* 70th Ed. (Boca Raton: CRC Press) p D-221
[12] Wolf A V 1966 *Aqueous Solutions and Body Fluids* (New York: Harper and Row)

IOP Conf. Series: Materials Science and Engineering **823** (2020) 012054 doi:10.1088/1757-899X/823/1/012054

Sensitivity analysis of fuzzy simple additive weighting to determine land suitability for corn in Kupang Regency

Dony Martinus Sihotang

Department of Computer Science, University of Nusa Cendana, Jl. Adisucipto Penfui, Kupang, Indonesia

E-mail: dmsihotang@staf.undana.ac.id

Abstract. The change in geographical states, such as weather and climate, makes it difficult for farmers to prepare adequate land for corn cultivation. This research analyses the fuzzy simple additive algorithm used to determine land suitability in Kupang Regency and the use of sensitivity analysis to obtain information on the most sensitive variable(s). Two testing models were conducted in this research, namely, accuracy and sensitivity tests. The result of the accuracy test showed 80% precision, while sensitivity showed that rainfall, soil depth, and C-Organic were highly sensible with a percentage of 100%, followed by slope and soil texture at 66%. Furthermore, the irrigation system, spring, and level of disaster had the same percentage of 33.33%.

1. Introduction

Approximately 70% of the East Nusa Tenggara (NTT) populations are predominantly corn farmers because it is a staple food for the community [1]. Corn commodities are also one of the outstanding opportunities for agribusiness development due to an increase in market demand because NTT will be among the top 12 food-producing regions in Indonesia [2]. There are three corn-producing zones in NTT: Flores, Sumba, and Timor which produced 109 tons in 2017 [3].

A technique used to increase corn productivity is through land evaluation, and this is achieved by comparing the characteristics of plants and nature of land [4]. The Food and Agriculture Organization [5] has updated evaluation indicators to include suitability, a comparative advantage with required inputs, technology, features, as well as the evaluation of its physical and socio-economic conditions. Furthermore, the indicators are called the FAO model [6].

Agricultural land suitability is an interdisciplinary approach that also involves major decisions at various levels starting from choosing a land type, criteria, organization, and preferences (both qualitative and quantitative) [7]. The relative importance of these parameters can be well evaluated to determine the suitability by multi-criteria technique [8] with the Simple Additive Weighting mainly utilized [9]. However, these methods have limitations, with several parameters or criteria linguistically [10]. The transition from uncertainty by linguistic values such as "Very Low", "Low", "High", and "Very High" can be quantified using fuzzy set theory [11]. Therefore, many researchers have attempted to use fuzzy in multi-criteria evaluation method [12].

The weight of the criteria is the biggest contributor to uncertainty in the multi-criteria evaluation method [13]. Sensitivity Analysis (SA) procedures help reduce uncertainty and stability of its outputs by illustrating the impact of introducing a small change to specific input parameters on outcomes [14].

This study aimed to determine the level of land suitability for corn plants in 10 plants alternative locations in Kupang Regencies namely Tarus, Noelbaki, Oebelo, Manusak, Oesao, Naibonat, Raknamo, Sulamu, Pariti, and Nonbes. In addition, it was also conducted to measure the sensitivity level of the criteria (land characteristics) such as rainfall, irrigation systems, water availability, land slope, soil texture, soil depth, C-organic, and level of disaster.

2. Methodology

There are four processes in this study, namely data collection, conversion to fuzzy numbers, F-SAW process, testing of accuracy and sensitivity.

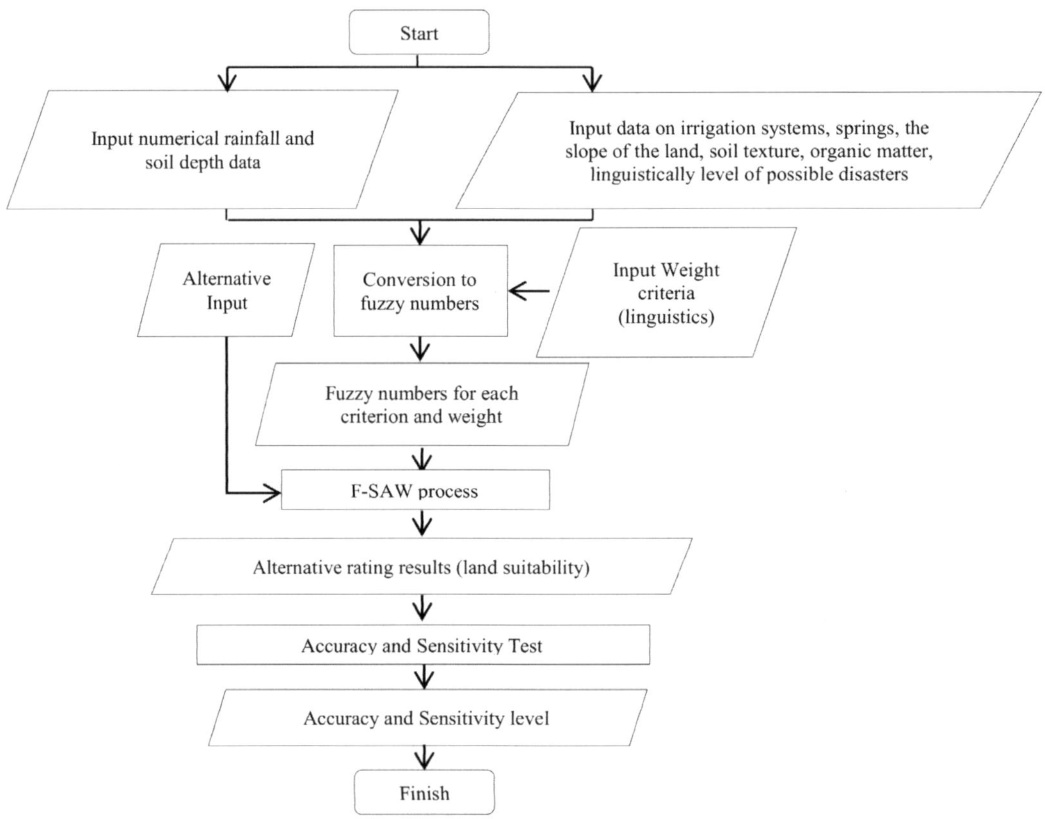

Figure 1. Research methodology

2.1. Data collection

Data was collected through interviews, observation, and documentation. The data, therefore, were divided into two models, namely crisp value and linguistic. Crisp value data was also acquired in the form of rainfall and soil depth at the Agricultural and Forestry Extension Agency (BP4K) of Naibonat, Kupang Regency. The linguistic data (irrigation system, springs, land slope, soil texture, organic matter, and level of disaster) were retrieved from observations that were made using questionnaires sent to 2 farmers groups in alternative locations.

Table 1. Alternative data, criteria, and weight of criteria.

Location	Rainfall	Springs	Land slope	Soil texture	Soil depth	C-organic	Level of disaster
Tarus	448 mm	VG	G	F	46 cm	G	G
Noelbaki	449 mm	VG	G	G	48 cm	F	G
Oebelo	453 mm	VG	VG	G	50 cm	G	VG
Manusak	1.043 mm	VG	VG	VG	65 cm	VG	VG
Oesao	1.040 mm	F	VG	VG	68 cm	VG	G
Naibonat	1.045 mm	F	VG	VG	65 cm	VG	G
Raknamo	485 mm	VG	G	G	65 cm	G	VG
Sulamu	420 mm	VG	G	VG	48 cm	VG	VG
Parity	425 mm	F	VG	VG	50 cm	G	VG
Nonbes	783 mm	F	VG	VG	65 cm	F	VG
Weight	VG	P	P	G	F	G	F

2.2. Conversion to fuzzy value

After data collection, those data were pre-processed by converting crisp and linguistic values (Table 1) into fuzzy values. The conversion process for variable rainfall and soil depth each used a membership function as in table 2. The membership function was built based on the requirements of land use for corn [15].

Table 2. Fuzzy membership function for crisp input

Variable	Membership function
Rainfall	$\mu_{rainfall}(x) = \begin{cases} 0; \ x < 300 \ atau \ x > 1600 \\ \frac{x-300}{500-300}; \ 300 \leq x \leq 500 \\ 1; \ 500 \leq x \leq 1200 \\ \frac{1600-x}{1600-1200}; \ 1200 \leq x \leq 1600 \end{cases}$
Soil depth	$\mu_{soil\ depth}(x) = \begin{cases} 0; \ x \leq 25 \\ \frac{x-25}{60-25}; \ 25 \leq x \leq 60 \\ 1; \ x \geq 60 \end{cases}$

The triangular fuzzy numbers are used to convert the irrigation system variables, springs, land slope, soil texture, organic matter, level of disaster to fuzzy values (Figure 2).

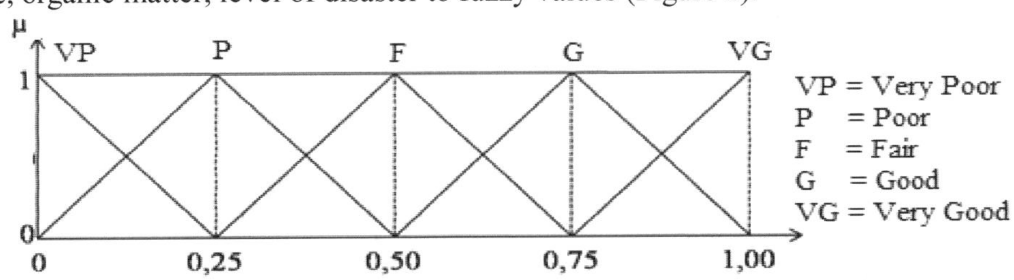

Figure 2. Fuzzy membership function for linguistic input

2.3. F-SAW method

Fuzzy set theory can be used to represent uncertainty by giving intervals from 0 to 1. F-SAW is a combination of the SAW method and Fuzzy mathematical logic with differences in the implementation of values. The F-SAW comparison matrix is represented by three variables (a, b, c) called Triangular Fuzzy Numbers (TFN), which will produce three values, according to three sequential weights in the triangle membership function [16]. The algorithm is as follows:

a. The rating of each criterion for alternatives was made in the fuzzy value through fuzzy membership function.
b. The fuzzy values for linguistic variables were calculated by the defuzzification equation as follows:

$$e = \frac{(a+b+c)}{3} \tag{1}$$

where:
e = defuzzification value
a = smallest fuzzy number
b = moderate *fuzzy* number
c = biggest *fuzzy* number

c. The fuzzy weight of each criterion was determined by the following equation:

$$W_i = \frac{e_i}{\sum_{i=1}^{n} e} \tag{2}$$

where:
W_i = weight for the-*i* criterion
e_i = defuzzification value for the-*i* criterion
$\sum_{i=1}^{n} e_i$ = the total defuzzification value of each criterion

d. A decision matrix (alternative to criteria) was produced, where the matrix element is a fuzzy value from the previous calculation.
e. The decision matrix is normalized by the following equation:

$$r_{ij} = \begin{cases} \dfrac{x_{ij}}{\underset{i}{MAX}(x_{ij})} \\ \\ \dfrac{\underset{i}{MIN}(x_{ij})}{x_{ij}} \end{cases} \tag{3}$$

where:
r_{ij} = normalized matrix element.
x_{ij}= decision matrix elements that have not been normalized
$MAX(x_{ij})$ = the biggest value of alternative *i* with respect to criterion *j*, for profit criteria
$MIN(x_{ij})$ = the smallest value of alternative *i* with respect to criterion *j*, for cost criteria

f. A normalized matrix of each criterion for each alternative is produced:

$$N = \begin{bmatrix} r_{11} & r_{12} & \cdots & r_{1j} \\ \vdots & & & \vdots \\ r_{i1} & r_{i2} & \cdots & r_{ij} \end{bmatrix}$$

where:
N = normalized matrix

g. The total value of each alternative is determined by the following equation:

$$V_i = \sum_{j=1}^{n} w_j . r_{ij} \tag{4}$$

The bigger V_i value showed that the A_i alternative is more chosen.

3. Result and Discussion

This section explains the results of the process: converting data to fuzzy values, F-SAW, and testing processes using the level of accuracy and sensitivity analysis.

3.1. Fuzzy value conversion

Fuzzy values (μ) are obtained by entering data (Table 1) into membership functions (Table 2 and Figure 1). The defuzzification equation (equation 1) is also used for linguistic data conversion.

Table 3. Fuzzy memberships function for crisp input.

Alternative	Rainfall	Irrigation system	Springs	Land slope	Soil texture	Soil depth	C-organic	Level of disaster
Tarus	0.740	0.916	0.250	0.750	0.500	0.600	0.750	0.750
Noelbaki	0.745	0.916	0.250	0.750	0.750	0.657	0.500	0.750
Oebelo	0.765	0.916	0.250	0.916	0.750	0.714	0.750	0.916
Manusak	1.000	0.916	0.750	0.916	0.916	1.000	0.916	0.916
Oesao	1.000	0.500	0.916	0.916	0.916	1.000	0.916	0.750
Naibonat	1.000	0.500	0.750	0.916	0.916	1.000	0.916	0.750
Raknamo	0.925	0.916	0.250	0.750	0.750	0.571	0.750	0.916
Sulamu	0.600	0.916	0.250	0.750	0.916	0.657	0.916	0.916
Pariti	0.625	0.500	0.916	0.916	0.916	0.714	0.750	0.916
Nonbes	1.000	0.500	0.916	0.916	0.916	1.000	0.500	0.916
Fuzzy weighted	0.916	0.250	0.916	0.250	0.750	0.500	0.750	0.500

3.2. The alternative ranking using F-SAW

After the fuzzy value is obtained, the next step is to calculate the normalized rating performance (r_{ij}) for each alternative and each criterion. The decision matrix and weight of criterion were normalized using (5) as shown in matrix *N* and *W* as follows.

$$N=\begin{bmatrix} 0{,}74 & 1{,}00 & 0{,}272 & 0{,}818 & 0{,}545 & 0{,}60 & 0{,}818 & 0{,}818 \\ 0{,}745 & 1{,}00 & 0{,}272 & 0{,}818 & 0{,}818 & 0{,}657 & 0{,}545 & 0{,}818 \\ 0{,}765 & 1{,}00 & 0{,}272 & 1{,}00 & 0{,}818 & 0{,}714 & 0{,}818 & 1{,}00 \\ 1{,}00 & 1{,}00 & 0{,}818 & 1{,}00 & 1{,}00 & 1{,}00 & 1{,}00 & 1{,}00 \\ 1{,}00 & 0{,}545 & 1{,}00 & 1{,}00 & 1{,}00 & 1{,}00 & 1{,}00 & 0{,}818 \\ 1{,}00 & 0{,}545 & 0{,}818 & 1{,}00 & 1{,}00 & 1{,}00 & 1{,}00 & 0{,}818 \\ 0{,}925 & 1{,}00 & 0{,}272 & 0{,}818 & 0{,}818 & 0{,}571 & 0{,}818 & 1{,}00 \\ 0{,}60 & 1{,}00 & 0{,}272 & 0{,}818 & 1{,}00 & 0{,}657 & 1{,}00 & 1{,}00 \\ 0{,}625 & 0{,}545 & 1{,}00 & 1{,}00 & 1{,}00 & 0{,}714 & 0{,}818 & 1{,}00 \\ 1{,}00 & 0{,}545 & 1{,}00 & 1{,}00 & 1{,}00 & 1{,}00 & 0{,}545 & 1{,}00 \end{bmatrix}$$

$$W = [\ 0{,}1895\ \ 0{,}0517\ \ 0{,}1895\ \ 0{,}0517\ \ 0{,}1552\ \ 0{,}1034\ \ 0{,}1552\ \ 0{,}1034\ ]$$

The last step of F-SAW calculation is ranking the preference value (*Vi*) obtained from summing the vector multiplication vector *W* and normalized matrix *N* (Pers. 4). The result of the calculation can be seen in Table 4.

IOP Conf. Series: Materials Science and Engineering **823** (2020) 012054 doi:10.1088/1757-899X/823/1/012054

Table 4. Result of alternative ranking (*Vi*).

Alternative	Rank
Manusak	0.9655
Oesao	0.9577
Naibonat	0.9232
Nonbes	0.9060
Pariti	0.8476
Sulamu	0.7413
Raknamo	0.7377
Oebelo	0.7315
Noelbaki	0.6513
Tarus	0.6444

3.3. Accuracy test
Accuracy testing is conducted to determine the precision of the F-SAW system produced. In this test, the result of the system-based ranking and the expert knowledge-based ranking have been compare.

Table 5. Comparison of rank.

Sub-dist	System-based rank	Expert-based rank
Tarus	10	10
Noelbaki	9	9
Oebelo	8	8
Manusak	1	1
Oesao	2	2
Naibonat	3	3
Raknamo	7	7
Sulamu	5	6
Pariti	6	5
Nonbes	4	4

From Table 5, it can be seen that there are 8 similar ranking results and 2 different rankings between the system and the expert so that the accuracy value is obtained:

$$Accuracy = \frac{8}{10} \times 100\% = 80\%$$

$$Error\ rate = \frac{2}{10} \times 100\% = 20\%$$

3.4. Sensitivity analysis
Sensitivity analysis aims to determine the criteria level utilized. In this section, the weights of preference were made on each criterion.

Table 6. Result of sensitivity analysis.

Variable	Default weight	Weight change	Number of ranking change	Count of change	Percentage of change
Rainfall	VG	F ; P ; VP	2 ; 2 ; 4	3	100%
Irrigation system	P	VG ; F ; VP	0 ; 0 ; 2	1	33.33%
Springs	VG	F ; P ; VP	0 ; 0 ; 4	1	33.33%
Land slope	P	VG ; F ; VP	3 ; 2 ; 0	2	66.67%
Soil texture	G	VG ; F ; P	0 ; 4 ; 5	2	66.67%
Soil depth	F	VG ; G ; P	2 ; 2 ; 2	3	100%
C-Organic	G	VG ; F ; P	2 ; 4 ; 5	3	100%
Disaster level	F	VG ; G ; P	0 ; 0 ; 2	1	33.33%

Table 6 shows that the rainfall criteria have an initial "Very Good (VG)" weight, which changes to "Fair (F)", "Poor (P)", and "Very Poor (VP)". There are 2, 2, 4 changes in sequence on the ranking results when the weight is changed to F, P, and VP. It indicates that the ranking of rainfall criteria is very sensitive to change in weight.

4. Conclusion

The result of the accuracy level test shows two different ranks produced by the F-SAW method and expert assessment, so the accuracy value obtained is 80%. The second test scenario, which is the sensitivity level test, shows that the rainfall, soil depth, and C-organic at 100%, followed by land slope and soil texture with 66%, then the irrigation systems, springs, and level of disaster with 33.33%. From the obtained results, the most sensitive criteria were rainfall, soil depth, and C-organic.

Acknowledgment

This study was supported by Nusa Cendana University Faculty of Science and Engineering and the Agricultural and Forestry Extension Agency (BP4K) of Naibonat, Kupang Regency to whom the author expresses his gratitude.

References

[1] Suryati 2015 *Pembelajaran Kontekstual Berbasis Inquiry Training Pendidikan Vokasional Pertanian pada Kompetensi Pemanfaatan Potensi Lokal* (Bandung: Universitas Pendidikan Indonesia) p 5

[2] Setiawan K and Fallo F A I 2017 Development prospects agroindustry processed corn in Kupang Regency *J. Partner.* **2** 172-80

[3] Statistic of Kupang Regency 2018 *Kupang Regency in Figure 2018* (Kupang: Statistic of Kupang Regency) p 158

[4] Bandyopadhyay S, Jaiswal R K, Hedge V S and Jayaraman V 2009 Assessment of land suitability potentials for agriculture using a remote sensing and GIS based approach *Int Journal of Remote Sensing* **30** 879-95

[5] Food Agriculture Organzation of The United State Nations 2007 *Land Evaluation* (Rome: FAO) pp 23-30

[6] Sitorus S R P 2010 Land capability classification for land evaluation *J. Sumberdaya lahan* **4** 69-78

[7] Prakash T N 2003 *Land Suitability Analysis for Agricultural Crops: A Fuzzy Multicriteria Decision Making Approach* (Enschede: International Institute for Geo-Information Science

and Earth Observation, Netherland) p 3

[8] Ceballos-Silva A and Lopez-Blanko J 2003 Delineation of suitable areas for crops using a multi-criteria evaluation approach and land use/cover mapping: a case study in central Mexico *J. Agricultural Systems* **0308-521** 117-36

[9] Velasquez M and Hester P T 2013 An analysis of multi-criteria decision making methods *Int. Journal of Operations Research* **10** 56-66

[10] Sediyono E, Setiawan A and Kaparang D R 2013 Fuzzy simple additive weighting algorithm to determine land suitability for crop in Minahasa Tenggara *Int. Journal of Computer Applications* **84** 26-9

[11] Gupta S and Kamarthi S 2003 Evaluation of production facilities in a closed-loop supply chan: A fuzzy TOPSIS approach *Proc. of SPIE* **5262** 128

[12] Kaya T and Kahraman C 2011 Multicriteria decision making in energy using a modified fuzzy TOPSIS methodology *J. Expert System with Applications* **38** 6577-85

[13] Chen Y, Yu J and Khan S 2010 Spatial sensitivity analysis of multi-criteria weights in GIS-Based land suitability evaluation *J. Environmental Modelling & Software* **25** 1582-91

[14] Ravalico J K, Dandy G C and Maier H R 2010 Management Option Rank Equivalence (MORE) – A new method of sensitivity analysis for decision-making *J. Environmental Modelling & Software* **25** 171-81

[15] Wirosoedarmo R, Sutanhaji A T, Kurniati E and Wijayanti R 2011 Land suitability assessment of corn (Zea mays.L) using spatial analysis method *J. AGRITECH* **31** 71-8

[16] Chang N-B, Parvathinathan G and Breeden J B 2007 Combining GIS with fuzzy multicriteria decision-making for landfill siting in a fast-growing urban region *J. of Environment Management* **0301-4797** 1-15

IOP Conf. Series: Materials Science and Engineering **823** (2020) 012055 doi:10.1088/1757-899X/823/1/012055

Principal component analysis with successive interval in K-Means Cluster Analysis (Study case: Poverty data 2013 in East Nusa Tenggara)

A M Tamonob[1], A Saefuddin[2], A Wigena[3]

[1]Faculty of Agricultural Technology, ArthaWacana Christian University, Kupang
[2]Department of Statistics, Bogor Agricultural University, Bogor, Indonesia

E-mail: arista_marlin@yahoo.com

Abstract. K-Means Cluster is a cluster analysis for continuous variables with the concept ofdistance used is a euclidean distancewhere that distance is used as observation variables, which are uncorrelated with each other. The case with the type of data that is correlated categorical can be solved by making categorical data into numerical data by the method called the successive interval and then used Principal Component Analysis. Applied this method in poverty data of East Nusa Tenggara Province in K-Means clusterobtained that Principal Component Analysis with Successive interval obtained variables that take effect to the cluster formation are toilet, fuel, and job.

1. Introduction

In the social and behavioral science, researchers are often confronted with a large number of variables. Often, variables used has a different measurement scale; there are numerical and categorical scale. Cluster analysis is a method in multivariate to grouping a set of objects in the same group (cluster) are more similar to each other than to those in other groups (clusters). In general, variables used as basic cluster consist of two types, there are categorical variables (nominal and ordinal) and numerical variables (interval and ratio). In general, there are two methods in cluster analysis, hierarchical and non-hierarchical method. The hierarchical method is used if the number of clusters is previously unknown while non-hierarchical method is used if the number of cluster is already known. One of non-hierarchical cluster analysis commonly used is the analysis of K-Means cluster used for numerical data and used Euclid distance as a measure to see similarity and dissimilarity measured. Euclidean distance is used if the observedvariables are independent oruncorrelated with one another.One of the assumptions in cluster analysis is no multicollinearitybetween variables [1]. Euclid distance concept used for numerical data by transforming the variables using Principal Component Analysis [2]. Nominal and ordinal data can be an interval by using the method proposed by Hays [3] called successive interval. Principal Component Analysis can be used in the results of successive interval to reduce data and solve multicollinearity, so it can be analyzed by cluster analysis.Previous research related to cluster analysis for categorical data by Yuniato [4] was about the comparison between cluster analysis with nonlinear principal component analysis and two-step cluster in mix data and the result that the two-step cluster method is better and more specific than centroid linkage method with

IOP Conf. Series: Materials Science and Engineering **823** (2020) 012055 doi:10.1088/1757-899X/823/1/012055

principal component analysis transformation. This study focusses on principal component analysis with a successive interval for categorical data in K-Means cluster analysis using poverty data 2013 in East Nusa Tenggara Province. East Nusa Tenggara ranks fourth poorest province in Indonesia after Papua, West Papua and Maluku [5]. This cluster analysis aims to find out the significant factors of poverty in East Nusa Tenggara so that it can be an input for the government to reduce and even overcome poverty in East Nusa Tenggara.

1.1 Categorical data

Based on the type of data, it can be divided into numerical data (quantitative) and categorical data (qualitative). Numerical data is data expressed in numerical magnitude (numbers), for example, the data in per capita income, expending, price, etc., whereas categorical data is classified by category or a particular class. Categorical data consist of nominal data and ordinal data. Nominal data is data with an order or the value does not indicate the level merely as a label only, such as religion, gender, ethnicity or race. Ordinal data is data with the order of categories shows level or rank, such as education level, smoking habits. Categorical data can also be obtained by grouping continuous data, but having a risk of losing information. In practice, it is easier to record categorical data rather than continuous data; respondents are easier to answer sensitive questions; it is more practical significance. Presentation of categorical data can be either frequency, frequency tables and contingency tables [6].

In cases where the numerical variables are linked, correlation analysis is one option to see the relation between variables. However, if the two categorical variables are linked, then correlation analysis can not be used because the numbers on a category code arenot the only form of actual values. Another reason why the correlation analysis can not be used on categorical data is due to one type of categorical variable that can not be sorted as nominal category. Giving a different order will give different correlation values so that two people who calculate the correlation value likely give the same results. For this reason, the chi-squared analysis will be used to find relationships (associations) between the categorical variables. The analysis is based on the chi-square contingency table (often called cross-tabulations). A contingency table is a table whose cells contain the frequencies of the intersection of rows and columns. The general form of contingency tables with the first variable has m categories and the second variable has k categories. The hypotheses were:

H_0 :two variables are independent (associated)

H_1: two variables are dependent (not associated)

The statistical test used was:

$$\chi^2_{hitung} = \sum_{j=1}^{k}\sum_{i=1}^{m}\frac{\left[O_{ij} - E_{ij}\right]^2}{E_{ij}} \quad (1)$$

If χ^2_{hitung} value is greater than $\chi^2_{tabel(db;\alpha)}$ with the value of the degrees of freedom $(m-1)(k-1)$ then reject H_0. Strong or weak associations can be seen from the association between the interval -1 to 1. Values association equals to 1 which means there is a strong association among variables, if the association equals to the value of zero, it means there is no association among variables.

Table 1. Correlation test or the chosen association

Variable 1	Variable 2	Correlation test or the chosen association
Nominal	Nominal	Contingency coefficient, Lambda, Phi, and Cramer
Nominal	Ordinal	Contingency coefficient, Lambda
Ordinal	Ordinal	Spearman, Gamma, Somers'd
Ordinal	Numeric	Spearman
Numeric	Numeric	Pearson

1.2 K-Means cluster

K-Means cluster analysis has been developed by Mac Queen in 1967 and it is the most well-known non-hierarchical data grouping method that widely used in many fields because it is simple and easy to implement [7]. K-Means is a partitioning cluster method that separates data into different groups. The purpose of grouping the data is to minimize the variance within a group and maximize the variance between groups. Basic K-Means algorithm is as follows:

1. Determine the k value as the number of clusters to be formed
2. Generate the center point of the initial k randomly cluster
3. Calculate the distance of each data into each cluster center using Euclid distance
4. Group each data based on the distance between the nearest inter-data center
5. Determine new cluster center position by calculating the average value of the data existing on the same cluster center.

1.3 Principal component analysis

Principal component analysis is a multivariate method that aims to reduce the dimension of original variables to new variables (called principal component) that uncorrelated but have big information about original variables.

Principal component analysis is a weighted linear combination of the origin variables that are able to explain the data maximally.

The principal component j of p variables can be expressed as follows:

$$Y_j = a_{1j}x_1 + a_{2j}x_2 + \ldots + a_{pj}x_p = \underline{a}'\underline{x} \quad (2)$$

and the variance of principal components j is

$$Var(Y_j) = \lambda_j;\ j = 1, 2, \ldots, p \quad (3)$$

$\lambda_1, \lambda_2, \ldots, \lambda_p$ is eigen values that obtained from the equation:

$$\left|\Sigma - \lambda_j \mathrm{I}\right| = 0 \quad (4)$$

Where $\lambda_1 \geq \lambda_2 \geq \ldots \geq \lambda_p \geq 0$. Eigenvector $\underline{a}$as the weight of transformation linear of origin variables obtained from the equation:

$$\left(\Sigma - \lambda_j \mathrm{I}\right)\underline{a}_j = 0 \quad (5)$$

Variance total of principal component is

$$\lambda_1 + \lambda_2 + \ldots + \lambda_p = tr(\Sigma) \quad (6)$$

and the percentage of variance total data explained by the principal component j is:

$$\left(\frac{\lambda_j}{tr(\Sigma)} x100\%\right) \quad (7)$$

The percentage of total variance is considered sufficiently representative of variance if data is 75% or more.

1.4 Successive interval

Methods to convert data from nominal or ordinal data to interval data [8].

Steps that have to be carried out are:

a. Calculate the frequency of each data category
b. Calculate the proportion based on the frequency of each category
c. Calculate the cumulative proportion
d. Calculate the value of for each cumulative proportion
e. Calculate the value of the density function
f. Calculate the value scale (the average interval) for each category
g. Calculates caling

2. Material and Method

2.1 Data

The data used is the core data SUSENAS (National Socio-Economic Survey) 2013 of East Nusa Tenggara Province. Variables used consisted of 15 variables categorical: Education (X_1), Floor Area (X_2), Floor Type (X_3), Wall (X_4), Roof (X_5), Toilet (X_6), Water (X_7), Fuel (X_8), Assets (X_9), Literacy (X_{10}), Employment (X_{11}), Work Status (X_{12}), Credit (X_{13}), Poor Rice (X_{14}), Handphone (X_{15}). Groups in this analysis consist of poor and non-poor households.

2.2 Method

In this research, steps to analyze are:

1. Data exploration
2. Checked multicollinearity among variables
   Applied NLPCA to data
3. Carried out the Successive interval to data
4. Applied PCA to successive interval data
5. Applied K-Means Cluster Analysis on principal component scores of PCA with Successive interval

3. Results

3.1. Data description

Description SUSENAS core data of East Nusa Tenggara Province in 2013 to 10422 households. The following three variables were taken for describing the data:

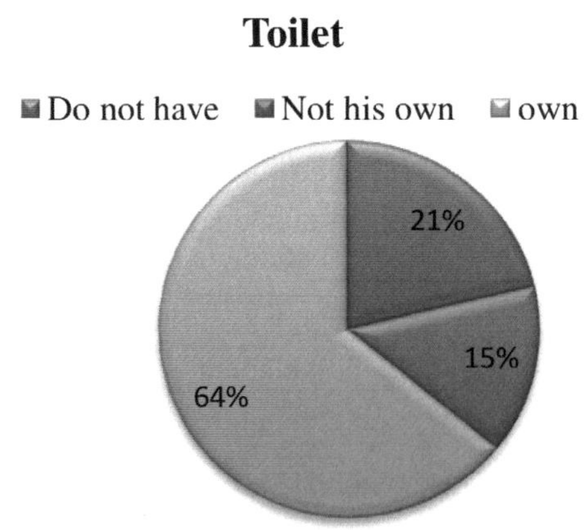

Figure 1. Percentage of toilet ownership

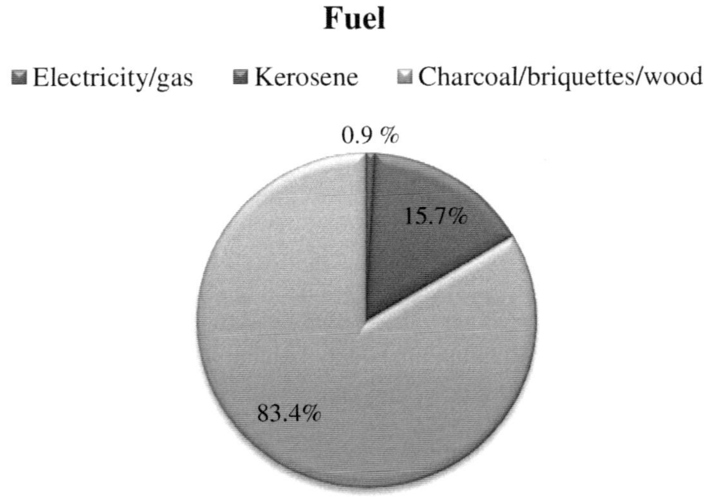

Figure 2. Percentage of the type of fuel used for cooking

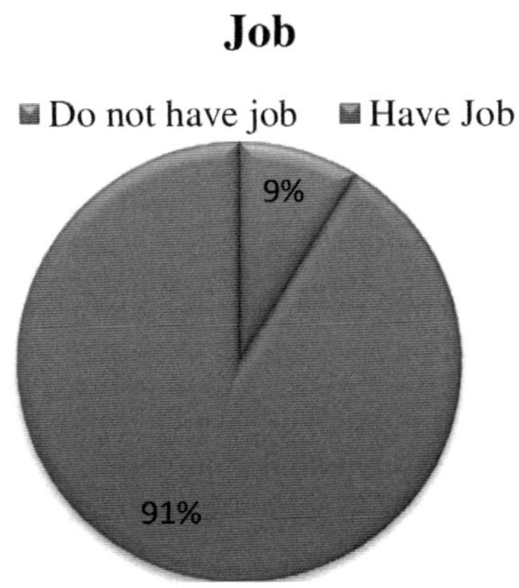

Figure 3. Percentage of job

From Figure 1, we can see 64% of household heads in East Nusa Tenggara Province 2013 had their own toilet, 21% did not have toilet, and 15% not has their own. From Figure 2, we see that household fuel is still dominated by charcoal / briquettes / wood by 83.4%, electricity / gas by 0.9% and 15.7% kerosene. Figure 3 shows that 91% ofthe households had a job while 9% did not have job.

3.2 Checking multicollinearity

From the 15 independent variables of nominal and ordinal scale, the test multicollinearity assumptions were done by looking at the association between variables in the contingency tables. Test of the association among the variables was large enough association between the variables of literacy and education by 0.815, between jobs and education by 0.656, between the working status and education by 0.567, and between jobs and working status by 0.813.

3.3 Principal component analysis with successive interval

Results of the principal component analysis showed that when the two main components were taken to mean, only 35.2% could explain the variance of the initial data. The selection of the main components to be used in the K-Means cluster is based on the percentage of the cumulative variance of data about 75% or more [9]. To be able to represent the variance of the data in this study were taken 9 main components with atotal variance of 79.1%. The final results of the principal component analysis area score of major components used for the analysis of K-Means cluster.

Table 2. Result of principal component analysis of successive interval data

Principal Component	Eigen Value	Variance Proportion	Cummulative Variance Proportion
1	3.7140	0.248	24.8%
2	1.5729	0.105	35.2%
3	1.2984	0.087	43.9%
4	1.0467	0.070	50.9%
5	0.9811	0.065	57.4%
6	0.9432	0.062	63.6%
7	0.8481	0.057	69.3%
8	0.7543	0.050	74.3%
9	0.7160	0.048	79.1%
10	0.6941	0.046	83.7%
11	0.6606	0.044	88.1%
12	0.5827	0.039	92.0%
13	0.5545	0.037	95.7%
14	0.3616	0.024	98.1%
15	0.2808	0.019	100%

Table 3. Principal component score of successive interval data

	KU_1	KU_2	KU_3	KU_4	KU_5	KU_6	KU_7	KU_8	KU_9
X_1	0.372	-0.040	-0.378	-0.201	0.169	-0.121	0.134	-0.073	0.012
X_2	0.201	0.417	0.375	-0.063	0.069	-0.088	0.128	-0.076	-0.312
X_3	0.229	0.085	0.206	0.175	-0.154	-0.638	0.040	-0.288	0.237
X_4	-0.302	-0.174	-0.208	-0.214	0.037	0.132	-0.099	-0.065	-0.019
X_5	0.284	-0.287	-0.069	-0.082	0.104	-0.079	0.354	-0.099	0.519
X_6	-0.244	-0.296	-0.070	0.115	-0.150	-0.302	-0.007	-0.578	-0.336
X_7	-0.229	-0.065	0.076	-0.264	0.234	-0.563	0.244	0.578	-0.166
X_8	-0.328	0.213	0.058	-0.228	0.070	0.041	-0.184	-0.104	-0.349
X_9	0.163	-0.328	-0.319	0.304	-0.294	-0.207	-0.327	0.366	-0.291
X_{10}	0.236	0.159	-0.539	-0.330	0.263	-0.112	0.062	-0.238	-0.162
X_{11}	0.308	-0.426	0.277	-0.257	0.034	0.068	-0.070	-0.020	0.014
X_{12}	0.221	-0.430	0.358	-0.409	0.053	0.063	-0.137	-0.084	-0.137
X_{13}	-0.000	0.145	-0.088	-0.464	-0.830	0.023	0.239	0.066	0.032
X_{14}	0.235	-0.170	-0.022	0.292	-0.006	0.269	0.680	0.041	-0.304
X_{15}	-0.312	-0.136	0.081	0.061	-0.014	0.004	0.285	-0.093	-0.302

IOP Conf. Series: Materials Science and Engineering **823** (2020) 012055 doi:10.1088/1757-899X/823/1/012055

From Table 3 above, variables X_1 value of the first principal component is 0.372 meaning that the principal component gives a contribution 0.372 but has a positive influence on the observation unit; the second component is 0.040 and has a negative influence on the observation unit. This also applies to principal components and other variables.

3.4 Clustering results

Table 4. Cluster variance

Variance	Method
	PCA with successive interval
Within Cluster	9.548
Between Cluster	24287.65
Ratio	0.00039

From Table 4, it can be seen that the ratio between the variance in the cluster of method showed that the method of Principal Component Analysis method with successive interval has a ratio of 0.00039. We can see variance within a cluster smaller than between clusters. Ratio 0.00039 can be used if we want to compare two methods of clustering, the smaller ratio shows that the method better than the other method.

Table 5. ANOVA of K-Means cluster for principal component analysis with successive interval

Variables	Deviation standard within a cluster	Deviation standard between cluster	F	p-value
X_1	0.2527	0.0320	0.06	0.8083
X_2	0.2125	0.0409	0.15	0.7134
X_3	0.3213	0.0395	0.06	0.8140
X_4	0.1547	0.0064	0.01	0.9364
X_5	0.2048	0.1125	1.21	0.3142
X_6	0.1628	0.2004	6.06	**0.0489**
X_7	0.1420	0.0478	0.45	0.5260
X_8	0.2129	0.3993	14.06	**0.0095**
X_9	0.3599	0.0745	0.17	0.6932
X_{10}	0.2658	0.2270	2.92	0.1385
X_{11}	0.2062	0.2312	5.03	**0.0662**
X_{12}	0.4184	0.0629	0.09	0.7739
X_{13}	0.2207	0.0489	0.20	0.6733
X_{14}	0.2351	0.2200	3.50	0.1104
X_{15}	0.1839	0.0398	0.19	0.6802

ANOVA results show that the variables that differentiate in cluster formation are variable with p-value less than 0.01. In this case, the significant variables are X_6 (toilet), X_8 (fuel), X_{11} (job). Therefore, the factors need to be considered by the East Nusa Tenggara government to reduce problems. Toilet is related to public health and cleanliness, so every household must have it. The result of SUSENAS shows that about 36% of households did not have a toilet. From fuel, there are still many households that use charcoal/briquettes/woods about 83.4%, the government must solve these problems, one way is to distribute electricity to the village. From SUSENAS about 91% households had a job, but mostly the jobs not a proper job with a good salary, so there are many households classified to poverty households.

4. Conclusions

In this study, it was concluded that PCA with successive interval produced 9 principal components with variance total about 79%. In K-Means cluster analysis with PCA successive interval had variance ratio within a cluster and between cluster about 47.26725. Variables that have influenced the household clustering were X_6 (toilet), X_8 (fuel), X_{11} (job). From this research we can see that this method can be used for categorical data, in PCA we can transform it back way and from the successive interval we can get the numerical data, so we can use them in another analysis using numerical data.

References

[1] Hair J F, Black W C, Babin B J and Anderson R E 2010 *Multivariate Data Analysis* 7th Edition (New Jersey: Prentice Hall International Inc.)
[2] Kaufma L and Peter J R 2005 *Finding Groups in Data: An Introduction to Cluster Analysis* (New York: John Wiley & Sons Inc.)
[3] Hays W L 1976 *Quantification in Psychology* (New Delhi: Prentice Hall)
[4] Yunianto Y 2011 Perbandingan Metode Penggerombolan dengan Komponen Utama Nonlinier dan Gerombol Dua Langkah pada Data Campuran *Skripsi* (Bogor: Institut Pertanian Bogor)
[5] Badan Pusat Statistik 2013 *Nusa Tenggara Timur dalam Angka 2013* (Kupang: BPS Nusa Tenggara Timur)
[6] Agresti A 2002 *An Introduction to Categorical Data Analysis* (New York, US: John Wiley and Sons)
[7] Wichern D W and Johnson R A 2002 *Applied Multivariate Statistical Analysis* (New Jersey: Prentice Hall)
[8] Waryanto B and Millafati Y A 2006 Transformasi Data Skala Ordinal ke Interval menggunakan Makro Minitab *Informatika Pertanian* **15** 881-95
[9] Morrisson D F 1990 *Multivariate Statistical Methods Third Edition* (New York: McGraw-Hill Inc.)

IOP Conf. Series: Materials Science and Engineering **823** (2020) 012056 doi:10.1088/1757-899X/823/1/012056

Prediction of student learning outcomes using the Naive Bayesian Algorithm (Case Study of Tama Jagakarsa University)

Arini Aha Pekuwali

Department of Computer Science, Christian University of Wira Wacana Sumba, R. Soeprapto 35 Street, Prailiu, Waingapu – Sumba Timur

E-mail: arini_pekuwali@apps.ipb.ac.id

Abstract. Assessment of student learning outcomes is the most important part in the learning process. Student achievement can be determined based on the achievement of final grades in certain subjects. Student final grades can be used to evaluate and predict student achievement in the future. This research was conducted to analyze the Naïve Bayesian Classifier (NBC) algorithm in predicting the final grades of students in the future based on student final grade data in the previous semester. This study is useful for students to improve their grades, according to their predicted weaknesses through this research (wake-up calling). The results of this study indicate that NBC successfully classifies data with an accuracy of 94.2446%.

1. Introduction

The main purpose of higher education institutions is to provide quality education for students. One way to achieve the highest level of quality in the higher education system is to find knowledge from educational data to learn the main attributes that can affect student achievement.

Research using Data Mining in the field of education is increasing. It is a new field, called Educational Data Mining (EDM), which processes data from the educational context into information. The information obtained is useful for supporting decisions in the form of constructive recommendations to academic planners in higher education institutions, cutting failure rates, to better understand student behavior, and many other benefits.

The quality of higher education can also be seen from the value of student learning outcomes. In general, higher education already has an information system that is used to manage student learning outcomes data assessment. The information system manages values based on established assessment points. Assessment points used by each educational institution are governed based on academic regulations owned by the institution. Based on Government Regulation No. 49 of 2014 explains that the assessment of learning is a minimum criterion regarding the assessment of student learning processes and outcomes in the context of fulfilling graduate learning outcomes. Assessment is done based on the value of assignments, quizzes, homework assignments, midterms, final examinations. In its implementation, the assignment value is the combined value of several other tasks, such as quizzes, homework assignments, practical reports, and presentation values.

IOP Conf. Series: Materials Science and Engineering **823** (2020) 012056 doi:10.1088/1757-899X/823/1/012056

This study predicts the value of student learning outcomes in the current semester based on data on student learning outcomes in the previous semester. The previous semester's assessment data is called training data, and the data that will be predicted to its final value is called test data. The assessment data consists of a number of values namely attendance value, assignment value, Midterm Exam scores and Final Semester Exam scores of the students Tama Jagakarsa University.

The three algorithms commonly used in EDM are the Decision Tree, Neural Network, and Naïve Bayes Classifier (NBC) [1]. NBC provides the highest accuracy compared to the other two algorithms. The variables used in this classification are GPA, student demographics, high school background, scholarships, and social interaction [1]. The researcher proposes the use of the NBC Algorithm in this study. The NBC algorithm extracts patterns from the training data and the test data testing process.

The problems are:

- How to determine the assessment components of the student's final grade that can be used in predicting the value of student learning outcomes?
- How to extract and analyze the value of student learning outcomes?
- How is the accuracy of the algorithm Naïve Bayesian Classifier to predict the value of student learning outcomes?

The purposes of this study are :

- Knowing the assessment components that can be used in predicting the value of student learning outcomes.
- Extracting and analyzing the value of student learning outcomes
- Knowing the accuracy of the algorithm Naïve Bayesian Classifier to predict the value of student learning outcomes.

2. Literature Review

2.1 Data Mining

The concept of *data mining* (DM) is extracting hidden patterns and trace relations between variables in a large data set [2]. Patterns are found used as the basis of decision making for the future. An important characteristic of DM is very large data volume [3].

DM divided into several stages:

a. Data cleansing
b. Data integration
c. Transformation of data
d. Application of DM techniques
e. Evaluation of the pattern that was found
f. Presentation of knowledge

DM includes several methods for the process of data analysis, namely classification, clustering and association rules [2]. Classification is predicting data testing based on patterns from training data. Clustering is a process to classify data that has characteristics that equal to one group, and the other to another group.

2.2 Education Data Mining (EDM)

Research Data Mining in the Education field now increases day by day. Education Data Mining is processing data from the educational field to more useful information. The information is useful for supporting to make a decision in the form of recommendations for the academic planner in university. EDM functions are cut the rate of failure, to better understand students behavior, and many benefits other [4].

Decision Tree, Neural Network, and Naive Bayes Classifiers are ordinarily used in EDM. Naive Bayes Classifier provides the highest accuracy compared to the other two algorithms. Variables of classification are student's GPA, demographics of the student, scholarships and social interaction [1].

2.3 Naive Bayesian Classifier Algorithm

The classification method used in this study is *Naïve Bayesian Classifier* (NBC). The NBC method is one of the classification methods that assume all the attributes of the examples are independent of each other in the class context. Although in general, the assumptions are bad, in practice, the NBC method shows very good performance. Probability Bayes can be used to calculate the conditional probability, the chance of occurrence if an event is known [5]. This method can predict the likelihood of members of a class based on samples from members of the class [6]. Because it is assumed to be an independent variable, only the variance of a variable in a class is needed to determine the classification, not the whole of the covariance matrix.

The first step in this classification is to calculate the average value and standard deviation of the training data features in each class [7,8]. The average value and standard deviation are used to calculate the probability $P(x_i|C_i)$

$$P(x_k \mid C_i) = g\left(x_k, \mu_{C_i}, \sigma_{C_i}\right) = \frac{1}{\sqrt{2\pi}\,\sigma_{C_i}} e^{-\frac{(x_k - \mu_{C_i})^2}{2\sigma_{C_i}^2}} \tag{1}$$

where $g\left(x_k, \mu_{C_i}, \sigma_{C_i}\right)$ is the Gaussian density function for the attribute A_k, during μ_{C_i} and σ_{C_i} is the average and standard deviation, each of which gives the attribute value A_k for the training data from the class.

After obtaining the probability $P(x_k|C_i)$ for each feature in each class, the probability value is $P(x_k|C_i)$ multiplied $P(X \mid Ci) = \pi^{n}_{(k=1)} P(x_k \mid C_i)$ so we get the probability $P(X \mid C_i)$ for each class. The probability of $P(X \mid C_i)$ multiplied by the prior probability of each class will produce a posterior probability $P(C_i \mid X)$.

$$P(C_i \mid X) = \frac{P(X \mid C_i) \quad P(C_i)}{P(x)} \tag{2}$$

The class determination of the test data follows the rules of P (C $_i$| X)> P (C $_j$| X), for $1 \leq j \leq m, j \neq i$.
Accuracy for classification results can be searched using the formula:

$$\text{Akurasi} = \frac{\sum \textit{data uji benar}}{\sum \textit{data uji}} \text{ x } 100\% \tag{3}$$

2.4 An important variable in predicting student performance

Shahiri stated that the variables most often used in predicting the performance of the student are GPA and values of college [1]. Ten of twenty-three journals reviewed by Shahiri been using GPA as a primary variable in predicting the performance of students [1].

Internal assessment tools are assignments, quizzes, practice, Middle Test, and Final Test. Furthermore, student demographic variables and external assessments. Demographics students include gender, age, and senior high school background. Men and women have a different style in the process of learning [9, 10].

3. Research Method

Figure 1 shows the research framework.

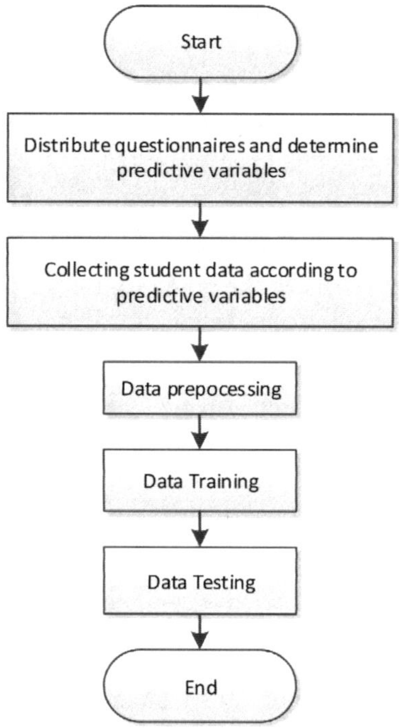

Figure 1. Research framework

3.1. Distributing questionnaires

Retrieval of data using a questionnaire is intended to know the point of view of Tama Jagakarsa University lecturers in assessing student performance. Student performance can be measured through GPA variables, internal assessment, student demographics, and external assessment. The questionnaire distributed was aimed to find out whether there were subjective factors that influenced the lecturers' assessment of students.

3.2. Collecting student data

Student data collection according to predictive variables. Student data were obtained from the academic section of Tama Jagakarsa University.

3.3. Data preprocessing and classification process

Prediction variables are normalized, so they have the same scale; this stage is called data *preprocessing*. *K-fold cross-validation* is used to divide the data into a data train and test data. The method is doing the loop as much as *k* times to divide a set of examples is randomly into *k* -subset are mutually freely. Each retest is left with a subset for testing, and the rest is used for training. The results of the *training* data *training* stage are the characteristic patterns of data that become the reference stages of *testing* the test data.

3.4. Research location

This research was conducted at Tama Jagakarsa University.

3.5. Software

The software which is used in the research:

- Microsoft Excel 2013
  Tabulation of data is done using the device software Microsoft Office Excel 2013.
- Matlab R2014b
  Processing of the data train and test data and the application of Naïve Bayesian classifier using Matlab R2014b.

3.6. Variable selection
The data value of student learning outcomes at the University of Tama Jagakarsa, Faculty of Engineering, Department of Engineering Informatics. Variables that are used as follows:

1. Value of attendance
2. Quiz Value
3. Practical Value
4. Task Value
5. Mid Test Value
6. Final Test Value

4. Results and Discussion

4.1 *Results*

Table 1. The recapitulation of the questionnaire

	Question	**Value**		
		TP	J	S
1	I will give good grades to students who always arrive on time at lecture hours	0	3	7
2	I will give fewer marks to students who often leave class without permission even though the lecture has not been completed	0	3	7
3	I will give good grades to students who are always on time in completing the assignments	0	5	5
4	I will give good grades to students who are active during lectures by either asking or answering questions	2	2	6
5	I will give good grades to students who help friends who have difficulty in lecturing material	1	4	5
6	I will give fewer marks to students who don't pay attention to lecture material	3	5	2
7	I will give good grades to students who always pay attention to lecture material	0	3	7
8	I will give good grades to students who are always polite in attending lectures	0	7	3
9	I will give good grades to students who always record responses from lecturers and classmates	0	4	6
10	I will give good grades to students who always permit when they want to leave the classroom and enter the classroom	0	8	2
	Total	6	44	50

Description:
TP: Never
J: Rarely
S: Often

4.2 Data processing stages
Data values are divided into the data train and test data using K-fold cross-validation. After that, the training data undergo a training process using the Naïve Bayesian Classifier algorithm. This process generates a training pattern. Then the process conducts a testing

IOP Conf. Series: Materials Science and Engineering **823** (2020) 012056 doi:10.1088/1757-899X/823/1/012056

process to test data. The process of testing this using a pattern that is generated by the training process. The testing process results in classification accuracy.

Tabel 2. Student grades

	Courses	2015/2016	2016/2017	Sum
1	Data Structure	23	20	43
2	Algorithm and Programming 1	43	42	85
3	Compilation Engineering	28	19	47
4	Web Programming 3	29	15	44
5	Database 1	47	42	89
6	Information Systems Modeling	26	16	42
7	Concept of Information Systems	29	42	71
8	Analysis and Design Information Systems	22	19	41
9	Algorithm and Programming 4	19	20	39
10	Computer Network	24	28	52
11	Algorithm and Programming 2	18	26	44
12	Assembly Language	23	41	64
13	Neural Network	21	15	36
	Total			697

Table 2 displays the amount of data students grades. There were 13 courses with 697 total data. The total data collected is separated according to letter values (Table 3).

Table 3. Student grade data according to letter grades

	Letter Grades	Sum
1	A	300
2	B	211
3	C	125
4	D	11
5	E	50
	Total	697

The number of data values the letter A is 300 the data or the amount of data most lots. Value letter D is the amount of data the bit, i.e., 11 data. Data grades students are divided into the data train and data test. Details of the amount of training data and data tests are presented in Table 4. Data sharing uses *K-fold cross-validation.*

Tabel 4. Data training and testing

	Data types	Sum
1	Data Training	558
2	Data Testing	139
	Total	697

Tabel 5. Data training divided by grades

	Grades	Sum
1	A	240
2	B	169
3	C	100
4	D	9
5	E	40
	Total	558

Tabel 6. Data testing divided by grades

	Grades	Sum
1	A	60
2	B	42
3	C	25
4	D	2
5	E	10
	Total	139

Data that has been divided, then processed using the Naïve Bayesian Classifier Algorithm. The results of the test data classification are shown in the Confusion Matrix (Table 7).

Table 7. Confusion matrix

		Prediction				
		A	B	C	D	E
Actual	A	58	2	0	0	0
	B	3	39	0	0	0
	C	0	0	24	0	1
	D	0	0	0	0	2
	E	0	0	0	0	10

The correct classification of grade A was 58 data and incorrect classification was 2 data. 39 data was classified into grade B, but 3 data was classified incorrectly to grade A. The correct classification of grade C was 24 data and incorrect classification was 1 data into grade E. There is no correct classification in grade D, and all of the data classified to grade E. All of data grade E had correct classification.

$$accuracy = \frac{Sum\_of\_true\_classified}{Sum\_of\_data} * 100\%$$

$$accuracy = \frac{131}{139} * 100\%$$

$$accuracy = 94.2446\%$$

So, the accuracy of the classification results is 94.2446%.

IOP Conf. Series: Materials Science and Engineering **823** (2020) 012056 doi:10.1088/1757-899X/823/1/012056

5. Conclusions and Suggestions

DHDNS component values consist of Presence, Tasks, Mid Test, and Final Test. The level of accuracy of the classification of 94.2446% by using a measurement of the effectiveness Confusion Matrix. For the development, can add the amount of data, so that the pattern of values is more varied.

Acknowledgments

Thank you to DPRM Higher Education through a grant DIKTI PDP has financed this research.

References

[1] Shahiria A M, Husaina W and Rashida N A 2015 The third information systems international conference a review on predicting student's performance using data mining techniques *Procedia Computer Science* **72** 414-22

[2] Ahmad F, Ismail N H and Aziz A A 2015 The prediction of students' academic performance using classification data mining techniques *Journal Applied Mathematical Sciences HIKARI* **9**(129) 6415–26

[3] Hermawati F A 2013 *Data Mining* (Yogyakarta: CV. ANDI OFFSET)

[4] Abuteir M M and El-Halees A M 2012 Mining educational data to improve students' performance: a case study *International Journal of Information and Communication Technology Research* **2**(2) 140-6

[5] Manning C D, Raghavan P and Schutze H 2009 *An Introduction to Information Retrieval* (Cambridge: Cambridge University Pr.)

[6] Rosen G, Garbarine E, Caseiro D, Polikar E and Sokhansanj 2008 Metagenome fragment classification using n-mer frequency profiles *Advances in Bioinformatics*

[7] Han J and Kamber M 2001 *Data Mining: Concepts and Techniques* (San Francisco: Morgan Kaufmann Publishers)

[8] Pekuwali A, Kusuma W A and Buono A 2018 Optimization of spaced k-mer frequency feature extraction using genetic algorithms for metagenome fragment classification *Journal of ICT Research and Applications* **12**(2) 123-37

[9] U. bin Mat, Buniyamin N, Arsad P M and Kassim R 2013 An overview of using academic analytics to predict and improve students' achievement: A proposed proactive intelligent intervention *IEEE 5th Conference on Engineering Education (ICEED)* 126–30

[10] Meit S S, Borges N J, Cubic B A and Seibel H R 2007 *Personality Differences in Incoming Male and Female Medical Students* Available from: https://files.eric.ed.gov/fulltext/ED490525.pdf

IOP Conf. Series: Materials Science and Engineering **823** (2020) 012057 doi:10.1088/1757-899X/823/1/012057

Simulation and parametric study of the innovated process to purify bioethanol with ethylene oxide hydration as an auxiliary reaction

I G P Wiratama[1], V Novianty[1]

[1]Department of Chemical Engineering, Faculty of Industrial Technology, Parahyangan Catholic University, Bandung, 40141, Indonesia

E-mail: pandega.wiratama@unpar.ac.id

Abstract. By the conventional process, to purify bioethanol from fermentation reactor or bioreactor is relatively high cost, since the purity of bioethanol in the bioreactor output is very low (just around 8-14%) and there are azeotropes between bioethanol with water. Therefore, in the conventional process, a combination of distillation-adsorption, distillation-pervaporation, or extractive distillation is needed in order to get bioethanol with purity of 99.6% or above. In this research, an innovative process to purify bioethanol in a reactive distillation column with an auxiliary reaction of ethylene oxide hydration was proposed. The auxiliary reaction broke the water/bioethanol azeotrope. Impacts of the reflux ratio and the bottoms rate were investigated by simulating the process in Aspen Plus software. The results revealed that the innovated bioethanol purifying process theoretically could produce bioethanol with a purity of 99.9%.

1. Introduction

Bioethanol is one of promising biofuels that is obtained from renewable sources, namely, biomass and is environmentally friendly [1]. Bioethanol is produced by fermentation of starch from biomass. However, the purity of bioethanol in the bioreactor output is very low (just around 8-14%), because above this concentration, inhibition of yeasts may occur that reduces their activity [1]. Meanwhile, based on PN A 79521 standard, bioethanol as fuel must have minimum purity of 99.6% [2]. Therefore, the bioethanol purification step after the fermentation process is an important issue.

The main components in the output stream of bioreactor/ fermentation process are bioethanol and water. There is azeotrope between these components as shown in Table 1. This azeotrope makes the separation of them difficult. In the conventional process, several methods can be used to separate bioethanol from water, such as the combination of distillation-adsorption, distillation-pervaporation, or extractive distillation [3]. However, the cost of these methods is relatively high since they need at least two columns and the addition of adsorbent, membrane, or solvent. Therefore, in this research, an innovative process to separate bioethanol from water in a single column of reactive distillation was proposed. The idea of this innovative process was to remove water by an auxiliary reaction of ethylene oxide hydration.

Ethylene oxide (EO) could be reacted with water and produce ethylene glycol (EG) as

$$C_2H_4O + H_2O \rightarrow (CH_2OH)_2 \quad (1)$$

IOP Conf. Series: Materials Science and Engineering **823** (2020) 012057 doi:10.1088/1757-899X/823/1/012057

The present of this auxiliary reaction is worthwhile, because the product, EG, is also a valuable chemical which is used as an antifreeze and coolant, in hydraulic fluids, and in the manufacture of low-freezing dynamites and resins [4]. There is also no azeotrope between EG and bioethanol (EtOH) as shown in Table 1, so the separation process is relatively easy and could be conducted in a single column of reactive distillation.

Table 1. System azeotrope

Temp (°C)	Mole Fraction			
	EtOH	Water	EO	EG
78.15	0.8952	0.1048	0	0

This study is a part of research about innovated bioethanol purification. In this study, the innovated bioethanol purification was simulated using Aspen Plus to investigate its feasibility and to study the effects of the process parameters in a steady-state condition. Based on the result of this study, the next parts of the research, which are about process design optimization, dynamic behavior, and process control, can be conducted.

2. Base Simulation

2.1. Kinetic model

The auxiliary reaction of EO hydration may also produce an unwanted by-product of diethylene glycol and triethylene glycol. However, the rate of the side reaction is much lower than that of the main reaction and consequently, the effect of this side reaction can be neglected [5]. The kinetic model of EG production from EO hydration is adapted as follows [5]

$$r(mol.cm^{-3}.s^{-1}) = 3.15 \times 10^{9} \exp\left(-\frac{9547}{T}\right) x_{EO} x_{water} \quad (2)$$

2.2. Process simulation

Innovated bioethanol purification was simulated in ASPEN Plus software, using a reactive distillation (RD) column as shown in Figure 1. The properties method that was used in this study was NRTL (Non-Random Two Liquid), which has been confirmed to be accurate in simulating a system containing EtOH, water, EO, and EG [5, 6].

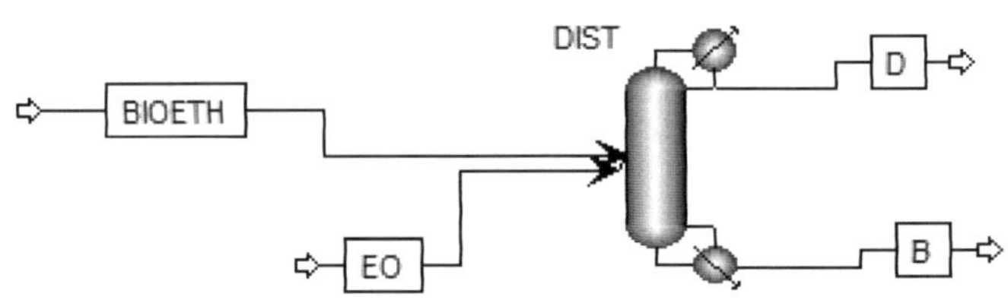

Figure 1. Reactive distillation for innovated bioethanol production

The process variables that were examined were reflux ratio and bottoms rate. The method of examination was by manipulating these two variables simultaneously with the range of 16-20 for reflux ratio and 6.6-8.6 kmol/h for bottoms rate. Other variables besides these two variables were set to constant at the initial condition of simulation. The initial conditions of simulation are shown in Table 2.

Reflux ratio and bottoms rate were examined their effects on the reaction, separation, and economic aspects of the process. The EO conversion was used to quantitatively describe the reaction aspect, while the economic aspect was described quantitatively by the reboiler duty, since it is connected to the operational cost. To describe separation aspect, two variables were used. They were the purity of EtOH in distillate stream and the purity of EG in bottom stream.

Table 2. Initial conditions of simulation

Parameters	Value	Units
Feed 1 (BIOETH)		
EtOH mole-flow	1.4	kmol/h
Water mole-flow	8.6	kmol/h
Temperature	25	°C
Feed 2 (EO)		
EO mole-flow	8.6	kmol/h
Temperature	25	°C
RD Column		
Enriching stages	1-4	-
Reactive stages	5-35	-
Stripping stages	36-40	-

Note that the initial conditions are not the optimum conditions. The optimum conditions will be covered in the next study.

3. Results and Discussion

3.1. The effect of reflux ratio and bottoms rate to the EO conversion

The simulation results from aspen were transferred to Matlab software to make a graph as shown in Figure 2. From the graph, it can be seen that when the reflux ratio was low, the increase in bottoms rate from 6.6 to around 8.1 kmol/h also increased the EO conversion. This was because the increase in bottoms rate forced much more EG to leave the column via bottom stream and maintained the water fraction in the column high. As can be seen in equation (4), the higher water fraction in the column increased the EO hydration rate. Therefore, the EO conversion was also increased.

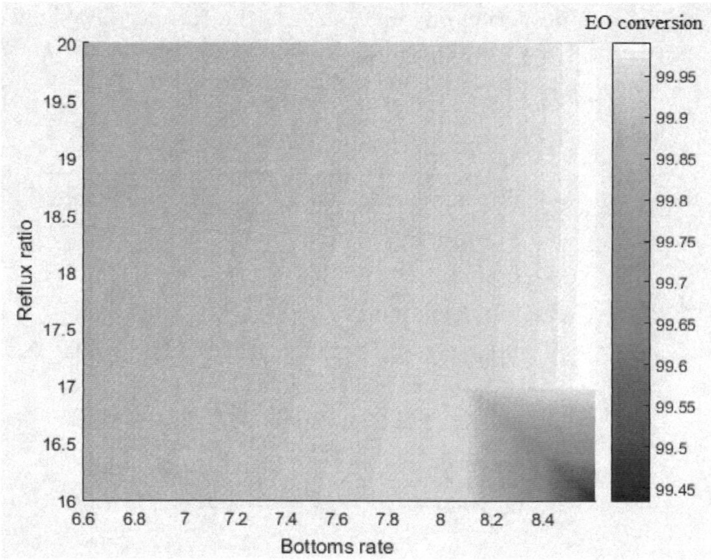

Figure 2. The effect of reflux ratio and bottoms rate to the EO conversion

However, if the bottoms rate was increased above 8.1 kmol/h when the reflux ratio was low, the EO conversion was decreased. This was because when the reflux ratio was low, there was less water that being returned into the column, resulting in the depletion of EO conversion or EG formation. The EG formation at this condition did not catch up with the value of bottoms rate that was 8.1 kmol/h or above. Therefore, water as the second heaviest component was also forced to leave the column with the EG via bottom stream. This phenomenon decreased the fraction of the water in the column, resulting in the further depletion of EO conversion.

At the higher reflux ratio, when the bottoms rate was increased, the EO conversion was also increased. When both of the reflux ratio and the bottoms rate were high, there was much water that being returned into the column and almost all EG was forced out from the column without dragged the water. This made the EO conversion became higher. The highest EO conversion (almost 100%) was achieved when the reflux ratio was 20 and the bottoms rate was 8.6 kmol/h.

3.2. The effect of reflux ratio and bottoms rate to the EG purity

It can be seen from Figure 3, that most of the variations in reflux ratio and bottoms rate made the EG purity reached 100%. This was because EG was the heaviest component in the system so it was naturally forced down to the bottom without dragged any other components, except when the reflux ratio was lower than 17 and the bottoms rate above 8.1 kmol/h.

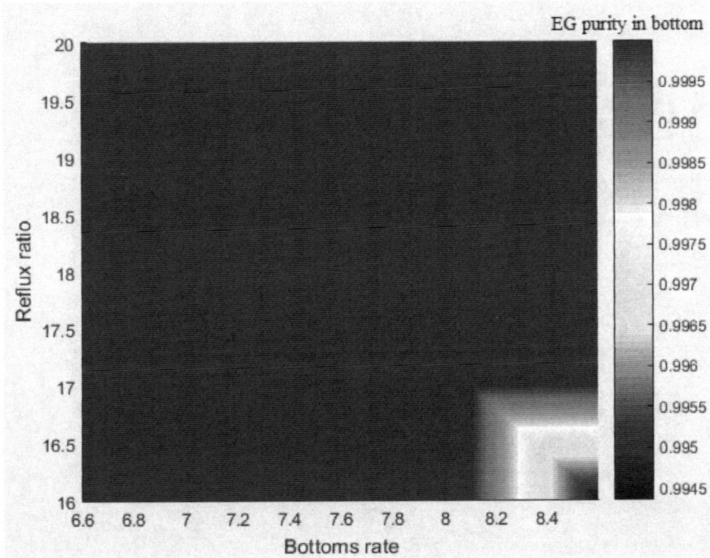

Figure 3. The effect of reflux ratio and bottoms rate to the EG purity in the bottom

As explained in the previous discussion of EO conversion, at the lower reflux ratio, the EG formation was also low and did not catch up with the value of bottoms rate that was 8.1 kmol/h or above. Therefore, water as the second heaviest component was also forced to leave the column with the EG via bottom stream, resulting in the depletion of EG purity in the bottom.

3.3. The effect of reflux ratio and bottoms rate to the EtOH purity

As shown in Figure 4, the reflux ratio did not significantly affect the EtOH purity in distillate, because EtOH was the second lightest component in the system, so even the reflux ratio was increased, the EtOH was evaporated again before it reached the bottom and came out as distillate product. This result was synchronous with the result in Figure 5, where the molar flow rate of EtOH at any variations of the reflux ratio and bottoms rate reached the value of 1.4 kmol/h. The EtOH in the feed stream also had a flow rate of 1.4 kmol/h. Therefore, it could be concluded that at any variations of the reflux ratio and bottoms rate, all of EtOH came out from the column via distillate stream.

In addition, within the range of variations, EO conversion reached above 99%, resulting in only small amount of EO that came out with the EtOH in distillate. Therefore, the EtOH purity in distillate was not affected by reflux ratio.

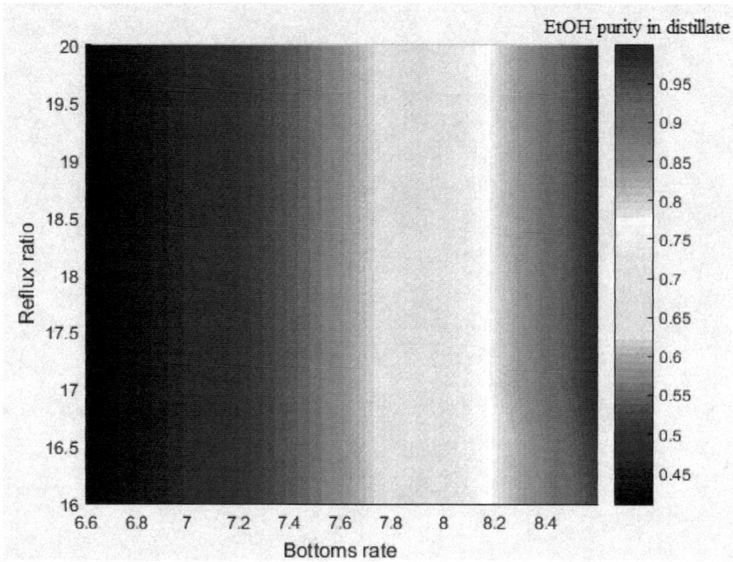

Figure 4. The effect of reflux ratio and bottoms rate to the EtOH purity in distillate

The EtOH purity was significantly affected by the bottoms rate. It can be seen in Figure 6, the decrease in the bottoms rate made some amount of EG was forced out from the column via distillate stream. This phenomenon decreased EtOH purity in distillate as it can be seen in Figure 4.

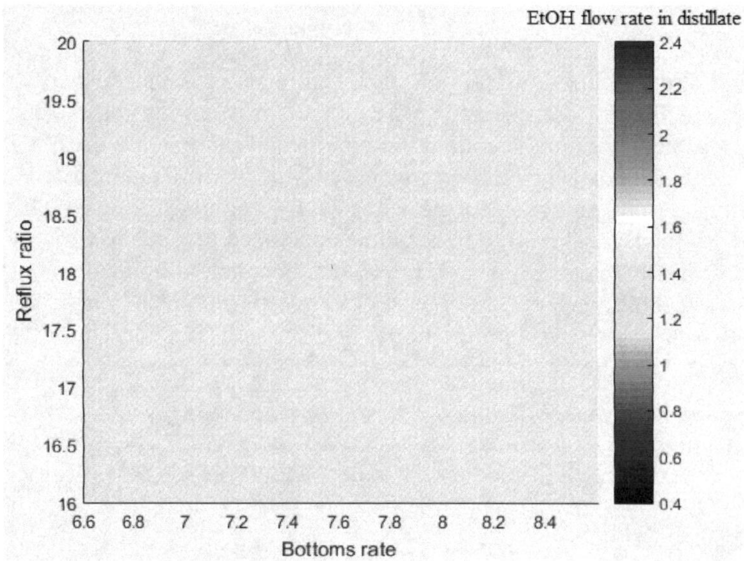

Figure 5. The effect of reflux ratio and bottoms rate to the EtOH molar flow rate in distillate

3.4. The effect of reflux ratio and bottoms rate to the reboiler duty

It can be seen in Figure 7, the increase in reflux ratio increased reboiler duty. This was because when the reflux ratio was increased, there was much more liquid that being returned into the column. Therefore, much more reboiler duty was needed to boil the liquid.

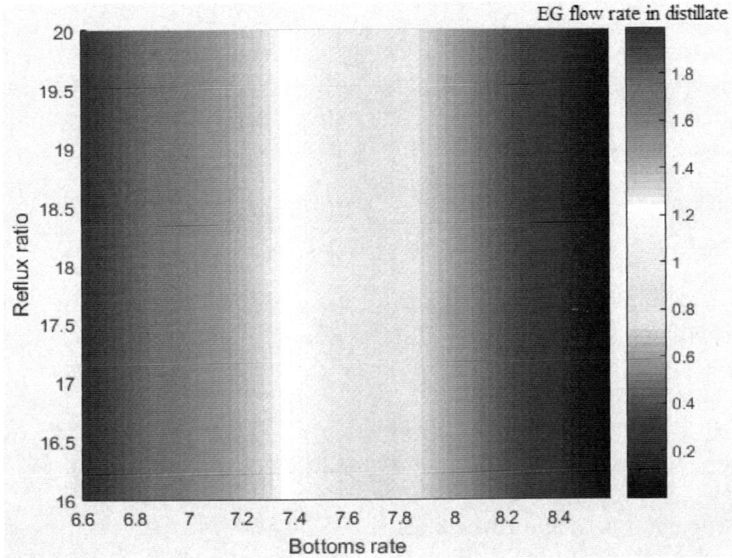

Figure 6. The effect of reflux ratio and bottoms rate to the EG molar flow rate in distillate

Meanwhile, the increase in bottoms rate decreased reboiler duty. This was because when the bottoms rate was increased, there was a less amount of liquid that needs to be boiled up. Therefore, the reboiler duty also decreased.

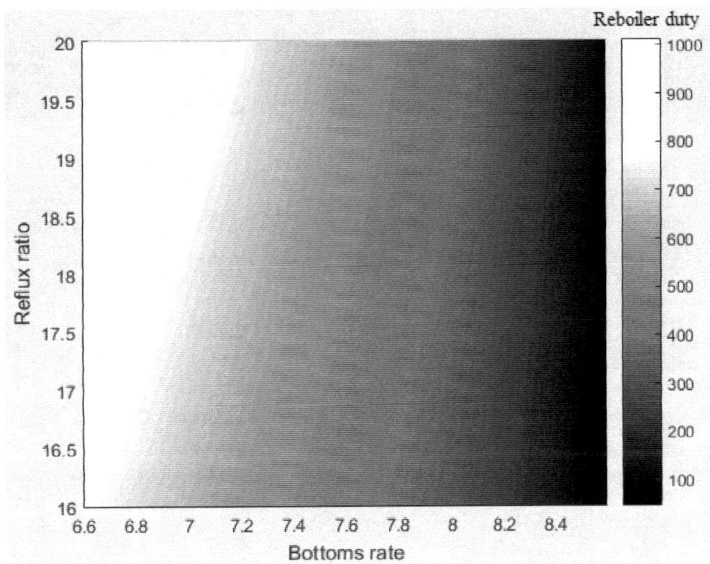

Figure 7. The effect of reflux ratio and bottoms rate to the reboiler duty

Based on the result in Figure 7, the best operating condition for this process was the reflux ratio of 20 and the bottoms rate of 8.6 kmol/h, because it gave the lowest reboiler duty, EO conversion of almost 100% and EtOH purity of 99.9%.

4. Conclusions

The simulation results revealed that the innovated bioethanol purification with auxiliary reaction of EO hydration in RD column is highly prospective. It gives very high hydration conversion (almost 100%) and very high purity of bioethanol (up to 99.9%). These results are achieved when the reflux ratio is 20 and the bottoms rate is 8.6 kmol/h.

5. References

[1] Vohra M, Manwar J, Manmode R, Padgilwar S and Patil S 2014 *J. Environ. Chem. Eng.* **2** 573-84
[2] Rutz D and Janssen R 2006 *Overview and Recommendations on Biofuel Standards for Transport in the EU* (Munchen: WIP Renewable Energy) p 8
[3] Frolkova A K and Raeva V M 2010 *Theoretical Foundations of Chemical Engineering* **44** 545-56
[4] National Center for Biotechnology Information *PubChem Database 1,2-Ethanediol, CID=174* Available from: https://pubchem.ncbi.nlm.nih.gov/compound/174 (accessed on Mar. 26, 2019)
[5] Tavan Y, Behbahani R M and Hosseini S H 2013 *Chem. Eng. Process. Process Intensif.* **73** 81-6
[6] Nabgan W, Saeh I, Abdullah T A T, Nabgan B and Mat R 2016 *International Journal of Environmental Research & Clean Energy* **1** 1-3
[7] Luyben W L and Yu C C 2008 *Reactive Distillation Design And Control* (Hoboken: John Wiley & Sons, Inc.) p 2-11
[8] Altiokka M R and Akyalcin S 2009 *Ind. Eng. Chem. Res.* **48** 10840-4
[9] Ali S H and Merchant S Q 2006 *Int. J. Chem. Kinet.* **38** 593-612
[10] Kumar A and Daoutidis P 2004 *AIChE. J.* **45** 51-68

IOP Conf. Series: Materials Science and Engineering **823** (2020) 012058 doi:10.1088/1757-899X/823/1/012058

$FeCl_3$ coagulant production from waste pickle liquor using electrolysis

T P Adhi[1,2], A Raksajati[1,2], A A Nugroho[2] and P Y Sitanggang[2]

[1] Department of Chemical Engineering, Faculty of Industrial Technology, Institut Technologi Bandung, 10th Ganeca street, Bandung 40135, Indonesia

E-mail: tpadhi@che.itb.ac.id

Abstract. Pickling is a typical method used in a cold milling plant to remove impurities from hot rolled coil by immersing the coil in strong acidic solution. The spent acidic solution or waste pickle liquor (WPL) requires certain waste treatment before it can be disposed to environment because of its metallic ion and acid content. WPL contains a number of metallic ions that can be utilized as ferric chloride coagulant. The purpose of this paper is to evaluate the kinetics of Fe^{2+} oxidation into $FeCl_3$ and the performance of the ferric chloride coagulant using the proposed electrolysis configuration. The electrolysis process was conducted in ambient condition with voltage 3 & 6 V, electrolysis time of 120 minutes, and addition of excess NaCl or HCl. The result shows that metal impurities in WPL and excess chlorine do not significantly affect the Fe^{2+} conversion. It can also be concluded that the optimum condition for electrolysis is using 6V of voltage, 90 minutes of electrolysis time, and excess chlorine at 0% stoichiometry which gave 60% conversion of Fe^{2+} in WPL. Due to the equilibrium between the Fe^{2+} dan Fe^{3+} which is formed during electrolysis process, the full conversion of Fe^{2+} is not possible yet, and the maximum conversion of Fe^{2+} is 60%. The presence of impurities in WPL is proven not affecting the Fe^{2+} conversion during the electrolysis process.

1. Introduction

The finishing stage of the steel industry consists of pickling, galvanizing and plating process, which aims to remove impurities in the form of dust, scale, and rust on the steel [4]. Pickling process is a process in which the steel is immersed in concentrated acid to remove rust from the steel surface [1]. The pickling process will produce liquid waste in the form of waste pickle liquor (WPL) in large quantities. WPL contains metallic ion and acid content that classified it as hazardous waste for the environment [6]. In this research, the WPL feed contains 120 g/L of Fe^{2+} and 40 g/L of HCl. To prevent environmental pollution, direct processing by steel industries or third parties is required before WPL can be disposed into the environment [11].

Commonly WPL treatment carried out by the steel industries aims to recover the acids and the dissolved metallic ions [8]. The recovery of the acids and metallic ions in WPL can be done through various methods. The treatment methods for WPL are pyrometallurgy, crystallization, separation using membrane, and liquid extraction [13].

Another alternative for WPL treatment is to use an oxidation method, specifically the oxidation of $FeCl_2$ in WPL that is produced from steel pickling process with HCl solution. To oxidize $FeCl_2$ in WPL,

IOP Conf. Series: Materials Science and Engineering **823** (2020) 012058 doi:10.1088/1757-899X/823/1/012058

oxidizer that can be used also diverse, but the most economical is to use Cl_2 gas. Oxidation of $FeCl_2$ to $FeCl_3$ by using Cl_2 gas is more economical compared to other oxidizing agents because WPL contains Cl compound which can be oxidized to Cl_2 gas to minimize treatment costs. Moreover, other oxidizer such as H_2O_2 is expensive and Cl_2 gas is quite difficult in handling process if used. Previous study from Pradana had proven the feasibility for electrolyzing WPL, but the dosage from the produced $FeCl_3$ are 155 and 300 ppm [11]. The aim of this paper is to determine the effect of metal impurities with Fe^{2+} conversion in WPL, determine the optimum equipment set up and condition for WPL electrolysis, and producing $FeCl_3$ coagulant with higher performance.

2. Methodology

This research consists of 5 stages, the first stage is making a feed solution to be electrolyzed, the second stage is electrolyzing feed solution, and the third stage is concentrating the sample through distillation. The fourth stage is determining the concentration of ferro (Fe^{2+}) and ferric (Fe^{3+}) ions from each sample, and the last stage is jar test to determine the quality of the $FeCl_3$ coagulant produced.

The materials required for the experiment are NaCl salt, H_2SO_4, H_3PO_4, HCl, $SnCl_2$, $HgCl_2$, $K_2Cr_2O_7$, $Na_2S_2O_3$ 0,1 N, and KI from Merck, mixed acid which was created from mix of H_2SO_4 and H_3PO_4 each at 5% with ratio of 1:1, and demineralized water.

Feed solution is made by mixing WPL with NaCl or HCl. NaCl and HCl are used to increase the content of chlorine compounds in the feed (excess). Therefore, the oxidation rate will not be affected by the concentration of chlorine compounds, and the oxidation rate can be assumed to be influenced only by Fe^{2+} concentration. In order to create the electrolysis feed, WPL is mixed with HCl or NaCl in a beaker and stirred for 15-20 minutes. The finished feed is then electrolyzed using the configuration in figure 1. The electrolysis process is done in 120 minutes with 10 minutes interval time for sampling.

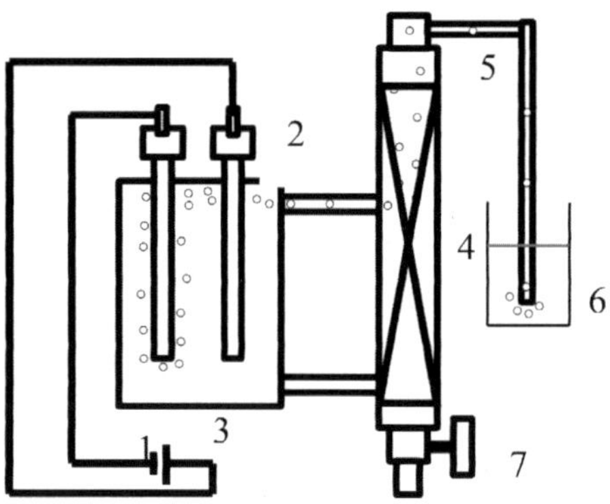

Notes:
1. DC electrical source
2. Platinum electrodes
3. Electrolysis reactor
4. Liquid gas contactor with *raschig ring*
5. Hose
6. Gas collector
7. Sample collector

Figure 1. The electrolysis configuration

The gas collector is used to dissolve the excess Cl_2 in water. The electrolysis samples are then distilled. The purpose of the distillation is to increase the $FeCl_3$ concentration in the sample to meet the commercial criteria for $FeCl_3$ coagulant. The distillation process is carried out using a simple distillation flask where the sample is heated in a boiling flask and the evaporated water is cooled with a condenser. The distillation process will reduce the volume up to 60% sample's volume. To analyze the content of Fe^{2+} and Fe^{3+} ions in the sample, an analysis of redox titration is performed using potassium dichromate for $FeCl_2$ and sodium thiosulfate for $FeCl_3$. The last stage is to perform jar test in order to determine the quality of $FeCl_3$ coagulant.

3. Result and Discussion

3.1. Electrolysis configuration

The proposed electrolysis configuration had been tested on various models. The process variables that had been studied were the amount of oxidation process needed, the use of contactor column, electrodes type, the source of excess chlorine, and whether or not the electrolyzed product is distilled.

3.1.1. Oxidation process. The oxidation process of Fe^{2+} compounds in WPL can be done through one or two stages of oxidation. The oxidant used in the one-stage-oxidation is Cl_2 gas produced by electrolysis of Cl^- ions in WPL (with excess chlorine). In two-stage oxidation, after oxidation with chlorine gas, the remaining Fe^{2+} is further oxidized by adding NaOCl to the electrolysis product. The coagulant performance from both methods can be seen in Table 2.

Table 1. One and two stages coagulant performance

Sample	Dose (ppm)	pH	TDS (µS/cm)	Turbidity (NTU)
One-stage	60	5.4	123.1	0.86
Two-Stage	110	5.81	169.6	1.6

From Table 1, $FeCl_3$ coagulant produce by a single-stage oxidation process not only able to reduce TDS and turbidity from the treated water in jar test lower than the two-stages, but also has lower dose than the two-stage.

3.1.2. Contactor column. The function of the contactor column is to prolong the contact time between Cl_2 gas produced from the electrolysis process with $FeCl_2$ in order to increase Fe^{2+} conversion. The result of $FeCl_3$ coagulant which is produced with and without contactor column can be seen in Table 2 and Table 3. The coagulant produced using contactor column (filled by raschig ring) results in higher conversion and coagulant performance because of the packing increase both reaction time and the contact between chlorine gas with WPL.

Table 2. Effect of contactor column

Sample	Fe^{2+} (g/L)	Fe^{3+} (g/L)	Conversion (%)
Without column	51.00	72.02	58.54
With column	44.28	67.53	60.4

Table 3. Effect of packing

Sample	Dose (ppm)	pH	TDS (µS/cm)	Turbidity (NTU)
Without packing	250	5.62	206	1.4
With packing	200	5.4	166	1.5

3.1.3. Electrodes type. The type of electrodes in an electrolysis reaction needs to be considered since some electrodes are inert. The ability of certain electrodes in conducting electrons also affects the conversion of Fe^{2+} during electrolysis. Table 4 shows the performance of coagulants produced using iron and platinum electrodes after being electrolyzed for 10 minutes.

Table 4. Effect of electrode

Sample	Fe^{2+} (g/L)	Fe^{3+} (g/L)	Conversion (%)
Iron electrode	80.36	24.42	23.30
Platinum electrode	68.90	43.60	38.76

Based on Table 4, the conversion from platinum electrode is higher than iron electrode. In the electrolysis process using iron electrode, there is a side reaction of oxidation Fe into Fe^{2+}. This reaction has higher selectivity than the oxidation of Cl^-, so the amount of Fe^{2+} will accumulate in electrolysis reactor overtime. Though it seems like the amount of $FeCl_3$ produced will increase, but on the contrary, the number of electrons which is needed to oxidize Cl^- is decreasing and lower the Cl_2 gas production. In platinum electrode, this phenomenon does not occur because of its inert nature. Based on those reasons, platinum electrodes are selected as electrodes to be used in the electrolysis process.

3.1.4. Chlorine sources. Table 5 and 6 below show the effect of excess chlorine from HCl and NaCl.

Table 5. Effect of chlorine sources and feed solution pH

No	Sample	Fe^{2+} (g/L)	Fe^{3+} (g/L)	Conversion (%)
-	WPL + NaOH	80.26	31.27	28.04
-	WPL+ HCl 100 ml	62.03	43.42	41.18
1	WPL + HCl 320 ml	39.47	40.32	50.53
2	WPL +NaCl excess 10%	73.31	51.32	41.18
-	WPL +NaCl excess 75%	69.36	40.32	36.76

Table 6. Coagulant performance from excess HCl and NaCl

Sample Number	Dose (ppm)	pH	TDS (μS/cm)	Turbidity (NTU)
WPL only	350	3.14	590	2.05
1	300	3.20	482	2.9
2	250	3.61	257	2.8

Based on Table 5, the addition of 100-320 ml HCl has a positive effect on Fe^{2+} conversion, while the addition of NaCl above 10% excess decreases the Fe^{2+} conversion. The addition of base (NaOH) to WPL reduced the conversion of Fe^{2+}. The conversion reduces because according to the Pourbaix diagram, in order to form Fe^{3+} the feed solution must be in acidic condition [2], and the base addition will make it harder for $FeCl_3$ to be formed. As shown in Table 6, both TDS and turbidity increases from the jar test result. This indicates that the addition of excessive chlorine sources will leave residual in the coagulant product. The phenomenon should be avoided to prevent the coagulant performance from decreasing due to the impurity which is carried. The effect of excessive addition of HCl is that the coagulant product will be too acidic, while excessive NaCl increases the TDS on the treated water.

3.1.5. Product concentration. Table 7 shows the difference on performance from electrolyzed samples with and without product distillation. The distillation process is able to increase the concentration of $FeCl_3$ in the product, hence it requires a lower dose than the non-distilled samples in the jar test.

Table 7. Effect of product concentration

Sample	Dose (ppm)	pH	TDS (µS/cm)	Turbidity (NTU)
Feed 1	250	5.62	206	1.4
Feed 2	200	5.40	166	1.5
Distilled feed 1	80	5.65	147.4	1.2
Distilled feed 2	60	5.40	123.1	0.86

3.2. Kinetic study

3.2.1. Kinetic profile of Fe^{2+} oxidation. Figure 2 shows the kinetic profile from the oxidation process of Fe^{2+} into Fe^{3+}.

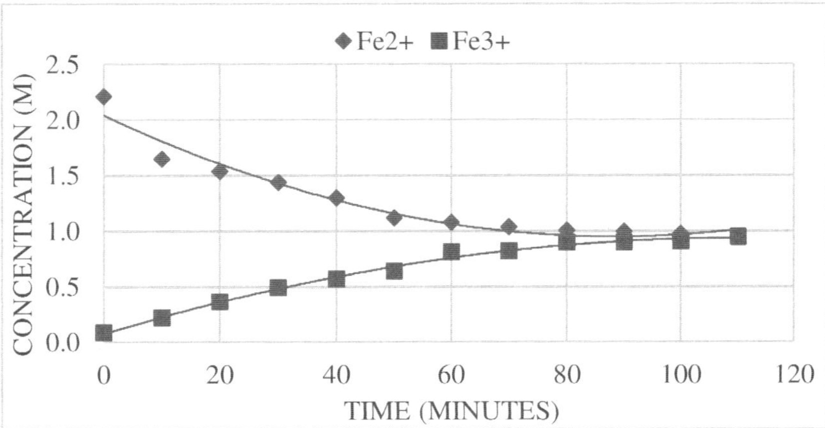

Figure 2. The kinetic profile from Fe^{2+} oxidation

As shown in Figure 2, the kinetic profile of Fe^{2+} oxidation represents a 1st order reaction. The concentrations of Fe^{2+} and Fe^{3+} do not change after the electrolysis process is carried out for more than 90 minutes. This can be seen from the non-intersecting Fe^{2+} and Fe^{3+} concentration curves, in addition for samples which are electrolyzed for more than two hours, the conversion is similar with the result from the two-hour electrolyzed samples. It means that the oxidation process will reach an equilibrium at certain Fe^{2+} and Fe^{3+} concentrations, so the 100% conversion of $FeCl_2$ into $FeCl_3$ cannot be achieved. In addition, experiments to electrolyze WPL in more than 2 hours with higher voltage had been carried out, however, the highest conversion of $FeCl_2$ that can be achieved was only around 60%. Figure 3 below shows the relation between voltage and Fe^{2+} conversion.

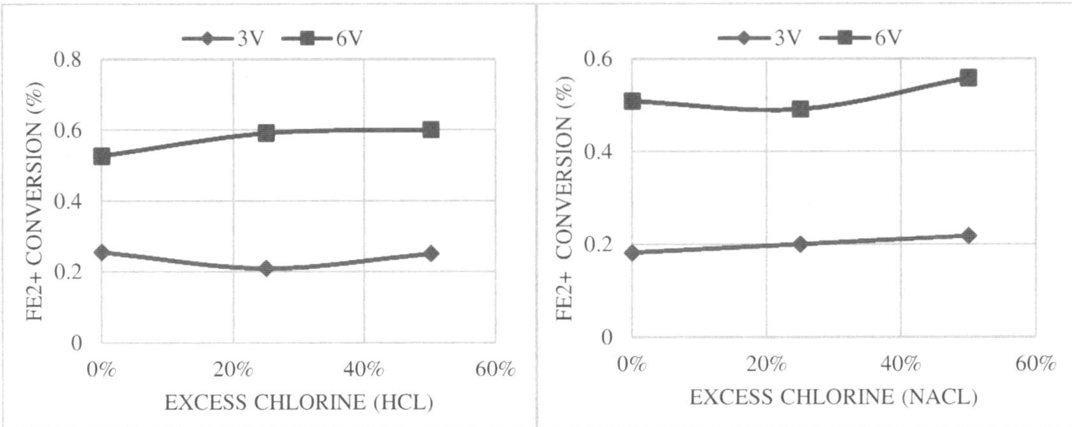

Figure 3. The effect of voltage to Fe^{2+} conversion

Higher voltage during electrolyzing process gives higher conversion since more electron are transferred and more Cl_2 created. However, higher voltage also means more thermal loss due to thermal dissipation whether from the DC generator, or the electrodes.

3.2.2. Excess chlorine effect. The data from the main experiments in this research is shown in Table 8.

Table 8. Effect of product concentration

Voltage	Excess Variation	Conversion
3 V	0% NaCl	18.2%
	25% NaCl	20.0%
	50% NaCl	21.8%
	0% HCl	25.5%
	25% HCl	20.9%
	50% HCl	25.0%
6 V	**Synthetic WPL 0% NaCl**	51.8%
	0% NaCl	50.9%
	25% NaCl	49.1%
	50% NaCl	55.9%
	0% HCl	52.7%
	25% HCl	59.1%
	50% HCl	60.0%

As shown in Table 8, the excess variation of 0%, 25%, and 50% HCl for 3 and 6V voltage have relatively similar conversion. This shows that the conversion of $FeCl_2$ during the electrolysis process is not significantly affected by the excess chlorine condition. The optimum excess chlorine more likely in saturated chlorine condition (0% excess) rather than 25 or 50% excess stoichiometry. The amount of chlorine needed to saturate a WPL solution is influenced by temperature and the HCl content in WPL [10].

3.2.3. Effect of metal impurities in WPL. Alongside with Fe ions, there are other ions such as Sn, Ni, Cu, and others. The impurities in WPL do not affect the Fe^{2+} conversion, as shown by results in Table 9.

Table 9. Effect of impurities

Voltage	Excess Variation	Conversion
	Synthetic WPL 0% NaCl	51.8%
6 V	0% NaCl	50.9%
	0% HCl	52.7%

The synthetic WPL was made from pure $FeCl_2$ which is dissolved in demineralized water. HCl was then added to duplicate the concentration from the Fe^{2+} ions and the acidic concentration. The result proves that the presence of impurities in WPL does not affect the Fe^{2+} conversion. If the impurities in WPL inhibit the oxidation reaction, the conversion of the synthetic WPL should be lower than the saturated WPL.

3.3. Coagulant performance

Samples chosen to be tested in jar test were the excess 25% HCl, 50% HCl, 50% NaCl, and the synthetic WPL. The concentration of $FeCl_3$ coagulant in each sample is shown in Table 10.

Table 10. $FeCl_3$ concentration in samples

Variation	$[Fe^{3+}]$ before distillation (g/L)	$[Fe^{3+}]$ after distillation (g/L)
50% HCl	72.86	121.85
25% HCl	73.91	112.31
50% NaCl	63.86	60.78
Synthetic WPL	45.37	75.58

The coagulant concentration in samples increases due to the distillation except for 50% NaCl sample. When the distilled sample was cooled, the NaCl solubility decreased and the dissolved NaCl in the sample crystallized. Because of the crystalized NaCl, the separation process was difficult for the distilled sample even after using a vacuum filter. To aid the separation process, demineralized water is used. Therefore, there was a decrease in the concentration of the Fe^{3+} coagulant in the 50% NaCl sample. The result from the jar testing on those samples is shown in Table 11.

Table 11. Results of jar test

Sample	Dose (ppm)	pH	TDS (μS/cm)	Turbidity (NTU)
Target	-	7-8	<250	<3
Dirty water	-	7.52	304	43
50% HCl	125	2.61	2.10	27
25% HCl	125	2.55	2.32	30
50% NaCl	100	2.50	2.33	10
Synthetic WPL	25	5.70	3.01	<1

From the jar test analysis, the $FeCl_3$ coagulant samples had been able to reduce TDS in dirty water in accordance with the required target of a commercial coagulant. However, as shown in Table 11 there is no sample from original WPL which able to meet the turbidity standard. The turbidity standards can only be met by coagulant sample produced from synthetic WPL. In addition, the treated waters pH was

IOP Conf. Series: Materials Science and Engineering **823** (2020) 012058 doi:10.1088/1757-899X/823/1/012058

around 3 except for synthetic WPL samples. The low pH from the treated water was caused by HCl residuals which were present in the distilled $FeCl_3$ coagulant. To overcome this problem further separation is needed to separate the HCl residuals from the $FeCl_3$ coagulant.

4. Conclusion

The optimum configuration to electrolyze WPL is one-stage oxidation using platinum electrodes and contactor column which packed by raschig ring. The electrolysis process parameters are 6V voltage, electrolysis time is 90 minutes, and excess chlorine at 0% using NaCl as the chlorine source. In order to increase the coagulant performance and lower the usage dose of the coagulant, distillation needs to be done to the electrolyzed product.

Due to the equilibrium between the Fe^{2+} dan Fe^{3+} which is formed during electrolysis process, the full conversion of Fe^{2+} is not possible yet, and the maximum conversion of Fe^{2+} is 60%. Though the excess of chlorine does not affect the conversion significantly, the source of chlorine must be considered because it will affect the coagulant performance. The presence of impurities in WPL was not affecting the Fe^{2+} conversion during the electrolysis process and though produced $FeCl_3$ has lower the dosage, further study is required to enhance the $FeCl_3$ coagulant because the pH and turbidity have not been met.

References

[1] Archana, Agrawal and Sahu K K 2009 An overview of the recovery of acid from spent acidic solutions from steel and electroplating industries *Journal of Hazardous Materials* **171** 61–75
[2] Channei 2017 Aqeous and surface chemistries of photocatalytic Fe-doped CeO_2 nanoparticles *Catalysts* **7** 2073-4344
[3] Crabtree 1996 The oxidation of Iron(II) by chlorine *Inorganic Chemistry* **5** 1348-51
[4] Devi and Singhal 2014 A study on treatment methods of spent pickling liquor generated by pickling process of steel *Clean Technologies and Environmental Policy* **16** 1515–27
[5] Farrington and Schaefer 1961 Coulometric titration of Iron(II) with chlorine *Analytical Chemistry* **33** 0003-2700
[6] Fitriani 2015 Estimasi Abatement Cost dan Efisiensi Pengolahan Limbah Cair Reject Treatment Plant (RTP) Cold Rolling Mill (CRM) PT. Krakatau Steel Cilegon *Skripsi Sarjana* (Bogor: Institut Pertanian Bogor)
[7] Green 1907 The effect of ferric salts on the rate of oxidation of ferrous salts and on the catalytic action of the latter *The Journal of Physical Chemistry* **12** 0092-7325
[8] Ozdemir 2006 Treatment of waste pickling liquors: process synthesis and economic analysis *Chemical Engineering Communications* **193** 548-63
[9] Pongchalermporn 2002 *Experimental Investigation on Ferric Chloride as Coagulant in Water Treatment Process* (Thailand: Asian Institute of Technology)
[10] Potter and Clynne 1980 Solubility of NaCl and KCl in aqueous HCl from 20 to 85°C *Journal of Chemical and Engineering Data* **25** 0021-9568
[11] Pradana 2017 Waste pickle liquor utilization as a coagulant for water treatment process in PT. Krakatau Steel (Persero) Tbk *SEAISI Conference Singapore*
[12] Prentice 1991 *Electrochemical Engineering Principles* (Englewood Cliffs: Prentice Hall)
[13] Regel-Rosocka 2010 Review on methods of regeneration of spent pickling solutions from steel processing *Journal of Hazardous Materials* **177** 57–69
[14] Zimmerman 1957 Equilibria and spectra of aqueous chlorine solutions *Journal of the American Chemical Society* **79** 0002-7863

AUTHOR INDEX

Institute of Physics
Dirac House, Temple Back
Bristol BS1 6BE UK

ISSN: 1757-8981
ISBN 978-1-7138-3608-7
9 781713 836087